International Conference on Physical Instrumentation and Advanced Materials (ICPIAM 2016)

Journal of Physics: Conference Series Volume 853

Surabaya, Indonesia
27 October 2016

ISBN: 978-1-5108-4244-1
ISSN: 1742-6588

Printed from e-media with permission by:

Curran Associates, Inc.
57 Morehouse Lane
Red Hook, NY 12571

CURRAN ASSOCIATES INC. proceedings.com

Some format issues inherent in the e-media version may also appear in this print version.

Printed by Curran Associates, Inc. (2017)

For permission requests, please contact the Institute of Physics
at the address below.

Institute of Physics
Dirac House, Temple Back
Bristol BS1 6BE UK

Phone: 44 1 17 929 7481
Fax: 44 1 17 920 0979

techtracking@iop.org

Additional copies of this publication are available from:

Curran Associates, Inc.
57 Morehouse Lane
Red Hook, NY 12571 USA
Phone: 845-758-0400
Fax: 845-758-2633
Email: curran@proceedings.com
Web: www.proceedings.com

Table of contents

Volume 853

International Conference on Physical Instrumentation and Advanced Materials
27 October 2016, Hotel Santika Premiere, Surabaya, Indonesia

Accepted papers received: 22 May 2017 Published online: 7 June 2017

Preface

Instrumentation Physics

Advanced Materials

IOP Conf. Series: Journal of Physics: Conf. Series **853** (2017) 011001 doi:10.1088/1742-6596/853/1/011001

Preface

The International Conference on Physical Instrumentation and Advanced Materials (ICPIAM) took place in the Hotel Santika Premiere Surabaya, on October 27, 2016 and was organized by the Department of Physics, Faculty of Science and Technology, Airlangga University, Surabaya Indonesia.

This conference will conduct biannual regularly. The main objective of this conference is to bring together students, researcher and practitioners of physical instrumentation and advanced material to share and disseminate the research result.

We are fortunate to have world-known keynote speaker: Prof. Tamio Oguchi and Prof. Yoshitada Morikawa (Osaka University), Prof. Sulaiman Wadi Harun (Malaya Uniersity), Prof. Moh. Yasin and Febdian Rusydi, Ph.D. (Airlangga University), who are willing to share their perspectives on the current status of researches.

We would like to thank the member of local organizing committee who help with all the preparations required to make the conference success. Of course, special thanks also go to the reviewers, the authors and all of the participants for their contributions in making ICPIAM 2016 successful.

In addition, we would also thanks to Science and Technology Faculty and Institute of Research and Innovation Airlangga University in which, without their support we would not able to organize this conference. I believe that we will have high quality discussion in order to increase our excellence in scientific area of physical instrumentation and advanced material. I also hope that this will become a stepping stone of potential collaboration between academicians and professionals in Indonesia in the future.

Finally, the success of this conference lies not only in the quality of papers presented but also to a large extent upon the dedicated team efforts of the many volunteers, in particular members of the Organizing Committee and International Advisory Committee. Their dedicated contribution and encouragement have been exemplary. We would also like to thank the staff at Integrated Meetings Specialist, who have given their best to ensure the smooth running of the conference.

Last but not least, I would like to acknowledge with sincere thanks to our sponsors and supporters. To all our delegates, I hope the 1st ICPIAM 2016 will be memorable not only from the scientific perspective but in the joy of meeting old friends and making new ones.

Dr. Khusnul Ain
General Chair of ICPIAM 2016

IOP Conf. Series: Journal of Physics: Conf. Series **853** (2017) 011002 doi:10.1088/1742-6596/853/1/011002

Peer review statement

All papers published in this volume of *Journal of Physics: Conference Series* have been peer reviewed through processes administered by the proceedings Editors. Reviews were conducted by expert referees to the professional and scientific standards expected of a proceedings journal published by IOP Publishing.

IOP Conf. Series: Journal of Physics: Conf. Series **853** (2017) 012001 doi:10.1088/1742-6596/853/1/012001

Modeling of electrical impedance tomography to detect breast cancer by finite volume methods

K. Ain[1,*], R A Wibowo[2], S Soelistiono[1]

[1]Biomedical Engineering - Airlangga University, Surabaya – Indonesia
[2]Physics - Airlangga University, Surabaya – Indonesia

[*]Corresponding email address: k_ain@fst.unair.ac.id

Abstract. The properties of the electrical impedance of tissue are an interesting study, because changes of the electrical impedance of organs are related to physiological and pathological. Both physiological and pathological properties are strongly associated with disease information. Several experiments shown that the breast cancer has a lower impedance than the normal breast tissue. Thus, the imaging based on impedance can be used as an alternative equipment to detect the breast cancer. This research carries out by modelling of Electrical Impedance Tomography to detect the breast cancer by finite volume methods. The research includes development of a mathematical model of the electric potential field by 2D Finite Volume Method, solving the forward problem and inverse problem by linear reconstruction method. The scanning is done by 16 channel electrode with neighbors method to collect data. The scanning is performed at a frequency of 10 kHz and 100 kHz with three objects numeric includes an anomaly at the surface, an anomaly at the depth and an anomaly at the surface and at depth. The simulation has been successfully to reconstruct image of functional anomalies of the breast cancer at the surface position, the depth position or a combination of surface and the depth.

1. Introduction

Based on data Globocan 2008, it was estimated that there were 12.7 million cancer cases and 7.6 million died of cancer in 2008 in the world. Breast cancer is one type of cancers that is most often diagnosed in women and at most cause of death in women, with the number of each about 23% of the total cancer cases and 14% of the total deaths from cancer [1]. Based on data from the American Cancer and society, data showed that the number of breast cancer increases with age. The more aged women, the higher the risk for breast cancer [2]. Therefore stepping 40 years should check the condition of her breasts regularly every year [3].

The earlier the detection and treatment, more and more women with breast cancer will be saved [4]. Therefore WHO running two programs to be against breast cancer, namely education and screening program. Education means helping women manually to be able to recognize the signs of breast cancer and immediately reported to the hospital if found. While screening means early identification using medical equipment before the signs appear.

Screening requires a medical device, which capable to detect the presence of breast cancer. Some equipment has been built to detect the presence of breast cancer, such as x-ray mammography, breast ultrasound and MRI mammography [5].

Mammography is a medical instrument that is commonly used to detect breast cancer, but this equipment has several weaknesses which are often provides negative or positive information false and the dangers posed by radiation because it uses X-rays to the breast is exposed [6]. Early detection using mammography equipment will increase the risk almost doubled if it is conducted at the age of 40 to 50 years, and this risk will continue to increase until the age of 75 years [7].

The ability of ultrasound to detect breast cancer is still inferior compared to X-ray mammography, although ultrasound capability even increasing in recent years. However, because it is much more flexible and interactive, and users can scan the desired area repeatedly from different angles, the ultrasound is often used to check for suspicious anomalies near the surface of the skin or cyst. It is generally used to check the type of tumour with diagnostic purposes or as a monitor to assist the placement of a needle biopsy [8].

MRI mammography is rarely used for the detection of breast cancer because it is not economically. In general, MRI is used only to verify the diagnosis. MRI mammography is only used for early detection when X-ray mammography is not possible to use, for example to young women who generally have dense breasts or a woman is at high risk if exposed to X-ray radiation [8].

The electrical impedance of body tissue has long been an interesting study, because changes of the electrical impedance of organs are strongly associated with physiological and pathological properties. Both physiological and pathological properties are closely related to the medical application information [9]. Some experiments shown that breast cancer has a lower impedance compared to normal breast tissue, so that the imaging based on impedance can be used as an alternative to detect breast cancer [10]. It is also supported by Zoa and Caves studies that in future electrical impedance tomography (EIT) will replace the role of X-ray mammography as a medical device for the early detection of breast cancer [11].

2. Methods

Electrical impedance tomography is the scanning technique that utilizes the electrical parameters to image the object cross section. The image is electrical impedance distribution which is one of the very important electrical parameters. Therefore, to understand the electrical impedance tomography, it would need to understand the concept includes forward problems, methods of data collection and inverse problem.

Forward Problem is the process of electrical potential calculation based on a mathematical model. If there is no current source inside the object and the distribution of electrical conductivity is known, then the distribution of electrical potential within the object will satisfy Laplace equation. Forward problem solutions can be obtained through the concept of conservation of current flow in each element of the object.

Furthermore, the current will flow and distributed to elements of its neighbor to the whole element. Thus, the potential distribution will follow the shape of the object. Determination of electric potential is obtained through the solution of equation (1) through the discretization of the object resistivity. In this study, the discretization is conducted by dividing the object into small elements rectangular.

As well as the admittivity discretization, the potential is made to be discrete into a number of specific points in the domainΩ. In 2D case, value and potential coordinates points are obtained using finite volume method through an integral approach of the surface $\boldsymbol{S}$ of the Laplace equation, thus it obtain the equation (1).

$$\iint_S (\nabla \cdot \gamma \nabla V) dS = 0 \tag{1}$$

By using divergence theorem, then apply

$$\iint_S (\nabla \cdot \gamma \nabla V) dS = \oint_l \gamma \nabla V \cdot dl \tag{2}$$

then the left side of equation (2) can be substituted into a closed integral along the trajectory l that surrounding elements, so it can be obtained the equation (6).

IOP Conf. Series: Journal of Physics: Conf. Series **853** (2017) 012001 doi:10.1088/1742-6596/853/1/012001

$$\oint_l \gamma \nabla V \cdot dl = 0 \quad (3)$$

Equation (3) states the sum of the current flux which penetrates normally to four sides that surround the rectangular surface. Equation (3) is a dot product of the current flux vector γ with a vector trajectory dl. This segment is not equal to zero if the two vectors are parallel. In other words, the current flux must penetrate perpendicular to the side of the element. Flux current flows through a potential difference between two points that form a line and cut perpendicular to the side of the border of two mutually neighboring elements. This flux flows stream as possible if the potential distribution represented as a set of potential value at the midpoint of the element. Thus, the model solutions developed through such an approach would produce a potential distribution point located at any point of the central element.

The numerical solution of the equation (3) produces equation (4).

$$\sum \gamma \nabla V \Delta l = 0 \quad (4)$$

with the current boundary $je(i)$ (current density on the element-e and ***i*** side) as follows

$$j_e(i) = \begin{cases} j_{injection}, & for\ element\ in\ injected\ boundary \\ 0, & for\ the\ element\ in\ the\ non\ injected\ boundary \end{cases}$$

this is a Neumann boundary. Whereas, the injection of current is done with by grounding the polarity of the current source, thus the potential on injection becomes equal to zero. This condition is a Dirichlet boundary condition.

If an element is at the coordinates (i, j), as Figure 1, the numerical solution of homogeneous equation (4) on the element produces equation (5).

$$\gamma_{ij}\left[\left(V_{i+\frac{1}{2}j} - V_{ij}\right)\Delta y - \left(V_{ij} - V_{i-\frac{1}{2}j}\right)\Delta y + \left(V_{ij+\frac{1}{2}} - V_{ij}\right)\Delta x - \left(V_{ij} - V_{ij-\frac{1}{2}}\right)\Delta x\right] = 0 \quad (5)$$

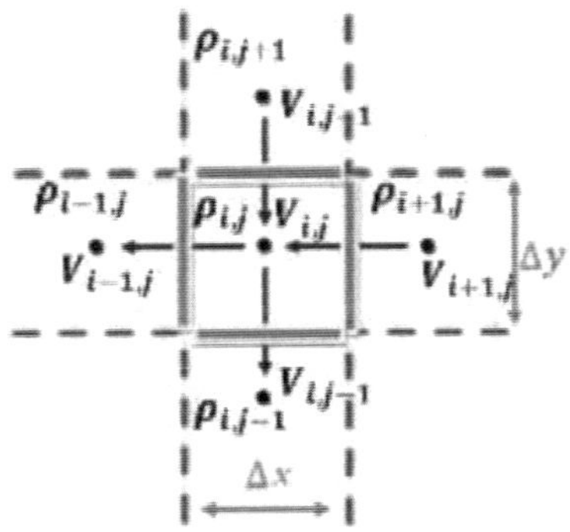

Figure 1. An element with 4 potentials point neighbor

boundary with a neighbor to the right, up, left and down:

$$\gamma_{i+1j}\left(V_{i+1j} - V_{i+\frac{1}{2}j}\right)\Delta y + \gamma_{ij}\left(V_{ij} - V_{i+\frac{1}{2}j}\right)\Delta y = 0 \quad (6)$$

$$\gamma_{ij+1}\left(V_{ij+1} - V_{ij+\frac{1}{2}}\right)\Delta x + \gamma_{ij}\left(V_{ij} - V_{ij+\frac{1}{2}}\right)\Delta x = 0 \quad (7)$$

$$\gamma_{i-1j}\left(V_{i-1j} - V_{i-\frac{1}{2}j}\right)\Delta y + \gamma_{ij}\left(V_{ij} - V_{i-\frac{1}{2}j}\right)\Delta y = 0 \quad (8)$$

$$\gamma_{ij-1}\left(V_{ij-1} - V_{ij-\frac{1}{2}}\right)\Delta x + \gamma_{ij}\left(V_{ij} - V_{ij-\frac{1}{2}}\right)\Delta x = 0 \quad (9)$$

Substituting any potential limit, i.e. $V_{i+\frac{1}{2}j}$, $V_{ij+\frac{1}{2}}$, $V_{i-\frac{1}{2}j}$, dan $V_{ij-\frac{1}{2}}$ into the equation (6, 7, 8 and 9), then obtained :

$$c_{ij}V_{i+1j} + d_{ij}V_{ij+1} + e_{ij}V_{i-1j} + f_{ij}V_{ij-1} + g_{ij}V_{ij} = h_{ij} \quad (10)$$

where,

$$c_{ij} = \left(\frac{2\gamma_{ij}\gamma_{i+1j}}{\gamma_{ij}+\gamma_{i+1j}}\right)\frac{\Delta y}{\Delta x} \qquad d_{ij} = \left(\frac{2\gamma_{ij}\gamma_{ij+1}}{\gamma_{ij}+\gamma_{ij+1}}\right)\frac{\Delta y}{\Delta x}$$

$$e_{ij} = \left(\frac{2\gamma_{ij}\gamma_{i-1j}}{\gamma_{ij}+\gamma_{i-1j}}\right)\frac{\Delta y}{\Delta x} \qquad f_{ij} = \left(\frac{2\gamma_{ij}\gamma_{ij-1}}{\gamma_{ij}+\gamma_{ij-1}}\right)\frac{\Delta y}{\Delta x}$$

$$g_{ij} = -\left(c_{ij} + d_{ij} + e_{ij} + f_{ij}\right)$$

Then equation (10) is applied to all elements and produces a number of linear equations that connect the potential point of an element with all four elements of the potential value of its neighbors. Through merger and rearrangement of linear equations, the potential point of each element can be arranged in the form of the vector into the equation (11).

$$[G]_{(NxN)}[V]_{(Nx1)} = [C]_{(Nx1)} \tag{11}$$

where, $[G]$ is admittance matrix, $[V]$ is potential matrix, $[C]$ is current matrix and N is number of element.

The electrical potential can be obtained when a current density is given on an element at the object boundary. There are several ways to put the position of the electrode as current source. Some ways to inject current and to capture potential data in the collection data are namely neighboring, opposite, multi-references, and adaptive methods. The study used neighboring data collection methods with 16 electrodes are placed on the surface of the object shown in Figure 2.

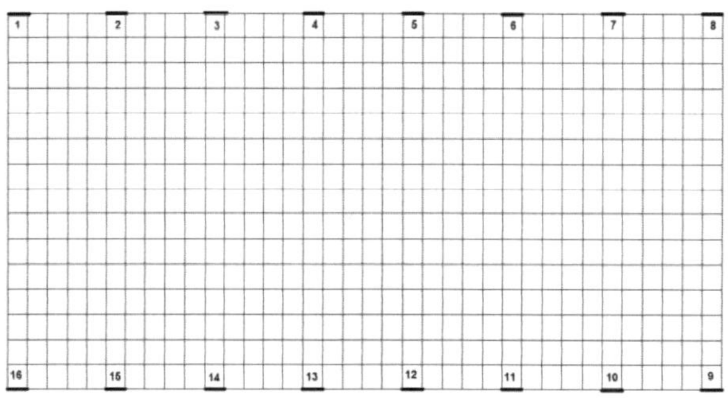

Figure 2. The methods of neighbor collection data

Data collection method was conducted by injecting an electric current into a pair of adjacent electrodes then measured potential difference between electrode pairs. For example, an electric current was injected into the electrode numbers 1 and 2, and then the potential difference was measured at electrode pair number 3 and 4, 4 and 5, and so on until the electrode pair number 15 and 16. The same procedure was repeated by injecting an electric current to the electrode pair number 2 and 3, 3 and 4, and so on until the electrode pair number 16 and 1.

Inverse Problem is the process to obtain the conductivity distribution of the object from limited voltage data. Several methods of different approaches have been proposed by some researchers that can generally be grouped into two categories, namely linear and non-linear [12]. The Linear reconstruction method will produce relative image that provides information about the relative differences to the reference image which to reflect changes of physiological function [13].

Nonlinear reconstruction methods will produce static image that gives information about the distribution of the absolute conductivity. The success of the non-linear method is determined by the match between the model geometry of forward problems and potential data from experiment [14]. Nonlinear Reconstruction consumes computing time because it requires iterative process, but it will produce a more accurate image reconstruction. Linearization method assumes that the potential change boundary is a linear function of the change in conductivity [14, 15]. The linearization reconstruction methods can be written as

$$[\delta\sigma] = ([S]^T[S] + \alpha I)^{-1}[S]^T\,[\delta V] \tag{12}$$

where $[\delta V]$ is change of boundary potential, $[S]$ is sensitivity matrix, α is regularisation parameter, I is identity matrix and $[\delta\sigma]$ is relative conductivity. The relative imaging methods require potential data as a reference data. The references can be obtained from the potential data from uniform conductivity or object conductivity from different frequency ω.

Potential data from uniform conductivity is very difficult to obtain, but the potential data from different frequency is allowed to be obtained. Potential data when the conductivity changed because the frequency is called imaging of difference frequency. Imaging different frequency will produce functional image that can be used for monitoring physiological. The image is only image relative, not absolute image of value $\sigma_{t,\omega}$. In this way more real for imaging applications than absolute imaging $\sigma_{t,\omega}$. in different frequency imaging, data obtained from the two conditions of different frequencies to produce an image of change admittance. This way will effectively cancel the error together. the Imaging of difference frequency has shown potential to produce a functional image in several medical applications.

Review comprehensively [16] and measurement [17] electrical parameters (electrical permittivity and conductivity) was conducted on normal breast and breast cancer based on different frequency. From the review shows that with different frequencies obtained different electrical parameter between normal breast and breast cancer significantly. There is information declare that the admittance of various biological tissues show a change with frequency [18, 19], so that imaging based on the difference frequency promises to generate an image of admittance distribution.

Therefore, the purpose of this research is to design, solve the mathematical model, namely forward and inverse problems and design 2D multi-frequency electrical impedance tomography which can be used to detect early breast cancer.

Electrical parameters of body tissue varies with frequency. This difference causes a change an electrical conductivity of body tissue. The breast is a tissue that change its electrical resistivity when measured at different frequencies. the study assume that the breast tissue is homogeneous, it contains anomalies, namely cancer. Here is the electrical conductivity of breast tissue used.

Tabel 1. The conductivity of breast tissue [20]

Conductivity of breast tissue (mS/cm)	10 kHz	100 kHz	10 MHz
Tumor (inside)	3.98	4.28	7.24
Tumor (tissue surface)	1.68	1.70	2.70
Fat with tumor cells	0.65	0.67	1.05
The tissue that away from tumor	0.28	0.31	0.37
Normal tissue	0.22	0.22	0.25

The numerical phantom is 2D numerical object that represents a rectangular shaped breasts in which there are anomalies with the size of the two elements. Each numerical objects represents a condition of one anomaly in a depth, one anomaly on surface and two anomalies in a depth and on surface with two different frequencies, namely 10 kHz and 100 kHz. The Breast numerical object is made based on the data Table 1, so that obtained the image 3.

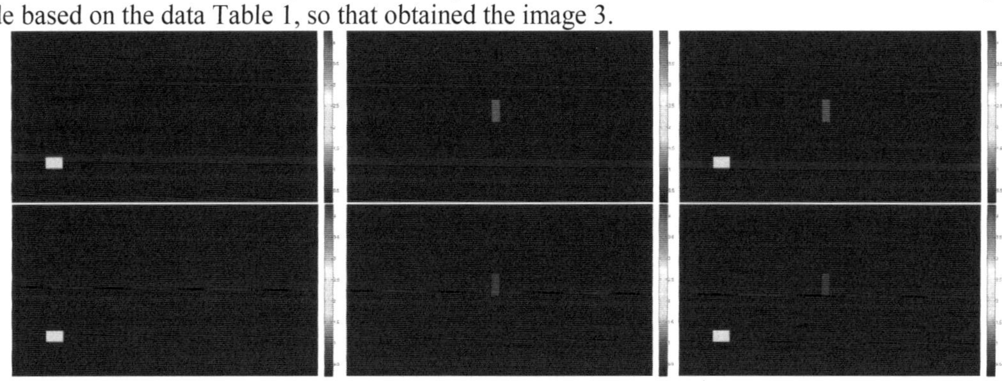

Figure 3. The breast numerical data with two element anomalies at a 10 kHz and 100 kHz

The study includes a development of a mathematical model of the electric potential field method 2D Finite Volume Method (FVM), solve forward problem and solve inverse problem by linear methods, developing the algorithm and program code to solve forward and inverse problems.

3. Results

Simulation was conducted by making the program code to solve the forward problem to obtain voltage data on the numerical object. The simulation used a numerical object of breast model, it shaped rectangular with neighbore method to collect data. Based on equation (13) and (14) code program has been compiled to solve forward problem by the finite volume method. The complete scan data using 16 channels in the numerical object of normal breast without anomaly was shown in Figure 7. The Scanning data in Figure 7 if taken at the surface where the electrodes position in the normal breast, containing anomalies at surface, depth and surface and depth at a frequency 10 kHz and 100 kHz, it will be resulting the voltage data shown in Figure 8.

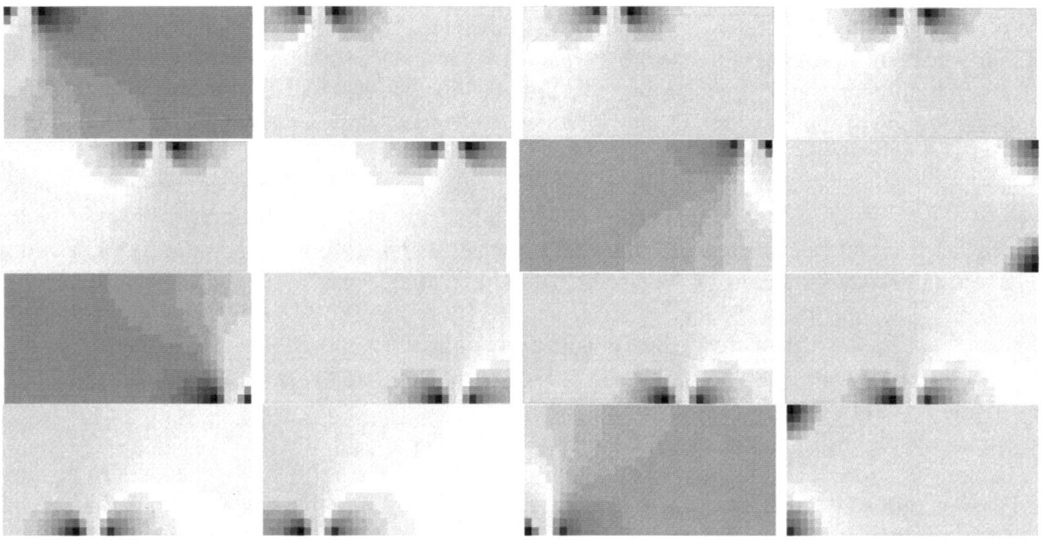

Figure 4. The complete Solution of forward problem of volume element methods on the rectangular uniform object

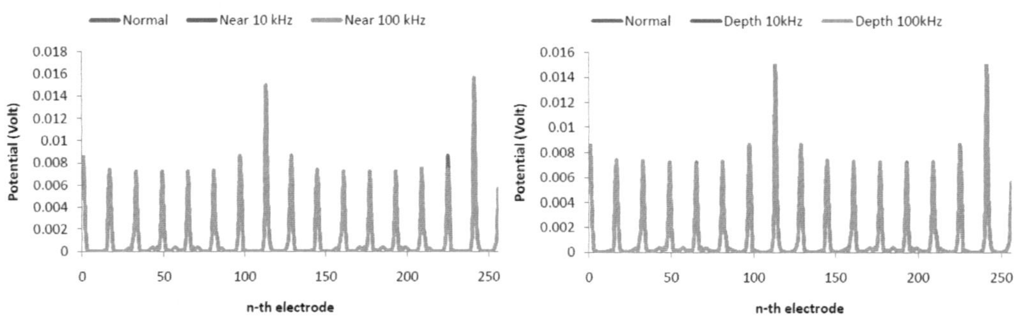

(a) (b)

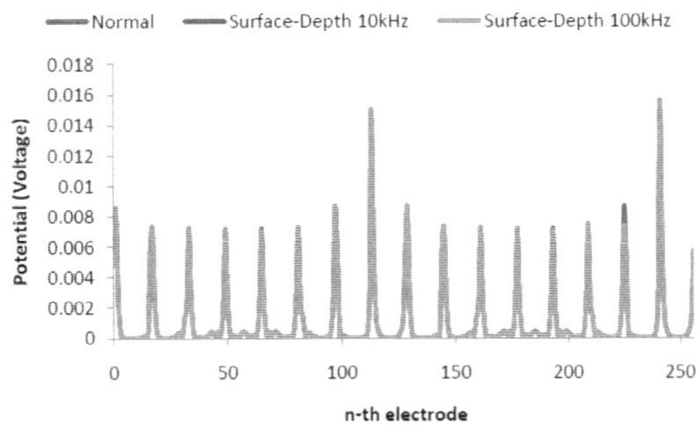

(c)

Figure 5. The scanning data from 16 channels elektrode by neighboring methode on uniform media and anomaly numerical object at 10 kHz dan 100 kHz (a) surface anomaly (b) depth anomaly (c) surface and depth anomaly.

Visually, Figure 6 does not show differences in the data between uniform and anomalies medium at 10 kHz and 100 kHz, but numerically has RMSE 0.188% and 0.189% toward the reference.

Based on equation (20), has been compiled a back projection program code to reconstruct voltage data in Figure 5 to obtain the image reconstruction in Figure 6. During the reconstruction process, the regularization parameters need to be given to obtain stable results. Regularization parameter selected by trial and error, to obtain optimal image reconstruction. The simulation voltage data from forward problem solution at Figure 5 then is reconstructed using back projection method with $\alpha = 5x10^{-5}$. By using data 10 kHz and a uniform object as a reference, so that obtained Figure 6.

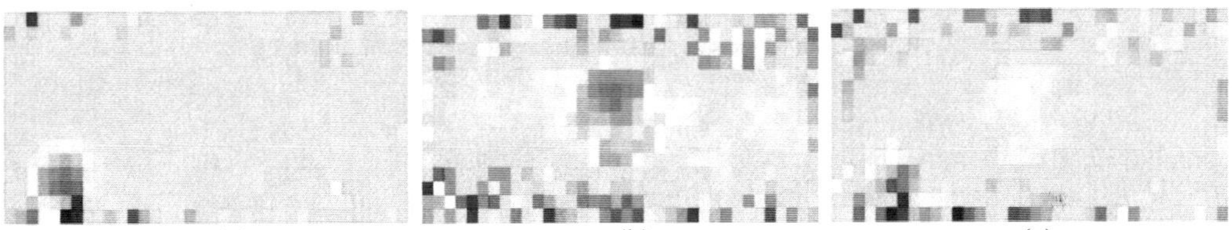

(a) (b) (c)

Figure 6. Reconstruction image from data V_f = 10 kHz and V_{ref} = uniform object by neighboring methods (a) surface anomaly (b) depth anomaly (c) surface and depth anomaly

The simulation voltage data from forward problem solution at Figure 5 then is reconstructed using back projection method with $\alpha = 5x10^{-4}$. By using data 100 kHz and data 10 kHz as references, Figure 7 is obtained. In Figure 7, the simulations have been able to reconstruct the image of functional anomalies of breast cancer at the surface, depth or a combination of surface and depth.

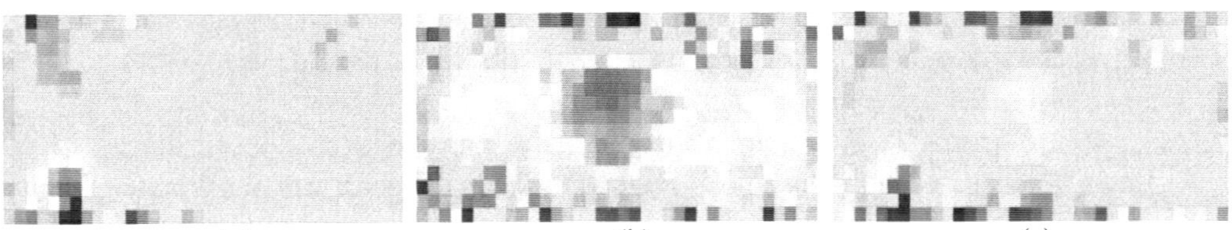

(a) (b) (c)

Figure 7. Reconstruction image from data V_f = 100 kHz and V_{ref} = 10 kHz by neighboring methods (a) surface anomaly (b) depth anomaly (c) surface and depth anomaly

4. Conclusion

It has conducted a modelling of Electrical Impedance Tomography to detect breast cancer by Finite Volume Methods (FVM). The research includes the development of a mathematical model of the electrical potential field by 2D Finite Volume Method, solving forward problem and inverse problem by linear method. Scanning is done by using 16 channels electrode by neighbor data collection method. Scanning is conducted at 10 kHz and 100 kHz with three anomaly objects. They are an anomaly in the depths, an anomaly at the surface and an anomaly at depth and surface. In simulations have successfully reconstructed image of functional anomalies of breast cancer at the surface position, depth position, or a combination surface and depth position.

Reference

[1] Jemal A, Freddy B, Melissa M, Jacques F, Elizabeth W, David F 2011 Global Cancer Statistics *CA Cancer J. CLIN* **61** (2) pp 69-90
[2] Breast Cancer, Facts and Figures 2011-2012 (http://www.cancer.org/acs/groups/content/@epidemiologysurveilance/documents/document/acspc-030975.pdf).
[3] Carol H *et al.* 2010, Breast Cancer Screening With Imaging: Recommendations From the Society of Breast Imaging and the ACR on the Use of Mammography, Breast MRI, Breast Ultrasound, and Other Technologies for the Detection of Clinically Occult Breast Cancer *J. American College of Radiology* **7** pp 18 -27
[4] Todd E, Keith D, Alex H, Sandra K, Steven P 2002 Electrical Impedance Spectroscopy of the Breast: Clinical Imaging Results in 26 Subjects *IEEE Transactions On Medical Imaging* **21** (6) pp 638 - 645
[5] Tyna A and Siân E 2004 Technology review: The Use of Electrical Impedance Scanning In The Detection of Breast Cancer, *Breast Cancer Res* **6** pp 69-74
[6] Shao W and Adams R 2012 Multi-Polarized Microwave Power Imaging Algorithm for Early Breast Cancer Detection *Progress in Electromagnetic Research* **23** pp 93 - 107
[7] Bijwaard H, Brenner A, Dekkers F, van Dillen T, Land C E, Boice J D 2010 Breast Cancer Risk From Different Mammography Screening Practices *Radiat Res* **174** pp 367–376
[8] David S 2005 Electrical Impedancetomography, Series in Medical Physics and Biomedical Engineering, Department of Medical Physics and Bioengineering, University College London.
[9] Gang H and Bin H 2011 Magnetoacoustic Imaging of Electrical Conductivity of Biological Tissues at a Spatial Resolution Better than 2 mm Vol **6** issue 8 (http://www.plosone.org/article/info%3Adoi%2F10.1371%2Fjournal.pone.0023421).
[10] Andrzej J, Stanislaw S, Stuchly J, Robin B, Arvind S 1988 Dielectric Properties of Breast Carcinoma and the Surrounding Tissues *IEEE Transactions On Biomedical Engineering* **35** (4) pp 257 - 263
[11] Zou Y and Guo Z 2003 A Review Of Electrical Impedance Technique For Breast Cancer Detection, Medical Engineering And Physics **25** pp 79-90
[12] Ain K, Kurniadi D, Suprijanto, Santoso O 2013 Lung Anomaly Detection By Filtered Back Projection Reconstruction Methods In Electrical Impedance Tomography *3rd international conference on instrumentation control and automation* pp 133-137
[13] Yaqin L 2010 A Novelty Dynamic Image Reconstruction Algorithm in Electrical Impedance Tomography Based on Nachman Theory *International Forum on Information Technology and Applications*
[14] Borsic A 2002 Regularisation Methods for Imaging from Electrical Measurements, A thesis of Philosophy Doctor, Oxford Brookes University.
[15] Chen X Y, Wang H X, Newell J 2011 Lung Ventilation Reconstruction by Electrical Impedance Tomography Based on Physical Information, *3rd International Conf. on Measuring Technology and Mechatronics Automation, IEEE Computer Society.*
[16] O'Halloran M, Conceicao R, Byrne D, Glavin M, Jones E 2009 FDTD Modeling of The

Breast: A Review *Progress In Electromagnetics Research B* **18** pp 1-24

[17] Kim T H and Pack J 2012 Measurement of Electrical CharacterisTics of Female Breast Tissues For The Development of The Breast Cancer Detector, *Progress In Electromagnetics Research C* **30** pp 189-199

[18] Gabriel C, Gabriel S, Carthout E 1996 the Dielectric Properties Of Biological Tissues I : literature survey *Physical Medical Biology* **41** pp 2231-2249

[19] Oh T I, Koo W, Lee K H, Kim S M, Lee J, Kim S W, Seo J, Woo E 2008 Validation Of Multi-Frequency Electrical Impedance Tomography (Mfeit) System KHU Mark I : Impedance Spectroscopy And Time Difference Imaging *Physical Measurement* **29** S295-S307

[20] Qiao G, Wang W, Wang L, He Y, Bramer B, AL-Akaidi M 2007 Investigation Of Biological Phantom For 2D And 3D Breast EIT Image *IFMBE proc.* **17** *13th International Conf. On Electrical Bioimpedance And 8th Conference On Electrical Impedance Tomography, Graz, Austria*

IOP Conf. Series: Journal of Physics: Conf. Series **853** (2017) 012002 doi:10.1088/1742-6596/853/1/012002

Dual frequency electrical impedance tomography to obtain functional image

Imam Sapuan*, Khusnul Ain, and Alif Suryanto

Department of Physics, Faculty of Sciences and Technology, Universitas Airlangga

*Corresponding email address: ispuan@yahoo.com

Abstract. Electric Impedance Tomography with two frequencies is a system to detect the anomalies. This system is expected to detect the presence of a cancer in the breast. In this study, the objects are modelled in a circle phantom within 13 cm diameter. Those objects are equipped with 16 electrodes of copperplate. The objects, carrots, are functioned as a cancer and water as a medium of the normal breast. This electrode works to inject the current and to measure the voltage at a certain point. The position of the electrode current injection is controlled by a de-multiplexer, whereas the measurement of voltage at the electrodes is controlled by a multiplexer. The electric current source utilized has two frequencies; 10 kHz and 100 kHz. This electric current is generated from a circuit of Voltage Controlled Current Source using an oscillator XR2206. The microcontroller is utilized to control the current injection through a de-multiplexer and the measurement of output voltage through a multiplexer.This research has produced three images. Two images are obtained from both frequencies of 10 kHz and 100 kHz. Those two images cannot be achieved in the reality. The object condition of normal breast cannot be measured, since the normal breast of a person is different from others. In this study, the two images can be obtained when the potential background of the phantom can be measured. The third image is obtained from the reconstruction of the electrical potential difference between the low and high frequencies. This image is called as a functional image. This functional image makes the EIT system can be implemented, since it can be obtained without measuring the potential background. This functional image reveals that the anomalies are more obvious than the single frequency image.

1. Introduction

Electrical impedance tomography (EIT) is a scanning technology using an electric current source to obtain an internal image of the object. The principle of tomography is to inject an AC current at the object through electrodes attached to the surface and to measure the electrical potential between the electrodes on the surface [1]. Electrical impedance tomography has the advantage to be safer for the body because it uses low electric current. The EIT system is also able to identify in the real time and is easy to be implemented because its creation is relatively simple [2].

The limitations of EIT system is firstly the low spatial resolution of the obtained image reconstruction. It is because of the limited number of electrodes that can be mounted on the surface of the object. The second drawback is that the injected current will be diffused into all directions, so the obtained potential data at the boundary is less sensitive or called as ill-pose conditions. The third drawback is the high noise of reached data measurement from the data signal cables, electrodes, and the movement of objects.

The third disadvantage can be overcome by making the scanning process at different frequencies. Data scanning at different frequencies can be reconstructed with a reconstruction method linearization. This reconstruction method produces significantly different conductivity other objects with other objects that are functional imagery.

One method of inverse problem can use linearization method. The linearization method assumes that the potential boundary alteration is a linear function of the conductivity alteration [3-4]. That linear function had been obtained by using Taylor series approach in $V(\gamma)$ around $V(\gamma_0)$ [5], namely:

$$V(\gamma) = V(\gamma_0) + V'(\gamma_0)(\gamma - \gamma_0) + O(\|\gamma - \gamma_0\|^2) \quad (1)$$

with $V'(\gamma_0)$ is Jacobion of $V(\gamma)$ on γ_0. By ignoring the last tribe, and then the equation (1) becomes

$$\delta V \approx J\delta\gamma \quad (2)$$

with $\delta V = V(\gamma) - V(\gamma_0)$, $J = V'(\gamma_0)$ and $\delta\gamma = \gamma - \gamma_0$.Jacobion is commonly referred as a matrix of sensitivity so that the equation (2) can be written in the form,

$$[\delta V]_{(q^2x1)} = [S]_{(q^2xp)}[\delta\sigma]_{(px1)} \quad (3)$$

with$[\delta V]$ is a potential alteration in the boundary, $[S]$ is the matrix sensitivity, and $[\delta\sigma]$ is the alteration of conductivity.

The relative imaging techniques require potential data comparator as a reference. References can be obtained from the data of potential object reference or the alteration of object conductivity since time t or object changes due to the frequency ω. The potential data of object reference is difficult to obtain, but the object of potential data during the conductivity changes related to the time or because the frequency is allowed to be enquiried. The potential data of object reference when the conductivity changes due to the time is called different time imaging, while the potential data of object reference when the conductivity changes because of the frequency is called imaging in the different frequency.

Different time imaging will produce functional image that can be used for physiological monitoring. The image is only relative image, not absolute image of $\sigma_{t,\omega}$ and $\omega\epsilon_{t,\omega}$. This way is more real for applications compared to absolute imaging. In this different time imaging, the data obtained from two different time conditions is reduced to produce an image of admittance distribution alteration. By effectively reducing the data, it would effectively omit the error.

Different time imaging has shown the potential as a functional image in several areas of medical applications. If the potential data of different time imaging is not obtained, breast cancer imaging for instance [6-8], so the frequency of different imaging can be performed. The admittance spectrum of various biological tissues shows an alteration toward frequency [9-10]. Thus, that different frequency imaging is ascertained to generate an image of alteration of admittance distribution.

In the image of different frequency, image is built from the inverted projection algorithm of the ratio of two potential data in the different frequencies [11-13]. The sensitivity matrix method for algorithm reconstruction with the potential difference in the two frequencies can also be used [14-15]. Both of these methods in principally use the potential difference in the two frequencies and linear algorithm reconstruction.

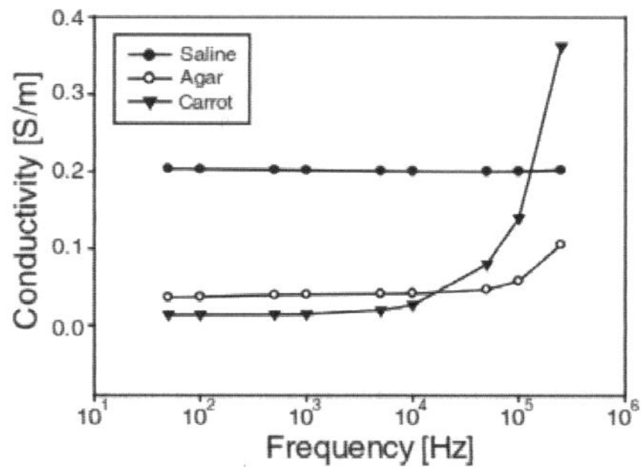

Figure 1.The conductivity of water (saline) with two objects, jelly and carrot [9]

Image with the different frequency using the potential difference weighted at two frequencies is easy to apply [17], for admittance spectrum of biological tissues alters toward frequency. It can be assumed that the image object with a frequency depends on the admittance of object reference. There are two different contrast mechanisms in the reconstructed image at different frequencies. First, there is a contrast in the admittance between the anomaly object and reference. The second, admittance distributions will alter toward frequency.

In this study, the carrot is used as the object of breast cancer, while the water represents the normal breast tissue. The carrot selection is based on the comparisons of different conductivity of carrot at frequency of 10 kHz and 100 kHz. According to research conducted by Zhao [16] at the frequency of 10 kHz and 100 kHz, the conductivity of carrots showed a very clear distinction.

Figure1. shows the different conductivity of carrot at 10 kHz and 100 kHz. Greater different conductivity between the carrots and water will produce higher contrast image. The frequencies used in this study are 10 kHz and 100 kHz. The source of electric currents is generated by current source circuit and the oscillator. The digital oscilloscope is used as an interface to display any potential difference of measurement results.

2. Research methods

In this research, the data collection process uses neighboring method. The phantom experiments use water and carrots as anomalies, elucidated in Figure 2. In this method, there are a pair of neighboring electrodes as flow injector and another pair of electrodes for measuring large voltages. Current injection is given by a pair of adjacent electrodes, while the potential difference is measured on all electrodes.

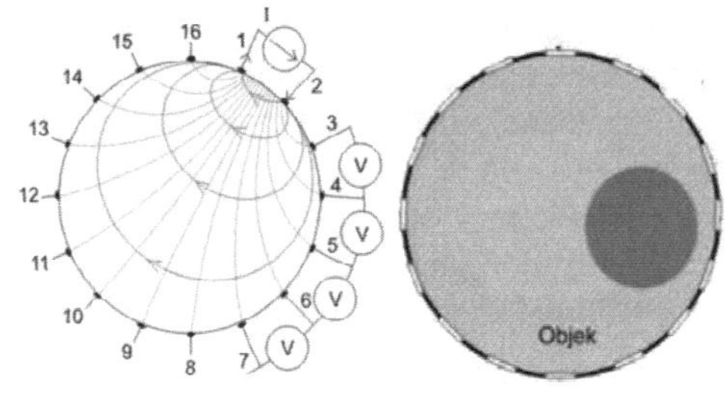

(a) (b)

Figure 2. (a) neighboring model (b) EIT Phantom model

The hardware of electrical impedance tomography system is composed of an oscillator circuit, a current source (VCCS), buffers, multiplexer, de-multiplexer and microcontroller. System hardware of electrical tomography system is revealed in Figure 3.

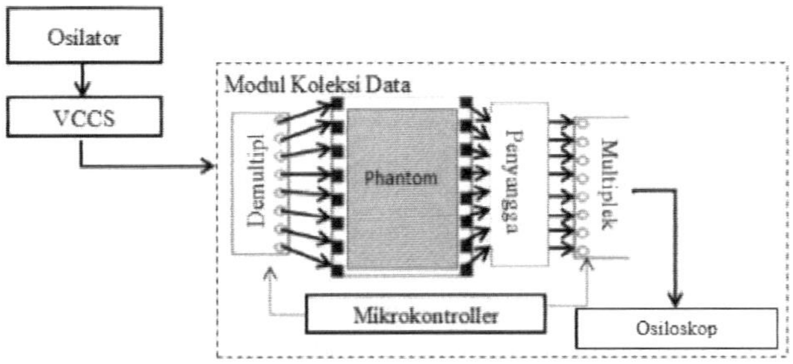

Figure 3. The electric impedance tomography system design

The oscillator is an electronic circuit that is able to generate an alternating current of electrical signal. IC oscillator circuit using XR-2206 can generate several types of waves, for instance, a square wave, triangular, and sinusoidal. XR-2206 circuit schematic is revealed in Figure 4.

By closing the switch of S_1, it will produce a sinusoidal signal on pin 2. The output frequency of this circuit can be determined by adjusting the value of the capacitor on pin 5 and resistor on pin 7. The frequency of the oscillator can be expressed by the following equation (4).

$$f = \frac{1}{RC} H \tag{4}$$

On pin 5 and 6, two capacitors are paralleled to adjust the frequency of the generated signal. Capacitor 470pF and 1nF is used for frequency 10 kHz or 100 kHz.

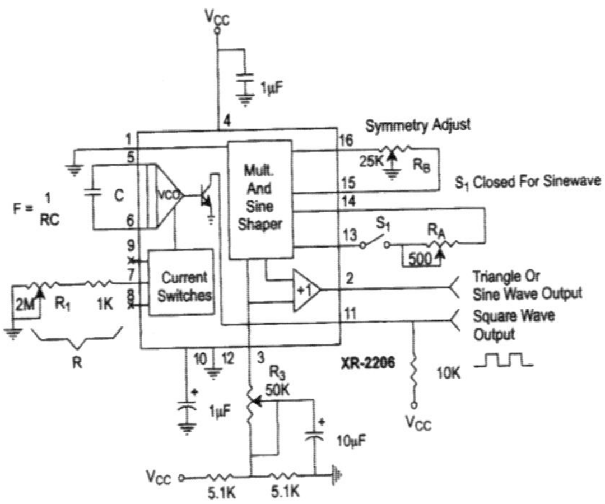

Figure 4. Schematic of IC XR-2206

Current source circuit or Voltage Controlled Current Source (VCCS) is a circuit that serves to maintain the stability of the current at a certain load range. This circuit serves to set the electrical current generated to be always constant when load is given. VCCS circuit schematic is shown in Figure 5.

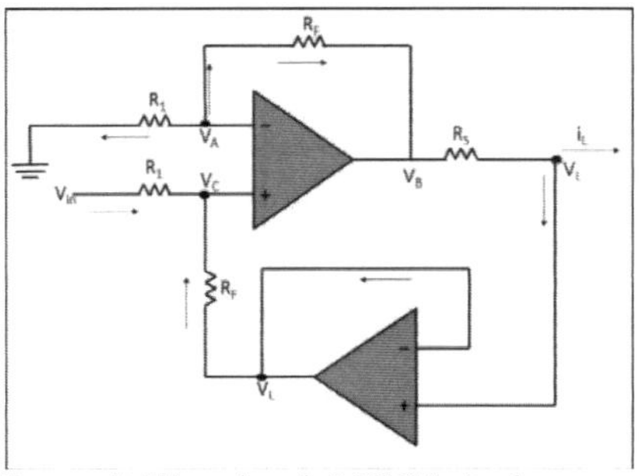

Figure 5. The circuit VCCS dual op-amp

VCCS circuit obtains the Vin input signal of the oscillator circuit and generates electric current Iout fixed I_{out}. This current will be injected into the object. The ideal I_{out} value is in accordance to the equation (5)

$$I_{out} = \frac{R_f}{R_S R_1} \times V_{in} \tag{5}$$

VCCS circuit uses the main component of op-amp of IC LM4562. This IC excellence compared to the op-amp IC LM741 or LM833 is more constant than current generated for greater load range. This is because the op-amp of IC LM4562 has very large output impedance. IC LM4562 has an impedance of 3000M Ohm while the LM741 has only 3M Ohm impedance.

The buffer circuit functions to maintain the measured voltage of the phantom, in order to avoid voltage drop when it is connected to the circuit in front of the circuit. Buffer circuit schematic is shown in Figure 6.

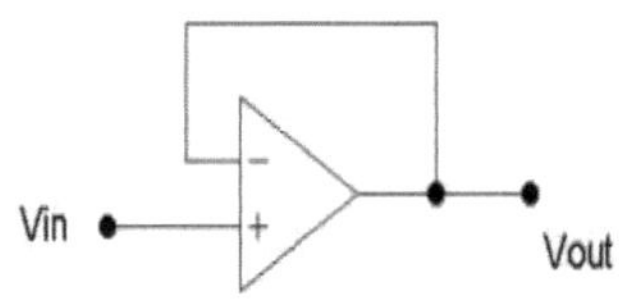

Figure 6. The simple buffer circuit

Due to the input voltage which is directly connected to the output, resulting output voltage will be equal to the input voltage (Vout = Vin).

Multiplexer can electronically be controlled to choose the activated path. One input of many inputs will be forwarded to the output. Selecting input is arranged by the control select. The multiplexer control the traffic signal of measurement result objects trough the electrode and the buffer to be forwarded to voltage measurement (digital oscilloscope) while the inversion of multiplexer is demultiplexer. Demultiplexer is used to regulate traffic flow injection to object through the connected electrodes.

Microcontroller is a device that is integrated into an IC to control electronic circuits. In this research, the type of microcontroller used is Atmega8535. This microcontroller controls in the current injection and voltage measurement via the multiplexer and demultiplexer. The minimum circuit system consists of four ports, namely ports A, B, C, and D. Table 1 shows the function of each - each port in this study.

PORT	Function	Interface
PA0 - PA7	Output	Demultiplexer
PB0 - PB7	Output	Multiplexer

Table 1. Pin Configuration Atmega8535 used

Image reconstruction is done by using Matlab software with the Filter Back Projection reconstruction methods. The algorithm method is based on the data collection through neighboring with circular geometry and based on equipotential. The algorithm of filtered Back Projection based on equipotential is initially proposed by Barber-Brown in matrix notation which can be expressed in the equation (6) [18].

$$[\delta\rho_n]_{(px1)} = [F]_{(pxp)}[B]_{(pxq^2)}[\delta V_n]_{(q^2x1)} \tag{6}$$

p is the number of elements and q is the number of electrodes, $[B]$ is the weight matrix of Back Projection, $[F]$ is the matrix representation of a filter, $[\delta Vn]$ are normalized limit of potential alteration, and $[\delta\rho n]$ is normalized conductivity distribution alteration.

Linearization method assumes that the boundary of potential change is a linear function of the change in conductivity (Ain in 2014), so that equation (6) can be written in the form of a matrix that can be seen in equation (7).

$$[\delta V]_{(q^2x1)} = [S]_{(q^2xp)}[\delta\rho]_{(pxp)} \tag{7}$$

$[\delta V]$ is a potential alteration of the boundary, $[S]$ is the matrix sensitivities, and $[\delta\rho]$ is the resistivity alteration. The step that could be done to acquire the matrix sensitivity $[S]$ is by doing $\delta\rho$ variation on all elements in the equation (7).

Equation (7) may be completed after the matrix $[S]$ is found by doing algebraic manipulations. However, since the matrix $[S]$ is not square then $[\delta\rho]$ cannot directly be obtained.

$$[S]^T{}_{(pxq^2)}[\delta V]_{(q^2x1)} = [S]^T{}_{(pxq^2)}[S]_{(q^2xp)}[\delta\rho]_{(pxp)} \tag{8}$$

$$[\delta V]_{(q^2x1)} = ([S]^T[S])^{-1}{}_{(pxp)}[S]^T{}_{(pxq^2)}[\delta\rho]_{(pxp)} \tag{9}$$

Generally $[S]^T[S]$ is a singular matrix so that the matrix has no inversion. To solve these problems, Tikonov regulation could be performed. Thus, the matrix will have the inversion.

$$[\delta V]_{(q^2x1)} = ([S]^T[S] + \alpha I)^{-1}{}_{(pxp)}[S]^T{}_{(pxq^2)}[\delta\rho]_{(pxp)} \tag{10}$$

α is a parameter of regulation and I is the identity matrix.

3. Results and discussion

The tomography hardware of electrical impedance with phantom experiments has been made as revealed in Figure 7

a

b

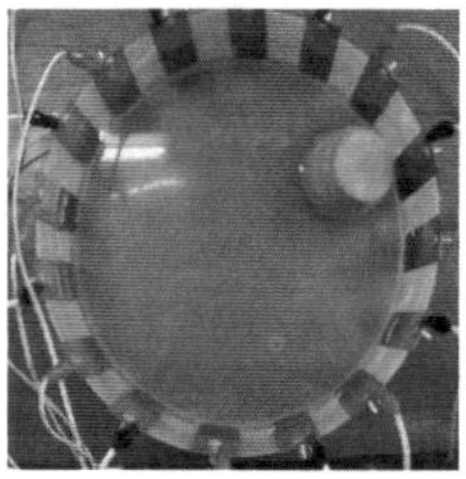

c

Figure 7. a. Electrical Impedance Tomography System
b. water reference c. phantom anomaly

3.1. Current injection module

Current injection module consists of an oscillator circuit and VCCS circuit. Based on the results having been obtained, an oscillator circuit is designed to fulfil the suitable frequency (10 kHz and 100 kHz) and required amplitude of 0.78 volts (2Vpp) with a sinusoidal output signal.

The constant current source (VCCS) that had been made is capable of producing a constant current 0.0057 A for the range of 155-560 ohm load and a frequency of 10 kHz. This testing was conducted on a VCCS input voltage of 0.7 volts (2Vpp). The data and graphics of current stability are elucidated in Table 2 and Figure 8.

This VCCS circuit is also capable of producing a constant current 0.0035 A for the range load of 300-621 ohm and a frequency of 100 kHz. The testing was conducted on a VCCS circuit with an input voltage of 0.35 volts (1Vpp). The data and graphics of current stability are elucidated in Table 3 and Figure 9.

The constant current source that had been made is suitable with the needs of current of the EIT system in milliamps. The resulted current is still safe when applied to the human body because it is below 10 mA.

Resistance (Ohm)	Voltage (Volt)	Current (mA)
101	0.7071	0.0070
155	1.0607	0.0058
243	1.4142	0.0058
303	1.7678	0.0058
364	2.1213	0.0058
436	2.4749	0.0057
498	2.8284	0.0057
562	3.1820	0.0057
687	3.5355	0.0051

Table 2. The data of circuit voltage test results with 10kHz frequency to the load

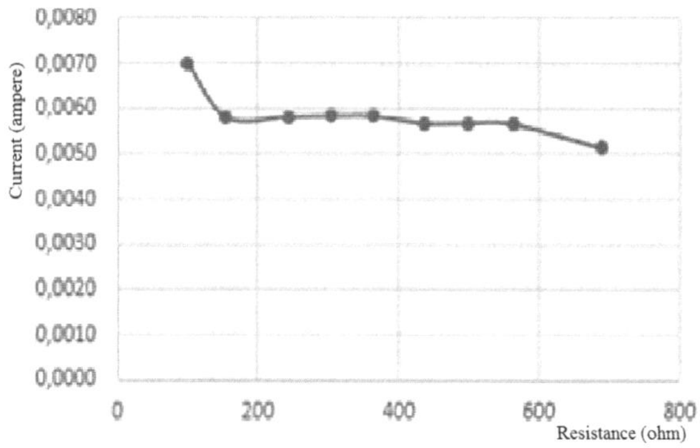

Figure 8. The graph of the relationship between output current to the load, at a frequency of 10kHz.

Resistance (Ohm)	Voltage (Volt)	Current (mA)
98	0.3536	0.0036
158	0.7071	0.0045
300	1.0607	0.0035
407	1.4142	0.0035
526	1.7678	0.0034
621	2.1213	0.0034
760	2.4749	0.0033

Table 3. The data from current source circuit testing at frequency of 100 kHz to the load

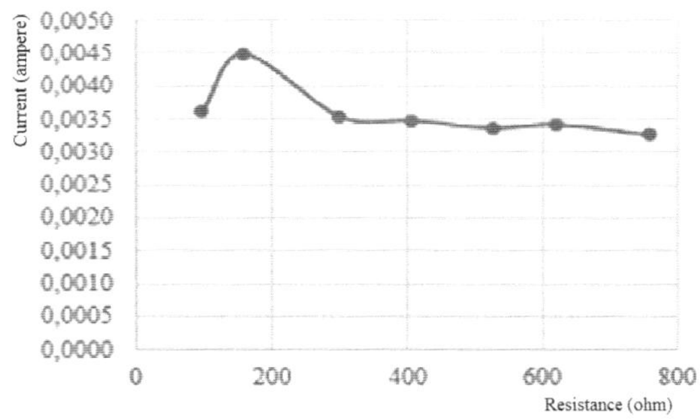

Figure 9.The graph of the relationship between the output current to the load at a frequency of 100 kHz

3.2. Voltage measurements
Hardware measuring the voltage that has been made is shown in Figure 7. This hardware consists of several components including a phantom, buffer circuit (buffer), multiplexer circuit, and microcontroller.

The process of voltage measurement in this research is controlled by a microcontroller circuit. This circuit serves to regulate the flow of the VCCS injection process via demultiplexer which is connected to the electrode. The microcontroller also regulates the voltage measurement process which is done through a multiplexer that is connected to the electrode.

3.3. Analysis of the collected data
In this research, the process of collecting data is done by using the electrode for 16 units. The first step starts from currents injected in the electrodes 1 and 2, where as a current injection electrode 1 and 2 as a ground electrode. Data retrieval is done on the potential difference electrode pairs 1 and 2, 2 and 3, 3 and 4, until the data to 16 is a pair of electrodes 16 and 1. This process is repeated until the current injection to 16 (the pair of electrodes 16 and 1), so that the amount of data obtained on measuring the potential difference of 256 data.

Potential measurements made on two State phantom, i.e. normal (water reference) and anomalies (water + carrots). Each measurement is performed in the same manner at a frequency of 10 KHz and 100 kHz. The graph of the results of the measurements of potential difference with 16 electrodes at water reference is shown in Figure 10. The graphs of measurement results of potential difference with 16 electrodes on the phantom breast cancer anomaly are as shown in Figure 11.

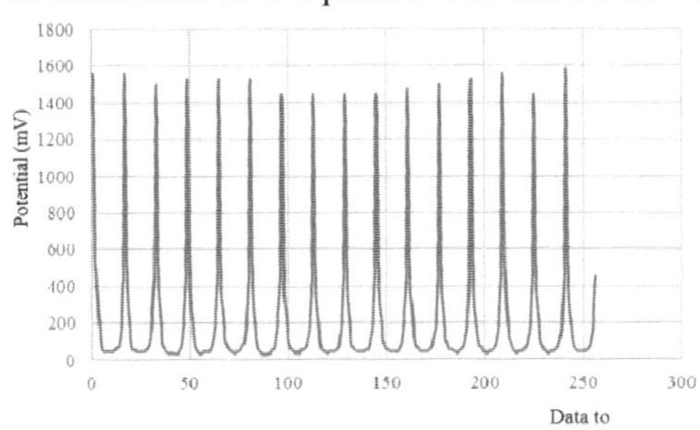

(a)

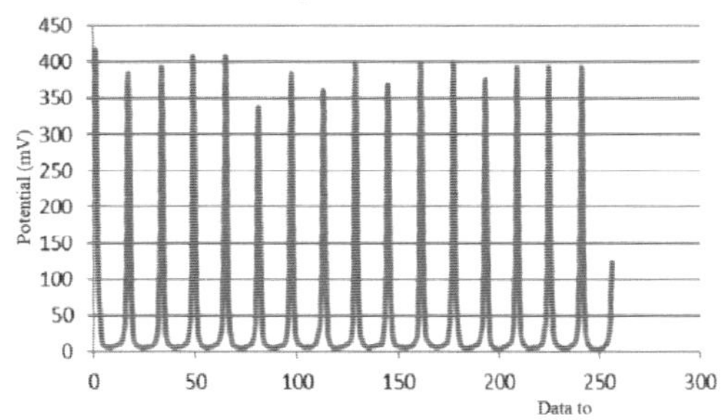

(b)

Figure 10. The graph is results of the measurements of potential difference with 16 electrodes at water reference. (a) The frequency of the10kHz, (b)The frequency of the 100kHz

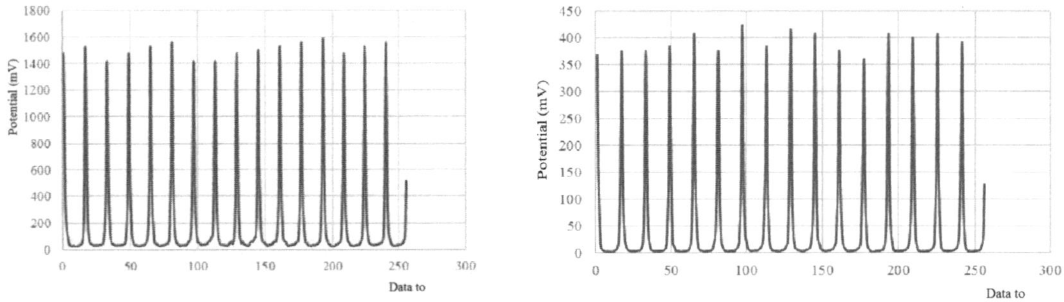

(a) (b)

Figure 11. The graph results on the measurements of potential difference with 16 electrodes at water reference. (a) The frequency of 10 kHz, (b) The frequency of 100 kHz.

Voltage measurements is performed on a phantom object made of water as a media and a normal object or as a background while the water as well as carrot acts as a representative of anomalous objects. Data collection was performed twice, for current injection frequency of 10 kHz and 100 kHz. These measurements performed at a frequency of 10 kHz produces a potential difference which is relatively higher compared with measurements made at a frequency of 100 kHz. This is because the magnitude of the potential difference measured is inversely proportional to the frequency used, as shown by the following equation.

$$V = I.Z = I\,.\sqrt{R^2 + \left(\frac{1}{2\pi f.C}\right)} \tag{11}$$

3.4. *Image reconstruction results*

There were three images obtained in the reconstruction process as shown in Figure 12. Image reconstruction method used equation (11). Two images were reconstructed from the data electric potential difference of the background and background + anomalies. Background electric potential is an electric potential measured in the water phantom object. Whereas, electric potential of background + anomaly is the electrical potential measured at the water + carrot phantom object.

Image (a) was obtained from the data reconstruction of electric potential difference between the background with background + carrot on the current frequency of 10 kHz. Whereas, Image (b) was obtained from the data reconstruction of electric potential difference between the background with background + carrot on the current frequency of 100 kHz

The both-images are relative images. Assuming that the change of potential boundary is a linear function of the change in conductivity, those two images cannot be achieved in the reality. The object condition of normal breast cannot be measured, since the normal breast of a person is different from others. In this study, the two images can be obtained if the potential background of the phantom can be measured.

Figure 12. Image Reconstruction resulted on (a) reconstructed image at frequency of 10 kHz between phantom with the water reference, (b) reconstructed image at frequency of 100 kHz between phantom with the water reference, and (c) the results of image reconstruction of the electric potential difference of water + carrot on 10kHz and 100kHz.

The third image was obtained from the reconstruction of the electrical potential difference between the low and high frequencies. This image is called as a functional image. This functional image makes the EIT system can be implemented, since it can be obtained without measuring the potential background. This functional image reveals that the anomalies are more obvious than the anomalies in the single free frequency image.

4. Conclusion

Based on the results of the conducted research, it can be summarized as follows:

1. Electrical Impedance Tomography systems (EIT) Dual frequency with several constituent components, like the circuit: oscillator, Voltage Controlled Current Source (VCCS), buffer, current injection and voltage measurement, as well as neighboring method for current injection process and measurement potential difference has successfully created.
2. Electrical Impedance Tomography System (EIT) has been successfully tested in water as medium and carrots as the object of breast cancer. Functional image reconstruction is obtained from the difference of electrical potential for low and high frequencies. With functional system image, EIT can be applied in the reality, because without doing the measurement the image background can be obtained. This functional image shows the anomaly clearly more than anomaly on the image of a single frequency.

Reference

[1] Ain K, Sapuan I, Wibowo R A, Kurniadi D2012 Electric Impedance Tomography System for Nondestructive test of Simple, University of Airlangga

[2] Setiyanto D 2011 Design and Buildingof Prototype Elektrical Impedance Tomography(EIT) with Single Frequency, Faculty of Mathematic dan Natural sciences, University of Indonesia

[3] Borsic A 2002 Regularisation Methods for Imaging from Electrical Measurements, A thesis of Philosophy Doctor, Oxford Brookes University

[4] Chen X Y, Wang H X, Newell J C 2011 Lung Ventilation Reconstruction by Electrical Impedance Tomography Based on Physical Information *3rd International Conf. on Measuring Technology and Mechatronics Automation IEEE Computer Society*

[5] Noor A 2007 Electrical Impedance Tomography at Low Frequencies *Thesis of Doctor Phylosophy* University of New South Wales

[6] Kukarni R, Boverman G, Isaacson D, Saulnier G J, Kao T J, Newell J C 2008 An analytical layered forward model for breast in electrical impedance tomography *Physical Measurement* **29** S27-S40

[7] Soni N K, Hartov A, Kogel C, Poplack S P, Paulsen K D 2004 Multi-frequency electrical impedance tomography of the breast : new clinical results *Physical Measurement* **25** S301-

IOP Conf. Series: Journal of Physics: Conf. Series **853** (2017) 012002 doi:10.1088/1742-6596/853/1/012002

S314

[8] Trokhanova O V, Okhapkin M B, Korjenevsky A 2008 Dual-Frequency Electrical Impedance Mammography For The Diagnosis of Non Malignant Breast Desease *Physical Measurement* **29** S331-S334

[9] Gabriel C, Gabriel S, Carthout E, 1996 The dielectric properties of biological tissues : I : literature survey *Physical Medical Biology* **41** pp 2231-2249

[10] Oh T I, Koo W, Lee K H, Kim S M, Lee J, Kim S W, Seo J K, Woo E 2008 Validation of Multi-Frequency Electrical Impedance Tomography (Mfeit) System KHU Mark I *Impedance Spectroscopy And Time Difference Imaging, Physical Measurement* **29** S295-S307

[11] Griffith H 1987 The Importance Of Phase Measurements In Electrical Impedance Tomography *Physical Medical Biology* **32** pp 1435-1444

[12] Griffith H and Zhang Z 1989 A Dual-Frequency Electrical Impedance Tomography System *Physical Medical Biology* **34** pp 1465-1476

[13] Schlappa J, Anesse E, Griffiths H 2000 Systematic Error In Multi Frequency Electrical Impedance Tomography Of The Breast *New Clinical Results, Physical Measurement* **21** S111-S118.

[14] Romsauerova A, Mc Ewan A, Horesh L, Yerworth R, Boyford R H, Holder D 2006 Multi-Frequency Electrical Impedance Tomography (EIT) Of The Adult Human Head : Initial Finding In Brain Tumor, Arteriovenous Malformations And Chronic Stroke, Development of An Calibration *Physical Measurement* **27** S147-S161

[15] Yerworth R J, Bayford R H, Brown B, Milnes P, Conway M, Holder D 2003 Electrical Impedance Tomography Spectroscopy (EITS) for Human Head Imaging *Physical Measurement* **24** S477-S489

[16] Zhao *et al.* 2012 High Density Trans-admittance Mammography Development and Preliminary Phantom Tests, Kyung Hee University, Yongin-si, Gyeonggi-do, 446-701

[17] Seo J K, Lee J, Kim S W, Zribi H, Woo E J 2008 Frequency Difference Electrical Impedance Tomography (Fdeit) Algorithm Development And Feasibility Study *Physical Measurement* **29** S929-S944

[18] Ain K 2014 Tomografi of Dual modality Electrical And Acoustical Properties To Increase Image Quality *Dissertation Department Of Engineering Physics ITB Bandung*

[19] Bera T K, Nagaraju J, Lubineau G 2016 Electrical Impedance Spectroscopy (EIS)-Based Evaluation Of Biological Tissue Phantoms To Study Multifrequency Electrical Impedance Tomography (Mf-EIT) Systems *Breast Cancer,Facts & Figures*

[20] Zhang J, Patterson R P, Koqenevsky A 2002 Comparison and Analysis of Electrical Impedance Tomographic Images Reconstructed Using Two Algorithms *Proc. of the Second Joint EMBS/BMES Conference*

IoT real time data acquisition using MQTT protocol

R A Atmoko*, R Riantini, and M K Hasin

Marine Electrical Engineering, Shipbuilding Institute of Polytechnic Surabaya (SHIPS-PPNS), Indonesia

*Corresponding email address: ra.atmoko@ppns.ac.id

Abstract. The Internet of Things (IoT) provides ease to monitor and to gain sensor data through the Internet [1]. The need of high quality data is increasing to the extent that data monitoring and acquisition system in real time is required, such as smart city or telediagnostic in medical areas [2].Therefore, an appropriate communication protocol is required to resolve these problems. Lately, researchers have developed a lot of communication protocols for IoT, of which each has advantages and disadvantages. This study proposes the utilization of MQTT as a communication protocol, which is one of data communication protocols for IoT. This study used temperature and humidity sensors because the physical parameters are often needed as parameters of environment condition [3]. Data acquisition was done in real-time and stored in MySQL database. This study is also completed by interface web-based and mobile for online monitoring. This result of this study is the enhancement of data quality and reliability using MQTT protocol.

1. Introduction

The Internet of Things (IoT) is a concept that aims to extend the benefits of continuous internet connection for various things such as data sharing, remote control, and data monitoring. IoT is the idea of researchers who want to optimize equipment's such as sensor, radio frequency identification (RFID), wireless sensor network, and all equipment's connected to the Internet network to communicate with humans. IoT's challenge is to bridge the physical world and the information world such as the processing of data obtained from electronic equipment via an interface between the user and the equipment. Sensors collect physical data such as temperature and humidity, and then send it to the server to be stored in database or be displayed on the application interface [4].

IoT technology has started being developed in manifestations such as the Smart City, which has been developed in many metropolitan cities of the world, Smart Factory which is used for production optimization, Telediagnostic which facilitates the process of the patient's health monitoring, weather information system which is used for weather prediction, and so forth. Two of the processes that are often used in IoT applications are process control and monitoring (acquisition) of data. Communication protocols for acquisition of data that can be used are HTTP and MQTT protocol. HTTP protocol is used to web data exchange where it can usually be used for handling data acquisition of hardware, but the ability of HTTP is less suitable when being used on IoT-based applications.

HTTP protocol uses a request/response model, which is currently the most common message exchange protocol. MQTT uses a publish/subscribe pattern [5]. HTTP protocol is not designed for pervasive network and can cause a decrease in performance, especially in bandwidth usage and battery durability. [6]. The protocol for IoT applications require less bandwidth, real-time response, and low energy use because it is commonly used for small appliances.

A study conducted by Vergara et.al showed the superiority of MQTT implemented on Android devices, which can decrease energy consumption compared to HTTP protocol [7]. MQTT and HTTP are both running over TCP. However, MQTT provides some advantages such as low energy, among others. [8]. Compared to IoT protocols such as COAP, MQTT also provides advantage to be used on client type of smartphone [9]. This study aims to implement the usability of MQTT protocol for temperature and humidity sensor data acquisition system interfaced on mobile-based and web-based applications.

2. HTTP protocol

Internet network is built to communicate via HTTP (Hyper Text Transfer Protocol). Various data, from images to texts, are sent over internet every day. HTTP is as a primary protocol interface to move a wide range of data quickly, easily, and stable from server to user devices such as browser. HTTP is built on TCP. HTTP ensures that data transmitted from one device to another will not corrupt so that the integrity of data transmitted is assured. HTTP is an open communication protocol that can be read by any devices that have been developed for HTTP protocol as browser or smartphone through browser application. An HTTP transaction consists of two parts: request command (request) sent from client to server, and response command (response) sent from server to client. The process of response and request is submitted using a data block with specific format known as HTTP Message. The messages are sent by HTTP which moves in one direction.

3. MQTT protocol

MQTT (Message Queuing Telemetry Transport) protocol is protocol specifically designed for "machine to machine" communication. MQTT protocol runs over TCP / IP and has a data packet size with low overhead minimum (> 2 bytes) so that consumption of the power supply is also small enough. This protocol is a data-agnostic protocol that can transmit data in various forms such as binary data, text, XML, or JSON and this protocol uses a publish/subscribe model rather than a client-server model.

Stack TCP/IP is now widely supported by microcontroller likeSTM32Fx7 series, as well as common market device board such as Wemos and Raspberry Pi. There are so many options for implementing MQTT protocol on devices. A common system of MQTT requires two main software components:

- MQTT Client is to be installed on device. A web platform, which uses Javascript, can use Client PAHO library of Eclipse.
- MQTT Broker serves to handle publish and subscribe data. A Linux platform can use broker that is available free such as mosquitto, HiveMQ etc.

The advantage of publish/subscribe system is that the data sender (publisher) and the data receiver (client) do not know each other because there is a broker between both. In addition, there is time decoupling which makes publisher and client unable to be connected simultaneously so that client will stay to receive delayed data previously.

3.1. Control signal

MQTT has 14 types of control signal, namely:

- CONNECT—Client request to connect to Server
- CONNACK—Connection Acknowledgement
- PUBLISH—A message which represents a new/separate publish
- PUBACK—QoS 1 Response to a PUBLISH message
- PUBREC—First part of QoS 2 message flow
- PUBREL—Second part of QoS 2 message flow
- PUBCOMP—Last part of the QoS 2 message flow

- SUBSCRIBE—A message used by clients to subscribe to specific topics
- SUBACK—Acknowledgement of a SUBSCRIBE message
- UNSUBSCRIBE—A message used by clients to unsubscribe from specific topics
- UNSUBACK—Acknowledgement of an UNSUBSCRIBE
- message
- PINGREQ—Heartbeat message
- PINGRESP—Heartbeat message acknowledgement
- DISCONNECT—Graceful disconnect message sent by clients
- before disconnecting

From those signals, there are only four main signals which are used directly by the client,namely PUBLISH, SUBSCRIBE, UNSUBSCRIBE, CONNECT. Other signals are part of the publish/subscribe mechanism.

3.2. Topic & Quality of Service (QoS)

In MQTT, it is as known topic that serves as filter for broker in sending message to every client connected and subscribed. MQTT provides quality level of service, which is called QoS. This level guarantees the reliability of message delivery. Level 0 message is sent only once. Messages are sent depending on the existence of network, and there is no attempt to transmit a message back. Level 1 messages are sent at least once so that if the subscriber does not recognize (acknowledge) the message, then the broker will send a message to publisher to receive the message acknowledgment status from the client. Level 2 is to make sure that message was received. With this level, it can be ensured that the message is certainly conveyed and avoided from duplication of messages sent.

4. System implementation

To implement usage of MQTT protocol, this study requires the following equipments:

1. Wemos D1 Mini & ESP 8266 Wifi Module
2. DHT 11 Temperature & Humidity Sensor
3. PC Server as MQTT Broker.
4. Android Smartphone

Wemos D1 Mini is a minimum system of controller that is equipped with ESP 8266 seriesWi-Fi module. The Wemos has 9 Input/Output digital and 1 Input/Output analogue. Wemos serves as publisher to transmit data of DHT 11 sensor to MQTT broker. DHT 11 is type of digital sensor of temperature and humidity. DHT 11 can measure temperature of 0^0-50^0 Celsius and humidity of 20%-90% RH. PC server is as MQTT broker to facilitate exchange of data between publisher and subscriber, or among hardware, mobile application, and web application.

Figure 1.DHT11 Temperature & Humidity Sensor

IOP Conf. Series: Journal of Physics: Conf. Series **853** (2017) 012003 doi:10.1088/1742-6596/853/1/012003

A DHT11 sensor was connected to digital pin number 4 at Wemos. The sensor read temperature and humidity and then the information was sent (publish) to the broker. The MQTT broker used in this study was mosquitto that is an open source broker application of MQTT from Eclipse product. The data, which entered on broker then, was subscribed by client application. Subscribe application was built to use Javascript that can install a library MQTT named PAHO, an open source application of Eclipse product.

Table 1.MQTT Parameters.

Parameter Used	
Broker	Mosquitto
Subscriber	PahoJavascript
Topic	teleppns/temphum
MQTT Port	1883
Websocket Port	9001

Figure 2.MQTT Broker &Arduino Serial Monitor.

The subscriber application for android platform was built with Phonegap using HTML and Javascript. The subscriber application of web platform was built using HTML and PHP. The subscriber application is added with a feature to save data to MySQL database. The subscriber application communicated to broker using Web Socket so that data exchange can happen in full duplex and real-time.

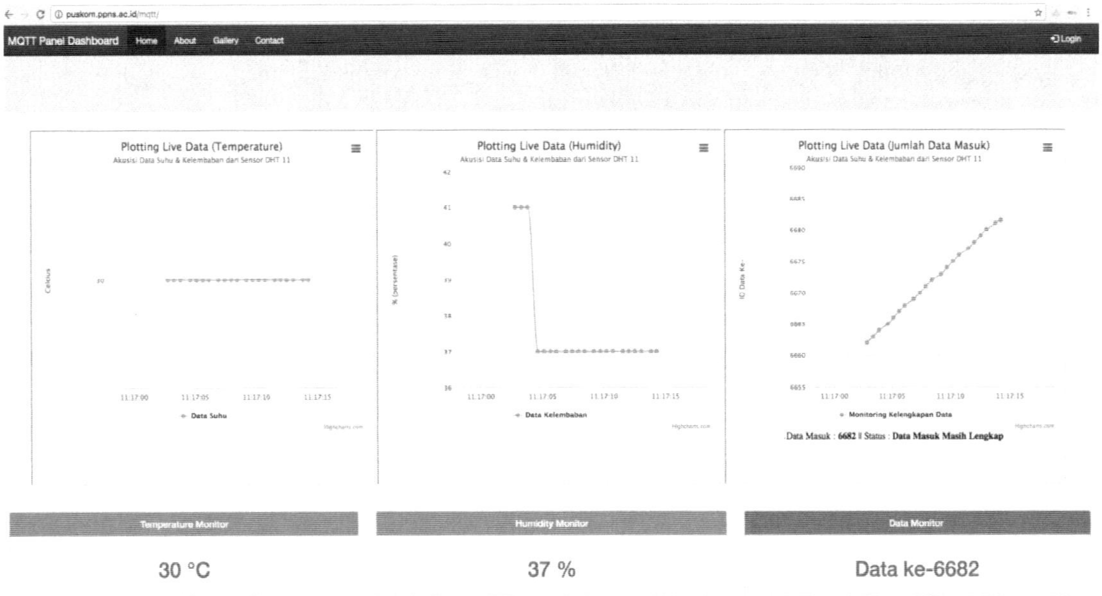

Figure 3. Web and Mobile Interfaces.

←T→	id_data	date1_data	date_data	suhu_data	kelembaban_data	nomer_data
Edit Copy Delete	11401	2016-10-19 18:09:46	09:46.237677097321	35	39	3496914
Edit Copy Delete	11402	2016-10-19 18:09:46	09:46.244874000549	35	39	3496916
Edit Copy Delete	11403	2016-10-19 18:09:46	09:46.244784116745	35	39	3496917
Edit Copy Delete	11404	2016-10-19 18:09:46	09:46.248136997223	35	39	3496915
Edit Copy Delete	11405	2016-10-19 18:09:46	09:46.374195098877	35	39	3496918
Edit Copy Delete	11406	2016-10-19 18:09:46	09:46.576997995377	35	39	3496919
Edit Copy Delete	11407	2016-10-19 18:09:46	09:46.578825950623	35	39	3496920
Edit Copy Delete	11408	2016-10-19 18:09:46	09:46.594352006912	35	39	3496922
Edit Copy Delete	11409	2016-10-19 18:09:46	09:46.594355106354	35	39	3496921
Edit Copy Delete	11410	2016-10-19 18:09:46	09:46.681370019913	35	39	3496923

Figure 4. Data Storage MYSQL.

5. Results and discussion

This study also reports the performance test of HTTP and MQTT protocol. The test was performed by taking data for 60 seconds. The following table shows the capability of each protocol to transfer data from hardware to server and store it into a MySQL database.

Table 2. Comparison HTTP & MQTT.

Data aqcuiring order	Amount of Success Data transfer via HTTP Protocol	Amount of Success Data transfer via MQTT Protocol
1	938 data	6560 data

Data aqcuiring order	Amount of Success Data transfer via HTTP Protocol	Amount of Success Data transfer via MQTT Protocol
2	930 data	6530 data
3	935 data	6501 data
4	931 data	6505 data
5	938 data	6505 data
Mean	934.4 data	6520.2 data

The performance test was conducted by calculating the successfully sent data from hardware to server and also the successfully inserted data into MYSQL database for 60 seconds. Each transmission from hardware has a sequential ID, so that when data has been entered into database, it can be checked for its completeness of data to determine whether a loss occurs. The test results show that the ability of MQTT is better in transferring data than HTTP. The above table shows that the MQTT has the ability to send data up to six times faster than HTTP. This is because MQTT architecture is designed to provide transfer speed by small data size that can reduce data transmission time. MQTT has a header data size smaller than HTTP.

6. Conclusion

This study has implemented the use of MQTT protocol to build data acquisition application of temperature and humidity sensor with a mobile interface, which is android and web-based. The test result indicates that MQTT protocol has the ability of transfer data faster than HTTP protocol, which can transfer amount of data 6 times of HTTP capability. MQTT usage can be an option for the hardware data acquisition real-time application based on the Internet of Things.

References

[1] Andrea Z, Nicola B, Angelo C, Lorenzo V, Michele Z 2014 *Internet of Things for Smart Cities, IEEE Internet of Things J.* **1** (1)

[2] Mitsa T 2003 An Evolvable Software Framework for an Internet-based Telediagnostic System *Information Technology Applications in Biomedicine 4th International IEEE EMBS Special Topic Conf.*

[3] Atmoko R A 2013 Sistem Monitoring dan Pengendalian Suhu dan Kelembaban Ruang pada Rumah Walet Berbasis Android, Web, dan SMS *Semantik* **3** (1) pp. 283-290 ISSN 979-26-02666

[4] Suresh P, Daniel J V, Aswathy R H 2014 A state of the art review on the Internet of Things *(IoT) History, Technology and fields of deployment*

[5] Lampkin V *et al.* 2012 Building smarter planet solutions with MQTT and IBM WebSphere MQ telemetry *IBM, ITSO*

[6] Colitti W, Steenhaut K, De Caro N 2011 Integrating Wireless Sensor Networks with the Web *Proc. IP+SN* Chicago, USA

[7] Vergara E J, Prihodko M, Nadjm-Tehrani S 2013 Mobile Location Sharing: An Energy Consumption Study *e-Energy* pp 289-290

[8] De Caro N, Colitti W, Steenhaut K, Mangino G, Reali G 2013 Comparison of Two Lightweight Protocols for Smartphone-based Sensing *IEEE 20th Symposium on Communications and Vehicular Technology in the Benelux (SCVT) Namur* pp. 1-6.

IOP Conf. Series: Journal of Physics: Conf. Series **853** (2017) 012004 doi:10.1088/1742-6596/853/1/012004

Simulation of leakage current measurement on medical devices using helmholtz coil configuration with different current flow

E Sutanto[1,*], **F Chandra**[1], **and R Dinata**[2]

[1] Department of Physics, Faculty of Science and Technology, Airlangga University, Surabaya 60115, Indonesia.
[2] Badan Pengkajian Fasilitas Kesehatan, Jalan Karang Menjangan No.22, Surabaya 60286, Indonesia.

*Corresponding email address: erwin_sutanto@fst.unair.ac.id

Abstract. Leakage current measurement which can follow IEC standard for medical device is one of many challenges to be answered. The IEC 60601-1 has defined that the limit for a leakage current for Medical Device can be as low as 10 μA and as high as 500 μA, depending on which type of contact (applied part) connected to the patient. Most people are using ELCB (Earth-leakage circuit breaker) for safety purpose as this is the most common and available safety device in market. One type of ELCB devices is RCD (Residual Current Device) and this RCD type can measure the leakage current directly. This work will show the possibility on how Helmholtz Coil Configuration can be made to be like the RCD. The possibility is explored by comparing the magnetic field formula from each device, then it proceeds with a simulation using software EJS (Easy Java Simulation). The simulation will make sure the concept of magnetic field current cancellation follows the RCD concept. Finally, the possibility of increasing the measurement's sensitivity is also analyzed. The sensitivity is needed to see the possibility on reaching the minimum leakage current limit defined by IEC, 0.01mA.

1. Introduction

The increasing number of medical devices being used in hospitals creates several basic things to be concerned. One of them is dealing with the patient's safety. It is common to see the use of electricity for medical device's power supply, so electrical safety is a mandatory standard. One of its requirements is to measure the leakage current. It is to prevent things that endanger patient's life caused by electrical over current. It may also prevent any electrical shocks.

1.1. Leakage current standard

The IEC (International Electro-technical Commission) published IEC60601 document that consists of collection of technical standard for safety and efficiency for medical devices. One of the standards talks mainly about leakage current standard. According to IEC, leakage current's limit can be categorized depending on the type of applied part.

The Applied Part itself is defined as part of a medical device which has a physical contact with patient. Following the type of contact, Medical Device is categorized into three types:

- B Type: There is no electrical contact with patient,
- BF Type: There is a part connected to patient but not directly to the heart,

- CF Type: There is a part connected to the heart of patient.

Following these applied part categories (B/BF/CF), a medical device must limit its leakage current to leakage current standard. The limit according to IEC 60601-1 is shown at Table I, the leakage current limits [1]. It is shown that the minimum limit is at CF Type, for Patient Leakage Current at Normal Condition (NC), is 0.01mA or 10 μA. The maximum limit is shown at most of categories. It is at 0.5mA or 500 μA.

Table 1. Leakage Current.

	B-Type		***BF-Type***		***CF-Type***	
Leakage Current	NC[a]	SFC[b]	NC[a]	SFC[b]	NC[a]	SFC[b]
Earth	0.5mA	1mA	0.5mA	1mA	0.5mA	1mA
Enclosure	0.1mA	0.5mA	0.1mA	0.5mA	0.1mA	0.5mA
Patient	0.1mA	0.5mA	0.1mA	0.5mA	0.01mA	0.05mA

[a]NC: Normal Condition.
[b]SFC: Single Fault Condition.

1.2. ELCB and RCD

ELCB (Earth-leakage circuit breaker) is a safety device which is used at electrical installation to prevent any electrical shocks. Later on, it is developed into the RCD (Residual Current Device). It improves the ELCB Voltage Based Measurement to become Current Operated Measurement. This current operated ELCB aka RCD is possible to measure leakage current directly. To understand its function and how it works, Typical RCD circuit [4] is shown at Fig. 1.

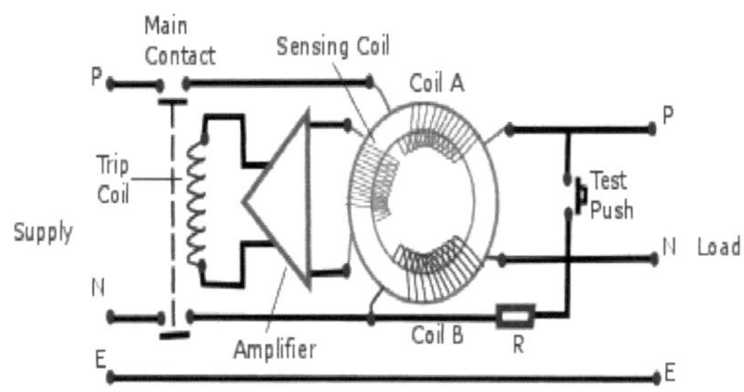

Figure 1. Typical RCD Circuit

The core are the toroid, two main coils (A and B), and one sensing coil. There is also circuit test to test the tripping mechanism. Lastly, there is tripping mechanism; a kind of relay, consisted of amplifier and trip coil.

Considering a possibility of leakage current, there are two paths on how the electrical current's flow through this RCD. They are:

- In normal condition the electrical current flow will be:
  Supply P (Phase) → Toroid's Coil A → Load P → (Load) → Load N → Toroid's Coil B → Supply N.
- When leakage current happens, the electrical current flow will be:
  Supply P (Phase) → Toroid's Coil A → Load P → (Load) → Load E (Earth) → Supply E.

Leakage Current is returning to supply through the earth path. It is different with normal path which use Toroid's Coil B and goes back to Supply N. If it is assumed that current flows through Toroid's Coil A is I1 and the current flows through Toroid's Coil B is I2, then the above case could be written as follow:

- In normal condition:

$$I_1 = I_2 \tag{1}$$

$$I_1 = I_2 + I_{leakage} \tag{2}$$

The measurement of leakage current is using the sensing coil by having induced e.m.f. During the leakage current case, ampere at Toroid's Coil A will exceed ampere at Toroid's Coil B. This leakage current will make alternating magnetic flux results in the toroid core. Then, this magnetic flux will be detected by the sensing coil, which is also wound at the toroid core.

Once there is enough induced e.m.f. (Electromagnetic Field), there will be a little current flow at sensing coil. This e.m.f. current will go through amplifier to drive trip relay coil. This is how the tripping mechanism happened. That is also how we can limit amount of leakage current before being cut off.

1.3. Helmholtz coil configuration with different current flow

Helmholtz Coil configuration is structured by two N-turn circular coils with radius R, each perpendicular to the y axis symmetrically, with a distance D. The original Helmholtz coil creates uniform magnetic field by having similar current flow direction on both circular coils. This will make similar magnetic field goes to the same direction. Its usage is well known for any experiments which may need to eliminate Earth Magnetic Field [2].

Here in this configuration, it has the same structure but different current flow. It is to differentiate magnetic field from those two circular coils. It is shown at Fig. 2, Helmholtz Coil with Different Current Flow. Looking at how the magnetic fields are against each other, it shows that it applies the same way like how the two main coils at toroid's RCD work. This concept can become the way to measure the leakage current after trying to explore more on theories of magnetic fields and later using simulation. There will be differences from Helmhotz Coil compared to RCD Toroid's Coil.

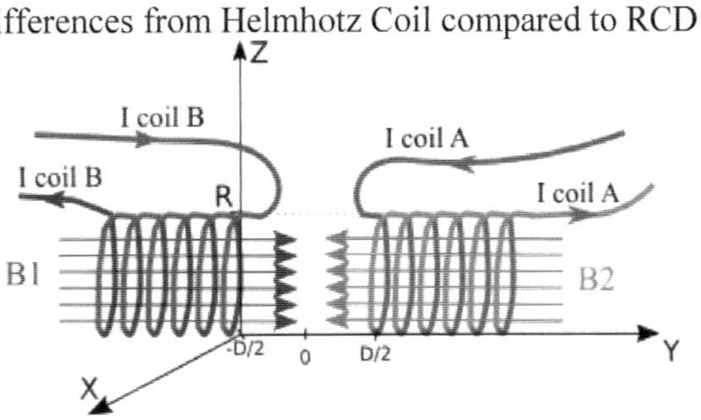

Figure 2. Helmholtz Coil with Different Current Flow

1.4. Helmholtz coil configuration with different current flow

Both devices, RCD and Helmholtz Coil with different current flow have main similarity on the use of magnetic field from two coils. That is why it is necessary to focus on this magnetic field measurement.

Thus, this work tries to explore more about the magnetic field and to make a model of that magnetic field. First is by using known formulas and then with making the simulation. Then it extracts the possible sensitivity measurement from the simulation.

2. Comparison between RCD and helmholtz coil with different current flow

The difference between existing RCD and the Helmholtz Coil Configuration is the medium for its coils or magnetic permeability. RCD is using toroid shape medium and most of used medium is a ferrite core. While Helmholtz Coil is simply free space of air. This difference has effect on the magnetic field's value and the equations.

IOP Conf. Series: Journal of Physics: Conf. Series **853** (2017) 012004 doi:10.1088/1742-6596/853/1/012004

2.1. Ampere law

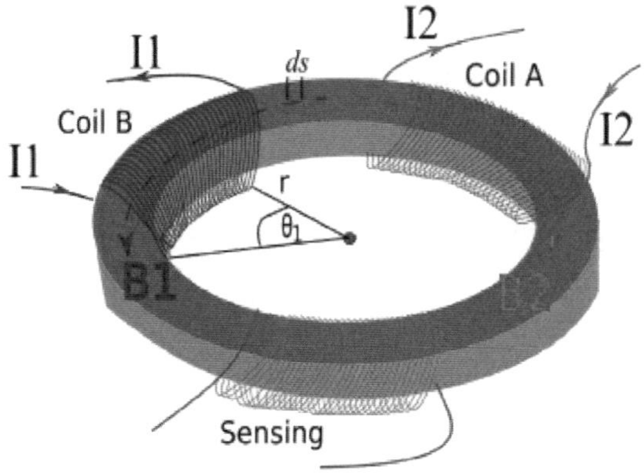

Figure 3. RCD Toroid's magnetic field

It is necessary to have a review of the basic formula. A magnetic field formula for a symmetrical current source is best calculated using Ampere's Law compared with Biot-Savart Law. The generalization Ampere's Law for an enclosed loop (enc) is written as:

$$\oint \vec{B} \cdot d\vec{s} = \mu_0 I_{enc} \tag{3}$$

Note: to use this formula, the system must have certain symmetry [3]. For RCD, this shape requirement is shown by solenoid wounded at toroid as part of circle.

2.2. RCD current equations

RCD is designed to get the magnetic field from the incoming current (I1) and subtract it with outgoing current flow (I2). Using the Ampere's Law for one of RCD's coil, the magnetic field (B1) can be calculated as follow:

$$\oint \vec{B}.d\vec{s} = B_1 \oint_{\theta_1} ds = B_1 \theta_1 r = \mu_r I_{enc} = \mu_r N_1 I_1 \tag{4}$$

$$B_1 = \frac{\mu_r N_1 I_1}{\theta_1 r} \tag{5}$$

The magnetic field's value depends on the toroid's medium μ_r , number of turns from coil B, enclosed part of circle ds1 in terms of θ_1 radian, and also the radius r as distance from center of the circle. First by assumption, there is no missing of magnetic field from the medium. Then, by having the same number of turns on Coil A and Coil B, $N_1 = N_2 = N$, $\theta_1 = \theta_2 = \theta$, and also using (2), total magnetic field's difference becomes:

$$\Delta B = B_1 - B_2 \cong \frac{\mu_r N}{\theta . r}(I_1 - I_2) = \frac{\mu_r N}{\theta . r} I_{leakage} \tag{6}$$

2.3. Biotsavart law

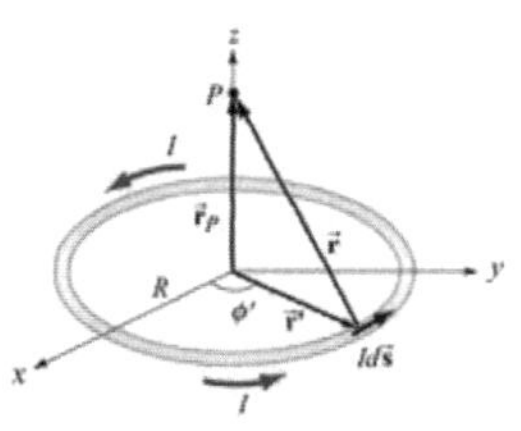

Figure 4. Point P at a distance |z| from Solenoid Coil [3]

For the Helmholtz coil configuration, the equation should be calculated using Biot-Savart Law by considering that the center point is outside the solenoid at a distance from the solenoid's coil. The Biot-Savart law expressed dB from a current source I_{ds}, as:

$$d\vec{B} = \frac{\mu_0}{4\pi}\frac{Id\vec{s}\times\hat{r}}{r^2} \tag{7}$$

μ_0 is the permeability of free space, $4\pi.10^{-7}$ (T.m/A). Looking at Fig. 4, components from (7) can be elaborated by considering the vector of r, as follow [3]:

$$\hat{r} = \frac{\vec{r}}{r} \tag{8}$$

$$r = |\vec{r}| = \sqrt{R^2 + z^2} \tag{9}$$

$$d\vec{s}\times\vec{r} = Rd\phi'[z\cos\phi'\hat{i} + z\sin\phi'\hat{j} + R\hat{k}] \tag{10}$$

Then (7) can be rewritten as:

$$d\vec{B} = \frac{\mu_0}{4\pi}\frac{Id\vec{s}\times\vec{r}}{r^3} = \frac{\mu_0 IR}{4\pi}\frac{z\cos\phi'\hat{i} + z\sin\phi'\hat{j} + R\hat{k}}{(R^2+z^2)^{3/2}} \tag{11}$$

To get amount of magnetic field B, this discrete magnetic field can be Integral. The value of component at axis vector; i and j will be zero. It is as a result of integral of sin and cos function respectively. Thus total B value becomes [3]:

$$B(z) = \frac{\mu_0}{4\pi}\frac{IR^2(2\pi)}{(R^2+z^2)^{3/2}} = \frac{\mu_0 IR^2}{2(R^2+z^2)^{3/2}} \tag{12}$$

2.4. Helmholtz coil with different current flow equation

The purpose of using this Helmholtz coil configuration is copying the previous RCD mechanism. It is to get the difference of magnetic field value from B1 and B2. As seen from Fig. 2, the direction of the two values is against each other on the same axis, axis y. Thus the equations, one can apply superposition on (12) and multiply it with number of coil's turn N_A and N_B, as follow:

$$\Delta B = B1 - B2 = \frac{\mu_0 R^2}{2}\left[\frac{N_B I_{CoilB}}{\left[(y-D/2)^2+R^2\right]^{3/2}} - \frac{N_A I_{CoilA}}{\left[(y+D/2)^2+R^2\right]^{3/2}}\right] \tag{13}$$

This equation shows that the magnetic field's value depends on its coil's radius (R), distance between the two coils (D), the coil's number of turns (N_A and N_B), and the current value on each coil ($I_{Coil\,A}$ and $I_{Coil\,B}$).

To simplify the equation, the calculation can focus only in the middle of those two coils. This can be calculated by setting the position of the middle point on y=0. Thus both coils will have a distance equals to D/2 from the middle point, assuming the number of both solenoid coils is the same, $N_A=N_B=N$. The (13) becomes (14).

$$\Delta B = \frac{\mu_0 NR^2}{2}\left[\frac{I_{CoilB}}{\left[(D^2/4)+R^2\right]^{3/2}} - \frac{I_{CoilA}}{\left[(D^2/4)+R^2\right]^{3/2}}\right] \tag{14}$$

Then, if it follows the default Helmholtz Configuration, the setting is also having D=R. That will make (14) as simpler, as follow:

$$\Delta B = \left(\frac{4}{5}\right)^{3/2}\frac{\mu_0 N(I_{CoilB} - I_{CoilA})}{2R} \tag{15}$$

IOP Conf. Series: Journal of Physics: Conf. Series **853** (2017) 012004 doi:10.1088/1742-6596/853/1/012004

This ΔB value is shown at the middle of Fig. 5, the magnetic field simulation with a text B=1.05E-24. Finally, it is possible to get the relation between magnetic field and $I_{leakage}$. It is by applying (2) into (15) for the leakage current, $I_{leakage,}$ which is shown at (16).

$$\Delta B \cong \left(\frac{4}{5}\right)^{3/2} \frac{\mu_0 N I_{leakage}}{2R} \tag{16}$$

3. Simulation and sensitivity

Fig. 5 is made using EJS (Easy Java Simulation). Many things are also simulated using this software. Some of the examples are common tools like vernier caliper and micrometer [5]. It shows how the magnetic field will behave. It also helps us to draw the equation.

3.1. Simulation

Easy Java Simulation or EJS is an *open source* software. It is designed to author and model science which allows user to create Java programs with minimal programming [2].By using this tool, it is easier for us to simulate the equations. It also helps us to confirm our possible core on designing how the leakage current measurement on medical devices will be.

The designed program has four possible inputs. It is shown at Fig. 5. The inputs are the possible components which are described at (15). They are the coil's radius (R), coil's number of turns (N), and the current value on each coil ($I_{Coil\ A}$ and $I_{Coil\ B}$).

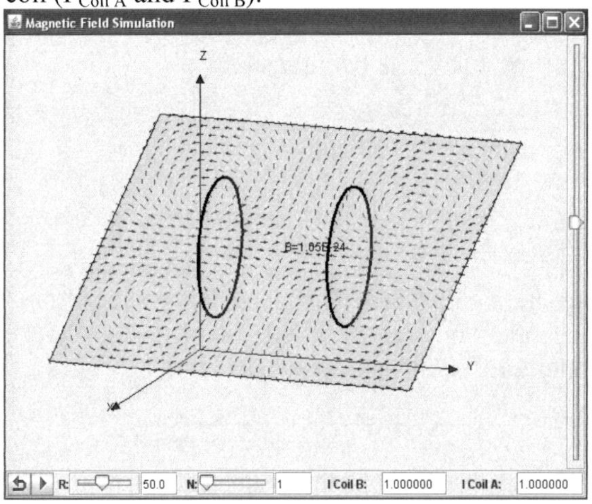

Figure 5. Magnetic Field Simulation

Figure 5 shows the magnetic field condition for a configuration of Helmholtz Coil with R=50, N=1, and there is no leakage current. It is seen from the value of I_{CoilA} and I_{CoilB}. They have the same value of 1.000000 A. At this condition, the magnetic field value at middle point is 1.05×10^{-24}. This value shows that there is very little flux at that position. It is almost zero. This value is not zero because the calculation step on this program is using numerical method.

The program's logic is not directly using (15). It is because each coil is an individual magnetic field source and simulated as it is. That is why the calculation of the middle point's magnetic field is actually subtraction of each individual coil's magnetic field.

3.2. Sensitivity

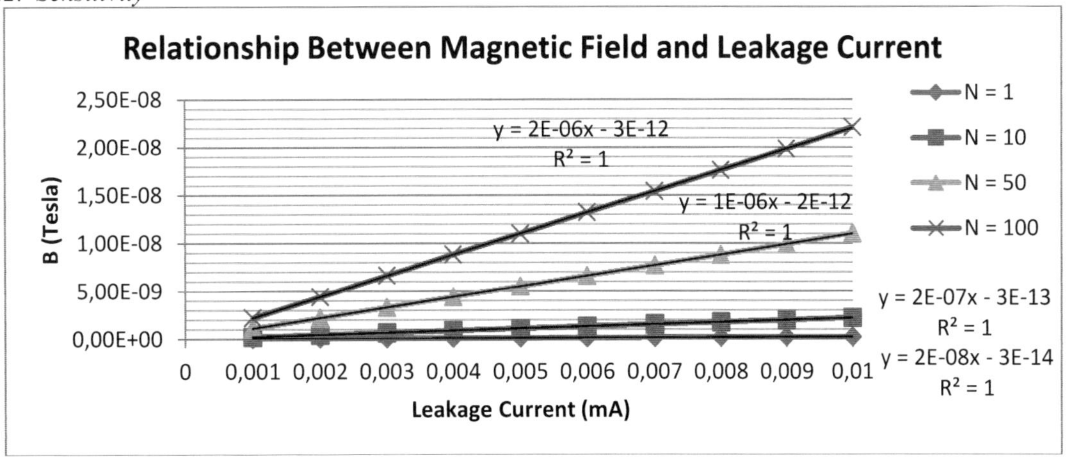

Figure 6. Magnetic Field vs. Leakage Current

From the simulation software, it is possible for us to draw the relation between the magnetic field's value at middle of two coils, ΔB, versus the leakage current, $I_{leakage}$. It is shown at Fig. 6, Magnetic field vs. leakage current.

Fig. 6 also shows possible different magnetic field value by increasing the coil's number of turns, N. It is shown at the figure as N=1, 10, 50, and 100, as examples. The lines at the figure are also showing the sensitivity of the configuration. The sensitivity is the gradient value of each line. They are $\approx 2\text{x}10^{-8}$, $2\text{x}10^{-7}$, $1\text{x}10^{-6}$, and $2\text{x}10^{-6}$ following the number of turns, N respectively. It shows that the magnetic field value is linear to coil's number of turns.

4. Conclusions

This work has shown the possibility of using Helmholtz coil configuration with different current flow as another form of RCD. From the explanation of this paper, some points can be listed:

- The RCD's Formula expressed at (6), and the Helmholtz Coil with Different Current Flow's Formula expressed at (16) shows similar relation between the leakage current and the resulted magnetic field with different value of *k* as in,
- The EJS software can be used to show the magnetic field of Helmholtz Coil with Different Current Flow. Its simulation shows how the magnetic fields from two coils are canceling each other.
- Either using resulted formula, or EJS software, it is possible to get the possible sensitivity of the magnetic field against the leakage current. That sensitivity is linear to coil's number of turns.

The future work will be investigating the sensing coil as shown at Fig. 3 RCD Toroid's Magnetic Field.

Acknowledgment

This work was supported by Faculty of Science and Technology, Airlangga University for Young Lecturer's Research Fund Program. Thanks to all team members for the supports.

References

[1] IEC 60601-1. 1995. *Medical Electrical Equipment-Part 1: General Requirements for Safety and Essential Performance.* (Geneva: International Electrotechnical Commission)
[2] Hwang F K, Wolfgang C, and Robert M 2014 OSP ComPADRE collection June.*Helmholtz Coils*. http://www.compadre.org/osp/items/detail.cfm?ID=8983
[3] Lewin W and John B 2007 *Physics II Electricity and Magnetism* MIT Open Course Ware,

Spring. http://ocw.mit.edu/courses/physics/8-02-physics-ii-electricity-and-magnetism-spring-2007/index.htm

[4] Withfield J 2008 *The Electricians Guide Book*. EPA Press. 16th ed, https://www.tlc-direct.co.uk/Book/5.9.2.htm

[5] Wee L K, Ning H 2015 Vernier caliper and micrometer computer models using Easy Java Simulation and its pedagogical design features--ideas for augmenting learning with real instruments. *Phys. Educ*. **49** issue 5

[6] John B 2007*A Practical guide to IEC 60601-1.* (United Kingdom: Rigel Medical). http://www.rigelmedical.com/downloads/Rigel-Medical-A-Practical-guide-to-IEC-60601-1.pdf

A simulation of surface plasmon resonance-based tapered fiber and sensing

Nina Siti Aminah[1,*], Siti Chalimah[2], Hendro[1], R Hidayat[2], and M Djamal[1,3]

[1]Theoretical High Energy Physics and Instrumentation, Institut Teknologi Bandung, Indonesia
[2]Physics of Magnetism and Photonics Group, Institut Teknologi Bandung, Indonesia
[3]Physics Department, Institut Teknologi Sumatera, Indonesia

*Corresponding email address: nina@fi.itb.ac.id

Abstract. This paper presents the fabrication and simulation of tapered optical fiber. To characterize the properties of these tapered optical fiber sensors, sugar solutions ranging from 0 to 20 in weight percent were utilized. The measured reflection spectra obtained from this SPR sensor shows the observation of a dip originated from the light absorption by the generated surface plasmon wave. The dip reflectance shifts with the increase in the refractive index. A simulation of the experiment was made using OptiFDTD. The simulation and the experiment showed a similar result. The result suggests that a compact sensor based on this structure may be useful for refractive index sensors.

1. Introduction

Surface Plasmon Resonance (SPR) is an interesting optical phenomenon that has been widely applied recently for the detection and analysis of bio molecular interaction. SPR can be observed when the wave vector of the incident light parallel to the conductor surface k_x matches the surface of the plasmon (SP) wave vector k_{sp}. The wave vector of the surface plasmon k_{sp} is sensitive to the change of the refractive index of the dielectric medium ε_s that is in contact with the surface of thin layer metal. The surface plasmon (SP) wave vector k_{sp} is given by

$$K_{SP} = \frac{\omega}{c}\left(\frac{\varepsilon_m \varepsilon_s}{\varepsilon_m + \varepsilon_s}\right)^{1/2} \quad (1)$$

where ε_m is the real part of the dielectric permittivity of the metal layer. When the resonance takes place, it will produce a dip in the measured reflection spectrum [1]. This wavelength and depth of the observed dip will then change when the refractive index of the dielectric medium change. In a sensor employing this SPR phenomenon, when the concentration of analyte molecules changes or the analyte mocecules are bound to the sensor surface, the refractive index at the surface changes, it leads to the shift of SPR dip.

In the past few years, optical sensor based on optical fiber technology has been a subject of intensive research. In order to improve the sensitivity, an enhancement through the inclusion of SPR phenomenon has also become a critical research issue. Several theoretical and experimental studies have been carried out. Tapered fiber structure seems offer an advantage because of its ease to be integrated in the existing optical fiber system. In addition, the formed evanescent field provided by tapered structure could be more intense, such as that it can be more effective to interact with the analyte molecules. However, the realization of this SPR sensor with tapered optical fiber structure is not easy as it is not easy to establish the resonance condition in this structure. In this paper, we report the preparation, characterization, and simulation of a SPR sensor with tapered optical fiber structure.

2. Experiment

The fabrications of tapered fiber have been demonstrated by using a wide range of techniques. In this paper, we used the flame heating technique to elongate the heated part of optical fiber, where the heated part then forms a tapered structure. This technique has proven to be one of the most versatile techniques, which can fabricate tapered fiber with good physical properties. The fabrication employed an oxy-butane torch.

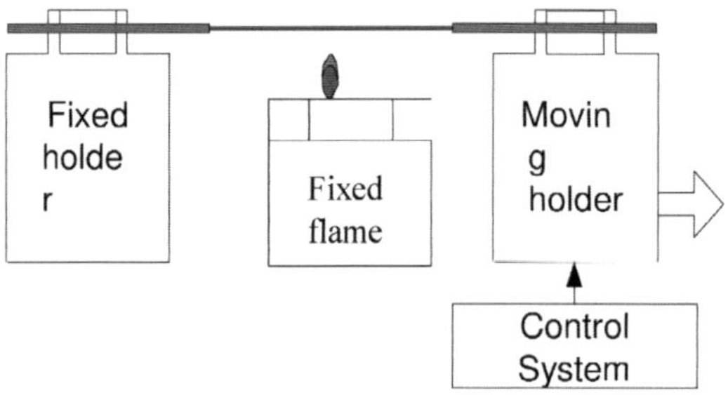

Figure 1. Fiber tapering rig.

Figure 1 shows the schematic diagram of fiber tapering rig used in this work. It mainly comprises of two fiber holders, an oxy-butane burner, a stepper motor, and the motor controller board. Two fiber holders were used for holding and stretching the optical fiber[2]. One was in a fixed. position while the other one was mounted above a motorized linear stage that can move back and forth. The elongated part was heated by an oxy-butane burner while the linear stage was moved to pull out the fiber slowly. For controlling the movement of motorized linear stage, namely controlling the speed of the optical fiber elongation, a high precision stepper motor (S1404) with a resolution of 1.8 /step were used. The tapering routine was implemented by using a computer program. A thin Au film of a thickness about 50 nm was then deposited to the resulted tapered fiber by sputtering technique.

Figure 2 shows a photograph of the resulted SPR sensor with tapered optical fiber structure. The taper angle was found dependent on the movement speed of the linear motor stage for a particular range.

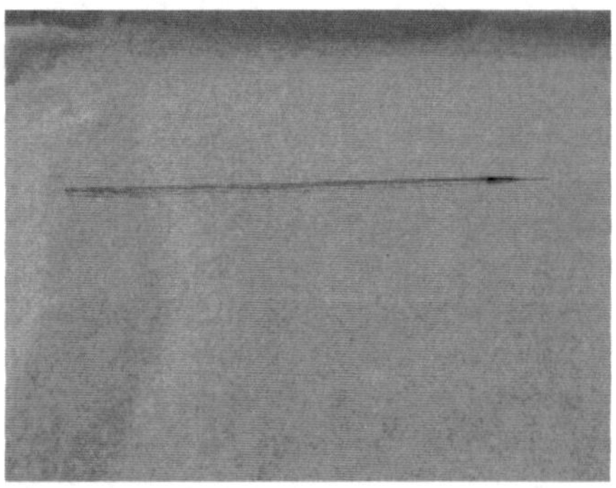

Figure 2. The resulted SPR sensor with tapered optical fiber structure.

The experimental set-up for evaluating the performance of the optical fiber sensors fabricated in this work is shown in Fig. 3.

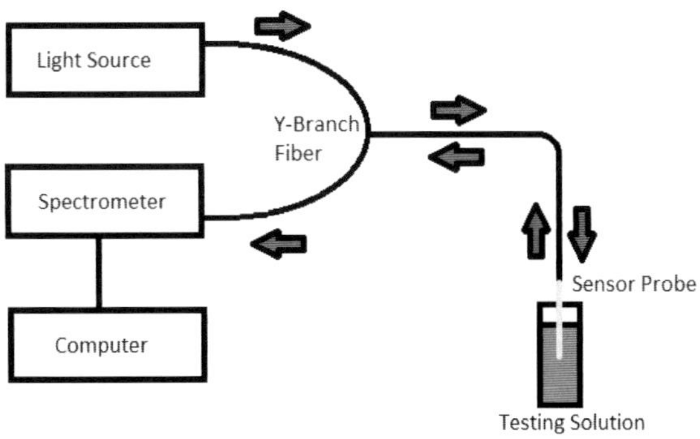

Figure 3. The resulted SPR sensor with tapered optical fiber structure.

The sensor probe was connected to one end of a Y-branch optical fiber and, it was illuminated by light from a tungsten lamp source connected to the other end of the fiber. In order to evaluate the sensor performance, the sensor probe was dipped into a testing solution with a known refractive index. Sugar solution was used as a convenient test sample, which provided refractive index variation by changing the sugar concentration [3]. The testing solutions were then prepared by dissolving sugar in water at different weight ratios. The reflection spectra were measured by a CCD based spectrophotometer which was connected to the other end of the fiber. The experimental set-up for this measurement is shown schematically in Figure 3. The reflection spectrum of air was recorded for comparison.

3. Simulation

The Finite Difference Time Domain (FDTD) is based on a numerical solution of the Maxwell's equation. In the 3D lossless media, Maxwell's equation takes the following form:

$$H_x\Big|^{n+1/2}_{i,j-\frac{1}{2},k-1/2} = H_x\Big|^{n-1/2}_{i,j-\frac{1}{2},k-1/2} + \frac{\Delta t}{\mu_0}\left[\frac{E_y\big|^{n}_{i,j-\frac{1}{2},k} - E_y\big|^{n}_{i,j-\frac{1}{2},k-1}}{\Delta z} - \frac{E_z\big|^{n}_{i,j,k-\frac{1}{2}} - E_z\big|^{n}_{i,j-1,k-\frac{1}{2}}}{\Delta y}\right] \quad (2)$$

$$H_y\Big|^{n+\frac{1}{2}}_{i-\frac{1}{2},j,k-\frac{1}{2}} = H_y\Big|^{n-\frac{1}{2}}_{i-\frac{1}{2},j,k-\frac{1}{2}} + \frac{\Delta t}{\mu_0}\left[\frac{E_z\big|^{n}_{i,j,k-\frac{1}{2}} - E_z\big|^{n}_{i-1,j,k-\frac{1}{2}}}{\Delta x} - \frac{E_x\big|^{n}_{i-\frac{1}{2},j,k} - E_x\big|^{n}_{i-\frac{1}{2},j,k-1}}{\Delta z}\right]$$

$$H_z\Big|^{n+1/2}_{i-1,j-\frac{1}{2},k} = H_z\Big|^{n-1/2}_{i-\frac{1}{2},j-\frac{1}{2},k} + \frac{\Delta t}{\mu_0}\left[\frac{E_x\big|^{n}_{i-\frac{1}{2},j,k} - E_x\big|^{n}_{i-\frac{1}{2},j-1,k}}{\Delta y} - \frac{E_y\big|^{n}_{i,j-\frac{1}{2},k} - E_y\big|^{n}_{i-1,j-\frac{1}{2},k}}{\Delta x}\right]$$

The locations of E and H components in the computational domain is shown below

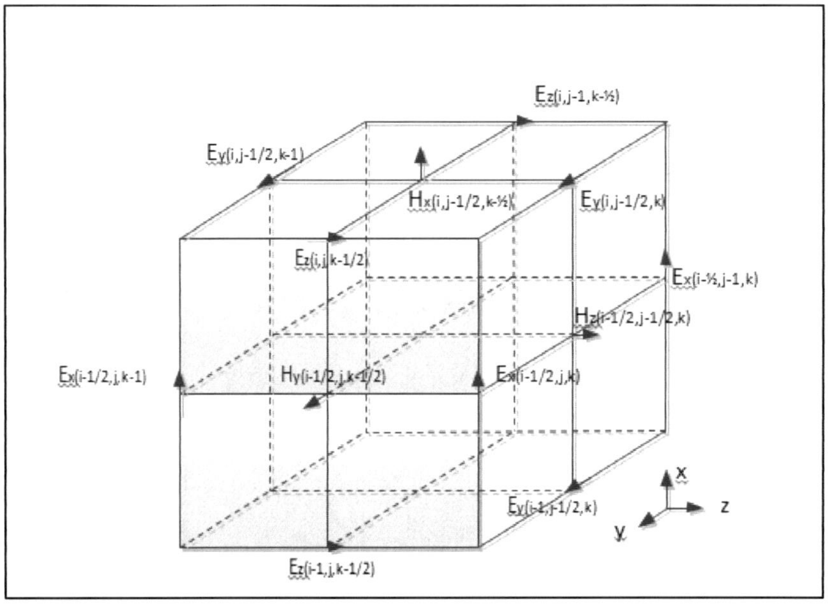

Figure 4. Location of the TM field components in the computational domain[4].

The resulting finite-difference equation are solved in software.

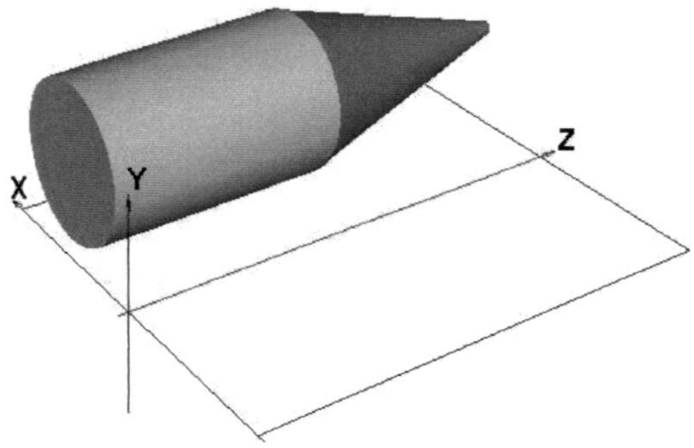

Figure 5. Location of the TM field components in the computational domain[4].

The structure and size of sensing probe used for simulation are adjusted to the sensing probe used in the experiment. We used unicone tapered fiber with pure silica core and fluorine doped silica cladding from Ocean Optics. Numerical aperture = 0.22, fiber core thickness (large end diameter) = 600μm, small end diameter = 10μm, gold as metal with thickness = 50nm.

4. Results and discussion

Fig. 6 clearly shows the reflection spectra obtained from the fabricated SPR sensor with tapered optical fiber structure when dipped into the testing solution with different refractive indexes

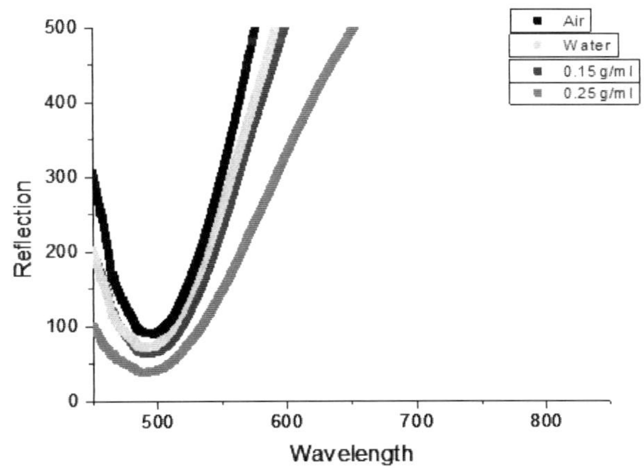

Figure 6. Overlapping reflection spectra of the tapered fiber refractive index sensors.

The comparison between the reflection spectrum measured in air (black), in water (green), in 0.15g/ml sugar solution (blue), and the reflection spectrum dipped in 0.25g/ml sugar solution (red) was shown is Figure 6. The figure shows the change of reflection spectra. The refractive index of air and water are 1.00 and 1.33, respectively, while the refractive index of sugar solution is larger than 1.33 depending on the sugar concentration. However, we also noted that the sugar solution also become darker with the increase of sugar concentration. Therefore, this leads to the increase in refractive index complex with the increase of sugar concentration.

It is noticeable that the dip wavelengths in the reflection spectra of these sensors were almost located at the same wavelength, namely at about 490 nm. With the increase of sugar concentration, we can observe the shift of the dip reflectance, although there is no significant change in the dip wavelength. Therefore, the absorbance of the dip in the reflection spectrum can be chosen as the sensitive parameter for this type of SPR sensor.

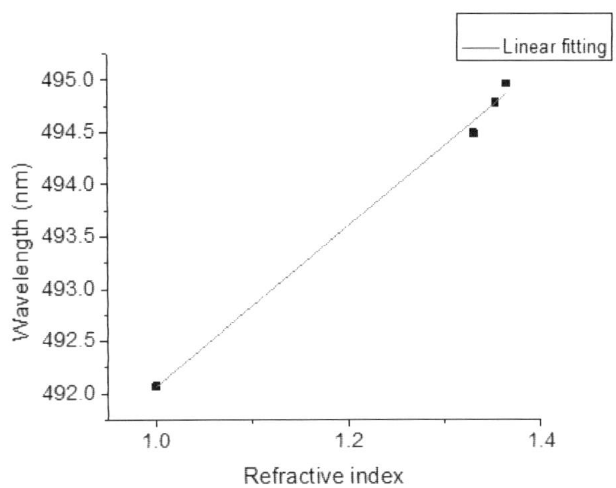

Figure 7. The fitting curve for wavelength versus refractive index curve experiment

In analyzing the data, we can find that the wavelength in which dip happened and the refractive index showed a relationship of similar linearity so we can get a fitting curve. The fitting curve is shown in figure 7. The fitting equation is as follows:

$$\lambda = 0.44024RI + 484.35606 \quad (3)$$

With RI means refractive index[5]. Sensitivity of the tapered fiber is 0.44024.

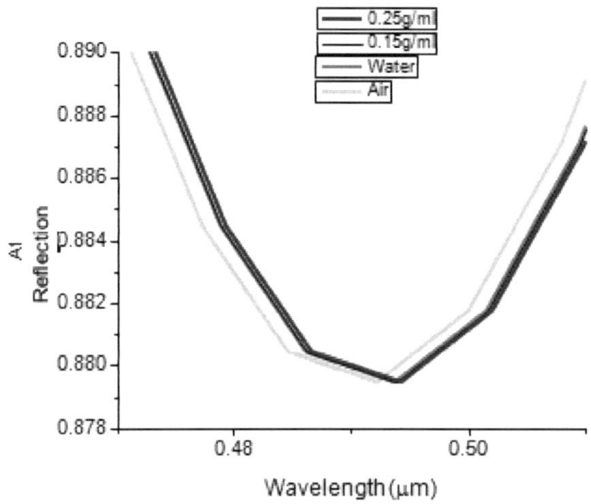

Figure 8. Overlapping reflection spectra of the simulation.

The experimental result showed small changes in profile and intensity. There might be some reasons for this. First, the tested tapered fiber sensor was not as ideal as we hoped for. Second, there might be physically adsorbed solution which also changed the properties of the tapered waveguide.

In the simulation, the shift of the reflection spectrum can be achieved by setting different media with different refractive index. Air and water were set at 13% and then sugar solution at 20% with refractive index 1.3525 and 1.3639 respectively [3]. Fig. 8 shows clearly the reflection spectra obtained from the simulation SPR sensor with tapered optical fiber structure when dipped into the testing solution with different refractive indexes. Fig. 9 shows the fitting curve.

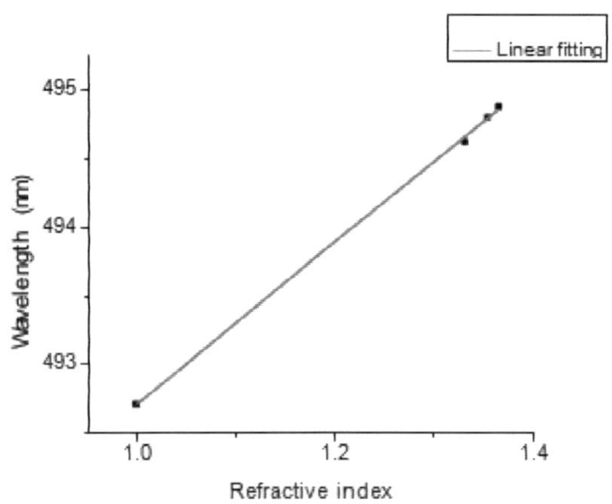

Figure 9. The fitting curve for wavelength versus refractive index simulation.

The fitting equation is as below:

$$\lambda = 0.13031\mathrm{RI} + 486.76028 \tag{4}$$

where RI means refractive index. The sensitivity of the tapered fiber using simulation is 0.13031. The simulation and the experiment showed similar result.

5. Conclusion

An SPR sensor with tapered optical fiber structure has been successfully fabricated by a simple technique. The characteristics of generated SPR waves in various medium with different refractive indexes, namely air, water, and sugar solutions, were able to be measured. A dip in the measured reflection spectrum was observed where its reflectance spectrum shifts with the increase in refractive index. The simulation of the experiment was made using OptiFDTD. The simulation and the experiment showed similar result. The result suggests that a compact sensor based on this structure may be useful for refractive index sensors.

Acknowledgments

The authors are thankful to Penelitian Unggulan Perguruan Tinggi ITB 2016. Authors would like to thank Dr. Yi Chen, Ocean Optics, for his help with information of fiber specification and Dr. Dwi Irwanto for his help with the world of computers.

References

[1] J Homola 2006 Surface plasmon resonance based sensors *Springer* Berlin
[2] S W Harun, K S Lim, C K Tio, K Dimyati, and H Ahmad 2013 Theoretical analysis and fabrication of tapered fiber *Optik* vol. **124** no. 6 pp. 538–543.
[3] Manual for Sugar solution Prism. Retrieved November 20 2015 from http://www.frederiksen.eu/fileadmin/user_upload/PDF/Export_Manuals/545920_E_sugarsolution.pdf.
[4] OptiFDTD technical program and tutorial Available: www.optiwave.com/download-1/optifdtd-10-technical-background-and-tutorials [August 20, 2016].
[5] L w Li and X h Sun 2012 Investigation on the tapered fiber evanescent-field sensor based on the comsol software *Photonics and Optoelectronics* (SOPO) pp. 1-3.

IOP Conf. Series: Journal of Physics: Conf. Series **853** (2017) 012006 doi:10.1088/1742-6596/853/1/012006

Development of TGS2611 methane sensor and SHT11 humidity and temperature sensor for measuring greenhouse gas on peatlands in south kalimantan, indonesia

I Sugriwan[1,*] and O Soesanto[2]

[1]Department of Physics, UniversitasLambungMangkurat, Jl. A. Yani KM 36 Banjarbaru Kalimantan Selatan 70714, Indonesia
[2]Department of Mathematics, UniversitasLambungMangkurat, Jl. A. Yani KM 36 Banjarbaru Kalimantan Selatan 70714, Indonesia

*Corresponding email address: iwansugriwan@unlam.ac.id

Abstract. The research was focused on development of data acquisition system to monitor the content of methane, relative humidity and temperature on peatlands in South Kalimantan, Indonesia. Methane is one of greenhouse gases that emitted from peatlands; while humidity and temperature are important parameters of microclimate on peatlands. The content of methane, humidity and temperature are three parameters were monitored digitally, real time, continuously and automatically record by data acquisition systems that interfaced to the personal computer. The hardware of data acquisition system consists of power supply unit, TGS2611 methane gas sensor, SHT11 humidity and temperature sensors, voltage follower, ATMega8535 microcontroller, 16 x 2 LCD character and personal computer. ATMega8535 module is a device to manage all part in measuring instrument. The software which is responsible to take sensor data, calculate characteristic equation and send data to 16 x 2 LCD character are Basic Compiler. To interface between measuring instrument and personal computer is maintained by Delphi 7. The result of data acquisition showed on 16 x 2 LCD characters, PC monitor and database with developed by XAMPP. Methane, humidity, and temperature which release from peatlands are trapped by Closed-Chamber Measurement with dimension 60 x 50 x 40 cm^3. TGS2611 methane gas sensor and SHT11 humidity and temperature sensor are calibrated to determine transfer function used to data communication between sensors and microcontroller and integrated into ATMega8535 Microcontroller. Calculation of RS and RL of TGS2611 methane gas sensor refer to data sheet and obtained respectively 1360 ohm and 905 ohm. The characteristic equation of TGS2611 satisfies equation $V_{RL} = 0.561 \ln n - 2.2641$ volt, with n is a various concentrations and V_{RL} in volt. The microcontroller maintained the voltage signal than interfaced it to liquid crystal displays and personal computer (laptop) to display result of the measurement. The result of data acquisition saved on excels and database format.

1. Introduction

Peatlands are a specific type wetland formed by the accumulation of incompletely decayed plant material by forming layers of peat. They are middling between terrestrial area and wetland where only found handful of standing water. The depth of peatlands is more than 30 cm to 40 cm with organic content more than 50% by weight [1].

Indonesia has tropical peatlands that spread out in the islands of Sumatera, Kalimantan, and Papua. Tropical peatlands are involved 10% of the coastal, sub-coastal and low altitude areas with estimates reach 16.8 to 27.0 million ha [2]. Peatlands in Kalimantan spans nearly 5.9 million ha, approach 11% of total terrestrial land area in Kalimantan [3]. Tropical Peatlands in South Kalimantan, Indonesia, could emission of three main greenhouse gases especially carbon dioxide, methane and nitrous oxide [4].

Methane, emitted by peatlands to the atmosphere, plays a major role in determining the climate of the atmosphere. The dominant sources of methane are single-celled archeae by manner inhibit anaerobic micro sites in the soil and water. Methane is produced by archeae via a group of methanogens. Environment parameters like humidity and temperature also take effect on production on methane [5].

Closed-chamber is a media that used to trap greenhouse gases that emitted by peatlands to atmosphere. The method of closed-chamber measurement is the most used to determine CH4 fluxes in peatlands and to measure rates of gas exchange between soil, vegetation and the atmosphere in terrestrial [6]. Mostly, in measuring the concentration of methane emitted from peatlands, gas chromatography such as gas chromatography (GC-8A) which consisted of a flame ionization detector maintained at 130 ^{0}C and a 2-m long activated carbon column (80/100 mesh) at 70 ^{0}C with pure nitrogen gas used as a carrier [7].

In this research, we develop microcontroller-based methane measurement with TGS2611 and humidity and temperature with SHT11. The system of hardware of data acquisition consists of powersupply unit, TGS2611 methane gas sensor, SHT11 humidity and temperature sensors, voltage follower, ATMega8535 microcontroller, 16 x 2 LCD character and personal computer. ATMega8535 module is a device to manage all part in measuring instrument. The software which is responsible to take sensor data, calculating characteristic equation and sending data to 16 x 2 LCD character are Basic Compiler. To interface between measuring instrument and personal computer is maintained by Delphi 7. The result of data acquisition which shown on 16 x 2 LCD characters, PC monitor and database with developed by XAMPP.

2. Material and methods

The data acquisition system for monitoring methane, humidity, and temperature in closed-chamber measurement on peatlands was conducted by pursuing block diagram presented in Figure 1. Two sensors, TGS2611 and SHT11, were connected with signal conditioning. TGS2611 was formed with the voltage divider, Figure 2, and then output voltage signals were taken from RL, whereas SHT11 configured with two wire interfaces. Output voltage signals connected with ATMega8535 microcontroller subsequently connected with LCD, personal computer, and database.

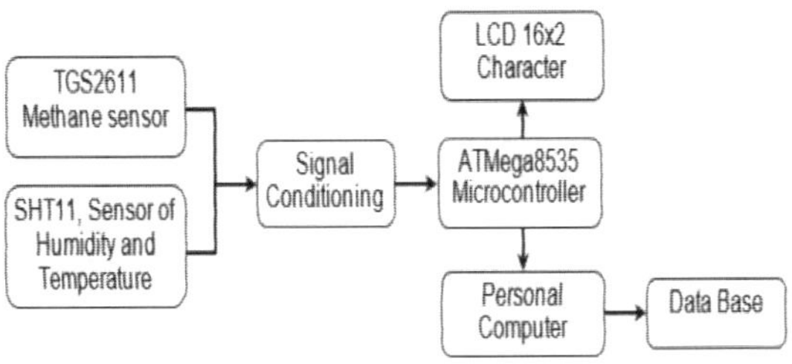

Figure 1. Block diagram of Instrumentations

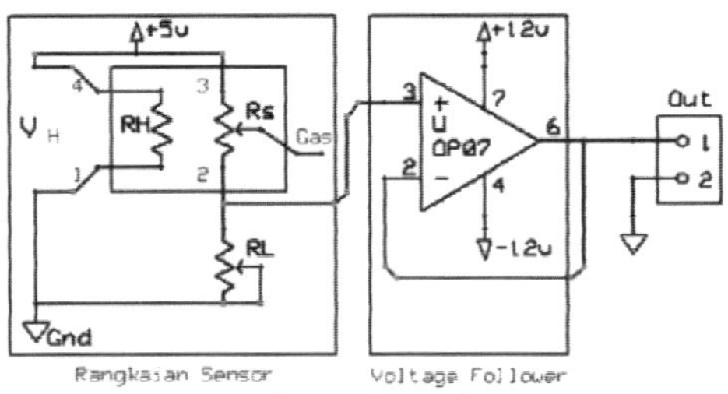

Figure 2. Configuration of TGS2611

TGS2611 is a methane gas sensor in the form semiconductor type which has low power consumption, very high sensitivity and long life. TGS2611 requires only 56 mA for heater current due to small size chip and housed in a standard TO-5 package [8]. The various concentration of methane displayed by TGS2611 has changes value of sensor resistance (RS). The sensor requires two voltage inputs that are the voltage of heater (VH) and voltage of the circuit (VC), shown in Figure 2. Power

IOP Conf. Series: Journal of Physics: Conf. Series **853** (2017) 012006 doi:10.1088/1742-6596/853/1/012006

supply circuit of TGS2611 was commonly applied +5 volt for both VH and VC. Pin 1 connected with +5 volt and pin 4 with the ground, similarly for pin 3 and pin 2. The load resistor (RL) which is assembled between pin 2 and ground and voltage signal (VRL) was accepted from RL whose value proportional to the concentration of methane were detected by TGS2611. The sensor resistance (RS), RL, VC, and VRL was calculated by using equation 1 and the power of dissipation using equation 2 in the following formula.

$$R_S = \frac{V_C - V_{RL}}{V_{RL}} \, x \, R_L \qquad (1)$$

$$P_S = \frac{(V_C - V_{RL})^2}{R_S} \qquad (2)$$

SHT11 has a fully calibrated, digital output, low power consumption, and long terms stability sensor of humidity and temperature. Internally, band gap sensor was applied to measure temperature while a unique capacitive element sensor was used to measure relative humidity. SHT11 pin assignment consisted eight pins; namely ground (GND), serial data (DATA), serial clock (CLK) and the source voltage (VDD) for pin 1 – 4, while pin 5 – 8 had no connection (NC) [9]. VDD and GND were connected to 5 volts and 0 volts (ground) from power supply unit. Output data signals from SHT11 were in form digital serial and then were interfaced with microcontroller by two wire serial interface which applies DATA and SCK respectively connected with port B.1 and port B.0 microcontroller.

Operational amplifier (op-amp) is a device with two input signals and one output signal which used to amplification output voltage signal from sensor device. Op-amp could be configured in five sharp: non-inverting amplifier, inverting amplifier, the differential amplifier, voltage follower and instrumentation amplifier. OP07 was a small scale integrated circuit to accommodate the op-amp [10]. In this research, OP07 are configured as voltage follower where inverting input (pin 2) directly connected to the output voltage (pin 6) of OP07 and non-inverting input receive the voltage signal from TGS2611. Output voltage signal was connected to the microcontroller, presented in Figure 3.

Figure 3. Component sides of TGS2611 and OP07

Figure 4. Connection ATMega8535 with OP07, LCD, and SHT11

The ATMega8535 is a heart of instrumentation systems. ATMega8535 is an 8-bit microcontroller with low power complementary metal oxide semiconductor (CMOS) by RISC architecture and part of Alf and Vegard RISC Processor (AVR) family [11]. Figure 4 showed the connection signal conditioning, SHT11, and LCD to ATMega8535. There were four input/output ports with each port have 8 bits, port A, port B, port C and port D. Port A.0 to A.7 are analog input port where internally include with analog to digital conversion (ADC) and the multiplexer. Port A.0 received the analog voltage from TGS2611 and port B.0 and B.1 received digital serial data from SHT11.

LCD is a thin film and flat screen device consisting amount of color pixel which composed in the front of light source. Each pixel consists of liquid crystal column that hangs up with two transparent electrodes and two filters polarization [12].

Basic Compiler (Bascom) is a software to write a program to the microcontroller that produced by MCS Electronics. Program language is based on the structure of basic language program. Bascom can execute the arithmetic command, logic, and special command to configure LCD pins [13]. Bascom was used to managing output data signal of TGS2611 which entered from port A.0 ATMega8535, to maintain two wire interface from SHT11 via port B.0 and B.1 ATMega8535, and interfaced between ATMega8535 and LCD.

Delphi is a visual program package in windows environment which using pascal compiler. Although Delphi uses pascal logic in programming, Delphi appearance is more interesting, less time consuming to write a long and complicated syntax for drawing, laying and adjusting of the image. Delphi also provides more choice interface component such as menu, drop down, pop up menu, text box, radio button, check box, TComPort and others. TComPort is a component on delphi to organize communication between peripheral devices and the personal computer via serial port [14]. Delphi is responsible for displaying the result data measurement from TGS2611 and SHT11 in the form of data array, graphs and excel. In addition, Delphi does not only display the result of data measurement, but also interface with the database.

Xampp is a database software package that includes with apache as a web server, MySql as a database server, and PHPMyAdmin. Xampp is used to storage the result data monitoring to the database [15]. The Xampp and MySql-ODBC-connect were installed to the personal computer to communication between Delphi and database systems. Recording data on MySql database can be displayed by online where MySql combined with web language program like HTML.

3. Discussion

3.1. Determination of characteristics equation of TGS2611

Data sheet of TGS2611 showed that the value of sensor resistance ranges from 0.68 to 6.8 kilo ohm and the span of methane concentration from 500 to 10000ppm. In other words, the value of RS was 0.68 kilo ohm equal to 500ppm of methane concentration as same as 6.8 kilo ohm equal to 10000ppm. The value of VRL was 2 volts when the concentration of methane is 2000ppm, so the value of RS was obtained by comparison with following formula:

$$R_S = \frac{2000\ ppm}{10000\ ppm} x6800\ ohm \tag{3}$$

$$R_S = 1360\ ohm$$

The value of RS = 1360 ohm used to compute the value of RL, which referred to equation 1 obtained that RL = 905 ohm. Typical sensitivity characteristics of TGS2611 could be determined by graph approximation on data sheets. For n was a various concentration of methane in ppm, VRL satisfies the equation:

$$V_{RL} = 0.561 \ln n - 2.2641 \tag{4}$$

3.2. Realizations of instrument systems

The instrumentation systems were realized to supporting measurement concentration of methane, humidity, and temperature on peatlands, presented in Figure 5. Two probes of sensors were embedded in closed-chamber measurement; the probe of TGS2611 was shown in Figure 1SHT11 in Figure 6. Closed-chamber was placed on peatland and methane emission accommodated in the chamber. Methane that released on peatlands was sensed by TGS2611 and SHT11 measure humidity and temperature in the chamber.

Figure 5. System of Instrument are implementes on Peatlands

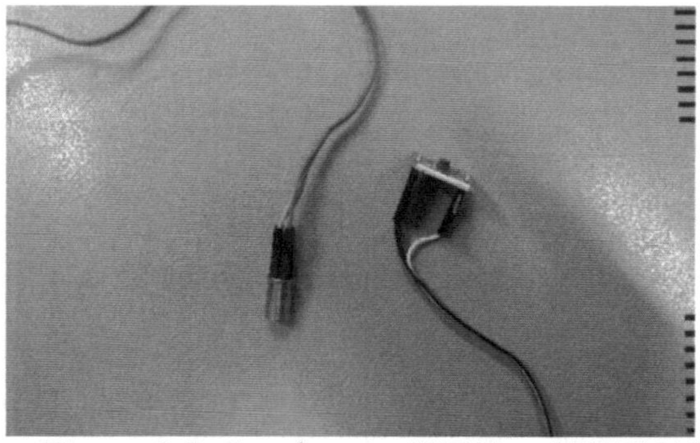

Figure 6. Probe of sensor SHT11 and TGS2611

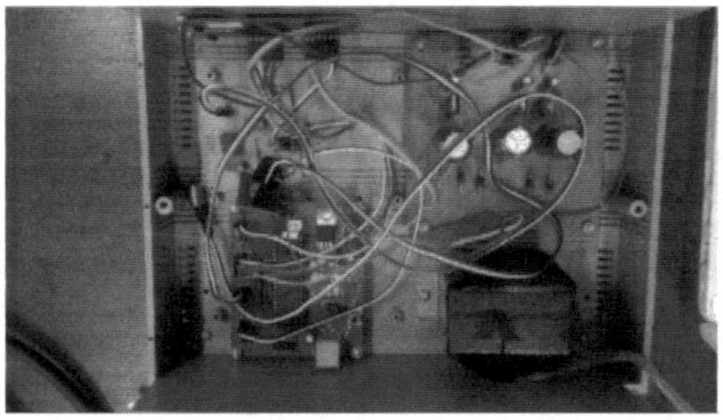

Figure 7. Realization of instrument system

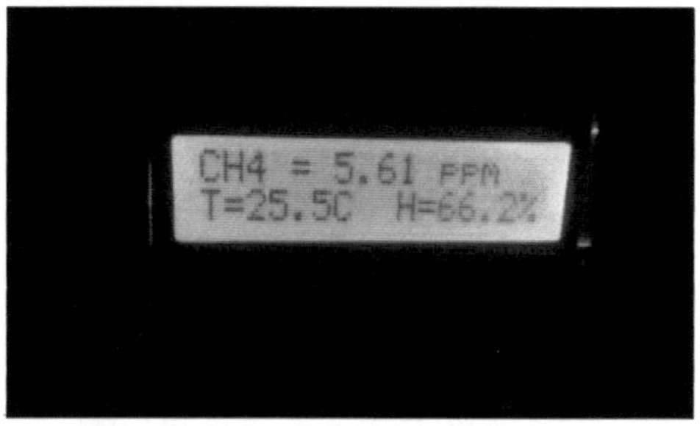

Figure 8. Displaying of LCD

All component in data acquisition system which consisted of TGS2611, SHT11, voltage follower with OP07, microcontroller ATMega8535 and LCD 16x2 character generated by the regulated power supply, shown in Figure 7. The power supply was designed with three output voltage signal, which were split supply ±12 volt for generates voltage follower and +5 volt for all microcontrollers ATMega8535, TGS2611 and SHT11. The voltage signal from TGS2611 was connected by voltage follower with OP07 then connected with port A0 microcontroller ATMega8535 to conversion by 10 bits ADC in ATMega8535. Microcontroller ATMega8535 realized to manage output voltage signal from TGS2611 and SHT11, to convert analog to digital conversion, to interface with LCD and personal computer. Figure 8 presented LCD which displaying three parameters which were CH4, T, and H.

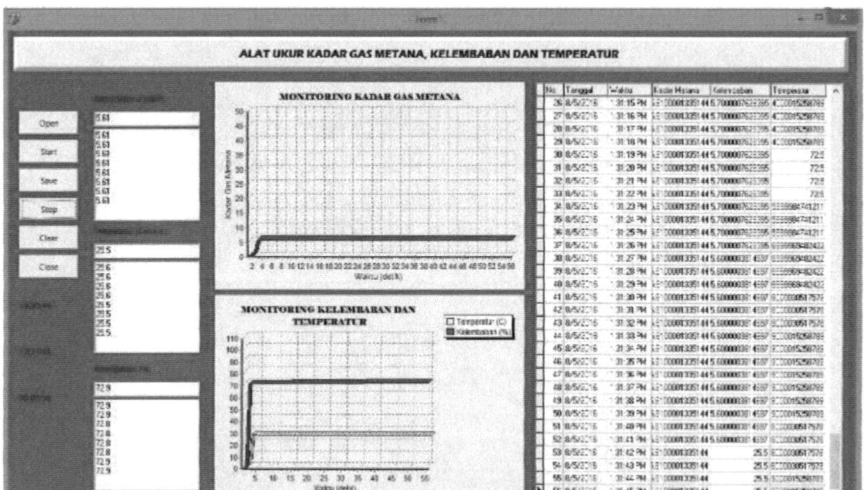

Figure 9. Display of data acquisition on PC/laptop

The result of data acquisition shown on the monitor screen of personal computer or laptop in Figure 9 is an application program to create the application developed by Delphi. On the left side on the form in Figure 9, monitoring windows were equipped with six buttons namely open, start, save, stops, clear and close. Each button running used to configure protocol interface, to start measurement of methane, temperature, and humidity, to storage the result data monitoring in excel form shown in Figure 10, to stop application program operation, to erase all value in text box and to close the window. Three textboxes showed respectively the content of methane, value of temperature and humidity. The value of physical parameters corresponded to two graphs that figured in the middle of windows. The upper one was graph referred to the content of methane were detected in closed-chamber measurement, the other one presentation of temperature and humidity. On the right side, the form of application program presenting all value of acquisition that is treated by the database.

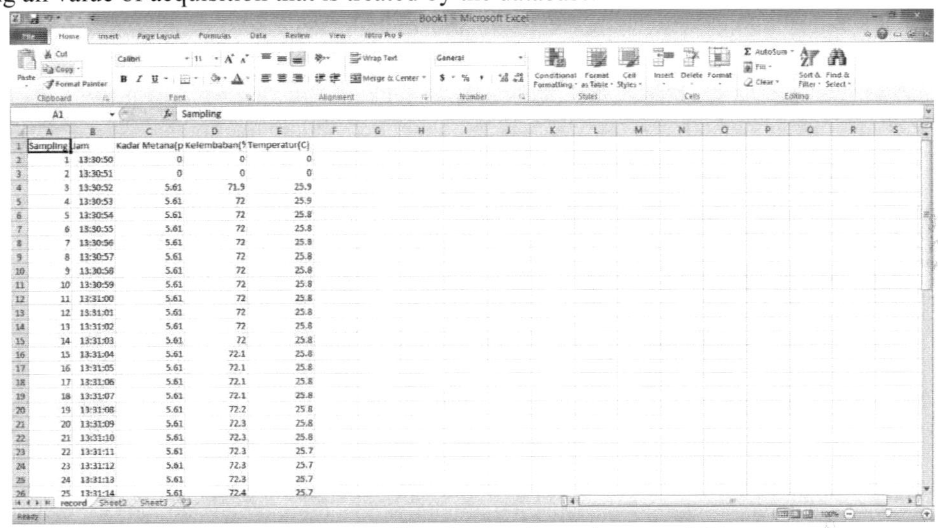

Figure 10. Recording data and saved on excel form

The display of data acquisition was shown on the database which developed by xampp. Database has contents six column that named *nomor* (number), *tanggal* (date), *waktu* (time), *kadarmetana* (concentration of methane), *kelembaban* (humidity) and *temperatur* (temperature). Each column presented real time data monitoring. The advantages in using a database in the measurement system are monitoring data can directly record, without the person to write the result of data acquisition, and data obtained can be processed and relocated to other media.

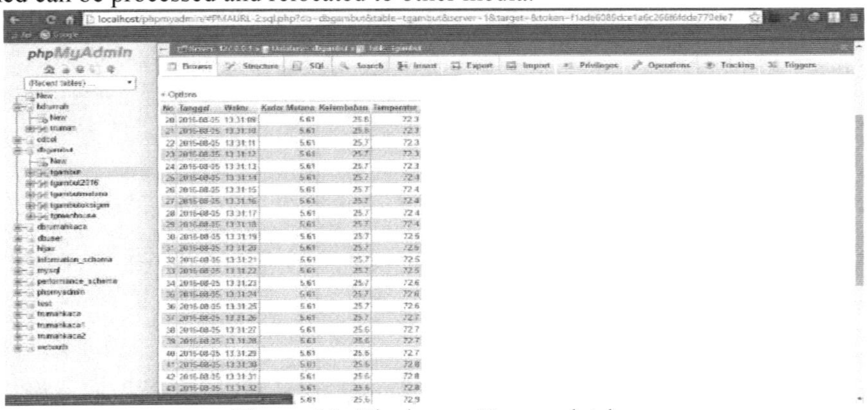

Figure 11. Display on Xampp database

4. Conclusion

Development of TGS2611 methane sensor and SHT11 humidity and temperature sensor gave a stable performance in monitoring methane, humidity, and temperature in peatlands. By monitoring system of methane, humidity and temperature on peatlands, methane production which emitted as greenhouse gas and change of humidity and temperature in closed-chamber measurement would be recorded always in real time and continuously. Portable dimension of the equipment gives another advantage which is easier to install in closed-chamber on peatlands.

References

[1] Charman D J 2009 *Peat and Peatlands* Elsevier pp. 541-548
[2] Jaenicke J, Rieley J O, Mott C, Kimman P, Siegert F 2008 Determination of the amount of carbon stored in Indonesian peatlands, *Geoderma* **147** (2008) 151–158
[3] Astiania D, Mujiman, Hatta M, Hanisah, Fifian F 2015 Soil CO2 respiration along annual crops or land-cover type gradients on West Kalimantan degraded peatland forest *Procedia Environmental Sciences* **28** pp 132 – 141
[4] Inubushi K, Furukawa Y, Hadi A, Purnomo E, Tsuruta H 2003 Seasonal changes of CO2, CH4 and N2O fluxes in relation to land-use change in tropical peatlands located in coastal area of South Kalimantan *Chemosphere* **52** pp 603–608
[5] Baldocchi D, Detto M, Sonnentag O, Verfaillie J, Teh Y A, Silver W, Maggi N K 2012 The challenges of measuring methane fluxes and concentrations over a peatland pasture *Agricultural and Forest Meteorology* **153** (2012) 177– 187
[6] Forbrich I, Kutzbach L, Hormann A, Wilmking M 2010 A comparison of linear and exponential regression for estimating diffusive CH4 fluxes by closed-chambers in peatlands, *Soil Biology & Biochemistry* **42** pp 507-515
[7] Hamada Y, Darung U, Limin S H, Hatano R 2013 Characteristics of fire-generated gas emission observed during a large peatland fire in 2009 at Kalimantan, Indonesia, *Atmospheric Environment* **74** 177e181
[8] Figaro Product Information, 2005, TGS 2611 - for the detection of Methane
[9] Data sheets SHT1x (SHT10, SHT11, SHT15) Humidity and Temperature Sensor IC, Version 5 – December 2011, www.sensirion.com
[10] Data sheets OP07, 2002, Ultralow Offset Voltage Operational Amplifiers, Analog Device
[11] Data sheets ATMega8535, 2006, 8-bit Microcontroller with 8K Bytes In-System Programmable Flash, Atmel Corporation
[12] Data Sheets character LCD 16x2, STN, Gray Background, White Backlight (#712-027)
[13] BASCOM AVR, Help Reference, Version 1.11.8.5 MCS Electronics 1995-2007
[14] Cantu M, Mastering Delphi 7, Sybex © 2003 (1011 pages)
[15] Dvorski D D, 2007, Installing, Configuring, and Developing with Xampp, Skills Canada – Ontario

IOP Conf. Series: Journal of Physics: Conf. Series **853** (2017) 012007 doi:10.1088/1742-6596/853/1/012007

Development of ozone sterilization system based microcontroller for *E. Coli* bacteria sterilization

Arfan Eko Fahrudin[1,*], Endarko[2], Amar Vijai Nasrulloh and Nurma Sari

[1]University of Lambung Mangkurat, Indonesia
[2]Sepuluh Nopember Institute of Technology, Surabaya, Indonesia

*corresponding email address : arfan_eko@unlam.ac.id

Abstract. It has been created microcontroller-based ozone sterilization system which is used for *E. Coli* bacteria sterilization. The created system consists of a neon sign transformer, corona electrode, ozone gas chamber container and ozone gas metering systems. Neon sign transformer and corona electrode are then assembled into an ozone generator that works on the principle of corona discharge. The use of ozone generators is controlled by microcontroller system. Ozone gas is produced by ozone generators which housed in a chamber and equipped by ozone gas indicator. The results of ozone gas effect testing in *E. Coli* with varied exposure of 15, 30 and 45 minutes showed the decrease number of *E. Coli* colonies of tested bacteria samples compared to the control sample without ozone gas exposure.

1. Introduction

In medical front, preventing hospital infection or pathogenic disease, for instance surgical instrument, sheet, food, and human body, has to be sterilized. These things cannot be applied in unitary sterilization technology. Therefore, the different sterilization methods are chosen depending on targets. For example, to sterilize small-sized and heat-resistant instrument, the autoclave apparatus, which utilizes heat, is used. However, this method cannot be applied to medical equipment which made of plastic or latex because their behaviour is destroyed by the heat. One of the applied methods for the cases is the plasma sterilization method [1]. Ozone is a kind of plasma sources used for sterilizing.

Ozone gas is the gas occurred from 3 atoms of oxygen gathering. The oxygen gas can burst to be oxygens free atoms (O). In addition, they are gathered in O_2 into ozone gas O_3 [2].Ozone is a practically transparent gas with very strong absorption in the region of a 245 nm wavelength. Besides, ozone is considered as a powerful disinfecting and oxidizing agent. Due to the reasons, it is used in a wide range of applications such as wastewater treatment, food processing, fire restoration, hospital disinfectant, etc. When ozone contacted with organic compounds or bacteria, the extra atom of oxygen will destroy the contaminant by oxidation. Furthermore, ozone will neutralize virtually all organic odors, specifically those containing carbons as their base element [3]. Ozone gas usually generated from corona discharge that arise from the specific electrode applied by a high voltage power supply [4].

Standard test methods typically include indicator test organism which represent of those found in surface to be sanitized. *E. Coli* is the most frequently used principal test organism for sanitizer studies. It is complemented by the selected secondary organism for the specific use recommended by sanitizer. For instance, *E. Coli* and *Staphylococcus aureus* are used for testing the same food-contact surface sanitizer; *E. Coli* and *Micrococcus caseolyticus* are often used for dairy sanitizers, and etc [5]. The

IOP Conf. Series: Journal of Physics: Conf. Series **853** (2017) 012007 doi:10.1088/1742-6596/853/1/012007

usage of sanitizer for *E. Coli* sterilization from previous research was done using ultraviolet light [6-7], Dielectric Barrier Discharge (DBD) air plasma [8], helium plasma and ozone gas [9]. This study developed ozone sterilizer based microcontroller and investigates the effect of ozone gas to the *E. Coli*.

2. Materials and methods

2.1. Ozone generator and microcontroller system

In this study, ozone generator system consisting neon sign transformer and corona electrode is made to produce ozone gas which used as sterilization process. Neon sign transformer serves as a high voltage source connected to the corona electrode. In addition, it results corona discharge in the electrodes and produce ozone gas. Corona electrode is made from two cylindrical electrodes with a glass as dielectric barrier between them. Corona electrode's picture is shown in Figure 1.

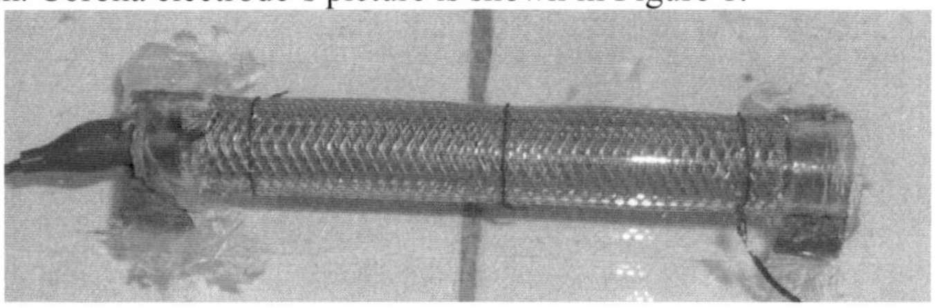

Figure 1. Corona electrode

Microcontroller-based controller system is developed to aid the use of ozone generator. Through this system, the use of ozone generator and ozone gas detector can be performed easily. Some components required to actualize the controller system, for instance microcontroller module arduino, MQ131 sensor, LCD 16x2, memory card module, Real Time Clock (RTC) module, and a rotary switch. The circuit of microcontroller system is presented in Figure 2.

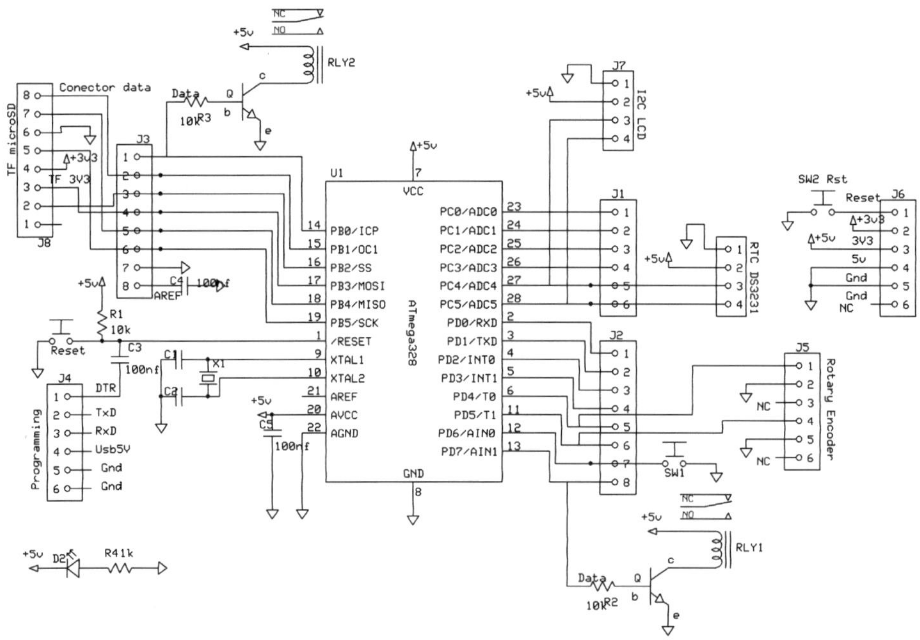

Figure 2. The circuit of microcontroller system

The algorithm is used for the microcontroller system to determine the time setting of ozone generator with the rotary switch considered as the first step. When the start button is pressed, the ozone generator will turn on according to the specified time setting and start to read the sensor data which automatically stored to the memory card. When the ozone generator stops working, the ozone detection continues to run until the reset button is pressed. The flow chart or flow diagram program from proposed algorithm is shown in Figure 3.

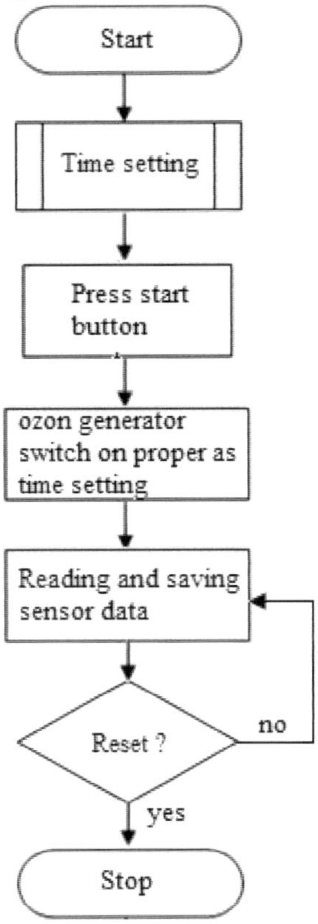

Figure 3. Flow chart algorithm of the microcontroller system

Generators electrode ozone and ozone sensors are placed in the chamber so that the ozone gas that will be used for sterilization accommodated in the chamber. Before the ozone generator is used, the pressure inside the chamber is set smaller than the outside air pressure. As the result, if the ozone gas generated by the ozone generator, there will be no ozone gas coming out during the use of ozone gas in the sterilization process.

2.2. Testing on the E. Coli bacteria

The survival rate of *E. Coli* was measured after ozone exposure under time variation to investigate the sterilizing effect of ozone gas. The cultured *E. Coli* bacteria on agar plates was used as indicator bacterium. During the sterilization treatment, the agar plate was placed in the chamber. After ozone treatment, colony condition of agar plate was observed for 24 hours in incubation.

3. Results and discussion

The prototype of microcontroller system to control the ozone generator was developed in this study as it is represented in Figure 4. In the LCD display output voltage, the sensor is shown as an indicator of the detection presence or absence of ozone gas. Automatically, if there is no detected ozone gas, the output voltage of the voltage sensor indicates nearly 5 volts (4.95 volts). Conversely, if the ozone gas is detected, the sensor output voltage will be decrease (approaching to zero volts). After the start button is pressed, the ozone generator will start to switch on and the valve opened, so that oxygen comes into the chamber and the process of ozone gas monitor begin to operate. Ozone sensor output voltage data will be stored automatically on the memory card in every second. The monitoring process of ozone gas will stop when the reset button is pressed. The test results showed the control system has worked properly in accordance with the created algorithm in the microcontroller program.

Figure 4. Prototype of microcontroller system

The concentration of ozone gas produced by ozone generator was determined by iodometric titration method [10]. There were several stages in iodometric titration as follows:

1. Prepared the required chemicals: 0.15 M, KI; 0.1 M, H_2SO_4; NaS_2O_3, 0.035 N and 1% starch solution.
2. Turned on the ozone generator for 30 seconds with oxygen flow rate of 1 litre (L) / minute, so that the formed ozone gas is accommodated in the chamber.
3. The ozone gas flow to a KI solution were placed in 20 ml Erlenmeyer. Furthermore, a KI solution will change into yellow indicating the reaction of ozone gas against a KI solution. In the experiment, discoloration occurred when ozone flow the volume of 240 ml.
4. dissolved the 4 ml of H_2SO_4 solution and 3 drops of starch to KI, so that KI solution turns into a deep blue colour.
5. Titrated with a NaS_2O_3 solution. The titration process was stopped when KI solution turned into clear and recorded the titrated volume of NaS_2O_3 solution. Mass of ozone trapped in KI can be found in this equation (1)[10].

$$mass\ of\ O_3 = 24\ x\ V_t\ x\ N_t \tag{1}$$

where : 24 = Conversion factor = 24000 me/L per 1000ml/L
V_t = Volume of sodium thiosulfate, in mL
N_t = Normality of s sodium thiosulfate, in mg/me

ozone concentration can be determined using equation:

$$Concentartion\ of\ O_3 = \frac{mass\ of\ O_3}{Volume\ of\ O_3} \tag{2}$$

From the experiment, the obtained concentration of O_3 = 0.84 mg/L.

The experiment results of *E. Coli* bacteria that have been exposed to ozone is figured in Table 1. Table 1 shows the decrease number of *E. Coli* bacteria on the medium exposed by ozone with three variations of exposure time (15, 30 and 45 minutes) compared to media without ozone exposure (control sample). When it was exposed by ozone for 15 minutes, the number of colonies was 6.7 x 10^6 CFU/ml. In addition, the colonies number were 5.3 x 10^6 CFU/ml with 30 minutes exposure and 3.4 x 10^6 CFU/ml with 45 minutes exposure. Meanwhile, bacteria media that was not exposed to ozone reached 8.1 x 10^6 CFU/ml of the number of colonies. The longer time of exposure influenced the amount of bacterial growth. It revealed the relation between ozone gas exposure time and the growth of the *E. Coli* bacteria. From the results, it was found out that ozone gas can be used for the bacteria sterilization, especially *E. Coli.*

Table 1. The number of *E. Coli* colonies condition after exposure and control sample without ozone exposure

No	Time of exposure	Colonies Number
1	15 minutes	6.7×10^6 CFU/ml
2	30 minutes	5.3×10^6 CFU/ml
3	45 minutes	3.4×10^6 CFU/ml
	Control sample	8.1×10^6 CFU/ml

4. Conclusion

A microcontroller-based sterilization system for *E. Coli* bacteria sterilization which created ozone generator and control system with ozone gas indicator has been developed successfully. The test results showed that the control system has worked properly in accordance with an algorithm created in the microcontroller program. Thus, the *E. Coli* bacteria test results indicated the decrease bacteria number the number on the medium of *E. Coli* bacteria with ozone exposure with three variations of exposure time (15, 30, 45 minutes) compared to bacteria media without ozone exposure (control sample). Based on the variations of exposure time, it is shown that the longer exposure time leads a little amount of bacterial growth.

Acknowledgments
This research is fully funded by Penelitian Kerja Sama Antar Perguruan Tinggi (PEKERTI) Grant from Ministry of Research, Technology and Higher Education.

References

[1] T Oshita, K Uehara, T Takamatsu, A Kawate, H Miyahara and A Okino 2013 Sterilization of *E. Coli* Using Temperature Controllable Plasma Jet. 31st ICPIG.
[2] S Ketkaew and A Jangwanitlert 2007 A Study of Ozon Gas Generating Technique Using a Hih Frequency Voltage DC Switching Power Suplly of High Ripple Voltage *IEEE* Computer Society.
[3] Pekárek S 2003 Non-Thermal Plasma Ozone Generation *Acta Polytechnica* **Vol43** No6
[4] A Suksri, K Karnchanalekha, K Tonmitra and P Apiratikul 2009 A Comparative Study on Suitable High Voltage Sources for Ozone Generation *IEEE* Computer Society
[5] Seymour S 2011 *Disinfection, Sterilization, and Preservation* Lippincott Williams & Wilkins
[6] Varna Kodoth, Meghan Jones 2015 The Effects of Ultraviolet Light on Escherichia coli *Journal of Emerging Investigators*
[7] Priyanka Verma, Shruti Chatterjee, Mayank Choudhary, Ramanathan K and Shanthi V 2014 Comparison of sterilization reaction rate for Bacillus subtilis and Escherichia coliusing UV-visible spectroscopy *Journal of Chemical and Pharmaceutical Research*
[8] Hu Miao, Guo Yun 2012 The Effect of Air Plasma on Sterilization of Escherichia coli in Dielectric Barrier Discharge *Plasma Science and Technology* **Vol.14** No.8, Aug.
[9] M Achen and AE Yousef 2001 Efficacy of Ozone Against Escherichia coli O157:H7 on Apples *Journal of Food Science* **Vol. 66** No 9
[10] Rakness, K Gordon, G Langlais, B, Masschelein W, Matsumoto, N Richard Y, Robson C R, Somiya, I 1996 Guideline for Measurement of Ozon Concentration in the Process Gas From an Ozone Generator *Ozone Cience & Engineering* **Vol. 18** pp 209-229.

Cholesterol detection using optical fiber sensor based on intensity modulation

Moh Budiyanto[1,2,*], **Suhariningsih**[3] **and Moh Yasin**[3]

[1]Doctoral Study Program of Mathematics and science, Faculty of Science and Technology, Universitas Airlangga, Surabaya 60115, Indonesia.
[2]Department of Natural Science, Faculty of Mathematics and Natural Science, Universitas Negeri Surabaya, Surabaya 60231, Indonesia.
[3]Department of Physics, Faculty of Science and Technology, Universitas Airlangga, Surabaya 60115, Indonesia.

*Corresponding email address: mohammadbudiyanto@unesa.ac.id

Abstract. The aim of the research is to detect the concentration of cholesterol by using the principle that a laser beam propagation is guided by optical fiber bundle in term of intensity profile through solution with vary concentrations of cholesterol from 0 to 300 ppm. The mechanism of cholesterol concentration detection is the propagation of He-Ne laser beam with wavelength of 632.5 nm through a fiber optic bundle and a solution of cholesterol, then is reflected by a flat mirror and enters receiving fiber. This signal is captured by a silicon detector (SL-818, Newport) in the form of output voltage. The result showed that the output voltage decrease linearly with the increase of concentration of cholesterol with a sensitivity of 0.0004 mV/ppm and the linearity more than 97%.

1. Introduction

Cholesterol is one component of the fat contained in the cell which is made bylow density lipoprotein (LDL), high density lipoprotein (HDL), total cholesterol and triglycerides. Cholesterol is actually one of the nutrients that are needed by the body in addition to other nutrients such as carbohydrates, proteins, vitamins and minerals. Cholesterol found in the bloodstream or body cells which needed for the formation of the cell wall and as raw material for some hormones. However, if there are excessive cholesterol levels in the blood, it can lead to diseases, including coronary heart disease and stroke. Normal cholesterol should be below 200 mg / dl. It is also a material basis for the formation of steroid hormones. Excessive cholesterol in the body will be buried in the walls of blood vessels and will cause a condition called atherosclerosis; a condition in which the arteries are narrowing or hardening. This kind of condition is the forerunner of heart disease and stroke.

Cholesterol is a compound which has a four-ring core siklopentano – phenanthrene, including fat with very little solubility in water. Levels in blood plasma 150-200mg / ml, are approximately 2x blood glucose levels. In blood plasma, 30% of it which bound to lipoproteins is able to increase solubility in blood. As many as 70% more blood cholesterol is in the form of cholesterol esters. Cholesterol does not dissolve in water but can be extracted from the tissue with chloroform, ether, benzene and alcohol heat. Cholesterol including steroid compounds of formula $C_{27}H_{45}OH$. Cholesterol is an important sterol which are many in nature. From the formula of cholesterol, it can be seen that

IOP Conf. Series: Journal of Physics: Conf. Series **853** (2017) 012008 doi:10.1088/1742-6596/853/1/012008

the hydroxyl groups are contained of C atom number 3 which has a beta position; therefore is connected by a full line.

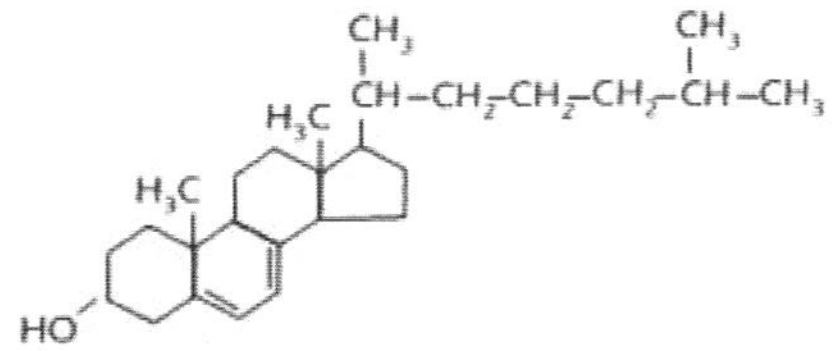

Figure 1. The compound cholesterol

There have been a lot of researches done on the detection of cholesterol levels as sensitive and selective electrochemical sensors use silver nanoparticles which modified glassy carbon electrode to determinate the bovine serum cholesterol [1]. Electrochemical sensor based on the molecular film printed on carbon nanoparticles modified electrode for the determination of cholesterol [2]. A layer of carbon nanotubes and gold nanoparticles bienzyme-based biosensor are usedto detect of cholesterol [3]. High-performance electrochemical biosensor could also be used for the detection of total cholesterol [4]. Cholesterol with amperometric detection using electrochemical oxidation method , electrochemical and optical polymer by using a biosensor are also to detect cholesterol [5]. The downside of sensors such as performance optimization of sensor sensitivity, linear range and detection limit is less than the maximum.

Another way to determine cholesterol levels can be determined by the character of the material through a laser beam fiber optic sensors. Fiber optic sensor method is more simple and inexpensive to be used to determine the concentration of a substance. One example of an optical fiber sensor to determine the salt content in a material is a simple instrument design, low cost, and high sensitivity measurement results [6]. Based on this description, there will be a developed fiber optic sensor research as a model instrument determines the concentration of cholesterol using fiber optic sensor bundle.

2. Experimental setup

The research method used to determine the cholesterol levels were determined by the character of the laser beam to the material through an optical fiber sensor by using a laser beam intensity modulation. Beam of light entering the optical fiber receiver and processed by an optical detector into electrical signals to be displayed on the computer. The experimental results would be obtained a linear relationship between the peak voltage output of the sensor as a function of variations in the concentration of cholesterol and measured parameters and sensor performance include sensitivity, linear range, and linearity.

The device consisted of a He-Ne laser (Thorlabs, 632.5 nm, 5 mW), Fiber bundle, an optical detector 818-SL (Newport), chopper and chopper controller (SR540, Stanford Research Systems, Inc.), Lock-in amplifier (SR510, Stanford Research Systems, Inc.), mirrors the wavelength of visible light (5101-Vis, New Focus), micrometer position (Newport), PC, and other supporting devices.

The first step was to characterize the shift sensor to determine the fiber bundle sensing channel shift to a flat mirror. Set-up characterization of the fiber bundle as sensor shift and subsequent detection of cholesterol concentration by making set-up as shown in the following figure 2:

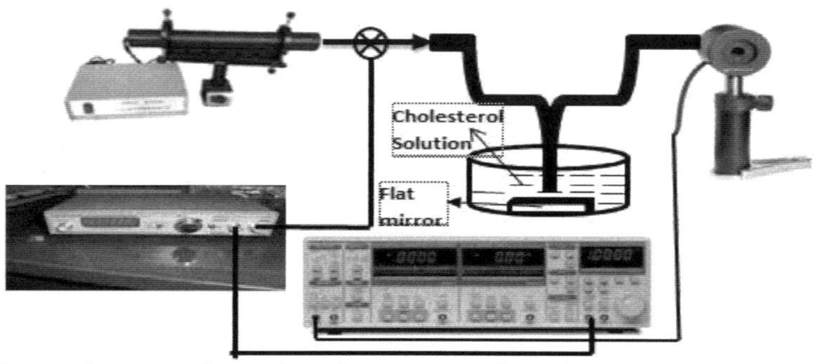

Figure 2.Setupfor fiber optic sensor determines the concentration of cholesterol using fiber bundle

Detection started when the fiber bundle was placed coincides with the mirror on the shift z = 0. The fiber bundle was placed on a micrometer position which was shifted every 50 μm. At each position shift detector output voltages were measured to obtain data such as output voltage detector as a function of shifting fiber bundle. The detection was done on several other variations of concentrations between 0 ppm, 50 ppm, 100 ppm, 150 ppm, 200 ppm, 250 ppm, 300 ppm.

3. Results and discussions

Figure 3 showed a graph of the output voltage detector as a function of shifting fiber bundle for each concentration ranging 0 to 300 ppm. The figure showed the relationship changes in cholesterol concentration variations of the maximum output voltage at each concentration; the higher the concentration of cholesterol, the lower the measured output voltage detector.

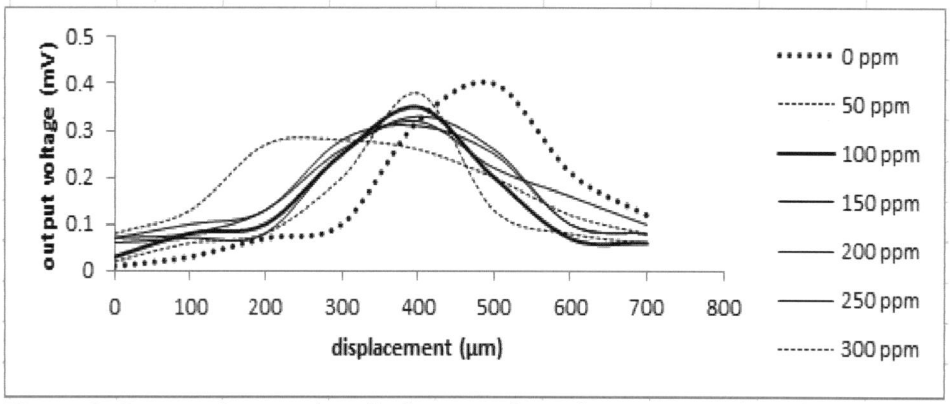

Figure 3. Output voltage against displacement for various concentration of cholesterol

Data detection output voltage detector began when fiber bundle placed the mirror coincided with the shift z = 0. The fiber bundle was placed on a micrometer position shifted every 50 μm. At each position shift detector output voltages were measured to obtain data such as output voltage detector as a function of shifting fiber bundle, so as to obtain the maximum output voltage of each of the various concentrations of cholesterol. The maximum output voltage value of variations in the concentration of cholesterol as in table 1.

Table 1.Maximum output voltage of the concentration of cholesterol

Data	Concentration (ppm)	output voltage(mV)
1	0	0.4
2	50	0.38

Data	Concentration (ppm)	output voltage(mV)
3	100	0.35
4	150	0.33
5	200	0.32
6	250	0.31
7	300	0.28

Figure4 showed that the peak voltages or light intensity showed a decrease linearly with the increase in the concentration of cholesterol with a sensitivity of 0.0004 mV/ppm and 97.5% of linearity.

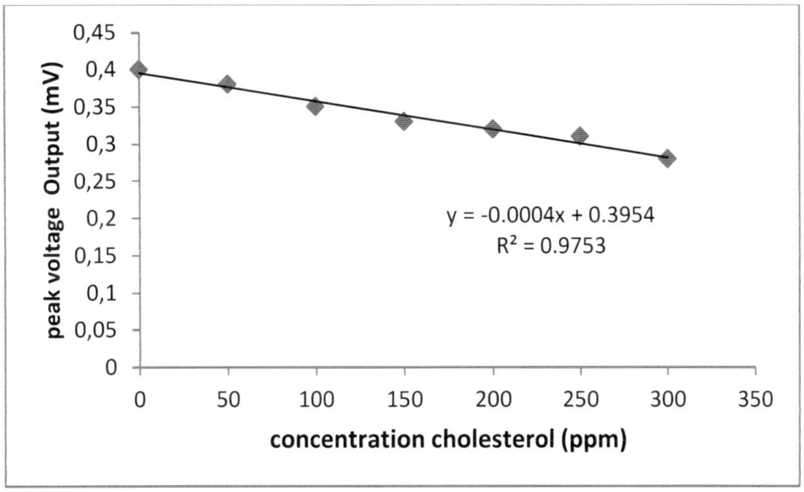

Figure 4. Peak voltage versus concentration of cholesterol

4. Conclusion

Optical fiber sensors based on intensity modulation for the concentration of cholesterol detection was presented. The results showed that the detection of the peak voltage decrease linearly with the increase in concentration of cholesterol with a sensitivity of the sensors 0.0004 mV/ppm and the linearity more than 97%.

References

[1] Siriwan Nantaphola, Orawon Chailapakul, and Weena S 2015 Sensitive and selective electrochemical sensor using silver nanoparticles modified glassy carbon electrode for determination of cholesterol in bovine serum *Sensors and Actuators* B **207**, 193–198.

[2] Jian Ji, Zhihui Zhou, Xiaolian Zhao, Jiadi Sun, Xiulan Sun 2015 Electrochemical sensor based on molecularly imprintedfilm at Au nanoparticles-carbon nanotubes modified electrode for determination of cholesterol *Biosensors and Bioelectronics* **66** 590–595.

[3] Xiaojun Cai, Xia Gaoa L W and Xianfu Lin 2013 A layer-by-layer assembled and carbon nanotubes/gold nanoparticles-based bienzyme biosensor for cholesterol detection *Sensors and Actuators* B **181**, 575– 583.

[4] Asieh Ahmadalinezhad and Aicheng Chen 2011 High-performance electrochemical biosensor for the detection of total cholesterol *Biosensors and Bioelectronics* **26** 4508 - 4513.

[5] Saniye Soylemez, Yesemin A U, Cevher Gundo H, Yavuz Ergun and Levent T 2015 Electrochemical and optical properties of a conducting polymer andits use in a novel biosensor for the detection of cholesterol *Sensors and Actuators* B **212**, 425–433.

[6] Yasin M, H Ahmad, K Thambiratnam, A A Jasim, S W Phang, and S W Harun 2012 Design of Multimode Tapered Fibre Sensor for Glucose Detection *Optoelectronics and Advanced Materials* Vol. **7**, No. 5 - 6. p. 371 - 376.

IIR digital filter design for powerline noise cancellation of ECG signal using arduino platform

Akif Rahmatillah* and Ataulkarim

Department of Biomedical Engineering, UniversitasAirlangga, Kampus C Mulyorejo, Surabaya, Indonesia, 60286

*Corresponding email address: akif-r@fst.unair.ac.id

Abstract. Powerline noise has been one of significant noises of Electrocardiogram (ECG) signal measurement. This noise is characterized by a sinusoidal signal which has 50 Hz of noise and 0.3 mV of maximum amplitude. This paper describes the design of IIR Notch filter design to reject a 50 Hz power line noise. IIR filter coefficients were calculated using pole placement method with three variations of band stop cut off frequencies of (49-51)Hz, (48 - 52)Hz, and (47 - 53)Hz. The algorithm and coefficients of filter were embedded to Arduino DUE (ARM 32 bit microcontroller). IIR notch filter designed has been able to reject power line noise with average square of error value of 0.225 on (49-51) Hz filter design and 0.2831 on (48 - 52)Hz filter design.

1. Introduction

ECG signal is a signal which indicates electrical activity of heart. It has been used in cardiovascular condition assessment [1]. On the frequency domain, its signal can be represented from 0,05 Hz to 150 Hz [2,12]. As a typical of bio potential signal, ECG signal can be interfered by noises. There are many interferences or noise sources on ECG signal measurement such as friction of an electrode with body (motion artifact), instrumentation noise, respiration movements, and power line noise. Interference is classified based on their frequency content [3]. The most significant interference of ECG signal measurement is at 50 Hz / 60 Hz power line noise [4, 12]. This noise can make an ECG signal difficult to be interpreted [11], and this is caused by a change in ECG signal shape as shown in Figure 1. For that reason, a filter which is able to reject a single frequency component of noise is needed. This refers to Notch Filter which can be applied using analog electronics circuit or digital filter design. However, analog filters have some disadvantages such as nonlinear phase shift of input-output filter signal and cut-off frequency which make it difficult to be applied due to its limitation of resistors and capacitors value and sensitivity of electronics circuit to temperature. Digital Filters can apply a precise value of cut - off frequency, and it is not sensitive to temperature or other environmental conditions [5].

IIR filter design is one of digital filter algorithm which applies linear time-invariant properties of filter system. A notch filter for Electromyography (EMG) signal has been used in IIR filter algorithm [6]. Coefficients of IIR filter can be calculated with many methods such as Pole Zero Placement method which has been used to calculate coefficients of IIR filter to be stable [7]. IIR filter design also has the ability to reduce the transient response of notch filter [10]. This method is applied to eliminate noises of ECG signal [5].

On the other hand, a computational platform to apply IIR filter algorithm is needed. An Arduino platform has been used to apply FIR digital filter [13]. Arduino DUE is one of low cost computational

IOP Conf. Series: Journal of Physics: Conf. Series **853** (2017) 012009 doi:10.1088/1742-6596/853/1/012009

platforms based on 32-bit ARM Microcontroller with 12-bit resolution of ADC and 13 kHz frequency sampling [8]. Although it is a "low-cost platform", it can be used as a hardware platform of IIR notch filter design for ECG referring to a resolution of ADC.

This study proposes a design of IIR notch filter algorithm, and the algorithm is applied to Arduino DUE platform.

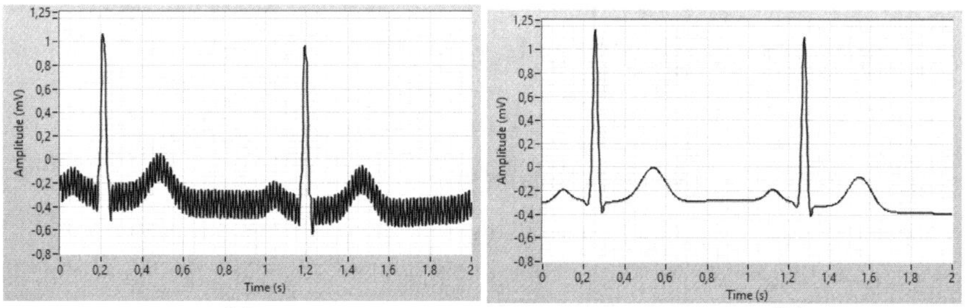

(a) (b)

Figure 1. (a) ECG Signal with 50 Hz Powerline Noise. (b) ECG Signal with 50 Hz Powerline Noise.

2. Background

IIR filter algorithm is characterized by recursive equation as shown in (1) in which $h[k]$ is filter impulsive response that it has an infinite value, and b_k and a_k are filter coefficients, while $x[n]$ and $y[n]$ are input and output filter.

$$y[n] = \sum_{k=0}^{\infty} h[k]x[n-k] = \sum_{k=0}^{N} b_k x[n-k] - \sum_{k=1}^{M} a_k y[n-k] \tag{1}$$

In Z domain, IIR filter transfer function is represented as in (2)

$$H(z) = \frac{b_0 + b_1 z^{-1} + \ldots + b_N z^{-N}}{1 + a_1 z^{-1} + \ldots + a_M z^{-N}} = \frac{\sum_{k=0}^{N} b_k z^{-k}}{1 + \sum_{k=1}^{M} a_k z^{-k}} \tag{2}$$

Therefore, it can be factorized as shown in (3)

$$H(z) = \frac{K(z - z_1)(z - z_2)\ldots(z - z_N)}{(z - p_1)(z - p_2)\ldots(z - p_N)} \tag{3}$$

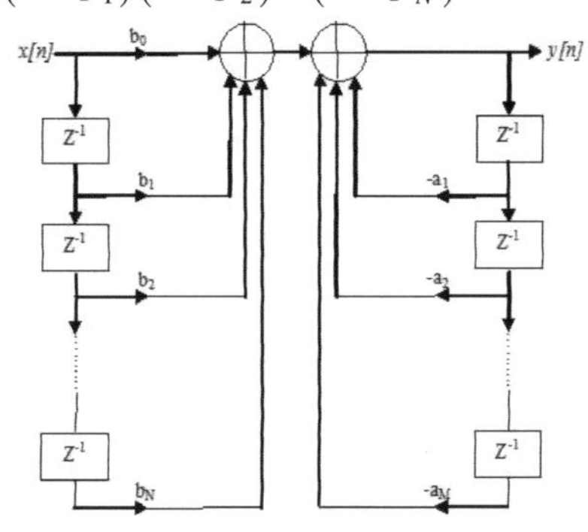

Figure 2. IIR Filter Data Flow Diagram.

Inwhich $z_1, z_2, ..., z_N$ are zeros, and $p_1, p_2, ..., p_M$ are poles of $H(z)$. The poles of $H(z)$ should contain the unit circle (i.e., $|z| = 1$) of which the filter is stable in the sense of Bounded Input - Bounded Output Stability. Thus, IIR filter may not be stable if a pole is outside of unit circle. On thc other hand, IIR filter is recursive. There is a feedback as shown in (1) that the coefficients of IIR filter are less than those of FIR filter. An Application of IIR filter is a computation algorithm which can be represented with data flow diagram as shown inFigure2.

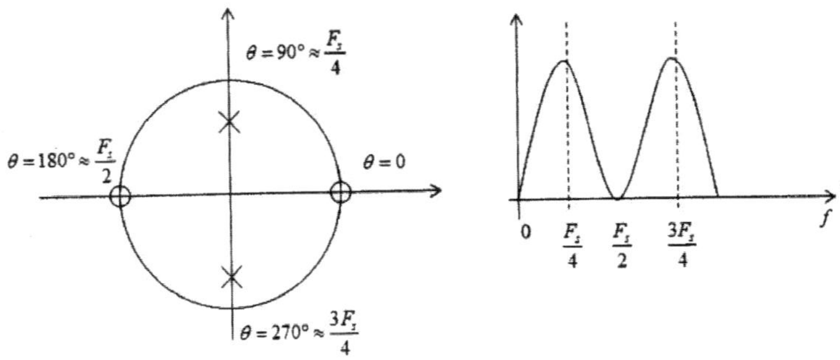

Figure 3. Figure with short caption (caption centred)[9].

Poles and Zeros Placement Method was used to calculate coefficients of notch filters. This method uses a property of poles and zeros position on Z domain. A frequency response of IIR filter value will be 0 on zeros position s and ∞ on poles positions as shown in Figure 3. In this method, there is a relationship between an angle (placement position of zeros and poles) and frequency [5]. The relationship is simple and follows linear comparison. Angle positions of poles and zeros can be calculated using (4)

$$\theta = \frac{f}{f_s} \times 360^o \tag{4}$$

In which f is stopband designed for frequency of filter, f_s is sampling frequency, and θ is angular position of poles and zeros. A radius of poles and zeros to the center of unit circles (r) is calculated using (5) in which BW is bandwidth of passband filter.

$$r = 1 - \left(\frac{BW}{f_s}\right)\pi \tag{5}$$

Notch filter is one of typical band stop filter with narrow stop band. Its common use is to reject a powerline noisewhich has a single frequency component to reject. Figure 4 shows the illustrationoffrequency response of notch filter in which ω_c is a frequency which is rejected.

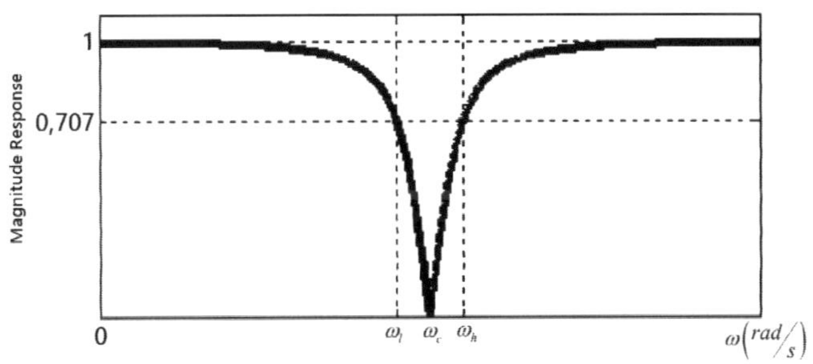

Figure 4. Frequency Response of Notch Filter.

IOP Conf. Series: Journal of Physics: Conf. Series **853** (2017) 012009 doi:10.1088/1742-6596/853/1/012009

3. Filter specification and design

For attenuating 50 Hz powerline noise, three variations of bandwidth rejection of notch filter were applied; they are 49 - 51, 48 – 52, and 47 - 53 Hz. Each variation had 50 Hz to be the center of stopband filter or the Q point. Using (4) and (5), filter coefficients were calculated and produced a transfer function of each design, and a sampling frequency has been designed at 200 Hz.

For 49 - 51 Hz bandwidth rejection is obtained that *BW* is 2 Hz. Using (4) and (5), angle position of θ is $\pm 90^{\circ}$ and *r* is 0,9686. Transfer function and difference equation of filter for this specification is shown in (6) and (7). Coefficients of filter are $b_0 = 1, b_1 = 0, b_2 = 1, a_1 = 0,$ and $a_2 = 0,9382$. The poles - zeros placement result and bode diagram response of filter are shown in figure 5.

$$H(z) = \frac{1 + z^{-2}}{1 + 0,9382 z^{-2}} \tag{6}$$

$$y[n] = x[n] + x[n-2] - 0,9382 y[n-2] \tag{7}$$

The second variation is 48 - 52 Hz which had 4 Hz of *BW*, $\pm 90^{\circ}$ for θ, and 0,9732 for *r*. Then, transfer function and difference equation of filter for this specification isshown in (8) and (9). Coefficients of filter are $b_0 = 1, b_1 = 0, b_2 = 1, a_1 = 0$ and $a_2 = 0,8783$. The poles - zeros placement result and bode diagram response of filter are shown in figure 6.

$$H(z) = \frac{1 + z^{-2}}{1 + 0,8783 z^{-2}} \tag{8}$$

$$y[n] = x[n] + x[n-2] - 0,8783 y[n-2] \tag{9}$$

The second variation is 47 - 53 Hz which had 6 Hz of *BW*, $\pm 90^{\circ}$for θ, and 0,9058 for *r*. Than transfer function and difference equation of filter for this specification are shown in (10) and (11). Coefficients of filter are $b_0 = 1, b_1 = 0, b_2 = 1, a_1 = 0$ and $a_2 = 0,8204$. The poles - zeros placement result and bode diagram response of filterare shown in figure 7.

$$H(z) = \frac{1 + z^{-2}}{1 + 0,8204 z^{-2}} \tag{10}$$

$$y[n] = x[n] + x[n-2] - 0,8204 y[n-2] \tag{11}$$

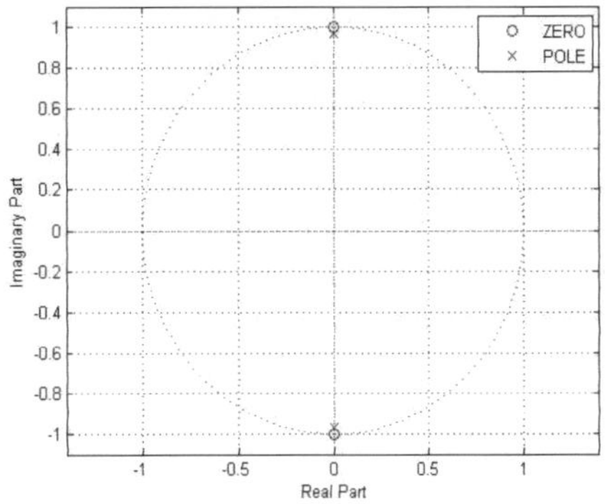

(a)

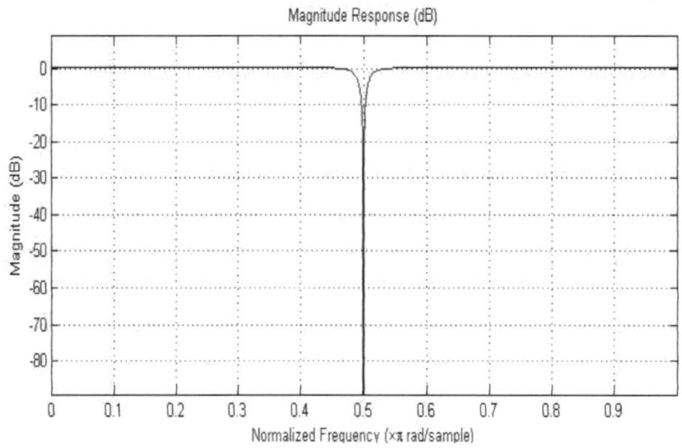

(b)

Figure 5. Notch Filter Design Result for Bandwidth rejection 49 - 51 Hz (a) Poles - Zeros Placement, and (b) Bode Diagram of Magnitude Response

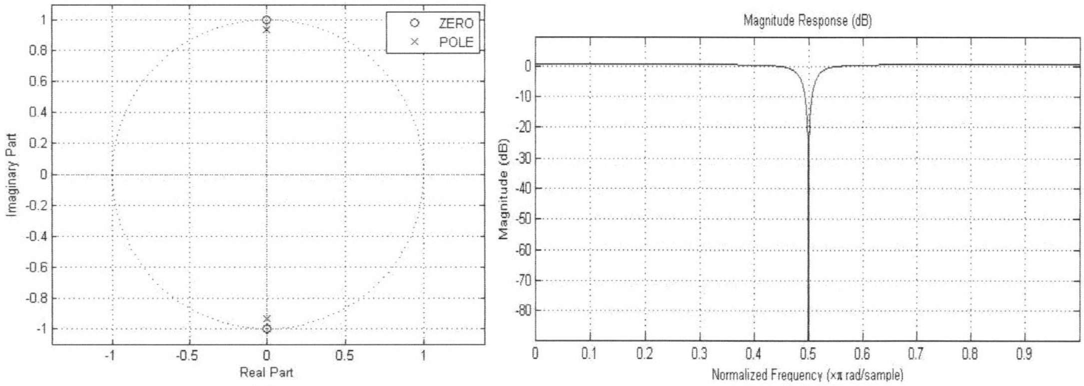

(a) (b)

Figure 6. Notch Filter Design Result for Bandwidth rejection 48 - 52 Hz (a) Poles - Zeros Placement, and (b) Bode Diagram of Magnitude Response

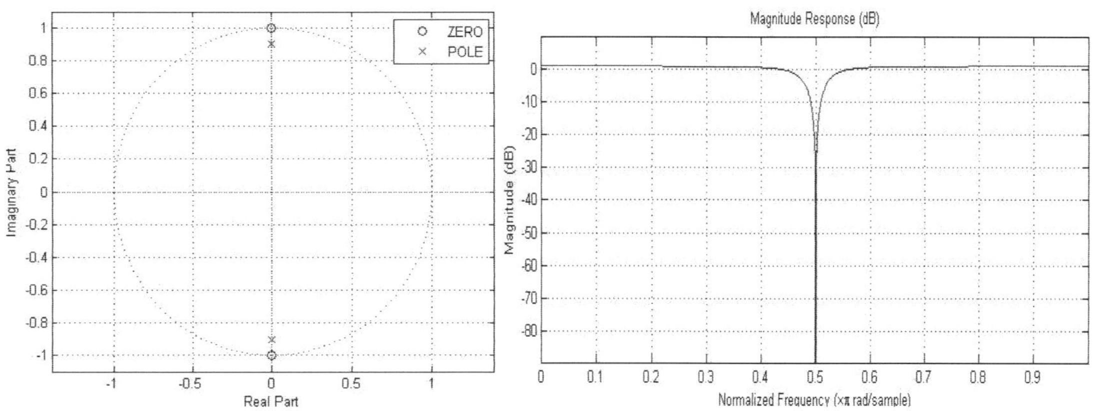

(a) (b)

Figure 7. Notch Filter Design Result for Bandwidth rejection 47 - 53 Hz (a) Poles - Zeros Placement, and (b) Bode Diagram of Magnitude Response

Each variation produced a second order filter. It happened because the poles - zeros placement method always placed two poles and zeros for every filter design.

4. Experimental result

To evaluate notch filter design, experimental setup was developed. LabVIEW 2013 Biomedical Toolkit was used to generate ECG simulated signal with powerline noise. The specification of power line noise was 50 Hz and 0,3 mV amplitude of noise signal. Noising ECG simulated signal generated on analog output of NI MyDAQ. Analog input of Arduino DUE which had been embedded on IIR Filter program received and filtered the signal. Denoised ECG signal was sent to Personal Computer to displaydenoised signal using serial communication. The experimental setup is shown on figure 8. Equation (7), (9), and (11) are represented by IIR Filter program written in C language and compiled by Arduino IDE 1.5.2.The program was written on Appendix.

The fourier transform of ECG signal before and after filtering is shown in Figure 9 (a) and (b), 10 (a) and (b), and 11(a) and (b). Notch filter on Arduino DUE had rejected the 50 Hz signal component which was power line noise signal. It was indicated by the amplitude of filter output signal which was zero on 50 Hz. Furthermore,the quality of ECG signal after filtering was measured with Root Mean

Square Error value. Error value was obtained by comparing signal after filtered to ECG to ideal signal shown in Figure 1(b). Table 1describes RMS values of each variation design of IIR Notch filters.

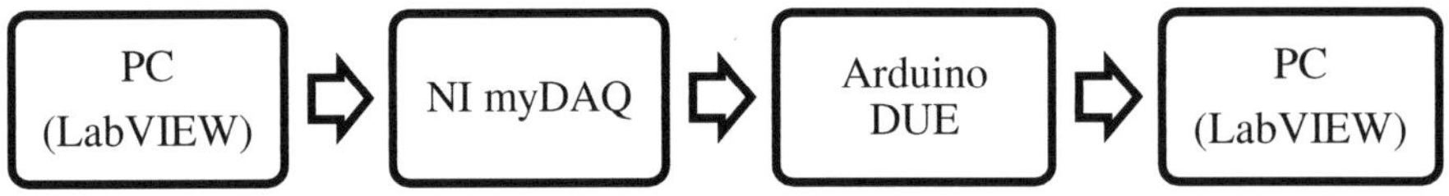

Figure 8. Experimental Set-Up.

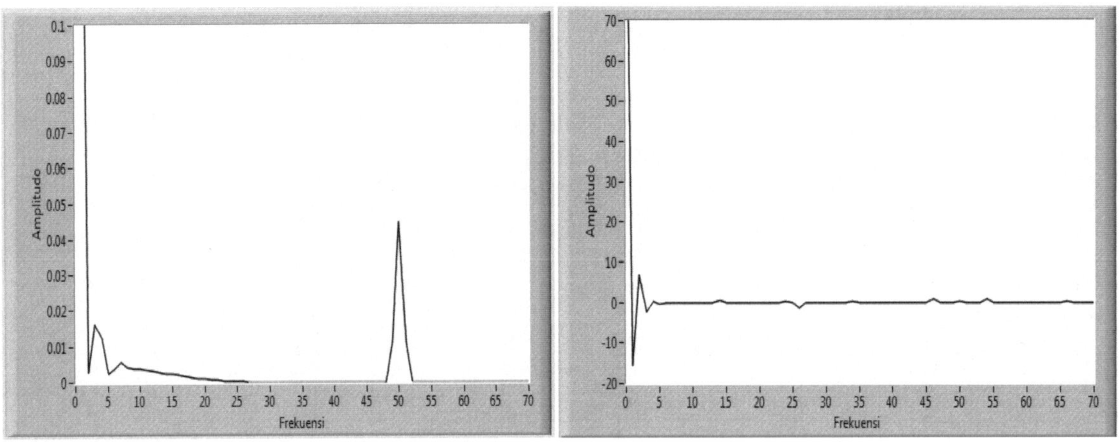

(a) (b)

Figure 9. IIR Filter Performance for Bandwidth Rejection 49 - 50 Hz (a) Before Filtering, and (b) After Filtering

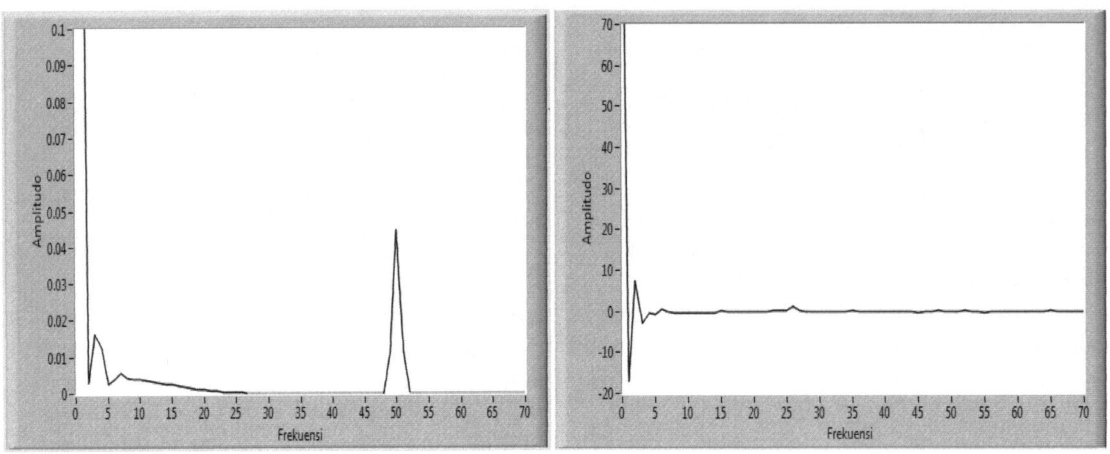

(a) (b)

Figure 10. Notch Filter Design Result for Bandwidth rejection 48 - 52 Hz (a) Poles - Zeros Placement. (b) Bode Diagram of Magnitude Response.

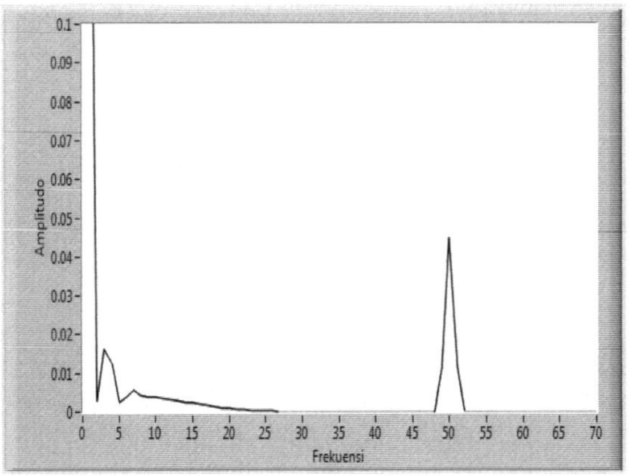

(a)

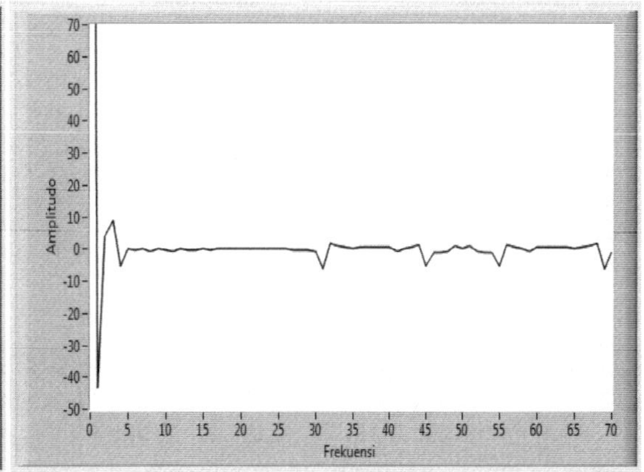

(b)

Figure 11. Notch Filter Design Result for Bandwidth rejection 47 - 53 Hz (a) Poles - Zeros Placement, and (b) Bode Diagram of Magnitude Response

5. Discussion

Design of IIR Notch filter using poles-zeros placement method was depended on bandwidth specification applied. From Figure 5 to Figure7, the width of a bandwidth correlated withpoles position. A narrow bandwidth made poles tend to keep away from thecenter of unit circler and vice versa. Then, a narrow bandwidth of notch filter gave perfect filtering only at 50 Hz (see Figure 9(b)). It can be compared with Figure 10(b) that the FFT result from after signal filtering gave ripplesat 30-70 Hz.

Table 1. RMSE Values of IIR Notch Filter for 50Hz Frequency Rejection

Bandwidth	RMSE Value
49 - 51 Hz	0.254
48 - 52 Hz	0.2831
47 - 53 Hz	1.5609

From table 1 it is shown a narrow bandwidth gave the filter betterqualitythan that of wide bandwidth. It is shown by RMSE value that the narrower bandwidth of notch filter is, the smaller the RMSE value. Small value of RMSE shown afterECG signal filtering was more like ECG signal without noise. Then, an Arduino DUE platform can be used as a hardware platform of Digital Signal Processing especially for IIR Notch filter design. This hardware platform can be embedded on IIR filter algorithm to perform filtering without any obstacle.

It can be concluded that IIR Notch filter whose coefficients calculated by poles-zeros placement method and applied to Arduino DUE platform was able to reject a 50 Hz powerline noise of ECG signal.

6. Appendix

IIR Notch Filter Program in C language and applied toArduino DUE Platform

```
float b0= ; //IIR Filter Coeficients
float b1= ;
float b2= ;
```

```
float a1= ;
float a2= ;
float x=0;//signal variable declaration
float x1=0.0;
float x2=0.0;
float y=0.0;
floatyl =0.0;
float y2=0.0;
intdatain[10000];//ECG data initialization
intdataout[10000];
int i=0;
void setup() {
// analogReadResolution(12);
Serial.begin(9600);
}
void loop() {
while (i<20000)
{
datain[i]=analogRead(0);
 x=datain[i];
 y=b0*x + b1*x1 + b2*x2 - a1*yl - a2*y2 - a3*y3;
 y2=yl;
yl=y;
x2=x1;
x1=x;
dataout[i] = y;
Serial.println(dataout[i]);
 i=i+1;
}
}
```

References

[1] Sornmo L and Laguna P 2006 Electrocardiogram (ECG) Signal Processing *Wiley Encyclopedia of Biomedical Engineering*
[2] Yeh Y C and Wang W C 2008 QRS Complexes Detection for ECG Signal: The Difference Operation Method *Comp. Methods and Prog. in Biomedicine* **91** 245
[3] Chavan M S, Agarwala R A, and Uplane M D, 2008 Supperssion of Baseline Wande and Powerline INterference in ECG Using Digital IIR Filter **2** 356
[4] Hutha J C and Webster J G 1973 60 Hz Interference of Electrocardiogram *IEEE Trans. Biomed. Eng.***20** 99
[5] Khaing H S and Naing Z M 2011 Quntitative Investigation of Digital Filters in Electrocardiogram with Simulated Noises *Inter. J. of Information and Electronics Eng.***1** 201
[6] Mello R G T Olivera L F and Nadal J 2007 Digital Butterworth Filter for Substracting Noise from Low Magnitude Surface Electromyogram *Comp. Methods and Prog. in Biomedicine***87** 28
[7] Hollman L J and Stevenson R L 2014Pole-zero placement algorithm for the design of digital filters with fractional-order rolloff*Signal Processing.* **107**218
[8] *https://www.arduino.cc/en/Main/ArduinoBoardDue*
[9] Khaing H S 2010 Powerline Interference Cancellation for ECG *Porc. Int. Conf. os Science and Engineering 2nd* (Yangon) 154

[10] Piskorowski J 2012 Powerline interference removal from ECG signal using notch filter with non-zero initial conditions *Proc. IEEE Int. Symp. on Medical Measurement and Application*
[11] Rana K P S Kumar V and Gupta A 2017 A Pole-Radius-Varying Notch Filter with Enhanced Post-Transient Perfomance *Biomedical Signal Processing and Control* **33** 379
[12] Rajan R K Kumar K Rani N Paul S K and Akashe S A Powerline Filter Circuit Design for Biomedical Applications *J. of Computational and Theoritical Nanoscience* **13** 3345
[13] Shajahan A H and Anand A 2013 Data Acquisition and Control Using Arduino - Android Platform : Smart Plug *International Conference on Energy Efficient Technologies for Sustainability* 241

Development of wireless sensor network for landslide monitoring system

Suryadi[1,*], Prabowo Puranto[1], Hendra Adinanta[1], Adrin Tohari[1,2] and Purnomo S Priambodo[1,3]

[1]Research Center for Physics – Indonesian Institute of Sciences Kawasan PUSPIPTEK Serpong, South Tangerang, Indonesia
[2]Research Center for Geotechnology – Indonesian Institute of Sciences Bandung, Indonesia
[3]Faculty of Engineering – University of Indonesia, Depok, Indonesia

[*]Corresponding email address: ment4r1@gmail.com

Abstract. A wireless sensor network has been developed to monitor soil movement of some observed areas periodically. The system consists of four nodes and one gateway which installed on a scope area of 0.2 Km^2. Each of nodehastwo types of sensor,an inclinometer and an extensometer. An inclinometer sensor is used to measure the tilt of a structure while anextensometer sensor is used to measure the displacement of soil movement. Each of nodeisalso supported by awireless communication device, a solar power supply unit, and a microcontroller unit called sensor module. In this system, there is also gateway module as a main communication system consistinga wireless communication device, power supply unit, and rain gauge to measure the rainfall intensity of the observed area. Each sensor of inclinometer and extensometer isconnected to the sensor module in wiring system but sensor module iscommunicating with gateway in a wireless system. Those four nodes are alsoconnectedeach other in a wireless system collecting the data from inclinometer and extensometer sensors. Module Gateway istransmitting the instruction code to each sensor module one by one and collecting the data from them. Gateway module is an important part to communicate with not only sensor modules but also to the server. This wireless system wasdesigned toreducethe electric consumption powered by 80 WP solar panel and 55Ah battery. This system has been implemented in Pangalengan, Bandung, which has high intensity of rainfall and it can be seen on the website.

1. Introduction

Nowadays, the development of communication system has been enhanced significantly. It is not only the increase of data transmission speed but also the absence of wire in communication system. The absence of wire in communication system is called wireless system. It leadsmany meritsto simplify the electronic system especially for data logger system that collect the data and send it to the database. Moreover, it reduces the complexity of installation of electronic system because of wire or cable for media communication and the wire installation cost. On the other hand, wireless systemalso has the disadvantage which needs more power consumption and data transmission stability. Unfortunately, the two problems takean important thing in every communication system. However, wireless system still becomesthebest choice for communication system in which the stability and power consumption has to be solved [1].

Related to the aspect above, Optoelectronics Research Group in Research Centrefor Physics has developed some sensors for Geotechnical Instrumentation [2]. These sensors are inclinometer, optical extensometer, and electrical extensometer. The extensometers and inclinometers have been used to measure the physical properties of a soil in slope. It can give the data of displacement of soil movement in a slope and the tilt or inclination of a reference structure respectively [3]. If the data can be processed and analysed correctly, it can providethe information about the stability of the slope itself. Nevertheless, in the reality,it is hard to install these sensors especially in a remote area whereasthere are many limitations to install the sensor connected to the wire. Therefore, the use of wireless system is needed to overcome this problem in order to simplify the installation in observation area [4,5].

In 2015, collaboration between Research Center for Physics and Research Center for Geotechnology has made the installation of landslide monitoring system in Pangalengan, Bandung. It begins with the report from the inhabitant that has seen many cracks in their village. Team of Research Center for Geotechnology has conducted pre-installation survey using geoelectrical equipment. This survey can analyse the underground condition to investigate the cause of those cracks. It also provides suggestion to improve the proper location for the landslide monitoring system installation where wireless system network is used in this system which presented in this paper.

2. Methodology

Landslide monitoring system is a system that consist of sensors, controllers, data logger and communication devices to measure the physical properties of observation area in order to monitor soil movement. This system is divided into some node units and a gateway module. Each node has one pair sensor that consistsof an extensometer and an inclinometer. This node is equipped by 20 WP solar panels, 10A solar charge controller, 12A/12V battery, data logger unit and Xbee module as the wireless communication device. Likewise, the gateway module has one rain gauge, 80WP solar panel and its controller, siren and rotary lamp, 55AH Battery, Xbee module as a wireless communication device, microcontroller, GSM Module, and relay controller. This system is shown in Figure 1.

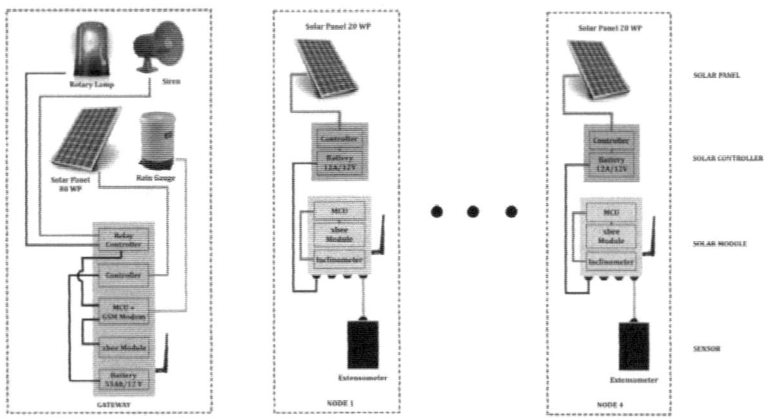

Figure 1. Scheme of wireless sensor network for landslide monitoring system

There are four nodes and one gateway to communicate each other. The gateway is a unit that send the instruction code to collect the data from each node whichhas an address. If the address sent is match to the node,this node will receivethe data from the sensor and send it to the gateway. It works in same way for each node. However, if a node position is too far from the gateway, this node still can end the data to the other closer nodes to it until the data send successfully to the gateway as the final data storage. After the gateway received completely the data from node 1 to node 4, this data is gathered and sent to the server via GSM/GPRS module.

The communication among nodes is provided by Xbee module. It is used as a wireless communication device that used to receive the instruction code and transmit the data. In this case, it uses Xbee Pro S2B that can be set in a mesh network topology and in a power save modes. Mesh network topology guarantees the data can be sent to the receiverin a long range yet takesthe short or medium range of Xbee module.This Xbee module can be seen in Figure 2.

Figure 2. Xbee Pro S2B Module [6]

Xbee module also can be set up in a power save mode. It is used to reduce the power consumption where the power supply has a limitation especially in remote area. This can be activated by setting the power down of Xbee in whichthe Xbee Pro S2B is the lowest current consumption. Thiscomparison between Xbee and Xbee ProS2B can be seen on the Table 1 as well.

Table 1. Specifications of the XBee/XBee-PRO2B RF Module[6]

Specification	Xbee	XBee-PRO (S2B)
Indoor/urban range	up to 133 ft. (40 m)	Up to 300 ft. (90 m), up to 200 ft (60m) international variant
Outdoor RF line-of-sight range	up to 400 ft. (120 m)	Up to 2 miles (3200 m), up to 5000 ft (1500 m) international variant
Transmit power output	2 mW (+3dBm), boost mode enabled 1.25 mW (+1dBm), boost mode disabled	63mW (+18 dBm) 10mW (+10 dBm) for International variant
RF data rate	250,000 b/s	250,000 b/s
Supply voltage	2.1 - 3.6 V	2.7 - 3.6 V
Operating current (transmit, max output power)	40 mA (@ 3.3 V, boost mode enabled) 35 mA (@ 3.3 V, boost mode disabled)	205 mA, up to 220 mA with programmable variant (@3.3 V) 117 mA, up to 132 mA with programmable variant (@3.3V),International variant
Operating current (receive))	40mA (@ 3.3 V, boost mode enabled) 38mA (@ 3.3 V, boost mode disabled)	47 mA, up to 62 mA with programmable variant (@3.3 V)
Idle current	15mA	15mA
Power-down Current	< 1 uA @ 25°C	3.5 pA typical @ 25°C

IOP Conf. Series: Journal of Physics: Conf. Series **853** (2017) 012010 doi:10.1088/1742-6596/853/1/012010

It can be seen that Xbee Pro S2B is a good choice for wireless device to use in landslide monitoring system that has longer range than regular Xbee. It also has good specification to be operated with the lowest current consumption when the power down mode is activated. To activate this mode,microcontroller controls each node and gateway. Beside Xbee Pro S2B, microcontroller and inclinometer sensors are the power down mode device which can be set in order to reduce their power consumption. For extensometer itself, the power consumption of extensometer cannot be reduced because it does not have power save mode. Moreover, it must not be switched off because it can cause the data loss.

In this system, the used sensorswere inclinometer and extensometer. Inclinometer is developed by using MEMS accelerometer SCA100T-D01 where The SCA100T Series is a 3D-MEMS-based dual axis inclinometer family which provides instrumentation grade performance for levelling applications [7]. It has resolution of 0.0035° and measuring range ±30°. It can measure the tilt of object referenced to gravity axis. On the other hand, extensometer is built using rotary encoder E40S6-2000-3-T-5 from Autonics. It has a resolution of 5000 pulses per revolution that is used to measure the displacement. In this case, the translational displacement can be converted from the counter of encoder pulse [8].

3. Result and discussion

3.1. Sensor characterization

There were two characterizationsdone in this research. Firstly, inclinometer characterization is presented in Figure 3

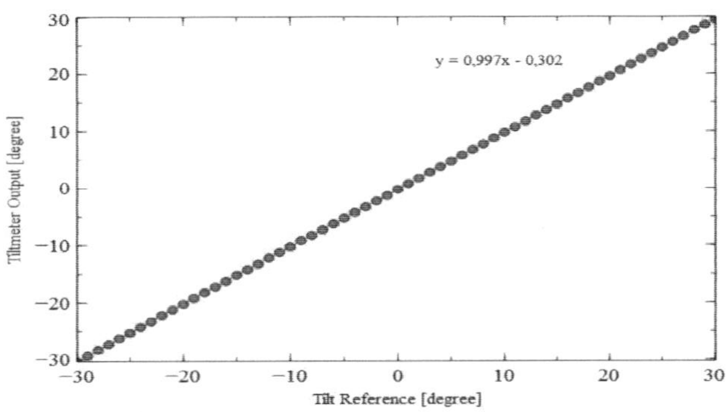

Figure 3. Inclinometer Characterization

In this graph, it is shown the result of inclinometer characterization where the x axis is the tilt references while the y axis is tiltmeter output. The range of measurement is limited at -30º to 30º and it gives very high precision of the tilt measurement. Thecalculated equation from the measurement result was y=0.997x-0.302. This equation referred to the 0.997 of sensor sensitivity of and 0.302º of sensor offset . It can be seenthat this inclinometer has good linearity to measure the inclination.

Secondly, the characterization of extensometer is presentedin Figure 4.

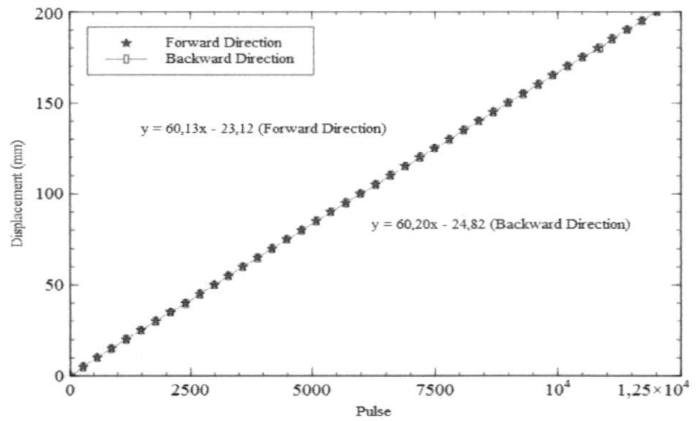

Figure 4. Extensometer Characterization

From the Figure 4, x axis is pulse of encoder while y axis is displacement of translational movement. This extensometer wasdesigned to measure the displacement for 1 m range. But in this characterization, it is onlymeasured by 200 mm range to represent the actual displacement range. They weretwo kinds of characterization that has been tested for extensometer. The first kind of characterization was forward direction with the equation of $y = 60.13x - 23.12$. The value of 60.13 figured out the sensitivity of extensometer sensor with-23.12 of offset value. Another characterization was backward direction with the equation of $y = 60.20x - 24.82$. Both equations between forward and backward direction werealmost similar inwhich the sensor is not affected by the direction.

3.2. Landslide monitoring system implementation

As it is mention before, this system has been implemented in Pangalengan, Bandung. This system consists of four nodes and one gateway unit that can be seen in Figure 5.

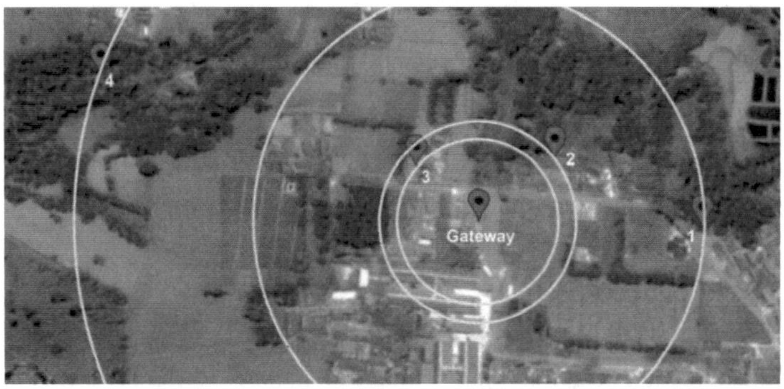

Figure 5. Location of Landslide Monitoring System Implementation

From the Figure. 5, it can be seen that four nodes is distributed in different pointsof area. The longest distance from the gateway is node 4 and the shortest distance is node 3. The distance of node 4 to the gateway is 274 meter while the other nodeshave theaverage of distance is 60 meter. This node can communicate each other to send data from the sensor ofeach node to the gateway.

3.3. Measurement result of landslide monitoring system

This landslide monitoring has been tested in Pangalengan within August – October 2016. The data was presented on the website. The graphic can be seen in Figure 6 where it represented the data of extensometer. There are four nodes that can measure the displacement of soil movement every 2 minutes. The x axis is the time while the y axis is displacement in mm.

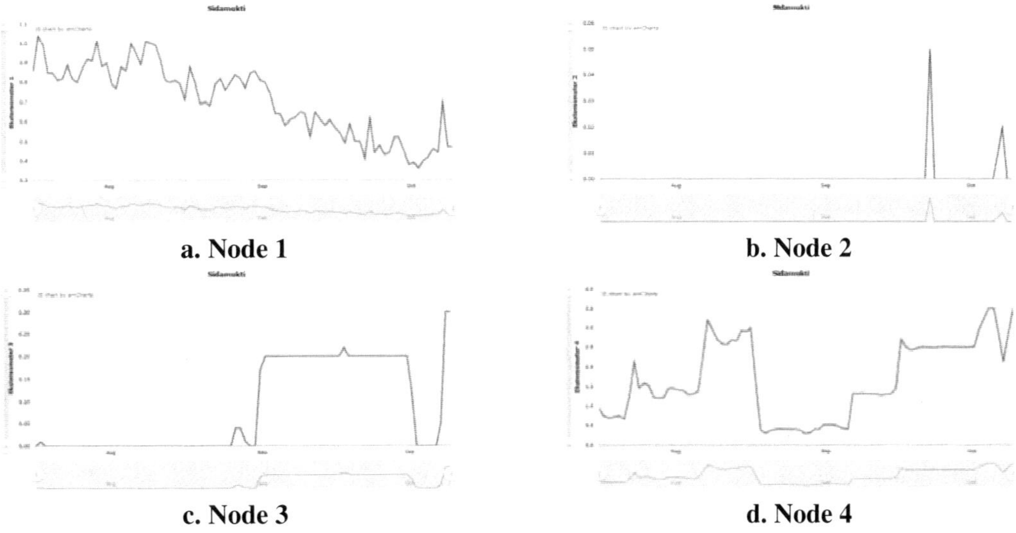

a. Node 1 **b. Node 2**

c. Node 3 **d. Node 4**

Figure 6. Extensometer Data

Another sensor is inclinometer that measures the tilt reference to gravity axis. The result can be seen in Figure 7 where the dual axis of x and y is measured in every 2 minutes. The x axis is the time stamps while the y axis is the tilt in degree. The x axis is in line to the soil movement direction in each node.

The interesting of this data was the trend of measurement result in every node. This might be caused by some factors like morphology, soil properties, humidity, and temperature of environment.The movement of soil can be in a shrinkage or expand that can be shown in graphic. The average of displacement was 0.5 to 1 mm. It can be concluded that it was still in normal condition.

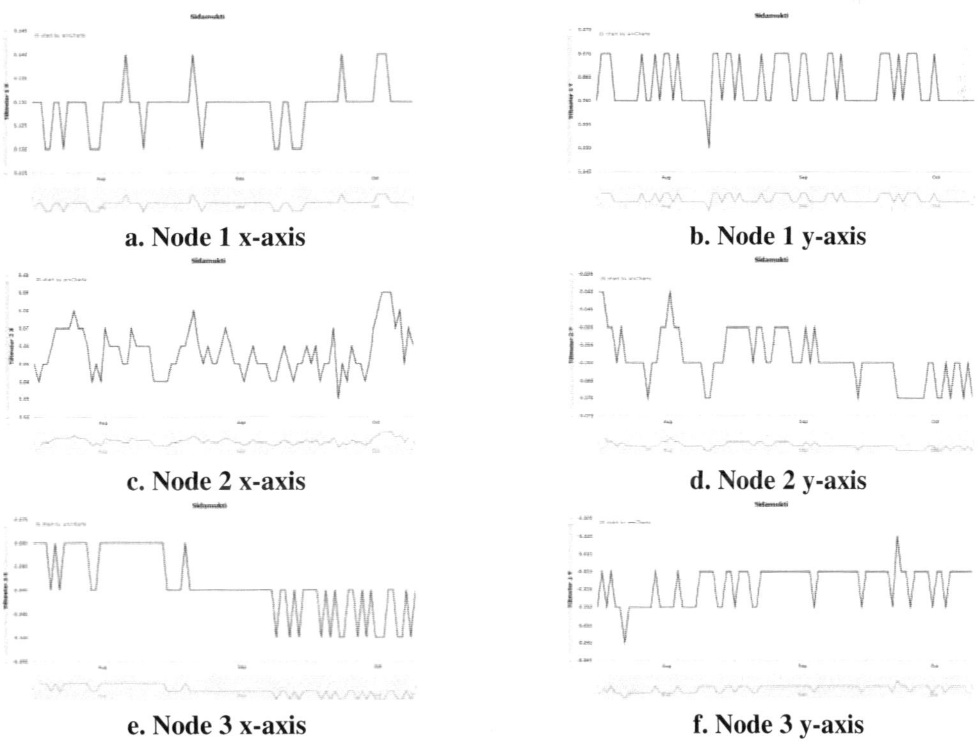

a. Node 1 x-axis **b. Node 1 y-axis**

c. Node 2 x-axis **d. Node 2 y-axis**

e. Node 3 x-axis **f. Node 3 y-axis**

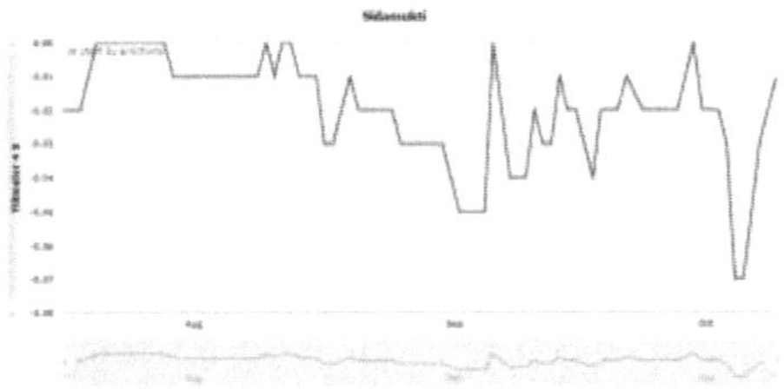

g. Node 4 x-axis

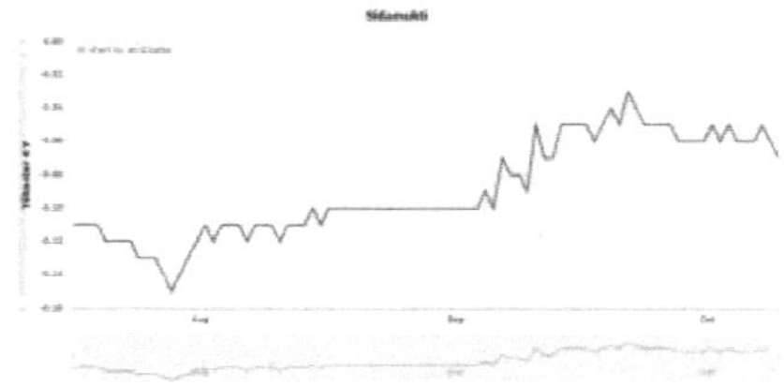

h. Node 4 y-axis

Figure 7. Inclinometer Data

4. Conclusions

The Landslide monitoring system has been installed successfully in the period of August to October 2016. It can measure the displacement and tilt in observation area. The sensor in each node was connected using wireless communication system. It is helpful to overcome the limitation of sensor system installation in remote area similar to landslide monitoring system. The inclinometer and extensometer has been made and characterized before it is installed.

References

[1] X Jian, Yuanhong Wang, Yu Zhang and Shushan Yang *Intelligent Early-Warning System for Landslides Based on The Zigbee Network* 2013 IEEE of 11th International Conference on Dependable Autonomic and Secure Computing

[2] H Dwi, B Widiyatmoko, B Hermanto, P Puranto and L T Handoko Real Time Inclinometer Using MEMS Accelerometer 2010 Proceedings of 5th International Conference On Physics And Its Applications Kentingan *Physics Forum* 2010 ISBN: 979-978-3456-68-5.

[3] Y Yan, HauyuanXie, Jie Wang and JinpingOu, Deflection Measurement Using Wireless Inclination Sensors for Bridge 2010 *IEEE of International Conference on Intelligent Controll and Information Processing*

[4] Y Yan, JinpingOu, Jun Zhang, Senior member, IEEE, Chunwei Zhang and Luyu Li April 2009 Development of Wireless MEMS Inclination Sensor System for Swing Monitoring of Large-Scale Hook Structures 2009 *IEEE Transaction on Industrial Electronics* **56** 4.

[5] D Rosanno JC, Jason Enriquez, Francis Gabriel Victorino Earl Anthony Mendoza, Marc Caesar Talampas and Joel Joseph Marciano Jr. 2009 Design, Development, and Evaluation of a Tilt and Soil Moisture sensor Network for Slope Monitoring Applications 2009 *IEEE of TENCON.*

[6] ZigBee RF Modules *DigiInc*[http://www.digi.com/]

[7] Datasheet SCA100T *Murata Inc*[http://www.murata.com/]

[8] Datasheet E40S6-2000 *AutonicsInc*[http://www.autonicsonline.com/]

Detection of acute lymphocyte leukemia using k-nearest neighbor algorithm based on shape and histogram features

Endah Purwanti and Evelyn Calista

Biomedical Engineering , Faculty of Science and Technology, Universitas Airlangga, Kampus C, Mulyorejo, Surabaya 60115, Indonesia

Corresponding email address: endah-p-1@fst.unair.ac.id

Abstract. Leukemia is a type of cancer which is caused by malignant neoplasms in leukocyte cells. Leukemia disease which can cause death quickly enough for the sufferer is a type of acute lymphocyte leukemia (ALL). In this study, we propose automatic detection of lymphocyte leukemia through classification of lymphocyte cell images obtained from peripheral blood smear single cell. There are two main objectives in this study. The first is to extract featuring cells. The second objective is to classify the lymphocyte cells into two classes, namely normal and abnormal lymphocytes. In conducting this study, we use combination of shape feature and histogram feature, and the classification algorithm is k-nearest Neighbour with k variation is 1, 3, 5, 7, 9, 11, 13, and 15. The best level of accuracy, sensitivity, and specificity in this study are 90%, 90%, and 90%, and they were obtained from combined features of area-perimeter-mean-standard deviation with k=7.

1. Introduction

The leukemia type will be used in this study is ALL. ALL is leukocyte cancer characterized by overproduction and continuous multiplication of malignant and immature leukocyte (lymphoblast or blast cell). One of the processes carried out in this study is obtaining the pattern image of normal leukocyte cells and blast cells. Therefore, the selection of features is essential in order to obtain information which distinguishes normal leukocyte cells from blast cells. Features used in this study were shape feature of the nucleus and histogram features. Shape features of the nucleus were adapted from those used by Kulkarni and Bhosale (2014), and the types of shape features that used were area, perimeter, eccentricity, form factor, and solidity [1]. Meanwhile, the histogram feature was adapted from that used by Scotti (2005), and it included mean deviation and standard deviation. The result of extraction features would be used as the input data in the next process, which was classification with k-Nearest neighbour. Euclidean distance was used to calculate the distance between testing data and reference data.

2. Methodology / experimental

2.1. Image dataset

Image dataset in this study used a public dataset ALL-IDB2 provided by Dr Fabio Scotti. Used imagery is a single image of peripheral blood cell with the size of 257 x 257 pixels.

2.2. Proposed system

Proposed system included pre-processing, nucleus segmentation, feature extraction, and classification. Figure 1 shows the proposed system algorithm in this study.

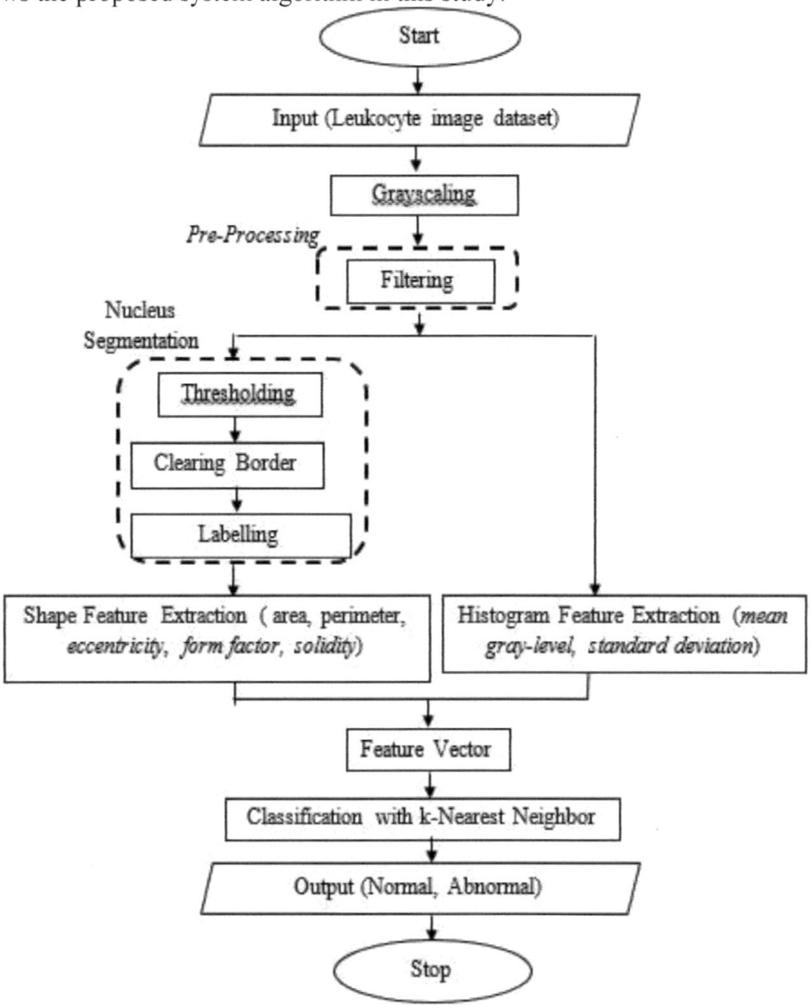

Figure 1. Proposed algorithm system

2.2.1. Gray scaling.

Gray scaling is an RGB conversion stage to a gray scale image by eliminating the hue and saturation information to maintain luminance [6]. This step was conducted to change RGB image with three-dimensional matrix into a gray scale image with two-dimensional matrix to make it easier to be processed.

2.2.2. Pre-processing.

In some cases, the preparation image encountered blurring, low contrast, and unwanted noise [7]. Therefore, pre-processing step was conducted to enhance the quality of the image. Median filtering was used in this step to remove unwanted noise.

2.2.3. Nucleus segmentation.

In this step, the segmentation used Otsu's thresholding method to take the nucleus of lymphocyte. The shape of nucleus is important to describe a blast cells. Lymphocyte cell has blue and regular shape of nucleus, whereas lymphoblast cell has irregular shape and spherical particles in nucleus [5]. To remove unwanted area which was still segmented with Otsu's thresholding method, clearing border and component labelling were used. Clearing border is one of morphological image processing as reconstruction application. It is used to remove objects touching the edge of image [8]. Component labelling is an image examination and classification of each pixel into a connected component according to the rules of connectivity [10].

2.2.4. Feature Extraction.
The extracted features were shape feature and histogram feature. Histogram feature was extracted using the first level of the nucleus image which had been created with preprocessing. Meanwhile, shape feature was extracted using an image which had been created with nucleus segmentation.

- Area: total number of nonzero pixel [5]
- Perimeter: total pixels of boundary image [5]
- Eccentricity: object's roundness with value 0 to 1. Perfect round object has 0 eccentricity and line segment has 1 eccentricity [5]
- Form factor: function of an area and perimeter of an object [5]
- Solidity: ratio of actual and convex hull area [5]
- Mean: the average intensity of the image
- Standard deviation: the deviation of mean

2.2.5. Classification.
Reference data used was 50 images for each class of normal and abnormal. Meanwhile, testing data used was 10 images for each class of normal and abnormal. This study was conducted by varying the parameter values of k=1 to k=15 [11] and also varying combinations of the image features. K-NN classifiers were used to look for a group of k objects on the training data set which was the closest to the test data and base the labelling of the dominant class on neighbouring regions [12]. According to current knowledge on the field, the selection of k optimal value depends on the amount of data that it requires different k values for different applications. This method is simple, but it is very effective in some cases [4]. Euclidean distance is used as a proximal distance calculation among neighbours. Equation 1 shows the Euclidean formula.

$$d_2(x,y) = \|x-y\|_2 = \sqrt{\sum_{i=1}^{r}(x_i - y_i)^2} \quad (1)$$

in which x, $y \in X$, and x_i, y_i are featuring values of $-I$ from x and y, and whereas r is the number of features in vector [9].

Determination of accuracy, sensitivity, and specificity is based on a comparison of the results of the test data output from the k-NN classification with the classification results of medical experts. Calculation used to get accuracy used ratio of correctly classified testing data and the number of testing data. Sensitivity was calculated from ratio of between true positive and the sum result of true positive and false negative, whereas specificity was calculated from ratio of true negative and the sum result of true negative and false positive.

TP (true positive) is the number of images correctly classified as positive on the test. TN (true negative) is the number of images correctly classified as negative in the test. FP (false positive) is the number of images classified as positive on the test while actually they are not. Meanwhile, FN (false negative) is the number of images that classified as negative image on the test while actually they are not [3].

3. Results and discussion
The image results of each step in the proposed system to segmenting nucleus are shown in Figure 2.

IOP Conf. Series: Journal of Physics: Conf. Series **853** (2017) 012011 doi:10.1088/1742-6596/853/1/012011

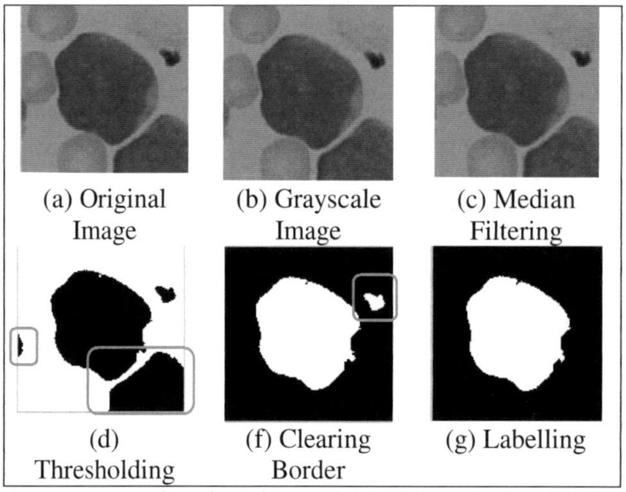

Figure 2. Image Results of Each Step in Forming Nucleus Segmentation

There are four combination features used in this study, such as (mean-standard deviation- area-perimeter-eccentricity-form factor-solidity), (area-perimeter-eccentricity-form factor-solidity), (mean-standard deviation), (area-perimeter), and (Mean-Standard Deviation-Area-Perimeter). Table 1 shows the highest accuracy, sensitivity, and specificity of each combination feature obtained with k-NN.

Table 1. Highest Accuracy, Sensitivity, and Specificity

Combination Features	*k*	Accuracy (%)	Sensitivity (%)	Specificity (%)
All features (mean-standard deviation-area-perimeter-eccentricity-form factor-solidity)	7	90	90	90
Shape features (area-perimeter-eccentricity-form factor-solidity)	7	85	80	90
Histogram features (mean-standard deviation)	5	75	80	70
	9	75	80	70
	11	75	80	70
	13	75	80	70
	15	75	80	70
Area-Perimeter	7	85	80	90
Mean-Standard Deviation-Area-Perimeter	7	90	90	90

Based on the information shown in Table 1, the highest accuracy, sensitivity, and specificity are obtained in combination (mean-standard deviation-area-perimeter-eccentricity-form factor-solidity) and (mean-standard deviation-area-perimeter) with k=7. However, the most effective combination features are mean-standard deviation-area-perimeter since the number of combinations is fewer than that of the combination features with similar accuracy, sensitivity, and specificity. Therefore, the computational load is also fewer than the others. In terms of application, it is advantageous for accelerating the working time of the program.

As a classification tool for medical experts to detect any abnormality in leukocytes led to acute leukemia, these applications still need to be developed, such as adding class classification. For example, the class is divided into more specific abnormalities namely acute lymphocytic Leukemia-L1 (ALL-L1), acute lymphocytic Leukemia-L2 (ALL-L2), and acute lymphocytic Leukemia- L3 (ALL-

L3). Furthermore, it also needs to be added with a feature which is able to count the number of abnormal leukocyte cells in the visual field preparations. Since to determine whether a person suffers from Leukemia and the type of Leukemia they suffer, it is required to calculate the number of abnormal leukocytes and any abnormal cell types found in the visual field preparations.

4. Conclusion

This study proposes a strategy and methodology to detect Acute Lymphocyte Leukemia (ALL). We have used combination made of shape and histogram features. The purpose of using k-NN algorithm is to classify the lymphocyte cells into two classes, namely normal and abnormal lymphocytes. The results show that the best combination features in this study are the combination of area-perimeter-mean-standard deviation with k=7. The values of accuracy, sensitivity, and specificity obtained are 90%, 90%, and 90%.

References

[1] Kulkarni T A dan Dilip S Bhosale 2014 A Robust Morphological Analysis of Normal and Abnormal Leukemic Cells Populations in Acute Lymphoblastic Leukemia, *National Conf. on Recent Innovations in Science Engineering & Technology* India

[2] American Cancer Society 2014 Leukemia-Acute Lymphocytic (Adults) www.cancer.org/Leukemia-acute-lymphocytic-pdf 10 September 2014

[3] Bain B J 2003 *Leukemia Diagnosis* (United Kingdom: Blackwell Publishing)

[4] Deore S G and Neeta Nemade 2013 Image Analysis Framework for Automatic Extraction of the Progress of an Infection *International Journal of Advanced Research in Computer Science and Software Engineering* **Volume 3 Issue 6** pp 703-6

[5] Joshi M D, Atul H Karode, S R Suralkar, 2013 White Blood Cells Segmentation and Classification to Detect Acute Leukemia *International Journal of Emerging Trends & Technology in Computer Science* **Volume 2 Issue 3** May-June 2013: 147-50

[6] Mathworks 2008 *Image Processing Toolbox™ User's Guide* (United States: The MathWorks, Inc.)

[7] Mokhtar N R, Nor Hazlyna Harun, M Y Mashor, H.Roseline, Nazahah Mustafa, R Adollah, H Adillah, N F Moh Nasir 2009 *Image Enhancement Techniques Using Local, Global, Bright, Dark, and Partial Contrast Stretching for Acute Leukemia* Images (London: WCE) July 1-3

[8] Prasetyo E 2011 *Pengolahan Citra Digital dan Aplikasinya Menggunakan MATLAB* (Yogyakarta: Penerbit Andi)

[9] Prasetyo E 2014 *Data Mining: Mengolah Data Menjadi Informasi Menggunakan MATLAB* (Yogyakarta: Penerbit Andi)

[10] Putra D 2010 *Pengolahan Citra Digital* (Yogyakarta: Penerbit Andi)

[11] Scotti F 2005 Automatic Morphological Analysis for Acute Leukemia Identification in Peripheral Blood Microscope Images *IEEE International Conference on Computationa;* Intelligence for Measurement Systems and Applications.

[12] Wu Xindong and Vipin Kumar 2009 *The Top Ten Algorithms in Data Mining* (United States: Taylor & Francis Group)

IOP Conf. Series: Journal of Physics: Conf. Series **853** (2017) 012012 doi:10.1088/1742-6596/853/1/012012

Color segmentation of multi variants tuberculosis sputum images using self organizing map

Riries Rulaningtyas[1,*], Andriyan B Suksmono[2], Tati L R Mengko and Putri Saptawati

[1] Biomedical Engineering, Department of Physics, Faculty of Science and Technology, University of Airlangga, Jl. Mulyorejo, Kampus C, Surabaya 60115,Indonesia
[2] School of Electrical Engineering and Informatics, Bandung Institute of Technology Jl. Ganesha, Bandung 40132, Indonesia

[*]Corresponding email address: riries-r@fst.unair.ac.id

Abstract. Lung tuberculosis detection is still identified from Ziehl-Neelsen sputum smear images in low and middle countries. The clinicians decide the grade of this disease by counting manually the amount of tuberculosis bacilli. It is very tedious for clinicians with a lot number of patient and without standardization for sputum staining. The tuberculosis sputum images have multi variant characterizations in color, because of no standardization in staining. The sputum has more variants color and they are difficult to be identified. For helping the clinicians, this research examined the Self Organizing Map method for coloring image segmentation in sputum images based on color clustering. This method has better performance than k-means clustering which also tried in this research. The Self Organizing Map could segment the sputum images with y good result and cluster the colors adaptively.

1. Introduction

Pulmonary tuberculosis is a deadly infectious disease occured in many countries in Asia and Africa. In Indonesia, many people with tuberculosis disease are examined in the community health centre. Examination of tuberculosis through sputum smear is a result from Ziehl – Neelsen staining in microscopy. From the results, Ziehl – Neelsen staining will give red color effect to the tuberculosis (TB) bacteria and blue color for background images. The first step of examination was to detect the presence of TB bacteria from the red side, then from the shape of the TB bacteria itself.

The Ziehl – Neelsen staining in sputum smear gives the complex color images, so that the clinicians have difficulty in doing slide examination manually. To assist the clinicians when reading the sputum smear slide, this research did color segmentation method for the sputum images which had been digitized. The purpose was to extract the TB bacteria images and eliminate the background images. The red TB bacteria after Ziehl – Neelsen staining provided the different brightness red levels, from brown, dark red, until to pink. The right color segmentation is necessary for the accurate classification techniques in order to distinguish the image of TB bacteria, the image of non TB bacteria, and the image of background.

Examination of patients with pulmonary tuberculosis began with an examination of sputum. The examination of sputum waa the main step, because it could identify the presence of TB bacteria so that the disease can be diagnosed as pulmonary tuberculosis. The Indonesian community health centers commonly use a light microscopic examination techniques using Ziehl - Neelsen staining for sputum observation. This staining gives a red effect for TB bacteria and blue color for the background of the sputum sample. The examination of the TB bacteria is still done manually by clinicians to perform manually counting the number of bacteria in each field of view of the microscope. For pulmonary TB disease diagnose becoming easy and fast, this research made a diagnosis of pulmonary tuberculosis automatically. The first step was the enhancement process for digital sputum images. After that, the color segmentation was done to sputum images for clearing up the background images and leaved the TB bacteria images. Some researchs which had been done correlated with the segmentation of TB bacteria images were [1],[2],[3]. Research [1] did the color segmentation using adaptive color thresholding in RGB color space while research [2] did color segmentation of the TB bacteria images using pixel classifier method. In addition, research [3] did color segmentation for TB bacteria using adaptive color thresholding in HSV color space. The researcher has opinion that the color segmentation has important rule to support the success of TB sputum diagnose. To improve the accuracy of TB identification in [1][2][3], this research already did color segmentation using self-organizing map (SOM). Previously, this research examined k-means cluster for color image segmentation, but there were some shortcomings to choose the best centroid values. It needs long computational time. With SOM method, the color could be segmented more fast and accurately. After segmenting background, identification is continued with the shape identification. Research [4] did TB bacteria identification in binary images using neural network based on geometry feature only without using color feature as information for classification process. SOM with clustering method to segment the color of TB images from background could support the identification TB bacteria images from sputum more accurately.

2. Material and Method

2.1. Image acquisition

The number of sputum smear slides used was about 1000 image of Ziehl-Neelsen sputum smear provided by Bandung Health Department in West Java Province, Indonesia. Images of sputum smear slide were captured using web camera integrated with ocular lens and mounted onto ocular field and saved in JPEG format in RGB 24 bits. The total of microscope magnification was 1000 times. The image size was 640x360 pixels.

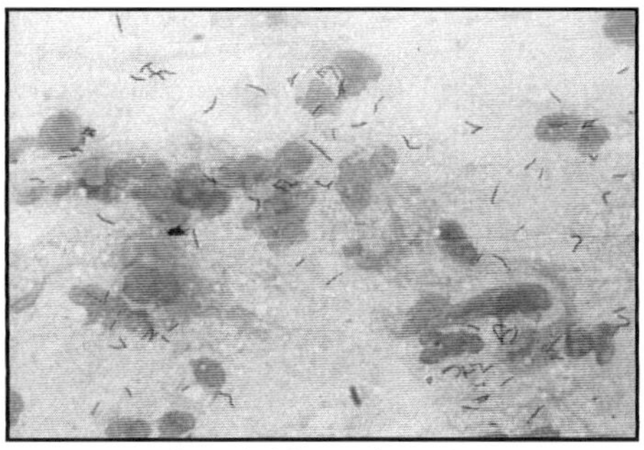

Figure 1. The Ziehl-Neelsen Sputum Image

2.2. K-Means clustering

K-means clustering is grouping data method which widely used. K-means clustering method has objective function based on the proximities of the data points to the centroid of cluster [8]. With $\boldsymbol{X} = \{\boldsymbol{x_1}, \ldots, \boldsymbol{x_n}\}$ be the data, $\boldsymbol{m_l} = \sum_{X \in C_1} \frac{x}{n_l}$ be the centroid of cluster $\boldsymbol{C_l}$, $\boldsymbol{1} \leq \boldsymbol{l} \leq \boldsymbol{k}$, where $\boldsymbol{n_l}$ is the number of the data objects in cluster $\boldsymbol{C_l}$, and k is the number of clusters [3]. The objective function of K-means clustering is then formulated as the sum of squared errors [5] :

$$F_k = \sum_{l=1}^{k} \sum_{x \in C_l} \|x - m_l\|^2 \quad (1)$$

With $d(C_p, C_q) = \sum_{x_i \in C_p} \sum_{x_j \in C_q} \|x_i - x_j\|^2$, the sum of all pair wise distances of data objects within K clusters as follows [3] :

$$D_k = \sum_{i=1}^{n} \sum_{j=1}^{n} \|x_i - x_j\|^2 = \sum_{l=1}^{k} d(C_l, C_l) + 2 \sum_{1 \le i \le j \le k} d(C_i, C_j) \quad (2)$$

2.3. Self Organizing Map

Self-Organizing Maps (SOM) have unsupervised training to get the best architecture. With the concepts of mapping input to the output based on similarity, so that SOM suitable to the clustering process. This research tried to use SOM for color image segmentation based on clustering method.

Kohonen SOM is designed to group a set of Q continuous - valued vectors $s^{(q)} = [s_1^{(q)}\ s_2^{(q)} \ldots s_N^{(q)}]$, q = 1, ..., Q, into M (< Q) clusters. The SOM consists of an input layer having N neurons, and an output layer having M neurons (cluster units) arranged in some predetermined fashion. Each neuron in the input layer is connected to a neuron in the output layer. Thus, output neuron (cluster units) j is connected to each of the input neuron through weights $w_{ij,} i = 1, \ldots, N$, which is referred to as the j-th weights vector. In vector or matrix notation, it is denoted the j-th weight vector by $W_{.j} = [w_{1j}\ w_{2j} \ldots w_{Nj}]$. It is given by the j-th column of the weight matrix. There is a total of M such weight vectors, one for each cluster units.

During the training process, each training vector is cyclically or randomly selected and presented to the network. Two vectors are considered the closest if the square of the Euclidean distance between them, $\|x - W_{.j'}\|^2$ was the smallest. The winning unit and its neighboring units (those located in its first and second neighborhoods) then update their weight vectors according to the Kohonen rule. This process is continued until the weight vectors changed by less than a preset amount. The square of the Euclidean distance between the input vector and each of the weight vectors is [6]:

$$d_j = \|x - W_{.j}\|^2 = \|x\|^2 + \|W_{.j}\|^2 - 2x.W_{.j} \quad (3)$$

The SOM algorithm:

1. Initialize M weight vectors. Set topological neighborhood parameters and learning rate, $\alpha(< 1)$
2. For step k = 1,2,..., do steps a-d by cycling through training set until weight vectors converge
   a. Set input vector $x = s^{(q)}$, one of the training vectors.
   b. Compute for each cluster unit j = 1, ..., M the Euclidean distance

$$d_j = \sum_{i\neq 1}^{N} (x_i - w_{ij}(k))^2 \quad (4)$$

   c. Find the index j' such that $d_{j'}$ is a minimum
   d. For all cluster units j within the specified neighborhoods of j', update the weight vectors

$$w_{ij}(k+1) = w_{ij}(k) + \alpha[x_i - w_{ij}(k)], \qquad i = 1, \ldots, N \quad (5)$$

3. Result

Firstly, color segmentation was done to TB bacteria identification. Color segmentation had a purpose to identify the red object as TB bacteria candidate. This research examined some methods based on clustering concepts for color image segmentation. They were k-means clustering and SOM.

IOP Conf. Series: Journal of Physics: Conf. Series **853** (2017) 012012 doi:10.1088/1742-6596/853/1/012012

Table 1. The color segmentation results

Before Segmentation	K-Means Clustering	SOM
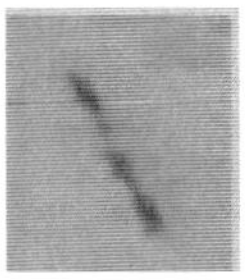	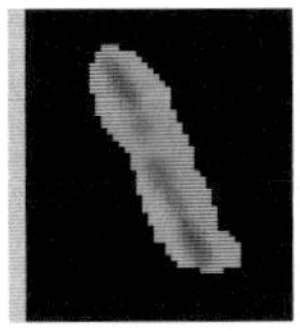	
	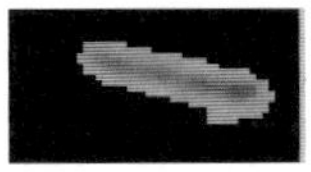	
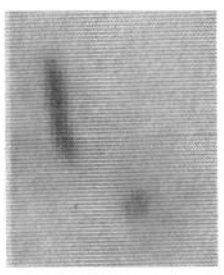		
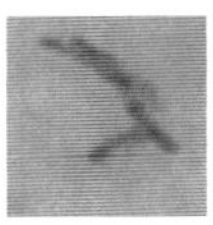		

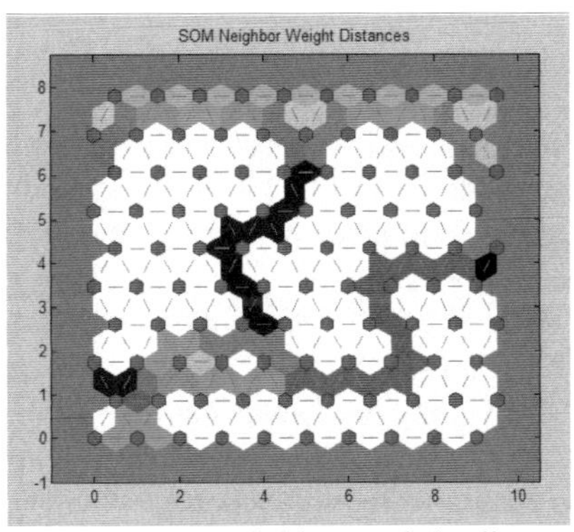

Figure 2. SOM neighbor weight distances

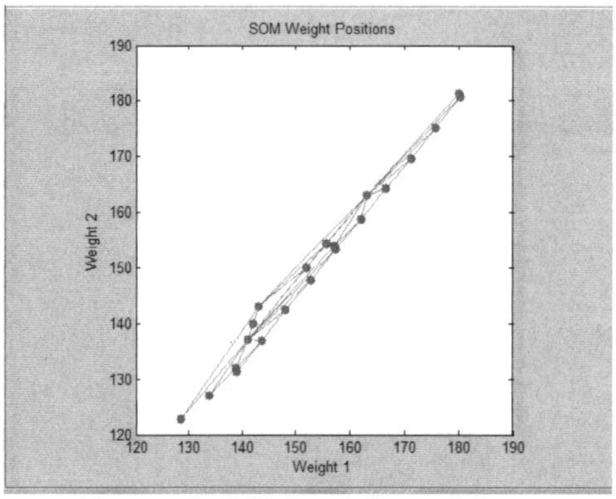

Figure 3. SOM weight positions

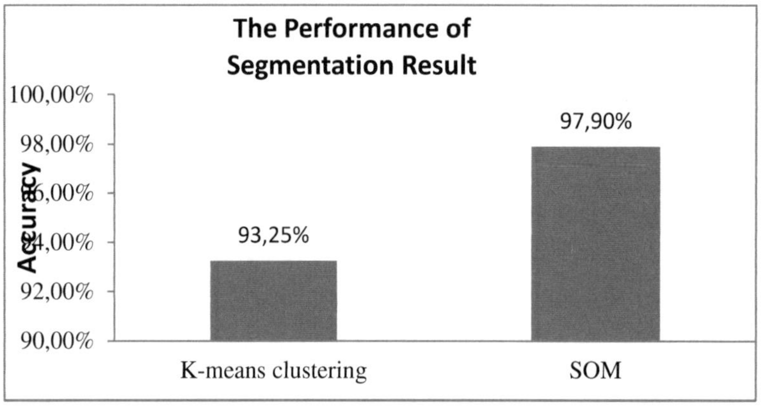

Figure 4. The Accuracy of Segmentation

4. Conclusion

SOM technique could do better segmentation in Ziehl-Neelsen sputum smear images which is very multi variants with the 97.90% of accuracy. The background images could be clear enough and remain TB bacilli candidate objects. Therefore, it could be easy of doing classification for the red objects to decide TB bacilli or not TB bacilli.

SOM technique could support the classification step to give the best accuracy in pulmonary disease diagnose. K-Means clustering still have some lacks in centroid finding. It remains use trial and error to find centroid and the data needs long computational time to achieve convergence. The SOM method could implement in real application for TB identification combining with the classification method.

Acknowledgment

This research was supported by the grant from Directorate General of higher Education Indonesia. We would like to thank to Indonesia Government who already supported and gave the fund for this research. Special thanks went to Bandung Health Department in West Java Province, Indonesia which already supported Ziehl-Neelsen sputum smear slide data.

References

[1] M K Osman, M Y Mashor, H Jaafar 2011 Combining Thresholding and Clustering Techniques for Mycobacterium Tuberculosis Segmentation in Tissue Sections *Australian Journal of Basic and Applied sciences* 5(12): 1270-1279 **ISSN 1991-8178**

[2] Khutlang, Rethabile, et al 2010 Classification of Mycobacterium tuberculosis in images of ZN-stained sputum smears *IEEE Transactions on Information Technology in Biomedicine* **14.4**

949-957.

[3] V Makkapati, R Agrawal, R Acharya 2009 Segmentation and Classification of Tuberculosis Bacilli from ZN-stained sputum /smear Images *IEEE Conference on Automation Science and Engineering* Bangalore India.

[4] Rulaningtyas R 2011 Automatic Classification of Tuberculosis Bacteria Using Neural Network *IEEE Conference Publications* Pages: 1 - 4 DOI: 10.1109/ICEEI.2011.6021502

[5] J Wu 2012 Advances in K-Means Clustering-Data Mining Thinking *Springer Theses, Recognizing Outstanding Ph.D Research* London.

[6] Leung KM 2007 *Kohonen Self-Organized Maps* Polytechnic University Department of Computer and Information Science

[7] G Guo, C R Dyer 2007 Patch-based Image Correlation with Rapid Filtering Proceeding/CVPR *IEEE Computer Society Conference on Computer Vision and Pattern Recognition* DOI:**10.1109**/CVPR.2007.383373.

[8] J Liu, YH Yang 1994 Multi resolution Color Image Segmentation *IEEE Transaction on Pattern Analysis and Machine Intelligence* Vol.**16**, No.7.

[9] K A Jain 1989 *Fundamental of Digital Image Processing* Prentice Hall

[10] MK Osman, MY Mashor, H Jaafar 2011 Tuberculosis Bacilli Detection in Ziel-Neelsen Stained Tissue Using Affine Moment Invariants and Extreme Learning Machine *IEEE 7th International Colloquium on Signal Processing and its Applications*

[11] MK Osman, MY Mashor, H Jaafar 2011 Improving Performance of Hybrid Multilayered Perceptron Network for Classification of Mycobacterium Tuberculosis in Tissue using Modified Recursive Prediction Error-Extreme Learning Machine *International Journal of Research and Reviews in Soft and Intelligent Computing (IJRRSIC)* vol.**1** No.3 ISSN : 2046-6412 Science Academy Publisher United Kingdom.

[12] P S Hiremath, P Bannigidad 2009 Automatic Classification of Bacterial Cells in Digital Microscopic Images *International Journal of Engineering and Technology* Vol.2 Num.4 Eashwar Publications.

IOP Conf. Series: Journal of Physics: Conf. Series **853** (2017) 012013 doi:10.1088/1742-6596/853/1/012013

Dose evaluation of organs at risk (OAR) cervical cancer using dose volume histogram (DVH) on brachytherapy

R Arif Wibowo[1,*], Bambang Haris[2], dan Inganatul Islamiyah

[1] Department of Physics, Faculty of Science and Technology, Airlangga University, Campus C Mulyorejo, Surabaya, Indonesia
[2] Department of Radiotherapy, dr Soetomo Hospital, Surabaya, Indonesia

*corresponding email address: aa.arifwr@gmail.com

Abstract. Brachytherapy is one way to cure cervical cancer. It works by placing a radioactive source near the tumor. However, there are some healthy tissues or organs at risk (OAR) such as bladder and rectum which received radiation also. This study aims to evaluate the radiation dose of the bladder and rectum. There were 12 total radiation dose data of the bladder and rectum obtained from patients' brachytherapy. The dose of cervix for all patients was 6 Gy. Two-dimensional calculation of the radiation dose was based on the International Commission on Radiation Units and Measurements (ICRU) points or called D_{ICRU} while the 3-dimensional calculation derived from Dose Volume Histogram (DVH) on a volume of 2 cc (D_{2cc}). The radiation dose of bladder and rectum from both methods were analysed using independent t test. The mean D_{ICRU} of bladder was 4.33730 Gy and its D2cc was4.78090 Gy. D_{ICRU} and D2cc bladder did not differ significantly (p = 0.144). The mean D_{ICRU} of rectum was 3.57980 Gy and 4.58670 Gy for D2cc. The mean D_{ICRU} of rectum differed significantly from D2cc of rectum (p = 0.000). The three-dimensional method radiation dose of the bladder and rectum was higher than the two-dimensional method with ratios 1.10227 for bladder and 1.28127 for rectum. The radiation dose of the bladder and rectum was still below the tolerance dose. Two-dimensional calculation of the bladder and rectum dose was lower than three-dimension which was more accurate due to its calculation at the whole volume of the organs.

1. Introduction

Cervical cancer is a deadly disease for women. This type of cancer is the most prevalent in the world. One way to cure cervical cancer is radiotherapy, either external radiotherapy irradiating the cancer tissue whose radiation source is outside the body or brachytherapy inserting the source into or around the cancerous tissue.

High dose rate (HDR) intracavitary brachytherapy is the most common method used to treat cervical cancer [1]. This therapy is able to give a radiation dose of more than 12 Gy/h[§] [2]. The process begins with the installation of brachytherapy applicator in the cavity of the cervix. This applicator serves as a radioactive source when irradiating the cervix. After the applicator is attached, the images of patient's organ are taken by radiograph devices. The next process called treatment planning system (TPS) can be carried out using the two and three-dimensional methods. Using two-dimensional radiograph of the patient, two-dimensional dose calculations are based on points of organ by the International Commission Radiation Units (ICRU) 38 in 1985. As the development of three-dimensional radiography technology, Groupe Européen de Churietherapie / European Society for Therapeutic Radiology and Oncology (GEC Estro) in 2006 gave the recommended dose calculations based on three-dimensional image of the patient obtained from dose volume histogram (DVH) and the volume histogram connecting organ with the radiation dose received [3]. TPS is a very important process in brachytherapy in order to produce the optimum radiation dose to the tumor and minimum dose to surrounding healthy tissue. The radioactive sources are then inserted into the applicator and irradiate cancerous tissue within a specific period before eventually taken back then stored in a well-protected place.

In practice, the radiation is received by not only cervix but also the healthy organs near the cervix such as bladder and rectum called organ at risk (OAR) which extremely sensitive to radiation when receiving doses above the tolerance, which consequently need a change of direction file or change in dose [2]. Hence, a dose evaluation of the bladder and rectum in cervical cancer brachytherapy is required. This study aims to evaluate the radiation dose of the bladder and rectum using both 2-dimensional and 3-dimensional methods.

2. Materials and methods

2.1. Installation applicator
This study used 12 cervical cancer brachytherapy patient data. The method used in accordance with cervical cancer brachytherapy procedures. The process began with the installation of applicator. The applicator used was fletcer suitdelclos consisting the tandem and ovoid. Installation of the applicator was conducted in patients who had been given an anaesthetic.

2.2. CT scan
The radiographic images of patient's organs were produced by CT scan. Patients did the CT scan after install edapplicator. CT scan produced images that can be reconstructed to form a three dimension (3D) through digital reconstruction radiography (DRR). Furthermore, CT scan image was transferred to the HDR brachytherapy for the treatment planning system via DICOM (Digital Imaging and Communications in Medicine) for defined tumours and coloured it by different colours for different organs.

2.3. Treatment Planning System (TPS)
TPS uses Brachyvision software both two-dimensional method and three-dimensional. Two-dimensional method based on points on two-dimensional radiographic image of the patient. The points are A, B, rectum and bladder as ICRU 38 recommendations. ICRU recommendations are made based on Three-dimensional image. Three-dimensional method uses dose volume histogram (DVH) to calculate radiation dose of organs. DVH summarize information in three-dimensional dose distribution. This histogram usually displayed as percent by volume of the total volume on the ordinate axis and the dose on the abscissa. The prescribed dose is 6 Gy for all patients. In accordance with the recommendations American brachyteraphy society (ABS), the dose of the rectum and bladder must be below 80% of the prescribed dose to the cervix.

2.4. Evaluation of dose
Dose evaluation using two-dimensional method performed at the ICRU 38 points of the bladder and rectum while three-dimensional ones obtained from 2 cc volume of organ of DVH (D_{2cc}).The radiation dose obtained from both methods are analysed using SPSS independent t test ($\alpha = 0.05$).

3. Results and discussion
Figure 1 is the result of two-dimensional method (D_{icru}) which shows point A, point B, bladder and rectum point. D_{icru} was obtained by drawing a line from the applicator to the ICRU points while three-dimensional method was obtained from volume 2cc of DVH as shown in Figure 2.

The mean Dicru for bladder is 4.33730 Gy and the mean $D_{2cc}$for bladder is 4.78090 Gy. As shown in Table 1, for bladder, the mean of D_{2cc} is greater than the mean of D_{icru} with ratio 1.10227. Based on independent test t test, two and three-dimensional methods dose of bladder did not differ significantly (p = 0.144). The mean D_{icru} of rectum was 3.57980 Gy while the mean D_{2cc} of rectum was 4.58670 Gy. Based on table 1, the mean D2cc is greater than D_{icru} with ratio 1.28127. The differences between the mean D_{2cc} and D_{icru} of rectum are significant based on the statistical test results (p = 0.000).

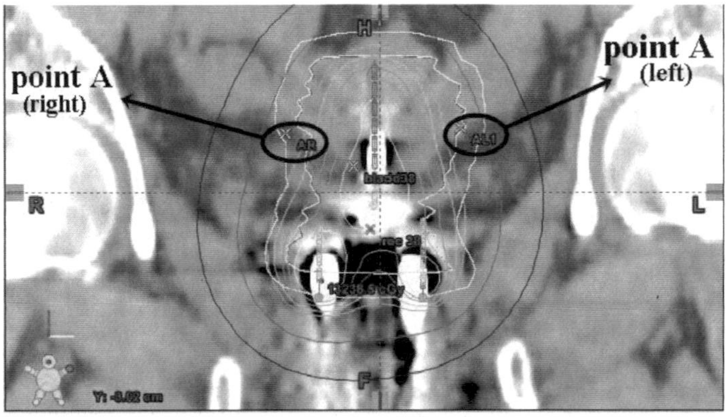

Figure 1. Supine position of the patient

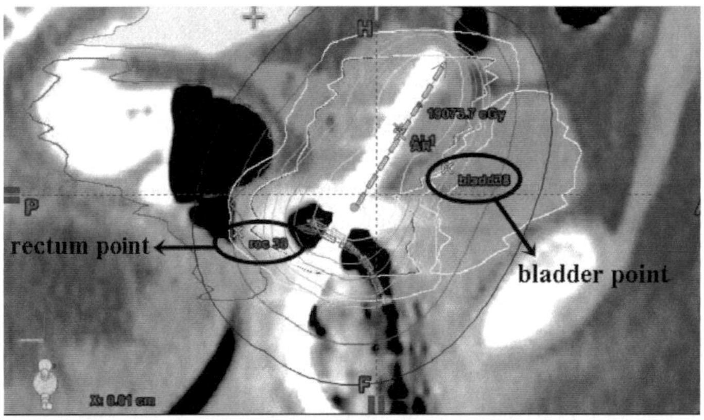

Figure 2. Oblique position of the patient

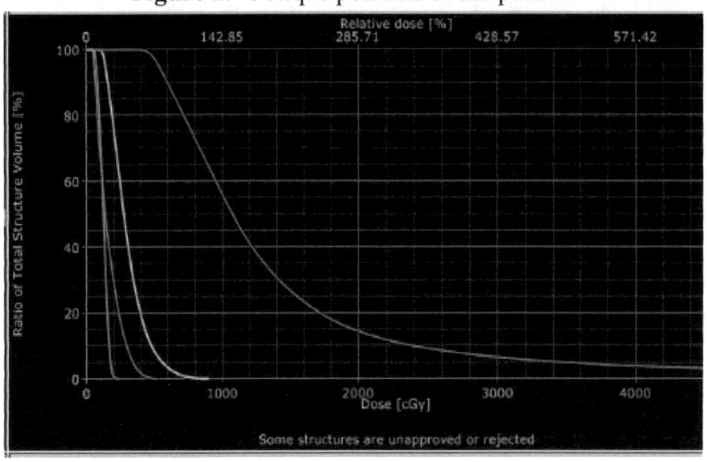

Figure 3. Dose Volume Histogram

Table 1. The mean D_{icru} and D_{2cc}

Organ	D_{icru} (Gy)	D_{2cc} (Gy)
Bladder	4.33730	4.78090
Rectum	3.57980	4.58670

As it is shown in Table 1, the average radiation dose bladder methods of 2-dimensional and 3-dimensional still below 4.8 Gy. At the same table, it also indicates the average radiation dose rectal method of 2-dimensional and 3-dimensional does not exceed 4.8 Gy. Both doses of radiation are in accordance with the recommendations of ABS. It means that the radiation dose is not harmful to the patients' bladder and rectum. But, there were some patients received radiation doses exceeding 4.8 Gy. The result was also performed by Sharma et al. This was caused by the distance of the bladder or rectum distance that was too close to the applicator so that received large doses. In accordance with the study of Zhang et al., the closer organs with the radioactive source in the applicator, the greater the dose of radiation received by these organs. Conversely, the distant organ of the applicator, the smaller the dose of radiation received.

It also found several differences in radiation dose between the bladder and rectum from one patient to another caused by patient anatomy. Based on the two-dimensional method, anatomy bladder and rectum belonging to different patients cause the difference in distance between the organs of the applicator and the radiation dose among patients' OAR. In addition, based on 3-dimensional method, the patients' anatomy lead to differences in organ volume and DVH generated as well.

The bladder filled with urine was able to influence the radiation dose to expand and push the applicator away from the surrounding organs. Thus, the radiation dose of bladder urine charged varies with the radiation dose when the bladder is empty. The rectum, with the same treatment filled with faeces slightly expands, leading to the differences in radiation dose between the empty and filled rectum.

Table 1 shows the average dose of radiation three-dimensional method is higher than two-dimensional one. The average ratio D_{2cc} bladder against D_{icru} is 1.10227 with insignificant differences in bladder radiation dose between 3- and 2-dimensional method ($p = 0.144$). This is consistent with the research conducted by Kirisits et al. and Hashim et al. However, some researchers, Jamema et al., Onal et al., And Tan et al. showed the opposite result, a significant difference between the dose of 3-dimensional and 2-dimensional bladder dose. Rectal radiation dose ratio using both methods was 1.28127 with differences radiation dose of those methods differing significantly ($p = 0.000$). Meanwhile, according to research Jamema et al., Tan., Et al., Onal et al., And Vinod et al., both doses were not different significantly.

As it is described above, the comparison results of the two methods are varied.But, with the similar indication that three-dimensional dose greater than the dose two dimensions which according to the research by Jamema et al., it can be caused by differences in radiographic and applicators. Besides, the latter method calculates the dose based on the distance of the applicator organ point compared to 3-dimensional method calculating the dose based on the volume of the organ. However, 3-dimensional method displays the more clear size of the tumor and OAR and the more accurate dose calculation. The size of the organ is obviously known that too large or small radiation doses can be avoided.

The different radiographic tools produced different images of the bladder and rectum. Based on the two-dimensional method, the OAR image differences lead to differences in ICRU point and dose. Meanwhile, according to three-dimensional method, the difference images of OAR produced different volume and DVH. Different applicator types produced different brachytherapy dose distribution related to the shape of the applicators. Various types of the applicator adapted to the location of the cervical tumour and the selection of the proper applicator can reduce the radiation dose of the bladder and rectum.

4. Conclusion

This study concluded the mean radiation dose of bladder from two-dimensional method (Dicru) was 4.33730 Gy while three-dimensional methods (D_{2cc}) was 4.78090 Gy. The mean Dicru of rectum was 3.57980 Gy while its mean D_{2cc} was 4.58670 Gy. The mean radiation dose of the bladder and rectum were below the tolerance dose. Besides, the ratio of the mean bladder D_{2cc} to D_{icru} was 1.10227. Differences bladder radiation dose of both methods was not significance ($p = 0.144$). The ratio of the D_{2cc} rectum to its D_{icru} was 1.28127. Rectal radiation dose of these two methods were different significantly ($p = 0.000$).

Acknowledgements

We would like to thank radiotherapy of dr Soetomo Hospital Surabaya especially medical physicists, doctors and lectures of radiation physics laboratories of the Faculty of Science and Technology, Airlangga University who has supported the implementation of this study.

References

[1] Harmon G, Chinsky B, Surucu M, Harkenrider M and Small Jr W 2016 Bladder distension improves the Dosimetry of organs at risk during cervical intracavitary high-dose-rate brachytherapy *The Amer Brachy Society* **15** 30-34

[2] Podgorsak EB 2005 *Radiation Oncology Physics* (Austria : IAIE) p 454

[3] Pötter R, Haie-Meder C, Limbergen E V, Barillot I, Brabandere M D, Dimopoulos J, Dumas I, Erickson B, Lang S, Nulens A, Petrow P, Rownd J and Kirisits C 2006 Recommendations from gynecological (GYN) GEC Estro working group (II) Concepts and terms in 3D image-based treatment planning in cervix cancer brachytherapy-3D dose volume parameters and aspects of 3D image-based anatomy, radiation physics, radiobiology *Rad and Oncology* **78** 67-77

[4] Hashim N, Jamalludin Z, Ung NM, Ho G F, Malik R A and EePhua V E 2015 CT-Based 3-Dimensional Treatment Planning of Intracavitary Brachytherapy for Cancer of the Cervix: Comparison between Dose-Volume histograms and ICRU Point doses to the rectum and Bladder *Asian Pa. J Cancer Prev* **15** 5259-64

[5] Jamema , Swamidas V, Saju S, Mahantshetty U, Pallad S, Deshpande DD, Shrivastava SK and Dinshaw KA, 2008 Dosimetric evaluation of the rectum and bladder using image-based CT planning and orthogonal radiographs with ICRU 38 recommendations in intracavitary brachytherapy *J Phys Med* **33** 3-8

[6] Kirisits C, Lang S, Dimopoulos J, Berger D, Georg D and Pötter R, 2006 The Vienna applicator for combined intracavitary and interstitial brachytherapy of cervical cancer: design, application, treatment planning, and dosimetric results *Int Jour Radiat Oncol Biol Physics* **65** 624-30

[7] Kirisits C, Pötter R, Lang S, Dimopoulos J, Wachter-Gerstner N and Georg D 2005 Dose and volume parameters for MRI-based treatment planning in intracavitary brachytherapy for cervical cancer *Int Jour Radiat Oncol Biol Physics* **62** 901-11

[8] Onal C, Arslan G, Topkan E, Pehlivan B, Yavuz M, Oymak E and Yavuz A 2009 Comparison of conventional and CT-based planning for intracavitary brachytherapy for cervical cancer: target volume coverage and organs at risk doses *Jour Exp and Clin Can Research* **28**:95

[9] Sharma B A and Tomcha Singh 2010 The rectum and bladder doses in intracavitary brachytherapy for cervical cancer *Jour Cont Brachytherapy* **4** 153-156

[10] Tan Y I, Choo B A and Lee K M 2010 2D to 3D Evaluation of organs at risk doses in intracavitary brachytherapy for cervical cancer *Jour Cont Brachytherapy* **2** 37-43

[11] Vinod S K, Caldwell K, A, Lau A and Fowler A R 2011 A comparison of ICRU point and volumetric doses doses of organs at risk (OARS) in brachytherapy for cervical cancer *The Jour. Med. Imag. and Rad. Oncology* **55** 304-310

[12] Zhang, Winson, Bhatia S K, Sun W, Modrick J M and Kim Y 2014 Bladder and rectal dose

estimations on digitized radiographs for vaginal brachytherapy after hysterectomy *Clin Med Jour Appl Physics* **15** 240-246

Effect of window length on performance of the elbow-joint angle prediction based on electromyography

Triwiyanto[1,3,*], Oyas Wahyunggoro, Hanung Adi Nugroho, Herianto[2]

[1]Department of Electrical Engineering & Information Technology, Universitas Gadjah Mada, Yogyakarta, Indonesia;

[2]Department of Mechanical & Industrial Engineering, Universitas Gadjah Mada, Yogyakarta, Indonesia;

[3]Department of Electromedical Engineering, Health Polytechnic of Surabaya, Ministry of Health, Indonesia

*corresponding author : triwiyanto123@gmail.com

Abstract. The high performance of the elbow joint angle prediction is essential on the development of the devices based on electromyography (EMG) control. The performance of the prediction depends on the feature of extraction parameters such as window length. In this paper, we evaluated the effect of the window length on the performance of the elbow-joint angle prediction. The prediction algorithm consists of zero-crossing feature extraction and second order of Butterworth low pass filter. The feature was used to extract the EMG signal by varying window length. The EMG signal was collected from the biceps muscle while the elbow was moved in the flexion and extension motion. The subject performed the elbow motion by holding a 1-kg load and moved the elbow in different periods (12 seconds, 8 seconds and 6 seconds). The results indicated that the window length affected the performance of the prediction. The 250 window lengths yielded the best performance of the prediction algorithm of (mean±SD) root mean square error = 5.68%±1.53% and Person's correlation = 0.99±0.0059.

1. Introduction

Nowadays EMG signal has important role in the development of the devices based on myoelectric control such as prosthetic device, exoskeleton for assistive or rehabilitative purpose, and sport activity monitoring. High accuracy in the prediction of the joint angle is significant in the devices based on myoelectric control. EMG is a signal which is generated when the muscle is in contraction. The characteristics of frequency and amplitude of EMG signal are 20 to 500 Hz and 0 to 10 mV [1] respectively. EMG can represent the activity of muscle when the limbs are in motion. By analysing the EMG signal, we can predict the position or angle of the joint which is under the study. The algorithm to predict the joint angle can be classified into two methods [2], namely e a non-pattern recognition and pattern recognition. A pattern recognition method uses classifiers to recognize the features. The classifiers, which are used often in the related research, are the artificial neural network [3], support vector machine [4], and neuro-fuzzy [5]. In the pattern recognition method, the classifiers can be assumed as a black box model which has inputs from feature extraction. The weakness of this method is the need of a training time to learn the pattern of the feature. A non-pattern recognition method uses a modelling function with the Hill-based method [6], a combination of feature extraction and low pass filter [7], and an optimization of the feature extraction using Kalman filter [8]. Feature extraction is an important part in the prediction algorithm which is used to predict the elbow joint angle. Previous researchers more concerned using time domain feature extraction than other domain (frequency and wavelet). This is due to the less complexities of processing time than those frequency and wavelet domain. Less complexity in the feature extraction is very important when the algorithm is applied to the real system.

Some parameters in the feature extraction need to be adjusted in order to obtain an optimum performance of the prediction. The parameters are window length, type of the segmentation (disjoint or overlap segmentation), and frequency sampling. The window length is essential in the feature extraction because it determines the accuracy of the prediction and the time processing of the system. In a real-time system, the window length cannot be too long because it introduces a lag time in the myoelectric control system. Smith [9] investigated the optimal window length for pattern recognition based myoelectric control. In his research, the classifier was used to discriminate seven patterns of motion by varying window length (50 to 550 milliseconds). Subasi [10] used window length of 256 milliseconds for feature extraction. Chu [11] suggested that in the devices based myoelectric control, the window length should be less than 300 milliseconds. Those previous studies concerned to find the optimal window for the best performance of the classifier. The study of window length which is related to the prediction of the elbow joint angle is rarely found in several literatures. Thus, studies in the window length, such as the Pearson's correlation coefficient and linear regression, which relate to the performance of the prediction algorithm are still promising.

The optimum of the window length needs to be investigated in order to obtain the best performance of the prediction and the linearity of the prediction algorithm. In this study, first, we proposed a non-pattern recognition method to predict the elbow joint angle. This method used zero-crossing feature extraction and second order Butterworth low pass filter. Second, we evaluated the effect of the window length on the performance of the prediction.

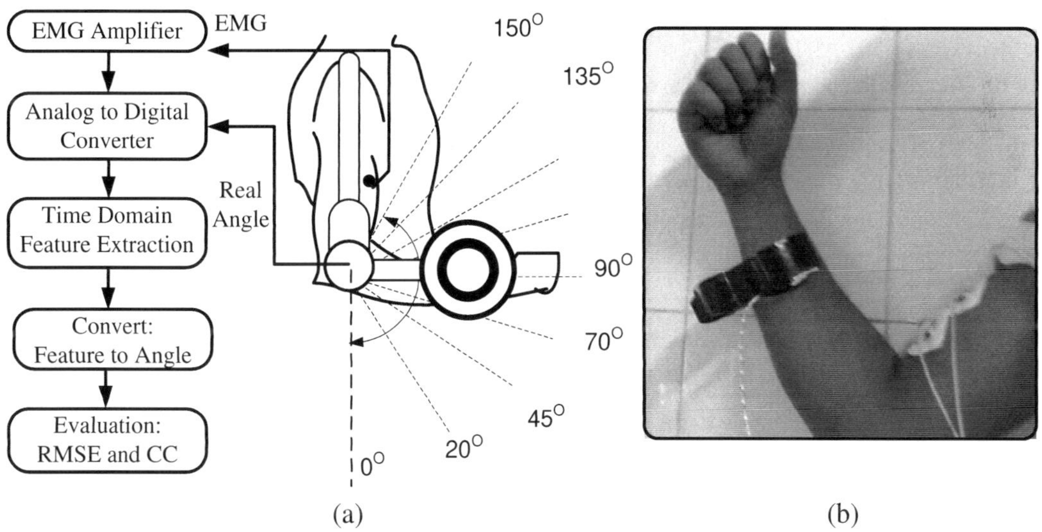

(a) (b)

Fig. 1 (a) Data acquisition (b) Electrode position

2. Materials and Method

2.1. Experimental Setup

EMG signal was collected from biceps muscle (Fig. 1(b)) using three disposable electrodes of Ag (AgCl). Two electrodes were placed in the biceps muscle, and one electrode was used as a common ground which was anatomically not related to biceps muscle. The distance of the electrodes was about

2 cm. Before placing the electrode, the skin surface was cleaned from dust and oil using alcohol 70%. This preparation was purposed to maintain the tape of the electrode, so that the electrode would not slip when there was a motion on elbow joint.

In this experiment, the elbow was moved in the flexion and extension motion. The period of the motion was varied with three different periods (12 second, 8 second and 6 second). The period of the motion was synchronized using a metronome application (Tempo Perfect). In the data collection (Fig. 1(a)) diagram, EMG was recorded while the elbow was moved in flexion and extension motion. The motion of the elbow joint derived from full extension (0°) to full flexion (150°) and then returned to full extension (0°). It was called as one cycle of motion. EMG signal was sampled using sampling frequency of 1000 Hz (time sampling equal to 1 milliseconds). An EMG amplifier was used to amplify the EMG signal to some level which was fit to the A/D converter. A 12 bit A/D converter was used to convert from analog to digital format.

2.2. Feature Extraction

Zero Crossing (ZC) is number of time that the signal crosses a certain threshold value. Chang [12], the first person who introduced the use of ZC as feature extraction, used this feature to determine the onset of muscle contraction. This feature can represent frequency information of the EMG signal [13]. The ZC feature is formulated as [14]

$$ZC = \sum_{i=1}^{N-1} [f(x_i \times x_{i+1}) \cap | x_i - x_{i+1} | \geq threshold$$

$$f(x) = \begin{cases} 1, if \rightarrow x \geq threshold \\ 0, otherwise \end{cases} \quad (1)$$

The x_i indicates the i-th EMG signal, N indicates the length of the signal, and threshold indicates a certain voltage value.

2.3. Infinite Impulse Response

Infinite Impulse Response (IIR) is a recursive filter which uses a previous output as an input of the filter. This filter is a mathematical process that is used to enhance or reduce a certain frequency band of the signal. This digital filter is usually implemented using a software programming [15]. The digital filter is written as following equation:

$$H = \frac{Y(z)}{X(z)} = \frac{b_0 + b_1 z^{-1} + b_2 z^{-2} + ... + b_N z^{-N}}{1 + a_1 z^{-1} + a_2 z^{-2} + ... + a_M z^{-M}} \quad (2)$$

$$y[n] = \sum_{k=0}^{N} b_k x[n-k] - \sum_{k=1}^{M} a_k y[n-k] \quad (3)$$

Where b_k and a_k are the (M+1) numerator and N denominator coefficient, respectively.

The coefficient of IIR filter was calculated using MATLAB program (version 2008). The coefficient was designed as a low pass filter second order Butterworth filter.

2.4. Data Processing

The data processing (Fig.2) was performed offline using a personal computer with Delphi programming and Microsoft Excel (version 2013). The Delphi programming was used to extract the raw EMG signal and to evaluate the accuracy of the prediction. The recorded EMG data, which were saved as text file, were opened using the Delphi programming. Zero crossing was selected as a feature extraction due to the previous study [16] which proved that the feature had better performance for root mean square error and Pearson's correlation coefficient to predict an elbow joint angle. For period of motion and window length of 12 second and 25 millisecond respectively, the feature extraction would result 480 features (12.000 millisecond ÷ 25 millisecond).

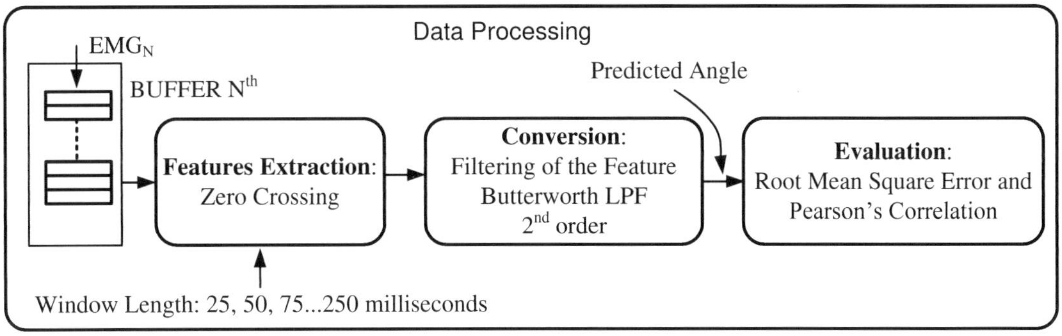

Fig. 2 Data processing of EMG signal

A low pass filter (LPF) was designed digitally using an infinite impulse response (IIR) as low pass Butterworth filter second order. The IIR was selected as a digital filter by considering the number of the coefficient of the filter which was less than the FIR filter. The calculation of IIR filter coefficient was performed using MATLAB application. The LPF was used to filter the ripple yielded from the output of the feature. To evaluate the performance of the prediction, the output of filtered features was normalized by multiplying with maximum angle so that the filtered feature had the same unit as the real angle. The maximum angle was accorded to the real angle of the elbow joint while EMG was recorded.

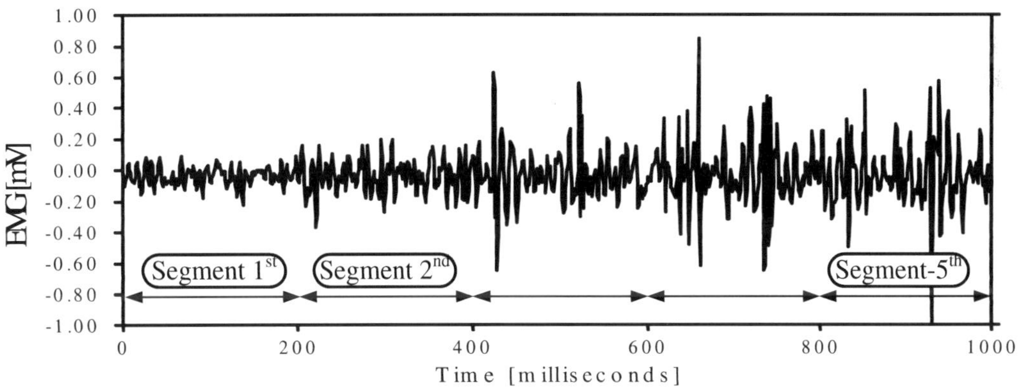

Fig. 3 Disjoint segmentation for window length 200 milliseconds and length of signal of 1000 milliseconds

2.5. Window Length

In regards to the aim of the study, each cycle motion was performed on period of motion of 12, 8, and 6 seconds.

It was then extracted with different window length from 25 to 250 millisecond with increment of 25 milliseconds in order to investigate whether there was any relationship between the window length and the performance of the prediction. The raw EMG data were segmented into several windows with disjoint segmentation. Fig 3 shows the illustration of a segmentation of the feature extraction process for window length of 200 millisecond. For time frame of 1000 milliseconds, the number of feature is five as shown in Fig. 4. The y-axis in the Fig. 4 is in degrees unit because the feature has normalized the angle unit by multiplying with the maximum value of the real angle.

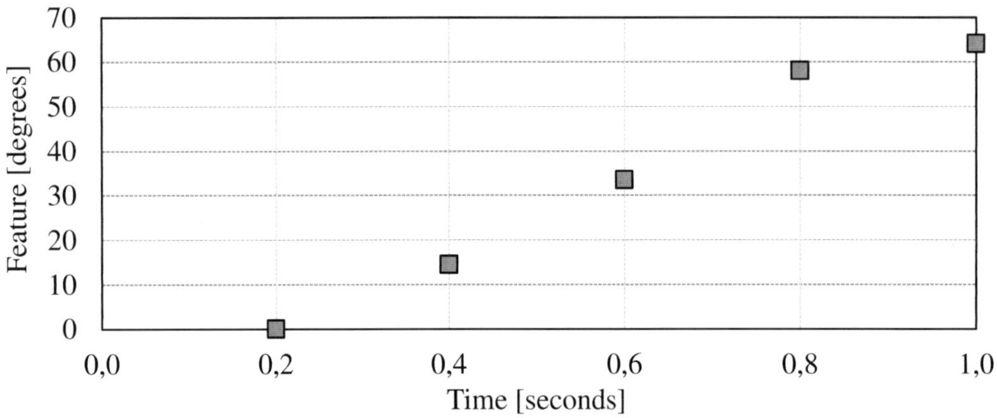

Fig. 4. Feature extraction results each 200 millisecond

2.6. Statistical analysis

The performance of the prediction algorithm, which was tested with some window length, was evaluated using root mean square error (RMSE). Several previous studies [17] [18] used this value (RMSE) to measure the accuracy of the prediction. The RMSE value can be calculated using the following equation [3]:

$$RMSE = \sqrt{\frac{1}{N}\sum_{i=1}^{N}(x_i - y_i)^2} \tag{4}$$

The Pearson's correlation coefficient (CC) is also often used as a complement of the RMSE value. The CC was used to measure the relationship between two parameters. In this case, the parameters were the predicted angle (x_i) and the real angle (y_i) and N indicated as the window length. Previous researchers [19] [20] used CC to measure the relationship between those parameters. The CC can be formulated as follows:

$$r_{xy} = \frac{\sum xy - \frac{(\sum x)(\sum y)}{N}}{\sqrt{(\sum x^2 - \frac{(\sum x)^2}{N})(\sum y^2 - \frac{(\sum y)^2}{N})}} \tag{5}$$

The significant different tests, between two or more condition, was performed using analysis of variance (ANOVA) of single factor. In this study, the effect of the window length would be tested to see whether any significant change of the performance of the prediction. As mentioned on several related work literatures, the significant confident was 95% (alpha =0.05).

3. Results and Discussion

3.1. Estimated Elbow Joint Angle

EMG signal shows a random [1] value with the amplitude of the EMG signal ranged from -1.5 to 1.5 millivolt (Fig.4(a)). A combination of feature extraction and a low pass Butterworth filter can estimate an elbow joint angle. The mean and SD of RMSE values (window length 200 milliseconds) for period of motion 12 seconds, 8 seconds and 6 seconds are 9.02%±2.19%, 7.17%±1.33%, and 8.58%±1.17% respectively.

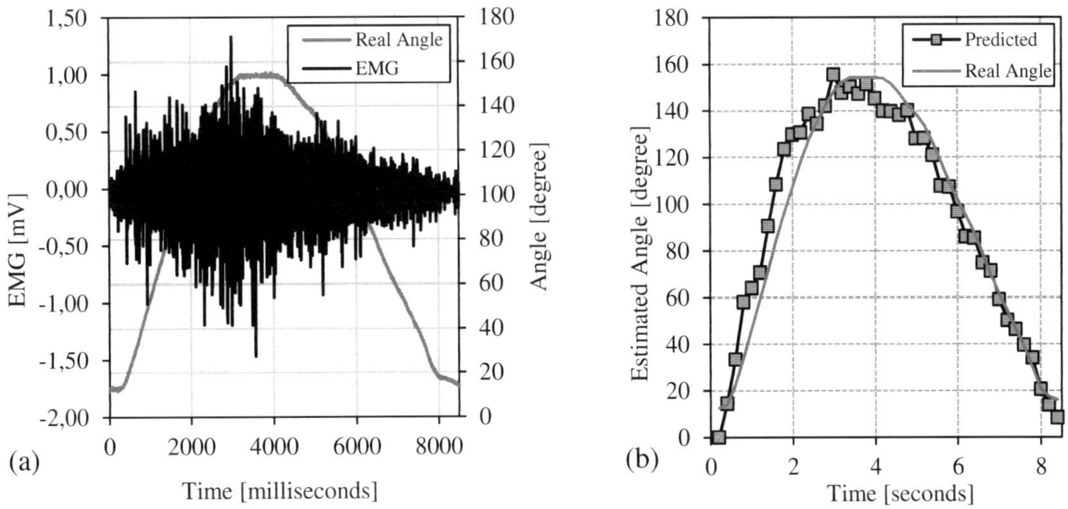

Fig. 5. (a) Responses of the EMG signal against the change of the elbow joint angle, (b) the predicted angle of the elbow joint

3.2. The Effect of Window Length on the Performance

Window length affected the performance (RMSE and Pearson's correlation) of the prediction. The performance of the prediction was calculated on each cycle of motion during five cycles for all of the periods of motion and window length. The highest performance was obtained in window length of 250 milliseconds and period of motion of 8 seconds (RMSE = 5.68%±1.53% and Pearson's correlation = 0.99±0.0059).

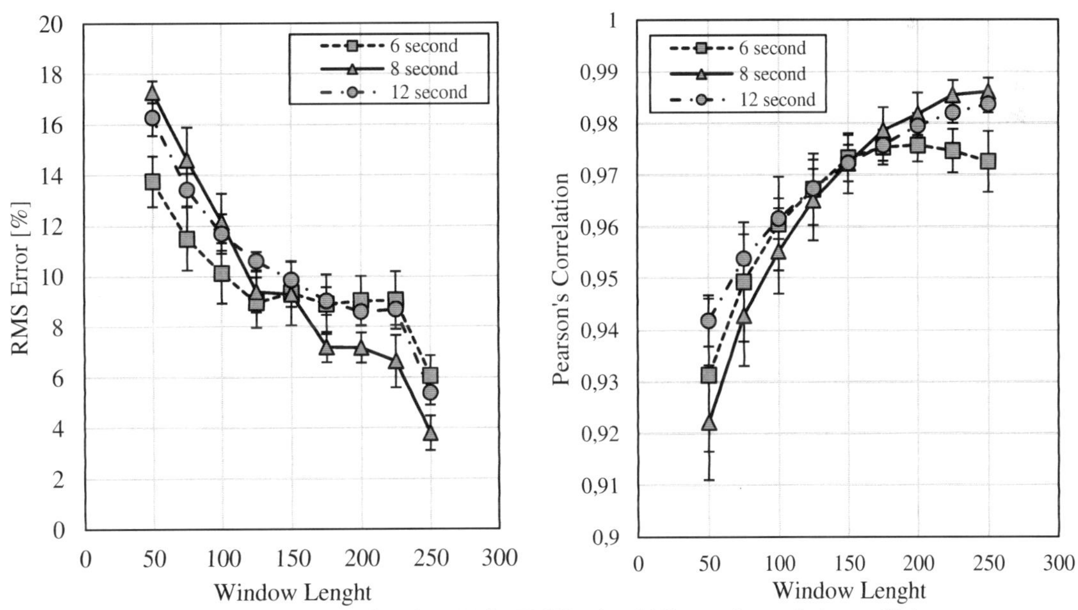

Fig. 6. Effect of window length to (a) the RMSE value (b) Pearson's correlation coefficient

Smith [9] reported that in the window length of 150 to 250 milliseconds, he obtained the best performance of the classifier. A similar result was also reported by Kamavuako [21] that the optimal window length was 250 milliseconds. In the window length of 50 milliseconds and period of motion 8 seconds, the performance was the lowest (RMSE = 17.30%±0.96% and Pearson's correlation = 0.92±0.025). Although, the RMSE and Pearson's correlation was varied for different periods of motion and window length, yet, the Pearson's correlation was higher than 0.9. It indicated that there was a high relationship between the predicted angle and real angle.

In the window length of 150 milliseconds, the RMSE and the Pearson's correlation value was almost coincided for all of periods of motion (12 seconds, 8 seconds and 6 seconds) namely 9.29%±1.13% and 0.98±0.009 respectively. In the window length of 50 to 150 milliseconds, the period of motion of 6 seconds has the lowest RMSE than other periods of motion, yet, after the window length was set to more than 150 milliseconds, the period of motion of 8 seconds has the lowest RMSE other periods of motion.

In ANOVA test, it was found that there was no significant difference on RMSE between window length of 50 and 75 milliseconds ($p>0.05$). Between window length of 50 and more than 100 milliseconds, there was a significant difference on RMSE ($p<0.05$).

3.3. The Effect of Window Length on Determination Coefficient

The linearity of the predicted angle is essential in the development of the devices based on myoelectric control. In this study, the linearity of the prediction angle was tested for each different windows length. The linearity was measured by plotting the linear regression of the output of the predicted angle against the real angle (Fig.6). The linear regression estimates the trend line of the predicted angle. The trend line shows the difference of the determination constant (R^2) for the different windows length. The highest determination constant was obtained when the window length was 250 milliseconds (R^2=0.96). The window length determined the determination constant. The longer the window length the more linear the predicted angle.

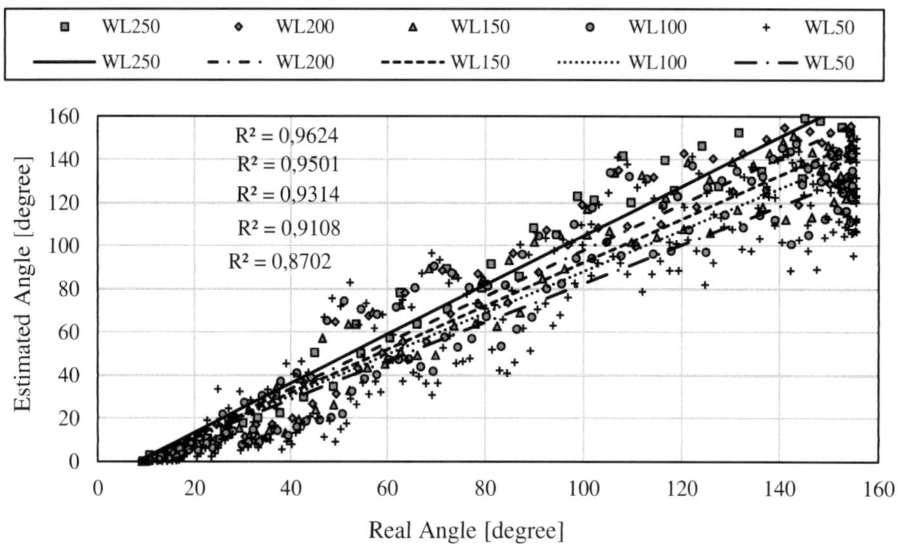

Fig. 7 The change of the linearity of the predicted angle for different windows length

4. Conclusion

In this paper, we demonstrated the effect of the window length to the performance of the prediction. The prediction algorithm was performed using the zero crossing feature extraction and filtering the output of the feature using a second order Butterworth low pass filter. The prediction algorithm offers a simple method to predict the elbow joint angles. The effect of the window length to the performance of the prediction was tested by varying some windows length (50 milliseconds to 250 milliseconds) in the feature extraction. Accuracy (RMSE and Pearson's correlation coefficient) of the prediction algorithm depended on window length. The results of this study suggested that the window length of

250 milliseconds has the highest performance (root mean square error and Pearson's correlation) to predict the elbow joint angle. The window length also affected the linearity of the predicted angle.

There were some limitations to this study which need to be considered in the next work such as muscle fatigue, the distance between the electrodes, the overlap segmentation, and the kinematic condition in the data collection. The results of this study can be used for other related studies based on electromyography such as sport activities, ergonomic tests, and devices based on myoelectric control.

References

[1] C J De Luca, Surface Electromyography : Detection and Recording, *DelSys Inc.*, **vol. 10**, no. 2, pp. 1–10, 2002.

[2] M Asghari Oskoei and H Hu, Myoelectric control systems-A survey, *Biomed. Signal Process. Control*, **vol. 2,** no. 4, pp. 275–294, 2007.

[3] Z Tang, K Zhang, S Sun, Z Gao, L Zhang, and Z Yang, An upper-limb power-assist exoskeleton using proportional myoelectric control, *Sensors (Basel).*, **vol. 14**, no. 4, pp. 6677–94, 2014.

[4] Z O Khokhar, Z G Xiao, and C. Menon, Surface EMG pattern recognition for real-time control of a wrist exoskeleton., *Biomed. Eng. Online*, **vol. 9**, p. 41, 2010.

[5] K Kiguchi and Y Hayashi, EMG-Based Control of a Lower-Limb Power-Assist Robot, in *Intelligent Assistive Robots, Springer Tracts in Advanced Robotics 106*, Springer, 2015, pp. 371–383.

[6] J Hashemi, E Morin, P Mousavi, and K Hashtrudi-Zaad, Surface EMG force modeling with joint angle based calibration, *Journal of Electromyography and Kinesiology*, **vol. 23**, no. 2. pp. 416–424, 2013.

[7] J Rosen, M Brand, M B Fuchs, and M Arcan, A myosignal-based powered exoskeleton system, *IEEE Trans. Syst. Man, Cybern. Part ASystems Humans.*, **vol. 31**, no. 3, pp. 210–222, 2001.

[8] Z Li, B Wang, F Sun, C Yang, Q Xie, and W Zhang, SEMG-based joint force control for an upper-limb power-assist exoskeleton robot, *IEEE J. Biomed. Heal. Informatics*, **vol. 18**, no. 3, pp. 1043–1050, 2014.

[9] L H Smith, L J Hargrove, B. a. Lock, and T a Kuiken, Determining the optimal window length for pattern recognition-based myoelectric control: Balancing the competing effects of classification error and controller delay, *IEEE Trans. Neural Syst. Rehabil. Eng.*, **vol. 19,** no. 2, pp. 186–192, 2011.

[10] A Subasi, Classification of EMG signals using combined features and soft computing techniques, *Appl. Soft Comput.*, **vol. 12**, no. 8, pp. 2188–2198, 2012.

[11] J U Chu, I Moon, and M S Mun, A real-time EMG pattern recognition system based on linear-nonlinear feature projection for a multifunction myoelectric hand, *IEEE Trans. Biomed. Eng.*, **vol. 53,** no. 11, pp. 2232–2239, 2006.

[12] G C Chang, W J Kang, L Jer-Junn, C K Cheng, J S Lai, J J J Chen, and T S Kuo, Real-time implementation of electromyogram pattern recognition as a control command of man-machine interface, *Med. Eng. Phys.*, v**ol. 18**, no. 7, pp. 529–537, 1996.

[13] B Hudgins, P Parker, and R N Scott, A New Strategy for Multifunction Myoelectric Control, *IEEE Trans. Biomed. Eng.*, **vol. 40**, no. 1, pp. 82–94, 1993.

[14] A Phinyomark, P Phukpattaranont, and C Limsakul, Feature reduction and selection for EMG signal classification, *Expert Syst. Appl.*, **vol. 39,** no. 8, pp. 7420–7431, 2012.

[15] Li Tan & Jean Jiang, *Digital Signal Processing: Fundamental and Applications*, 1st ed., vol. 70, no. 3. California: Elsevier, 2008.

[16] Triwiyanto, Oyas Wahyunggoro, H. A.Nugroho, and Herianto, Quantitative Relationship Between Feature Extraction of sEMG and Upper Limb Elbow Joint Angle, in Proceeding *2016 International Seminar on Application for Technology of Information and Communication*, 2016, pp. 44–50.

[17] J Hashemi, E Morin, P Mousavi, and K Hashtrudi-Zaad, Joint angle-based EMG amplitude

calibration, *Proc. Annu. Int. Conf. IEEE Eng. Med. Biol. Soc. EMBS*, no. 2, pp. 4439–4442, 2011.

[18] P K Artemiadis and K J Kyriakopoulos, EMG-Based Position and Force Estimates in Coupled Human-Robot Systems: Towards EMG-Controlled Exoskeletons, *Springer Tracts in Advanced Robotics*, **vol. 54**. pp. 241–250, 2009.

[19] P K Artemiadis and K J Kyriakopoulos, An EMG-based robot control scheme robust to time-varying EMG signal features, *IEEE Trans. Inf. Technol. Biomed.*, **vol. 14**, no. 3, pp. 582–588, 2010.

[20] G Li, Electromyography Pattern-Recognition-Based Control of Powered Multifunctional Upper-Limb Prostheses, in *Advances in Applied Electromyography*, Prof. Joseph Mizrahi, Ed. InTech, 2011, pp. 99–177.

[21] E N Kamavuako, D Farina, K Yoshida, and W Jensen, Relationship between grasping force and features of single-channel intramuscular EMG signals,*J. Neurosci. Methods*, **vol. 185**, no. 1, pp. 143–150, 2009.

IOP Conf. Series: Journal of Physics: Conf. Series **853** (2017) 012015 doi:10.1088/1742-6596/853/1/012015

Confocal microscope is able to detect calcium metabolic in neuronal infection by *toxoplasma gondii*

A D Sensusiati[1,*], T K S Priya and Y P Dachlan

[1]Medical Faculty of Airlangga University, Dr. Soetomo Hospital, Surabaya

*corresponding email address : radiologigdc1@gmail.com

Abstract. Calcium metabolism plays a very important role in neurons infected by Toxoplasma. Detection of change of calcium metabolism of neuron infected by Toxoplasma and Toxoplasma requires the calculation both quantitative and qualitative method. Confocal microscope has the ability to capture the wave of the fluorescent emission of the fluorescent dyes used in the measurement of cell calcium. The purpose of this study was to prove the difference in calcium changes between infected and uninfected neurons using confocal microscopy. Neuronal culture of human-skin-derived neural stem cell were divided into 6 groups, consisting 3 uninfected groups and 3 infected groups. Among the 3 groups were 2 hours, 24 hours and 48 hours. The neuron *Toxoplasma gondii* ratio was 1:5. Observation of intracellular calcium of neuron and tachyzoite, evidence of necrosis, apoptosis and the expression of Hsp 70 of neuron were examined by confocal microscope. The normality of the data was analysed by Kolmogorov-Smirnov Test, differentiation test was checked by t2 Test, and ANOVAs, for correlation test was done by Pearson Correlation Test. The calcium intensity of cytosolic neuron and *T. gondii* was significantly different from control groups ($p<0.05$). There was also significant correlation between calcium intensity with the evidence of necrosis and Hsp70 expression at 2 hours after infection. Apoptosis and necrosis were simultaneously shown with calcium contribution in this study. Confocal microscopy can be used to measure calcium changes in infected and uninfected neurons both in quantitatively and qualitatively.

1. Introduction

Many brain calcifications were found on congenital toxoplasmosis (Couvreur et al, 1962; Dunn et al, 1999). Neurons calcification as the result of *Toxoplasma gondii* infection is associated with important role of calcium for *Toxoplasma gondii*. *Toxoplasma gondii* as a part of apicomplexan parasite relies on calcium signaling to review different important functions including protein secretion, motility, cell invasion and differentiation (Negaunee, et al, 2008)[1]. In parasite infection, during invasion and persisting, the host cells need movement and gliding motility, also calcium scavenging (Dobrowolski and Sibey, 1996)[2]. All of this dynamic calcium change can be evaluated by measuring the calcium intra parasite and neuron as a host cells (Lovett and Sibey, 2003)[3]. Confocal microscope has the ability to capture the wave of the fluorescent emission of the fluorescent dyes used in the measurement of cell calcium. The purpose of this study was to prove the difference of calcium changes both in infected and uninfected neurons using confocal microscopy. In order to make the similar condition like infected neuron in neonatal toxoplasmosis, skin-derived neural stem cell was used.

2. Material and methods

A fully synchronized laser scanner in a single compact design for simultaneous laser stimulation and confocal observation (Olympus FV1000) was used to measure calcium intensity of the uninfected and *Toxoplasma gondii* infected neuron.

Two month baby's preputium was used to grow multi potent neural stem cell. The existence of its growth was confirmed by immunofluorescence of CD 133. After isolation, proliferation, and differentiation of neural stem cell from multi potent, neural stem cells (MNSC) became adult neuron. Beta III tubulin with immunocytochemistry was also confirmed for adult neuron.

Toxoplasma gondii isolate from RH lineage taken from Biotechnology Centre of Gajah Mada University Yogyakarta was used in order to increase the number of isolate intra peritoneal mouse injection of 10^3 in every mouse. After 4 days, the mice were sacrificed to take the intraperitoneal fluid. Parasites inside the peritoneal fluid were counted with haemocytometer, and the number of *Toxoplasma gondii* (tachyzoites stage) in every millimetre was obtained. Adult neurons were divided into 6 groups consisting of 3 uninfected groups and 3 infected groups. Adult neurons then were infected by tachyzoite through incubation with tachyzoite and fixation at 2 hours, 24 hours and 48 hours. The optimal neuron tachyzoites ratio was 1:5 (Rozenfeld, 2005)[4].

Fluorescent calcium indicators fura-2 was used to dye calcium. The excitation was set at 340 and 380 nm. Confocal laser scanning microscopy then used to measure the calcium intensity.

The normality of the data was analysed by Kolmogorov-Smirnov Test, differentiation test was checked by t2 Test, and ANOVAs, and for correlation test was done by Pearson Correlation Test.

3. Result

Laboratory experimental research has been carried out in vitro on the mechanism of deposition of calcium in neurons infected by *Toxoplasma gondii.* It was conducted through observation of neuronal cytosolic calcium, *Toxoplasma gondii* calcium by imunofluorense using fura-2.

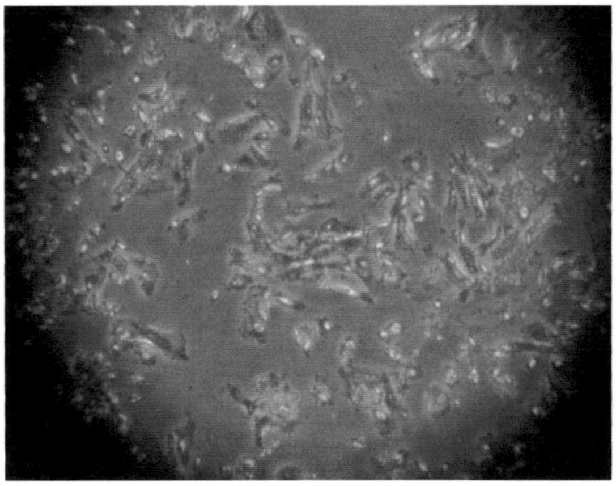

Figure 1. Neural stem cell at day 7. Blue arrow shows neural stem cell, red arrow shows fully differentiated neuron. Evaluation with 200x

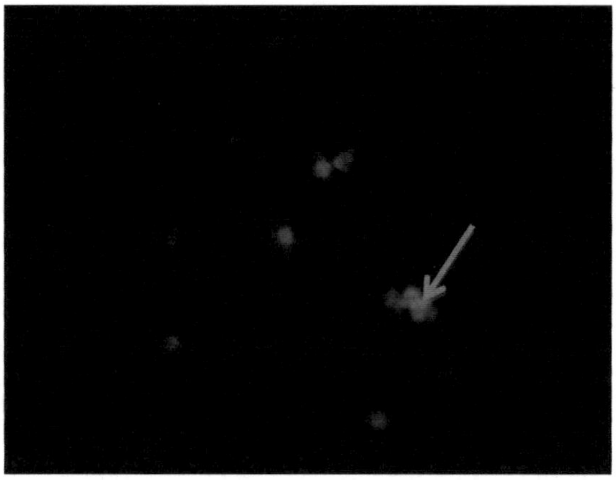

Figure 2. Identification and characterization of neural stem cell using fluorescent microscopy, the CD 133^+ expression can be detected. Green colour is the expression of neural stem cell.

As the main ingredient in this laboratory experimental research, neurons were obtained from the process of isolation and culture using discharged skin tissue from prepuce circumcision of a 2-month-old baby. Within 10 days after the process of proliferation and differentiation, fully differentiated adult neurons were formed as seen in Figure 1. Characterization of neuronal stem cells was observed by CD 133 using immunofluorescence. Figure 2 shows that the cells are checked neuron stem cells for expressing the CD 133^+.

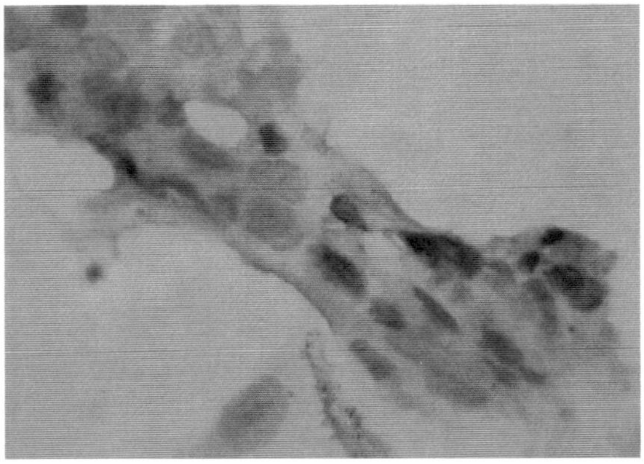

Figure 3. Fully differentiated neuron confirmed with anti-β-III tubulin, the specific marker for neuron. Expression of this marker is presented as chocolate colour.

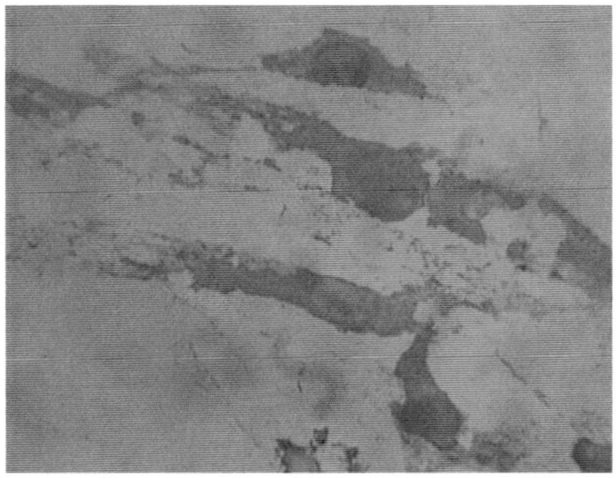

Figure 4. Immunohistochemistry test on fully differentiated neuron

Figure 5. Neuron culture from control group on 2 hours. Check with different interference contrast, CLM

Figure 6. Detection of neural intracellular calcium after fura-2 dye. Nucleus and cytosol can be well observed. Green fluorescence is the indicator of cytosolic and nucleus

Culture examination of neurons with fura-2, which was observed in confocal microscopy, showed the results in a 3-dimensional image based on different interference contrast (DIC) (Figure 5). Intracellular calcium and nucleus was showed in luminescence of green colour of fluorescent fura-2 after incubated in cultured neurons (Figure 6).

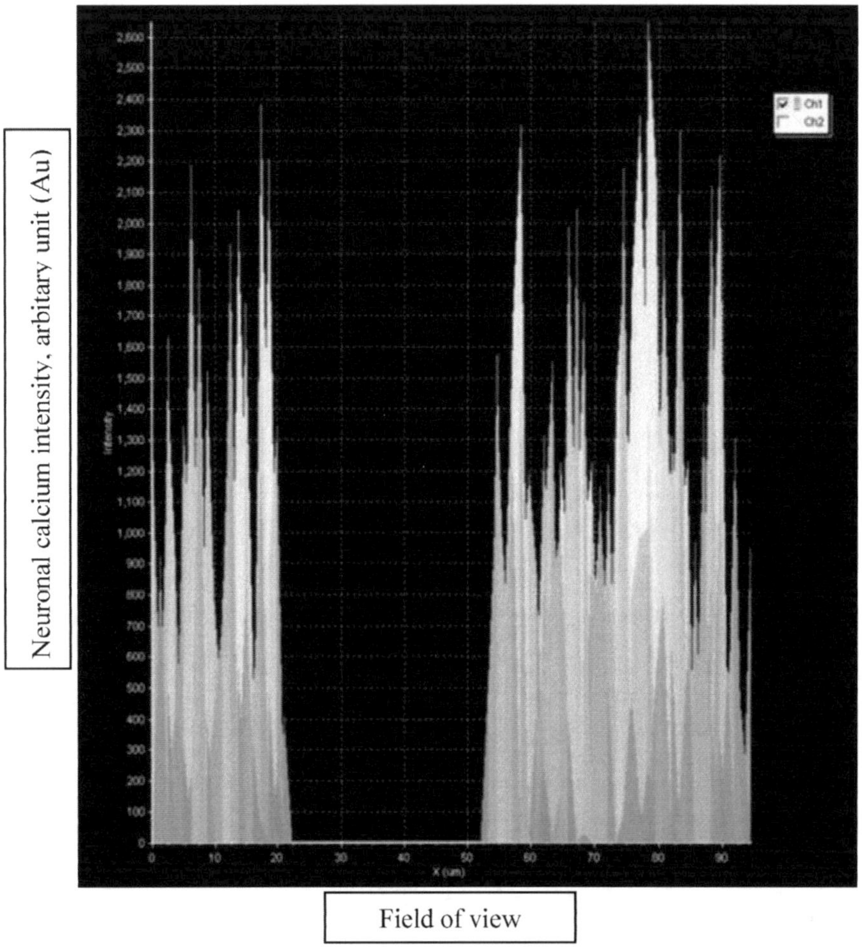

Figure 7. Calcium fluorescence intensity on control group was measured using confocal laser microscopy scanning

The measurement of calcium on experimental group was performed using the method of chemical fluorescent indicator fura-2 and antibodies (double staining, Figure 8 to 10) on *Toxoplasma gondii*. The result then was utilized to distinguish neurons intensity of cytosolic calcium and calcium intensity of *Toxoplasma gondii*. The observations indicated that the intensity of calcium in *Toxoplasma gondii* was gradually increasing in which its second-hour result was equal to the intensity of the control group in 2 hours. It was increasing at 24 hours and 48 hours while the intensity of calcium in neurons was gradually declining (Figures 11 to 17).

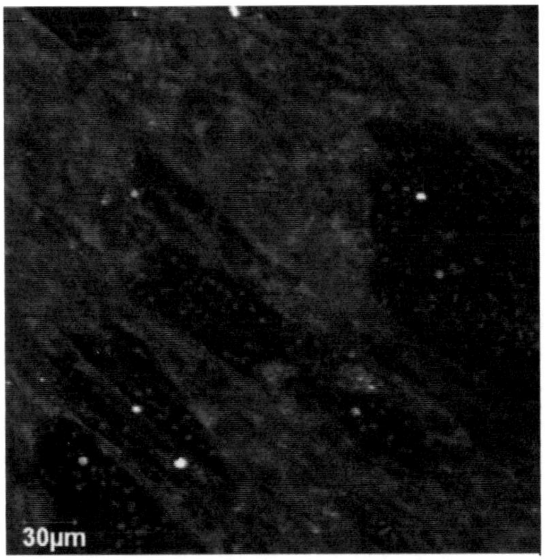

Figure 8. Neurons after infected by tachyzoites in two hours. The identification and characterization of tachyzoites used *Toxoplasma gondii* antibody, shown by red fluorescence, and neural intracellular calcium, characterized by green fluorescence.

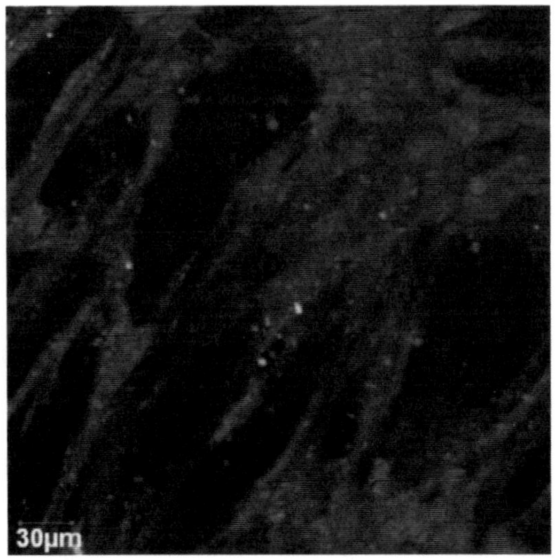

Figure 9. Twenty four hours after tachyzoites infection, red fluorescence was more prominent than 2-hour tachyzoites infection. The green fluorescence decreased.

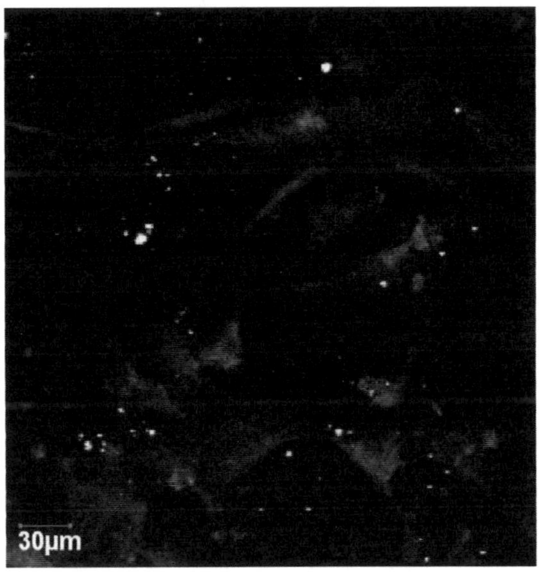

Figure 10. Forty eight hours after tachyzoites infection, the neuron population diminished, and tachyzoites intracellular calcium increased.

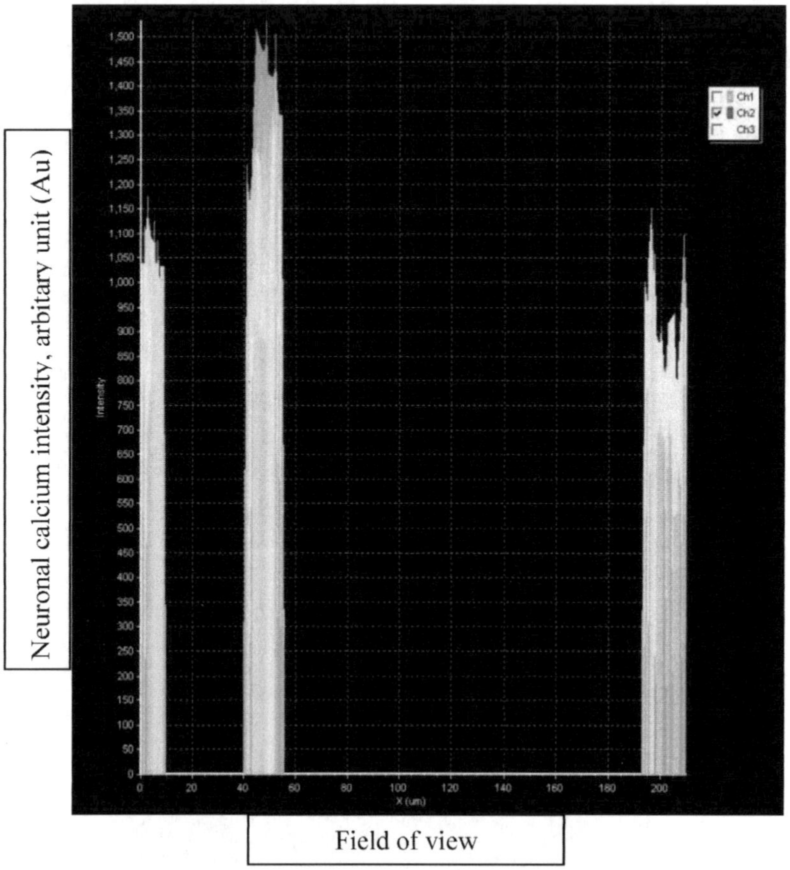

Figure 11. Calcium intensity measurement on neural cell in 24 hours after tachyzoites infection.

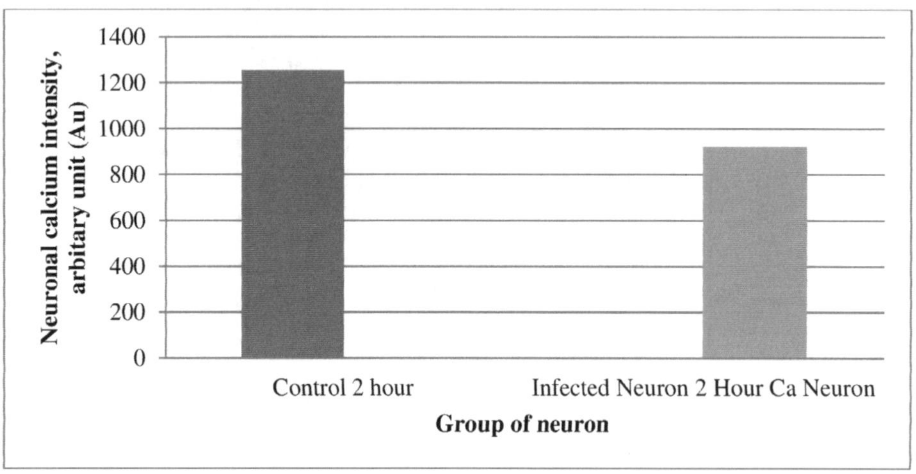

Figure 12. The comparison between the intensity of neuronal calcium of control group and after 2 hours tachyzoites infection.

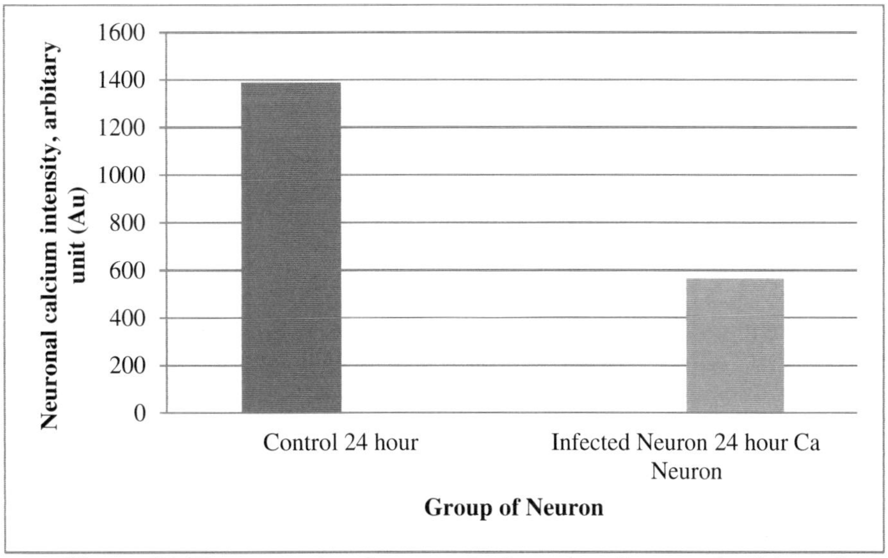

Figure 13. The comparison between the intensity of neuronal calcium of control group and after 24 hours tachyzoites infection.

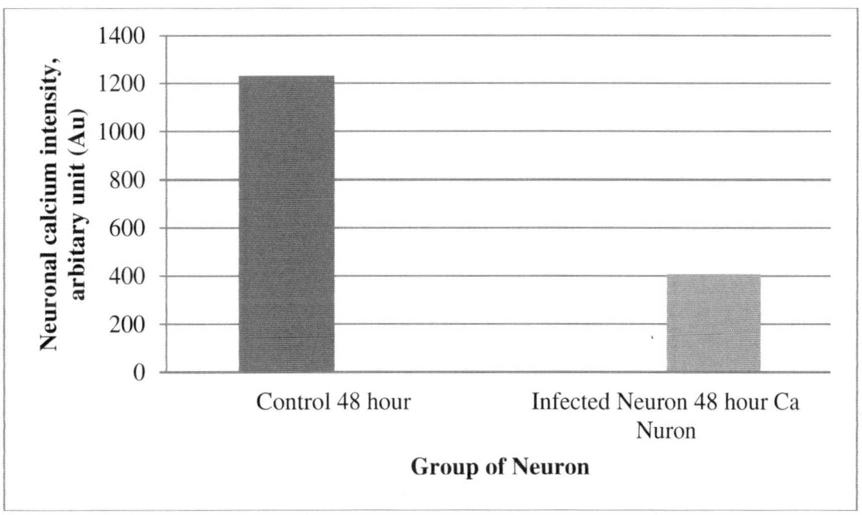

Figure 14. The comparison between the intensity of neuronal calcium of control group and after 48 hours tachyzoites infection.

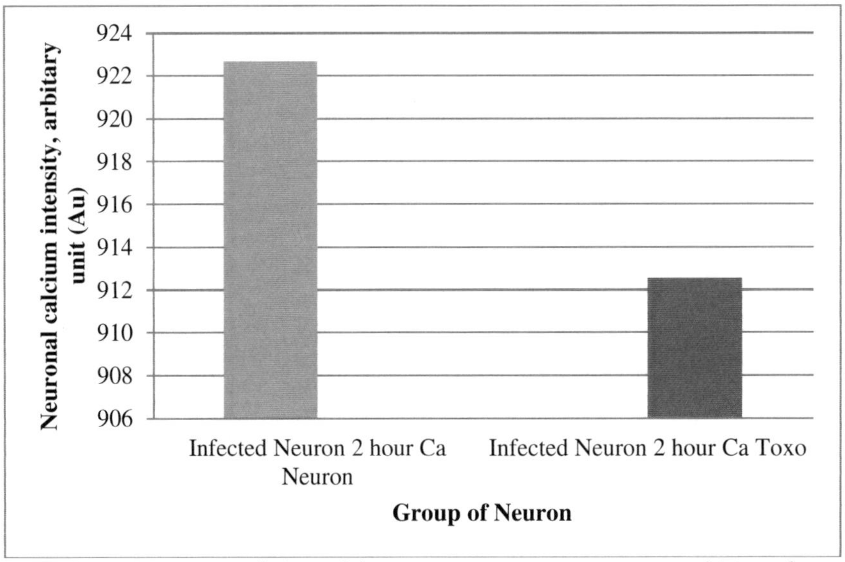

Figure 15. The intracellular calcium measurement on neuron and *Toxoplasma gondii* 2 hours after tachyzoites infection.

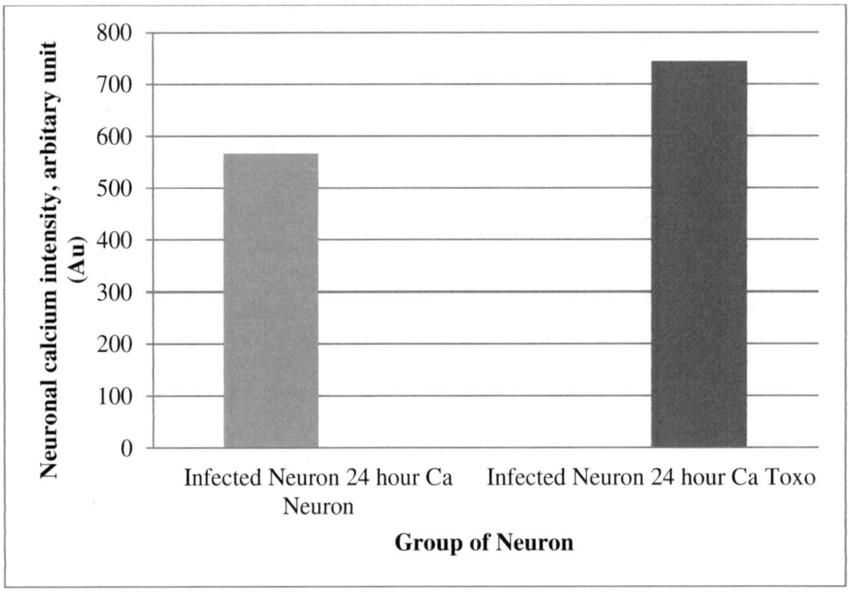

Figure 16. The intracellular calcium measurement on neuron and *Toxoplasma gondii* 24 hours after tachyzoites infection.

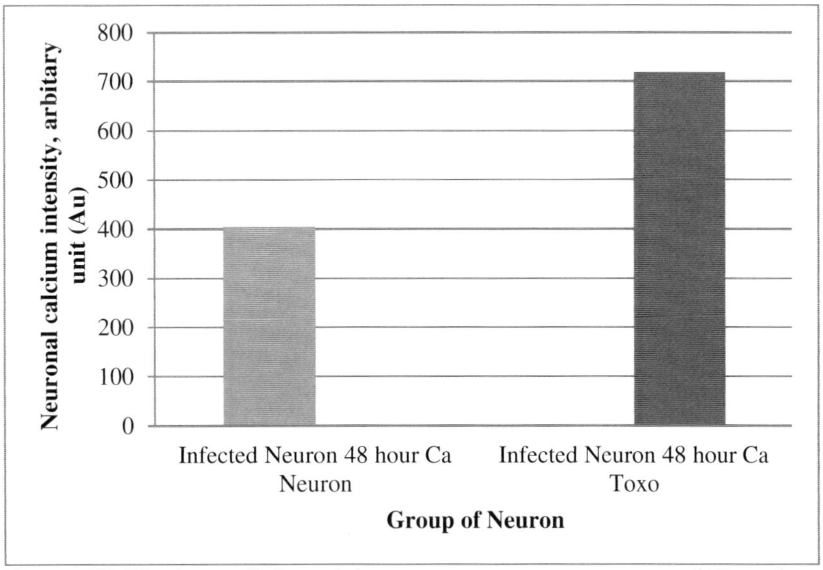

Figure 17. The intracellular calcium measurement on neuron and *Toxoplasma gondii* 48 hours after tachyzoites infection.

Table 1. Results of Data Normality Test Using One Sample Kolmogorov-Smirnov Test

Group	Significant value (p)	Description
Control 2 hour (Ca Neuron)	1.000	Normal Distribution (p > α)
Control 24 hour (Ca Neuron)	1.000	Normal Distribution (p > α)
Control 48 hour (Ca Neuron)	0.988	Normal Distribution (p > α)
Infected Neuron 2 hour (Ca Neuron)	0.764	Normal Distribution (p > α)
Infected Neuron 2 hour (Ca Toxo)	1.000	Normal Distribution (p > α)
Infected Neuron 24 hour (Ca Neuron)	0.946	Normal Distribution (p > α)
Infected Neuron 24 hour (Ca Toxo)	0.984	Normal Distribution (p > α)
Infected Neuron 48 hour (Ca Neuron)	0.942	Normal Distribution (p > α)
Infected Neuron 48 hour (Ca Toxo)	0.772	Normal Distribution (p > α)

Based on the observation of the control group and infected group, as shown in Table 1, and the analysis using abnormality test data by Kolgomorov-Smirnov's test, the result showed that the data was in normal distribution.

Table 2. Results of the Difference Test of Calcium between Neuron Control group and Infected Neurons Using t2 Free Sample Test

Groups	Mean ± Standard Deviation	Significant value (p)	Description
Control 2 hour (Ca Neuron)	1254.206 ±177.629	0.038	significant difference (p < α)
Infected Neuron 2 hour (Ca Neuron)	922.678 ± 177.173		
Control 24 hour (Ca Neuron)	1388.460 ± 110.720	< 0.0001	
Infected Neuron 24 hour (Ca Neuron)	566.320 ± 94.669		
Control 48 hour (Ca Neuron)	1231.460 ± 196.058	0.002	
Infected Neuron 48 hour (Ca Neuron)	405.523 ± 51.144		

Table 2. shows free sample t test on both control and infected group (2-hour, 24-hour, and 48-hour infection). The result showed a significant difference between control group and infected group.

Table 3. Results of the Difference Test between Calcium Neurons and Calcium *Toxoplasma gondii* using t2 Free Sample Test

Groups	Mean ± Standard Deviation	P	Description
Infected Neuron 2 hour (Ca Neuron)	922.678 ±177.173	0.934	Was not significantly different (p > α)
Infected Neuron 2 hour (Ca Toxo)	912.550 ± 152.719		
Infected Neuron 24 hour (Ca Neuron)	566.320 ± 94.669	0.026	significant difference (p < α)
Infected Neuron 24 hour (Ca Toxo)	743.130 ± 74.386		
Infected Neuron 48 hour (Ca Neuron)	405.523 ± 51.144	0.004	
Infected Neuron 48 jam (Ca Toxo)	717.635 ± 130.578		

Based on the result presented in Table 3 with 2 free sample t test, there was no significant difference between control group and after 2 hours infection, yet the significant difference presented after 24 and 48 hours infection.

IOP Conf. Series: Journal of Physics: Conf. Series **853** (2017) 012015 doi:10.1088/1742-6596/853/1/012015

4. Discussion

4.1. Confocal laser microscopy, a fluorescent method to study intracellular calcium

A laser-scanning microscope converges the laser beam into a small spot using an objective and scans the specimen in the X-Y direction using the laser beam. The microscope then captures fluorescent light and reflected light from the specimen using light detectors, and outputs the specimen image on an image monitor. The confocal optics incorporate a confocal aperture on the optically conjugate position (confocal plane) with the focus position to eliminate light from other parts than the focus position. This causes the extraneous light to be viewed as darkness in the observation image. It is possible to slice optically a specimen tissue that has thickness. On the other hand, an ordinary optical microscope, the light from other part than the focus position is overlapped with the imaging light of the focus position so the image is blurred in overall. Laser beam that has been transmitted through the specimen is detected by the transmitted light detector and provides the transmitted image, which is not a confocal image.

In regards of its importance, numerous methods for analysing the mechanism of cellular and/or sub cellular calcium activity have been established. Fluorescence dyes methods have been widely used in *Toxoplasma gondii* since it had high sensitivity (Grynkiewcz *et al.*, 1985; Tsien, 1989)[5]. Fura-2 loaded neuron was used in this study, for fura-2 measurement, excitation set at 340 and 360 nm and emission at 510 nm.

To study the calcium dynamic during *Toxoplasma gondii* infection on neuron, neural stem cell cultured from baby's preputium was used to grow adult neuron. Figure 1 to figure 4 presents the step of neural cell development.

Confocal microscopy detected intracellular calcium on control group as well as infected group, as shown on figure 5 to figure 11. To separate calcium content inside the neuron and inside *Toxoplasma gondii*, antibody for *Toxoplasma gondii* was used. The intensity of both cells can be measured.

4.2. The influence of toxoplasma gondii infection to host cytosolic calcium

Calcium plays important roles during attachment of tachyzoite to the host-cell surface and invasion (Viera and Moreno, 2000; Bouchot, et al., 1999; Mondragon and Frixione, 1996). The sequential secretion of parasite organelles marks the tachyzoite invasion process. Micronemes, rhoptries, and dense-granule contents are released by invading parasites, and participate in attachment, vacuole formation and intracellular survival (Carruthers and Sibbley, 1997)[6]. In invasive phase, tachyzoites-neuron attachment begins with protein secretion that functions as enzyme from secretory organelles microneme, rhoptry, and dense granules from unique microtubules cytoskeleton called conoid. They are active PLA, rhoptry proteins (ROPs), and microneme protein (MICs). These enzymes influence host's plasma membrane phospholipids to produce that cause local lysis, influence membrane fluidity or egress intracellular calcium store (Bonhomme, 1999; Lin, 1993)[7]. Arachydonic acid influenced calcium entry to host cell. MICs and ROPs also secreted from apical *Toxoplasma gondii*. After exocytosis MIC3, an *epidermal growth factor (EGF-) like* signal may activate inositol phosphate signal (IP), causing calcium to be released from inner pool, and increase calcium concentration. Figure 12 shows cytosolic calcium on infected neuron was lower than from uninfected group, but on tachyzoite was higher than from control and neural cytosol. It shows that increment of neural cytosolic calcium immediately present after invasion in 20 seconds (Morisaki et al., 1995) cannot be evaluated in this study because the observation was done in 2 hours. Higher calcium intensity on *Toxoplasma gondii* was due to calcium transfer from neural cytosol to *Toxoplasma gondii*. Calcium intensity was observed from 2 hours to 24 hours and 48 hours as shown on Figure 8, 9, 10 and may proof that cytosolic calcium in neuron decreased. The increment of calcium inside the tachyzoite and calcium transfer proceeded until 48 hours. T2 test resulted a significant difference between neural cytosolic calcium in control group than experimental group, proofing that calcium signalling worked on plasma membrane and extracellular influx to neuron.

4.3. Toxoplasma gondii's calcium intensity transformation during infection

Toxoplasma gondii needs calcium during motility, conoid extrusion, microneme secretion, attachment, and invasion to the host cell (Debrowlski, et al., 1997). Gliding motility depends on actomyosine motor and important for active invasion host cell (Debrowlski and Sibley, 1996), involving *Toxoplasma gondii*'s intracellular calcium. Conoid extrusion exists during invasion and egress from host cell, and it is stimulated by material that can enhance calcium (Mondragon and Frixione, 1996; Monteiro et al., 2001)[8]. Microneme is the first secretory organelle that produced from conoid. This microneme secretion needs intracellular calcium from *Toxoplasma gondii*, and host attachment triggers microneme release. (Brecht *et al*., 2001).

Free calcium intracellular concentration in the infected host cell is variable (Bouchot A., et al., 1999). When *Toxoplasma gondii* enters the host cell, calcium pump is activated to their environment. Calcium may freely diffuse across membrane because of the porus of Parasitoporus vacuole membrane (PMV), implicating that product secreted to the vacuole can bind the calcium. The candidate for this activity is dense granule protein GRA 1, which is highly related to tubovesicular network from parasitoporus vacuole (PV) and high affinity for calcium. This study showed that calcium decreased after 2 hours infection (Figure 15) and even more decreased after 24 hours infection (Figure 16). The consequence, 48 hours after infection, when the parasites have more replicated and built rosette, host cytosolic calcium decreased significantly (Figure 17). Host intracellular calcium increment and egress of parasite from host cell are seemed to be correlated. Calcium increment on PV was a signal that was triggered the *Toxoplasma gondii* egress and parasites push the host cell membrane during this moment. Rupture of plasma membrane happened, parasites came out from the cell, and continued by host cell collapse. Thus, regulation of host intracellular calcium and parasitic vacuole compartment has a close relationship with the growth and egress of parasites from host cell.

4.4. Calcium signalling during neural infection by Toxoplasma gondii

According to Krebs and Michalak (2007), eukaryotic cell is surrounded by free calcium media more than 1mM. Yet, it regulates cytoplasmic intracellular calcium concentration maintained four times lower. This low concentration conditions calcium precipitation prevention, and phosphate calcium sodium low solubility. This condition also prevents the energy loss that is needed for targeted calcium concentration transformation. Total intracellular calcium is higher than mM, but it is reduced until ionic sub-μM. First bounding with phospholipids membrane (acidic) to lower the metabolite molecular weight, to be inorganic ion such as phosphate, and then built complex with certain protein. This protein is divided into some classes: one consists membrane intrinsic protein that operates as calcium messenger on plasma membrane and on organelle membrane. This protein plays an important role on cellular calcium homeostatic, to relocate calcium into and so on between cytosol, where the highest targeted calcium was located, extracellular space or organelle luminal space.

Calcium signalling coordinates the motility, cell invasion, and egress from host cell of this apicomplexan parasite that until recently the mediators was not known. Lourido et al., (2012) states that *Toxoplasma gondii* calcium-dependent kinase 1 (TgCDPK 1) is needed during invasion, TgCDPK1 and TgCDPK3 during egress from host cell. During host cell invasion, *Toxoplasma gondii* secret PLA2 with microneme3, activates phosphate inositol signalling and continues with calcium release from internal pool. From statistical test, this study has proven significant difference of calcium measurement between control and infected group, and the calcium intensity transformation during 2 hours, 24 hours and 48 hours of infection. This result of this research supported transformation of neural intracellular calcium and during *Toxoplasma gondii* infection, and showed that calcium signalling as a first messenger and second messenger works in this process.

5. Conclusion

The confocal laser microscopy is able to detect intracellular calcium on control group and infected group. It also detects intracellular calcium by capturing fluorescent light and reflecting light from the specimen using light detectors. The outputs of the specimen image then presented on image monitor.

Intracellular neuron calcium intensity was significantly different between the control and *Toxoplasma gondii* infected neuron groups. It was because *Toxoplasma gondii* may alter neural calcium concentration. Among the 3 groups of infected neuron (2 hours infection, 24 hours infection and 48 hours infection), there was significant difference of calcium intensity inside the neuron and parasite cells. The calcium intensity in *Toxoplasma gondii* was gradually increasing while the intensity of calcium in neurons was gradually declining. This condition caused significant difference between control and *Toxoplasma gondii* infected groups on neuron intracellular neuron calcium intensity.

Acknowledgment
This study was supported by Medical Faculty of Universitas Airlangga.

References

[1] Nagamune K et al 2008 *Calcium Regulation and Signaling in Apicomplexan Parasites Molecular mechanism of Parasite Invasion* pp 70-81.
[2] Dobrowolski, JM, Carruthers, VB and Sibley LD 1997 *Participation of myosin in gliding motility and host cell invasion by Toxoplasma gondii. Mol. Microbiol. 26* pp 163-173.
[3] Lovett, JL, and Sibley L.D., 2003. *Intracellular calcium stores in Toxoplasma gondii govern invasion of host cells*. J. Cell Sci. 116 pp 3009-3016.
[4] Rozenfeld C 2005 Toxoplasma gondii Prevents Neuron Degeneration by Interferon-γ-Activated Microglia in a Mechanism Involving Inhibition of Inducible Nitric Oxide Synthase and Transforming Growth Factor-β1 Production by Infected Microglia *The Americal Journal of Pathology,* **vol. 167**.Issue4. October 2005 1021-1031.
[5] Tsien, RY 1989 *Fluorescent indicators of ion concentrations* Methods Cell Biol 30 pp 127-156.
[6] Carruthers V B and Sibley L D 1997 *Sequential protein secretion from three distinc organelles of Toxoplasma gondii accompanies invasion of human fibroblast* Eur. J. Cell Biol. 73 pp 114-123.
[7] Bonhomme A 1999 *Signaling during the invasion of host cells by Toxoplasma gondii FEMS Microbiol* Rev 23 pp 551-561.
[8] Mondragon R and Frixione E 1996 *Ca2+-dependence of conoid extrusion in Toxoplasma gondii tachyzoites*. J. Eukaryot. Microbiol. 43 pp 120-127.
[9] Bouchot A, Zierold K, Bonhomme A, Killian L, Belloni A, Balossier G, Pinon J and Bonhomme P 1999 *Tachyzoite calcium changes during cell invasion by Toxoplasma gondii Parasitol Res 85* pp 809-818.
[10] Brecht S, Carruthers V B and Ferguson D J 2001 *The toxoplasma micronemal protein MIC4 is an adhesin composed of six conserved apple domains J. Biol Chem 276* pp 4119-27.
[11] Couvreur J, Desmonts G, 1962 *Congenital and Maternal toxoplasmosis* A review of 300 congenital cases Dev. Med. Child Neurol. 4 pp 519-530.
[12] Dobrowolski J M and Sibley L D 1996 *Toxoplasma invasion of mammalian cells is powered by the actin cytoskeleton of the parasites Cell 84* pp 933-939.
[13] Dunn RB 1999 *Mother-to-child transmission of toxoplasmosis: risk estimates for clinical counseling*. Lancet. 353 pp 1829-1833.
[14] Grynkiewicz, G, Poeni M and Tsien RY 1985 *A new generation of Ca2+ indicators with greatly improved fluorescent properties*. J. Biol. Chem. 260 pp 3440-3450.
[15] Krebs J and Michalak M 2007 *Calcium : A Matter of Life or Death first edition* (Amsterdam:Elsevier) pp 5-12.
[16] Lin L L 1993 *cPLA2 is phosphorylatedand activated by MAP kinase Cell* 72 pp 269-278
[17] Lourido S, Tang K, and Sibley L D 2012 Distinct signalling pathways control Toxoplasma egress and host-cell invasion *The EMBO Journal* **31** pp 4524-34.
[18] Monteiro VG 2001 *Morphological changes during conoid extrusion in Toxoplasma gondii tachyzoites treated with calcium ionophore* J Struct Biol 136 pp 181-189.
[19] Morisaki J H, Heuser J E and Sibley L D 1995 *Invasion of* Toxoplasma gondii *occurs by active*

penetration of the host cell J Cell Sci 108 pp 2457–64.

[20] Viera M C, and Moreno S N 2000 *Mobilization of intracellular calcium upon attachment of Toxoplasma gondii tachyzoites to human fibroblasts is required for invasion* Mol Biochem. Parasitol 106 pp 157-162.

Performance weight sensor using graded index optical fiber on static test with UTM

R Khamimatul Ula[*] and Dwi Hanto

Research Center for Physics, Indonesian Institute of Sciences, Building 440-442, PUSPIPTEK, South Tangerang, Indonesia

[*]Corresponding email address: riniula@gmail.com

Abstract. Overloading the vehicle on a highway cause the damage to roads, accidents and harm other road users. Required a weight sensor has a high sensitivity, resistant to corrosion and electromagnetic wave interference. Graded index optical fiber is a kind of fiber that has the potential to be used as a deformation sensor. This research aims to optimize the load sensor has been developed previously to detect a load on a ton scale. The weight sensor-based micro bending graded index fiber and LED as a light source capable of detecting a load from 0.7 to 1.93 tons with a standard deviation of 1.18 and 99.45% accuracy level in a static text using UTM. This sensor has been able to be used to detect heavy vehicles such as water truck wheels 6 and fuel trucks. The study will be further developed in order to detect the load to more than 8 tons.

1. Introduction

Research on the application of optical fiber as the sensor has been developed[1][2]. One application is the use of fiber optics as the load sensor with the principles of micro bending fiber[3][4].Overloading the vehicle on a highway cause the damage to roads, accidents and harm other road users. Required a weight sensor has a high sensitivity, resistant to corrosion and electromagnetic wave interference. The weight sensor that can be planted in highway structures can have a large role to detect the presence of heavy vehicles with a load exceeding the normal capacity. Therefore, it is necessary based optical sensor is resistant to EM waves and water.

Refer to the results of our previous studies [5], that using LED 1310 nm as the light source on the weight sensor is more stable than LASER 1310 nm and 970 nm. Bending period is an important factor that determines the precision level of a sensor[6][7][8]. In the our previous study [9], the weight sensor with a micro bending deformer double side and bending period is 1.5 mm with step index optical fiber is able to detect the load until 100 kg with an accuracy value of 98.77%, equivalent to the load of the vehicle wheel two. This study aimed to optimizing the weight sensor based on fiber bent techniques be able to detect loads up on a ton scale.

1.1. Graded index optical fiber

Plastic Optical Fiber (POF) graded index capable as deformation sensing system[10][11]. In this research, would analysis the performance of the weight sensor using graded index multimode optical fiber types GIF50C with static test.

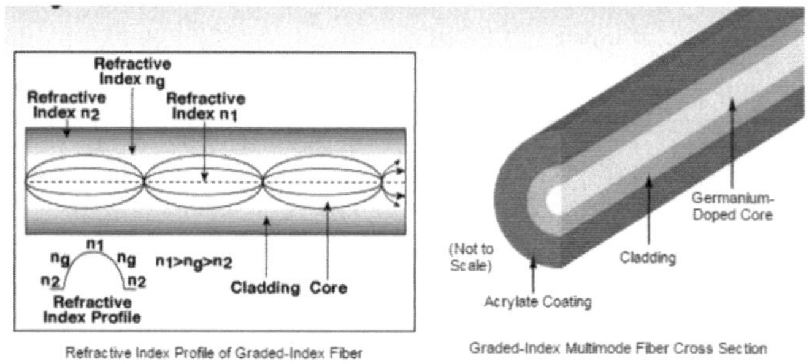

Figure 1. Graded index optical fiber type GIF50C

The refractive index of graded index optical fiber is a function of the distance to the optical axis. So that, the light will experience refraction storied away sheath and approaching the fiber core[12]. This is causes theoptical waves that propagate through multiple modes will arrive at the end of the fiber optic at the same time.

1.2. Micro bending fiber bent

Micro bending is thc bending of an optical fiber in micro sized and affects the optical power loss. This bending is caused by mechanical stress on the optical fiber in an axial direction. Based on studies [7], if the optical fiber has bent in axial direction of the optical axis, the transmission coefficient of the light will change with magnitude as this equation:

$$\Delta T = \frac{\Delta T}{\Delta X} A_p k_f^{-1} \Delta P \quad (1)$$

Transmission of light in an optical fiber would result in attenuation. It is caused weakening of energyso that the amplitude of wave would arrived at receiver smaller than those sent by the transmitter[13].

Deformer micro bending consists of two grooved plates and between plates are optical fibers. When the bend radius exceeded the critical angle, the light began into the leak to the cladding results in intensity modulation. When a small bent exposed fibers or disorder, a certain portion of the propagation of light in the fiber core are incorporated in the radiation mode and lost. Mode merger can be achieved by using a corrugated plate that changes the shape of fiber into a series of bends. Therefore, micro bending causes the light intensity decreases

2. Experiment

The weight sensor consists of LED 1310 nm as a light source which propagated in a graded index optical fiber GIF 50c which has a refractive index of the core material 1,479 at 1300 nm, Numerical Aperture 0.20 ± 0.015 and a core diameter of 50.0 ± 2.5 um. These fibers are enclosed in micro bending deformer has a length 50 cm are made of PVC material with a thickness of 1.5 cm and 1.5 mm bending period. The method of static pressure test by pressing a fiber bent using UTM. This machine could generate the pressure until 100 kN. Output optical power after a sensor in the press by UTM captured by detector optical power meter with a minimum capability -60 dBm. All equipment will be setting as a Figure 2.Press test is intended to determine the maximum load that can be detected by these sensors. Tests conducted by the repeated 5 times in each load.

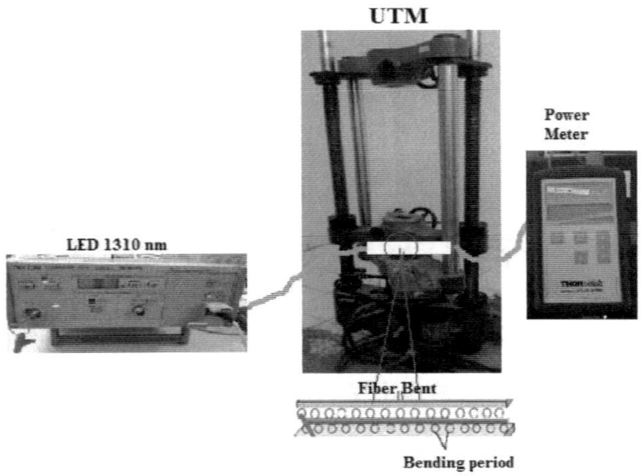

Figure 2. Set up experiment static pressure test

3. Result and discussion

Several of the most important to static characterization of sensor are repeatability, resolution, sensitivity, linearity and accuracy[14][15]. From experimental data obtained graphics are 1,2,3,4 and 5 represent the output optical power meter as a function of pressure of UTM as in Figure 3. Graphics 1 and 2 have different profile with graphic 3,4 and 5. This is due to the time of data collection is done shortly after the equipment assembled; do not wait for the stability of the optical wave propagation in the fiber. Nevertheless, this sensor has a good repeatability.

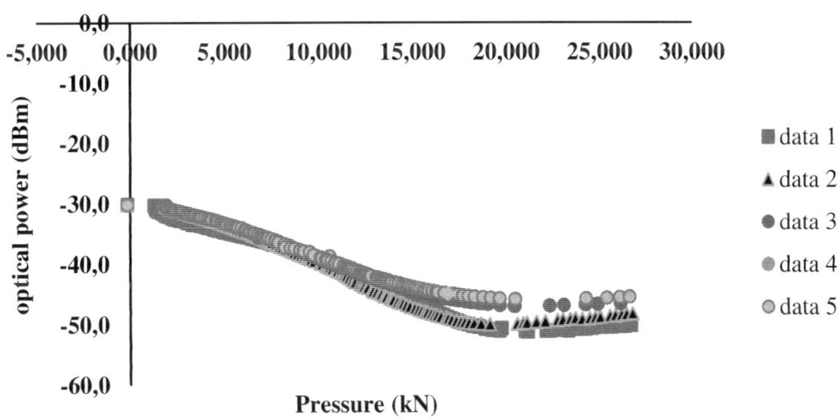

Figure 3.Response the optical power causes pressure on weight sensor

From 5 data of repetitions performed the calculation of average output optical power due to the pressure UTM and represented in the Figure 4.

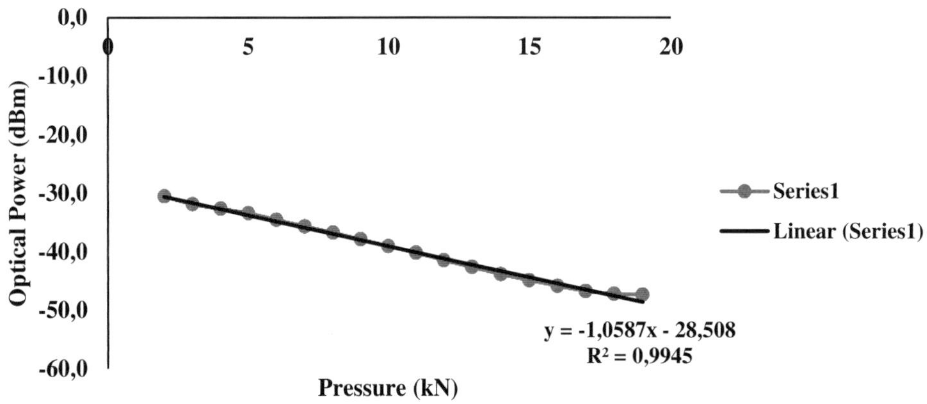

Figure 4.Fitting linear of average optical power on weight sensor

Linear graph in the Figure 4 representation the linearity have a equation y = -1,0587x - 28.508, that means for every 1 kN pressure exerted on the weight sensor will result in a decrease of -1.0587 dBm optical power. In other words, the resolution of this sensor is every 1 dBm on the detector shows the pressure on the sensor caused by load of 92.106 kg.

This sensor has been able to detect a load in the range of 0.7 - 1.93 ton at 99.45% accuracy level. This success can already be applied to detect load 6 wheel vehicles such as water vehicle and vehicle fuel.

4. Conclusion

Weight sensor using graded index optical fiber on deformer micro bending double side 1,5 mm and LED 1310 as a light source able to detection the load in the range of 0.7 - 1.93 tonnes with good repeatability. Has resolution 92.106 kg/dBm and the accuracy is 99.45%.

Acknowledgments

The work has been supported by RisetUnggulan LIPI 2016. The authors also thank to MuhamadSulhin, Fajar Tri Wibowo, Tari Rafika and AzkaZarirohfrom State University of Surabaya for helping this experiments

References

[1] X Yang, Z Chen, C Ser, M Elvin, L Hong, and Y Janice 2015 *Textile Fiber Optic Microbend Sensor Used for Heartbeat and Respiration Monitoring* **15** 2 pp. 757–761.

[2] P Wang, Y Semenova, Q Wu, and G Farrell *2010 A macrobending fiber based micro-displacement sensor* pp. 25–28.

[3] A R Mejía-Aranda, V I Ruiz-Perez, M A Basurto-Pensado, E E Antúnez-Cerón, D A May-Arrioja, P Li Kam Wa, and J J Sanchez-Mondragon Dec. 2013 Design of a pressure sensor of 0–7bar in fiber optic using MMI methodology Opt. - *Int. J. Light Electron Opt.* **124** 23 pp. 5927–5929,

[4] S F Knowles, B E Jones, S Purdy, and C M France 1998 *Multiple microbending optical-fibre sensors for measurement of fuel quantity in aircraft fuel tanks* **68** pp. 320–323,.

[5] R K Ula, D Hanto and B Widiyatmoko 2015 Analisis Perbandingan Penggunaan LASER dan LED sebagai Sumber Cahaya pada Sensor Berat Berbasis Optik, *Proseding SNF* UNESA.

[6] J. W. B. Iii 1998 *Microbend fiber optic sensors* **3** 7 pp. 1193–1199.

[7] N Lagakos, J H Cole, and J A Bucaro 1987 *Microbend fiber-optic sensor* **26** 11 pp. 2171–2180,

[8] F Luo, J Liu, N Ma and T F Morse 1999 *A fiber optic microbend sensor for distributed sensing*

application in the structural strain monitoring, pp. 41–44, 1999.

[9] R K Ula, D Hanto, A Hendra, and B Widiyatmoko *Design of Micro Bending Deformer for Application Weight Sensor Based on Optic* pp. 1–6.

[10] A B Ã and J Maryles 2007 *Graded-index plastic optical fiber for deformation sensing* 45 pp.757–760.

[11] Y He and F G Shi 2006 *A graded-index fiber taper design for laser diode to single-mode fiber coupling* 260 pp. 127–130.

[12] Keiser Gerd 1991 *Optical Fiber Communication*,New York, McGraw-Hill

[13] B E A Saleh, M C Teich and C J Wiley 1991 *Fiber optics* 8.1 5.

[14] J McGrath Michael and Scanail Ni Ci *2013Sensor Technologies* Frenos Apres

[15] Kalantar Zadeh 2013 *Sensors An Introductory Course* Springer

IOP Conf. Series: Journal of Physics: Conf. Series **853** (2017) 012017 doi:10.1088/1742-6596/853/1/012017

Investigation of michelson interferometer for volatile organic compound sensor

Marzuarman*, Muhammad Rivai, Tri Arief Sardjono and DjokoPurwanto

Department of Electrical Engineering, Institut Teknologi Sepuluh Nopember, Surabaya, Indonesia

*Corresponding email address: marzuarman14@mhs.ee.its.ac.id

Abstract. The sensor device is required to monitor harmful gases in the environments and industries. Many volatile organic compounds adsorbed on the sensor material will result in changes of the optical properties including the refractive index and the film thickness. This study designed and realized a vapor detection device using the principle of Michelson Interferometer. The laser light beamed with a wavelength of 620 nm was divided by using a beam splitter. Interference occurredwhen the two separated lights were recombined. The phase difference between the two beams determined whether the interference would destruct or construct each other to produce the curved fringes. The vapor samples used in these experiments were ethanol and benzene. The results showed that the ethanol concentration of 1611-32210 ppm produced a fringe shift of 197 pixels, while the concentration of benzene of 964-19290 ppm produced a fringe shift of 273 pixels.

1. Introduction

Volatile organic compounds are widely used in the industrial manufacture of paints, cosmetics, fragrances and others. However, in short term, the exposure to the vapors can be harmful at high concentrations due to its effect on causing health problems such as shortness of breath, dizziness, and eye irritation. While in the long term, it can even cause cancer.

Currently, sensors have been developed to detect volatile organic compounds, including crystal resonator, surface acoustic wave (SAW), fiber optic sensors, and the likes. This study uses the Michelson interferometer because it has a sharper and clearer interference pattern and narrower distance between the interferometer fringes [1]. In addition, this type of interferometer has several advantages including high sensitivity, fast response, low power consumption, stable in hot temperatures, and resistant to electrical or magnetic field interference [2][3][5].

The experiment was conducted by observing the fringe shift when exposed to various concentrations. Differences in travel time of two beams of light when one of the light passed through a vapor sample would result in a fringe shifts [3][4]. In addition, the Michelson interferometer was also expected to provide high sensitivity by using image processing as data acquisition method for measuring the concentration of gas. The type of sensor can be developed into a gas sensor that is effective, efficient, high accuracy in a long term usage.

2. Method

The measurement of concentrations of the volatile organic compound with Michelson interferometer was conducted by observing the interference pattern shift. The vapor concentration changed the value of the refractive index. When light passes through a medium, in which its refractive index value is greater than the air's one, that will decrease its speed; there is the phase change between the two beams of light [2]. The observance of magnitude of the fringe shift is shown in Equation 1.

$$\Delta m = \Delta m + 1 - \Delta m = \frac{s\gamma}{a} \quad (1)$$

Where Δm is the shift of the fringe, s is the distance the beams splitter to the screen, a is the different optical path [2]. The number of fringe shift is affected by changes in the refractive index of vapor, with the initial condition of the shift is expressed by Equation 2.

$$m = \frac{2d}{\gamma} \quad (2)$$

where m is the initial fringe, d is the thickness of the sample chamber (4 cm), and λ is the wavelength of the laser light (620 nm). Equation 3 expresses the change of m:

$$m = \frac{2d}{\gamma/n} \quad (3)$$

where n is the refractive index of vapor. The fringe shift is expressed in Equation 4.

$$\Delta m = m_2 - m_1 = (n-1)\frac{2d}{\gamma} \quad (4)$$

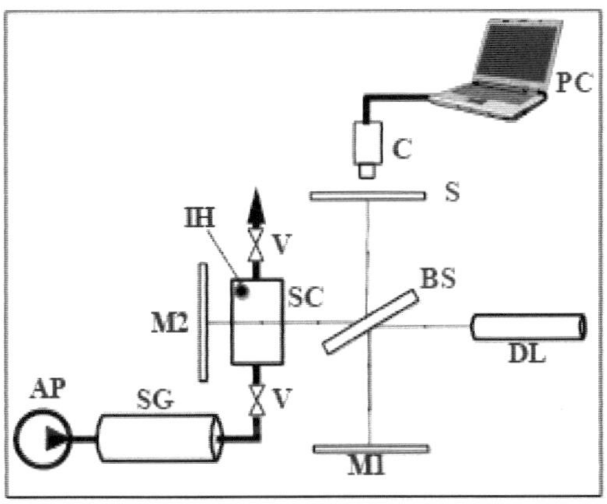

Figure 1. Experimental set-up. AP: Air Pump, SG: Silica Gel, V: Valve, SC: Sample Chamber, IH: Injection Hole, DL: Diode Laser, BS: Beam Splitter, M1: Mirror 1, M2: Mirror 2, S: Screen, C: Camera, PC: Personal Computer

The refractive index can be calculated by Equation 5.

$$n = 1 + \frac{\Delta m\gamma}{2d} \quad (5)$$

In this study, the light source wasa laser diode with a wavelength of 620 nm and the vapor samples were ethanol and benzene. The block diagram of the equipment is shown in Figure 1. The light source of DL was divided by the BS towards the M1 and SC. Reflected light from the M1 and M2 was recombined to form a pattern displayed on the S and captured by the camera.

3. Result and discussion

The vapor concentration was given by injecting sample fluid into the chamber. The fluid volume ranged between 0.05 to 1 mL. The number of fringe shifts was observed in the pattern and intensity is shown in Figure 2 and Figure 3 respectively. The results of the measurement of ethanol and benzene solvents are shown in Figure 4 and Figure 5 respectively.
The vapor concentration can be estimated by using Equation 6 [6].

$$C(ppm) = \frac{\rho V}{M} \frac{V_m}{V_c} 10^6 \tag{6}$$

where C is the vapor concentration in ppm, ρ is the liquid density (g / mL), V is the volume of fluid (mL), M is the molar mass of fluid (g / mol), Vm is the molar volume of ideal gas (22.4 L / mol) and Vc is the volume of the chamber (0.013 L).

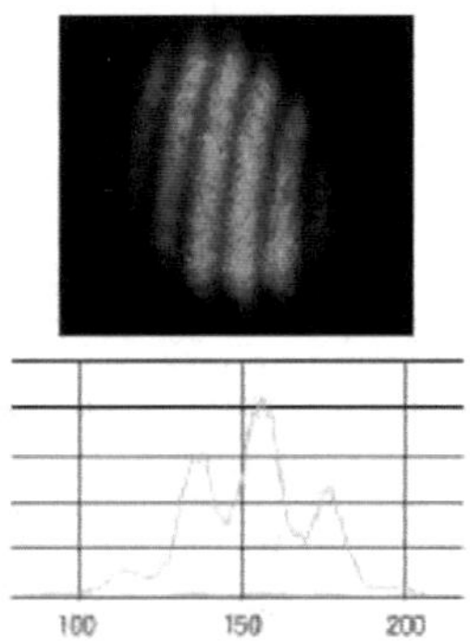

Figure 2. Fringe pattern before the phase shifts and intensity pattern

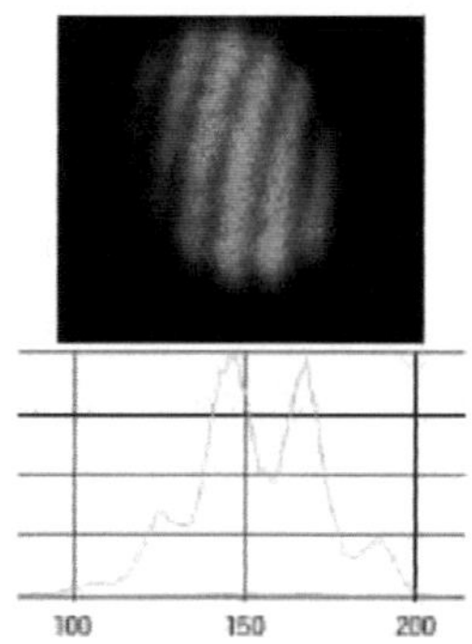

Figure 3. Fringe pattern after the phase shifts and intensity pattern

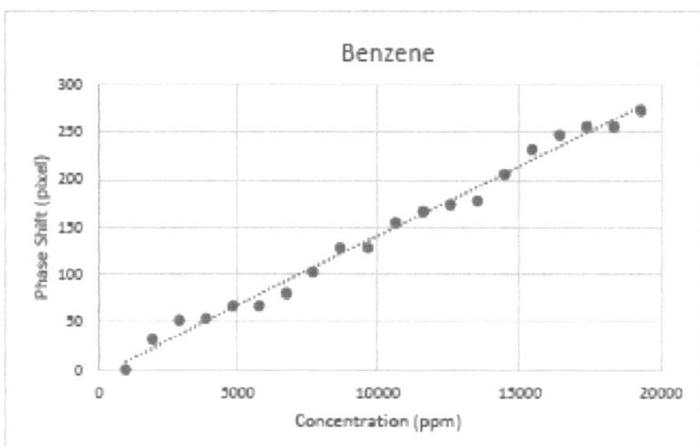

Figure 4. The fringe shift of the benzene vapor

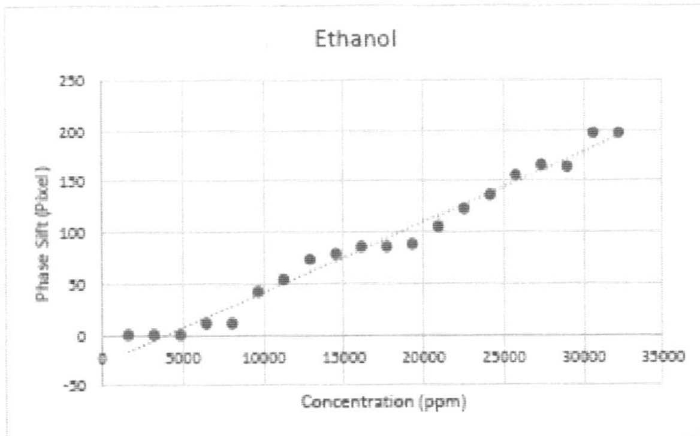

Figure 5. The fringe shift of the ethanol vapor

The results showed that the number of the fringe shift was proportional to the concentration. The ethanol concentration of 1611-32210 ppm produced a fringe shift of 197 pixels, while the concentration of benzene of 964-19290 ppm produced a fringe shift of 273 pixels.

4. Conclusion

This study has been made of the gas sensor based on Michelson interferometer. This type of sensor is used to measure the concentration of volatile organic compound vapor observed from the change in the number of fringe shifts. The results show that the number of the fringe shift is proportional to the concentration. The ethanol concentration of 1611-32210 ppm produces a fringe shift of 197 pixels, while the concentration of benzene of 964-19290 ppm produces a fringe shift of 273 pixels. The use of polymer materials is needed to increase the sensitivity and selectivity of this optical sensor. An array of polymers with different its polarities will result in a shift of the fringe pattern that is unique to each sample of vapor.

References

[1] Hall J L, Petropavlovskikh S, Nilsen O and Hacioglu B 2005 Detection and Discrimination of Low Concentration Gas Contaminants by means of Interferometrically-Sensed Polymers *Proc. IEEE Sensors* 2005 pp 1366-1369

[2] Halliday D and Resnick R 1999 *Physics* (version Pantur Silaban dan Erwin Sucipto) **2** 3 (Jakarta: Publisher Erlangga)

[3] Abbasian K and Abdollahi M H 2013 Electromagnetically Induced Transparency-based Gas Detector Design Using Michelson Interferometer *Journal of International Nano Letter s* **3** 34 pp 1-7

[4] Kachiraju S R, Don A and Gregory 2012 Determining the Refraction Index of Liquids Using a Modified Michelson Interferometer (*Optics & Laser Technology 44*) pp. 2361-2365

[5] Byeong H L, Young H K, Kwan S P, Joo B E, Myoung J K, Byung S R and Hae Y C 2012 Interferometric Fiber Optic Sensors *Sensors* vol **12** pp 2467-2486

[6] Feng L, Tang W, Zhao J, Yang R, Hu W, Li Q, Wang R and Guo X 2016 Unencapsulated Air-Stable Organic Field Effect Transistor by All Solution Processes for Low Power Vapor Sensing *Scientific Reports* p 8

IOP Conf. Series: Journal of Physics: Conf. Series **853** (2017) 012018 doi:10.1088/1742-6596/853/1/012018

The transformation of equation of the characteristic curves to widen the working area of the hall effect sensor

D Darmawan*, F Rahmawati, Suprayogi and A Suhendi

Engineering Physics, Telkom University, Bandung, Indonesia.

*Corresponding email address: dudiddw@telkomuniversity.ac.id

Abstract. The ability of a measuring instrument is determined by the characteristics of the sensor used. One of the important characteristics of a sensor is its area of work. The working area of a sensor is determined by a linear relationship between the physical quantities to be measured with the actual measured physical quantity. A narrow sensor working area will limit the use of the instruments. Transformation equation of sensor characteristic curve is expected to widen the working area of this sensor. In this study, the magnitude of a magnetic field is used as an example of the physical quantity to be measured. Hall Effect sensors are used as sensing elements and the transformation equation performed on the sensor characteristic curve. Hall Effect sensor which is used as the object of widening the area of work is UGN3503UA while the reference sensor is A1302UA. The result of this equation transformation is embedded inside a micro-controller. The interpolation value of the magnetic field, as the result of the characteristic curve transformation, which compared to the actual value of the magnetic field provides an average relative error of measurement results of 3.83 %.

1. Introduction

In recent years, the measurement of the magnetic field has become an important concern in the industrial world [1] [2]. One of the sensors used to detect magnetic fields is Hall Effect sensor.

Hall Effect sensors is mostly used because it provides lossless and isolated current-sensing solution, by sensing the current without dissipating the power that passive resistors do [3]. Their low maintenance, robust design and sealed Hall Effect devices are immune to vibration, dust and water [4].

Hall Effect sensors work based on the concept of Lorenz which posed the charged particles under the influence of magnetic fields [3]. Linear relationship between the electric potential generated by the magnetic field on charged particles, which polarized the particles, is used as the basic concept of measuring the magnetic field [4]. Interval linearity which is the characteristic of each sensor is referred to as the working area of the sensor. The working area of the sensor is one of the problem of the measurement devices.

The Hall Effect sensors used in this study are UGN3503UA and A1302UA. Both magnetic sensors are only a sample of sensors to implement the results of widening the working area of the sensor. Two magnetic sensors used in a Hall Effect sensor which have different work area. One sensor, which has a wider working range, is used as a reference. Meanwhile, the other sensor, which have a more narrow working area, is used as an object sensor that will be widened.

Sensor UGN3503UA is one of the Hall Effect sensors that is widely used and easily available in the market at a price that is not too expensive. The test results of this sensor in a previous study [5] found that the electrical potential linear response obtained in the magnetic field are given below 900 Gauss.

Meanwhile, in another study [6], A1302UA sensor can provide a linear response to electrical potential to the provision of a magnetic field of 1800 Gauss. Therefore, the sensor UGN3503UA has a work area that is narrower than A1302UA sensor.

Based on the data acquired from the results of these studies, we can make measurements from a magnetic field with a value greater than 900 Gauss using UGN3503UA sensor. Through this idea, the sensor UGN3503UA, with a narrow working area, can be used to measure the magnitude of the magnetic field that is greater than the maximum limit of the linear sensor. To be able to do that, the sensor characteristic curve equation UGN3503UA needs to transform to the sensor characteristic curve equation A1302UA. The equation of this transformation is then used to interpolate the value of the magnetic field of the electric potential measurements by sensors UGN3503UA. Through this process, the working range of the sensor UGN3503UA will be wider so that it can be used to measure the magnetic field to the maximum extent of A1302UA sensor. Equations transformation and interpolation step are then translated into the program and can be implemented into the micro-controller.

2. Research method

The study begins by characterizing the source of the magnetic field that will be used, namely neodymium magnet rods. This step is done by measuring the magnitude of the magnetic field as a function of distance. The magnitude of the magnetic field as a function of distance measurement produces a characterization curve. This curve is then used to calibrate and validate the results of the determination of the value of the magnetic field by the sensor under test.

The next stage is to characterize the second Hall Effect sensor. Through the characteristic curve of each sensor, the working area of the sensor and the characteristic equation can be determined. After the equation of each characteristic curve is obtained, then the transformation equation of the sensor characteristic curves is done. Transformation is done from the sensor with a narrow working area to the reference sensor characteristic curve which has a wider working range. Overview of the characteristic curve and the working area of each sensor are shown in Figure 1.

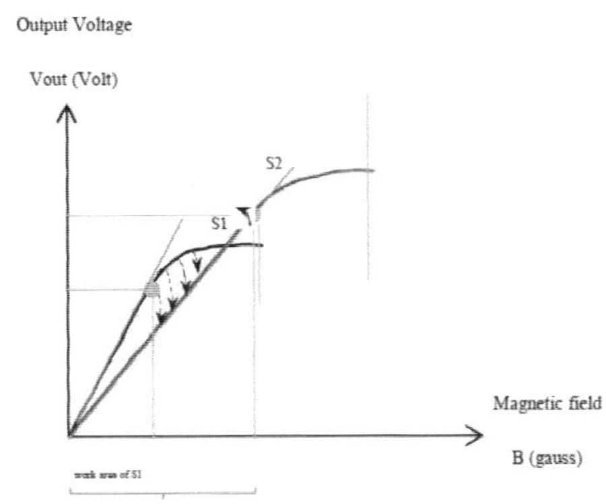

Figure 1. An overview of the sensor characteristic curve

Figure 1 shows the different working areas of both sensors. Both sensors provide a voltage output that is proportional to the magnetic field strength, the greater the magnetic field is given, the greater the output voltage measured by the sensor. Sensor 1 has a working range narrower than sensor 2.

Thus, the transformation of the characteristic curve equation will be conducted from sensor 1 to sensor 2. Through this transformation equations, interpolation value of the magnetic field may be based on electric potential value measured by sensor 1. It is expected that the sensor 1 will behave the same as sensor 2. Flowchart of research conducted seen in Figure 2.

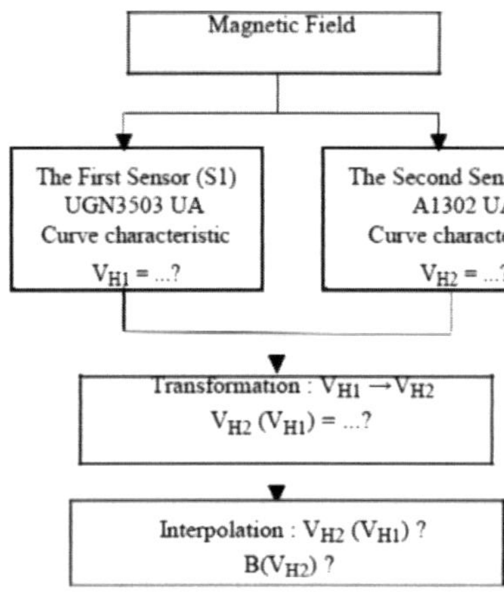

Figure 2. Flowchart of research

3. Results and analysis

3.1. Characteristic of magnetic field source

To test the Hall Effect sensor, a source of magnetic field measured and calibrated needed. Sources of magnetic fields have two important characteristics which are the flux density and polarity. To obtain the characteristics of the source of the magnetic field, neodymium magnet is used in this study. Specifications bar magnets used is 2 cm long, 1 cm wide and 0.5 cm high. Results of measurement of magnetic field strength showed a function of distance measurement, using teslameter 7010 models that have been calibrated, shown in Figure 3.

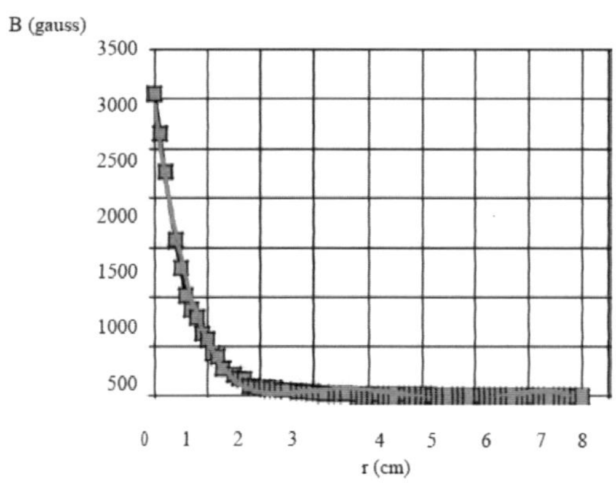

Figure 3. The characteristic curves of the magnetic field source rod neodymium

Figure 3 shows that the maximum magnetic field generated by the magnet rod is achieved at 3000 gauss neodymium magnet at the position approaching the probe teslameter (position 0 cm). Meanwhile, if the position is eliminated, the amount of neodymium magnet magnetic field decreases and reaches a value close to 0 gauss at a distance of about 8 cm. However, the linear relationship between the magnitude of the magnetic field and the distance measurement can be achieved between 0 cm and 2 cm. Therefore, the validation of Hall Effect sensors using neodymium magnet rods will only be done in the measurement range below 2 cm.

3.2. Characteristic of hall effect sensor

In the same way, to ensure the measuring area to be used, the characterization of the Hall Effect sensor needs to be done. Measurement of magnetic field strength neodymium rod do the same using both UGN3503UA hall Effect sensor and A1302UA, as shown in Figure 4. The measurement results of the second voltage sensor bar magnet distance changes were obtained as shown in Figure 5.

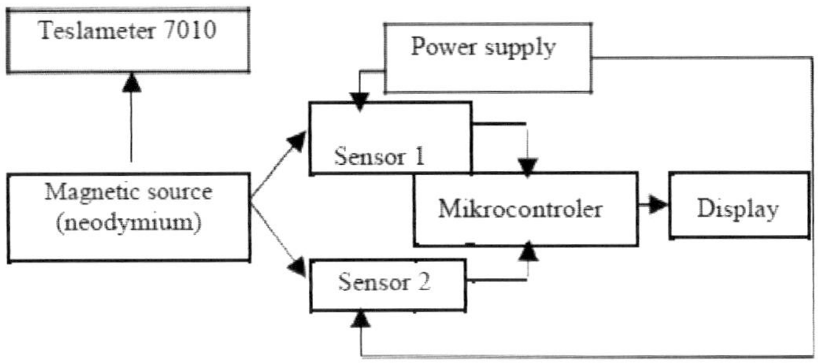

Figure 4. Measurement of the magnetic field

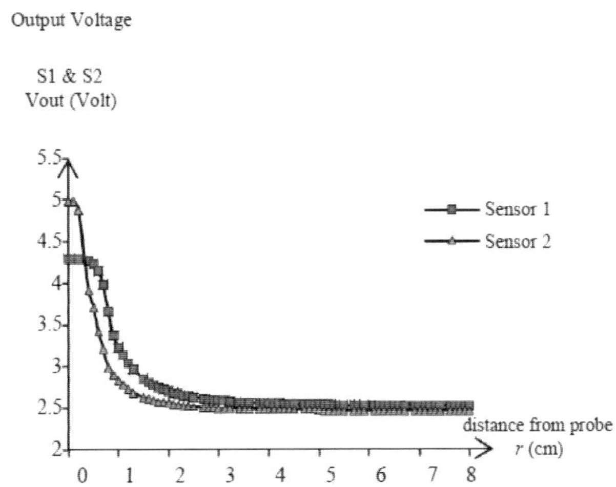

Figure 5. Graph comparison voltage output sensor test results 1 (UGN3503UA) and sensor 2 (A1302UA) with neodymium magnet rods against distance measurement

The output of the two sensors, in Figure 5, shows that the sensor responds to changes in the magnetic field is also given linearly at a distance of less than 2 cm.

Furthermore, by using a measuring instrument that has been calibrated (Teslameter 7010), the measurement of magnetic field strength on every point distance measurement between 0 and 2 cm is done. Results of measurement of magnetic field strength by teslameter and the electric potential arising from the Hall Effect sensor by a multi meter plotted become characteristic curve, as shown in Figure 6.

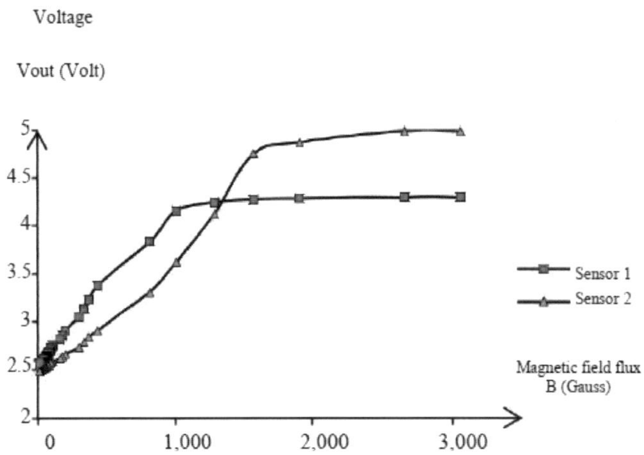

Figure 6. The second characteristic curves of sensors with neodymium magnet rods source

The relationship between the values of the magnetic field sensor output voltage 1 did not produce a linear curve entirely. For sensor 1, the curve is linear only in the range of 0 to 900 gauss. If the magnetic field exceeds 900 gauss is given then the sensor experience saturation. Similarly, in the second sensor, the resulting curve only produces a linear curve until approximately 1400 gauss. From the measurement data, sensor 1 can be targeted widening of the work area. Through this process, the sensor 1 is expected to work on the areas of work as well as sensor 2.

3.3. The equation of characteristic curve of hall effect sensor

At this stage, the characteristic curve equation will be determined by the method of interpolation. Equation characteristic curve for each sensor is broken down into two intervals in the range 0-900 gauss and 900-1400 gauss. Sampling and interpolation curves to measure data at the sensor characteristic curve 1 (UGN3503UA), within the range of 0-900 gauss, obtained as shown in Figure 7. Meanwhile, the equation interpolation on the curve characteristic of the sensor 2 (A1302UA), in the same range, is obtained as shown in Figure 8.

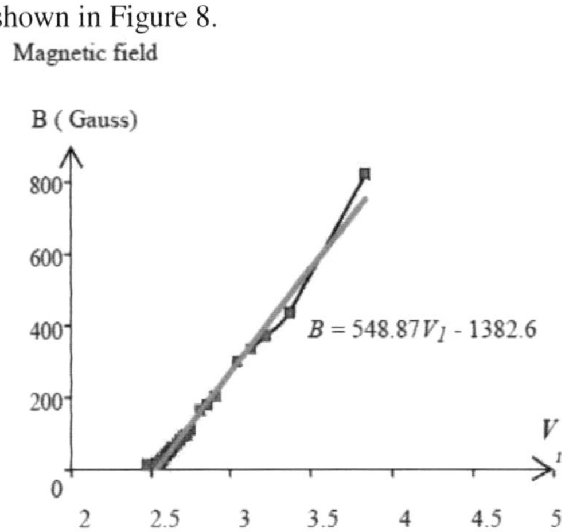

Figure 7. Sampling data and interpolated curves characteristic of the sensor 1 in the range of 0 to 900 gauss

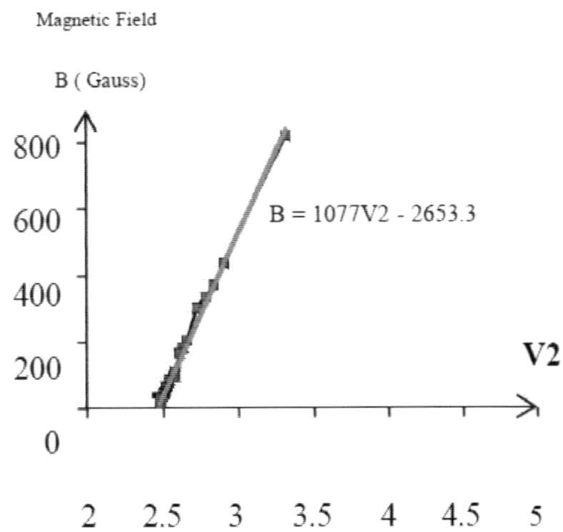

Figure 8. Sampling data and interpolated sensor characteristic curve 2 in the range of 0 to 900 gauss

From Figure 7 and 8, it can be obtained that equation characteristic curve for sensor 1 and sensor 2 is

$$B\,(V_1) = 548.87\,V_1 - 1382.6 \tag{1}$$
$$B\,(V_2) = 1077\,V_2 - 2653.3 \tag{2}$$

With
B is a large magnetic field (gauss) and
V_1 output voltage (volts) sensor 1
V_2 output voltage (volts) sensor 2

As for the magnetic field ranges between 900 to 1400 gauss, the characteristic curve is obtained as Figure 9 and 10.

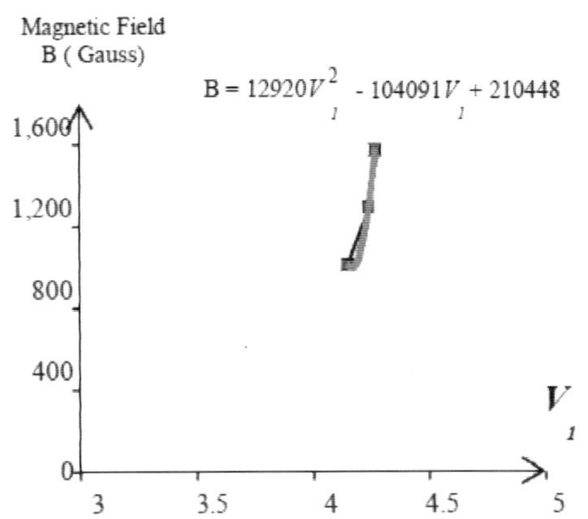

Figure 9. Sampling data and interpolated curves characteristic of the sensor 1 in the range 900 to 1400 gauss.

IOP Conf. Series: Journal of Physics: Conf. Series **853** (2017) 012018 doi:10.1088/1742-6596/853/1/012018

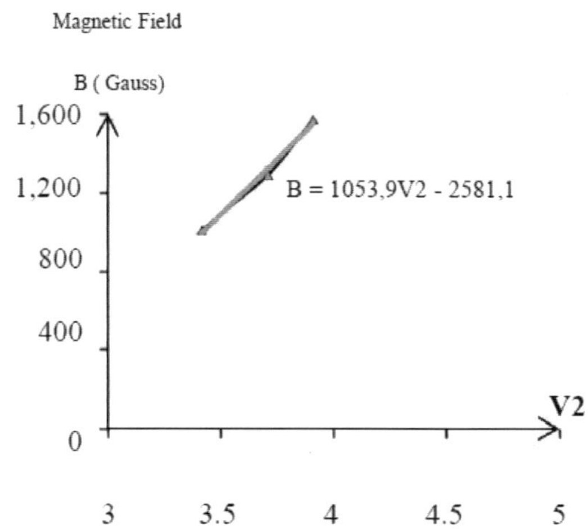

Figure 10. Sampling data and interpolated sensor characteristic curve 2 in the range of 900 to 1400 gauss

Equation characteristic curve for the range 900 to 1400 gauss is

$$B\,(V_1) = 12920\,V_1^2 - 10409\,V_1 + 210448 \tag{3}$$
$$B\,(V_2) = 1053\,V_2 - 2581.1 \tag{4}$$

3.4. *The transformation of characteristic curve equation*

Transformation equation is done in two ranges. For the range of 0 to 900 gauss, because the magnetic field measured by the sensor is the same value of the magnetic field, the value of equation (1) will be equal to the value of equation (2).

$$B\,(V_1) = B\,(V_2) \tag{5}$$

Completion of the equation (5) yields as a function of V1 V2, such as the equation (6)

$$V_2(V_1) = \frac{548.87\,V_1 - 1270.7}{1077} \tag{6}$$

Equation (4) is the transformation equations to get the same voltage as the voltage sensor 2. Then equation (6) is substituted into the equation (2) and will generate the equation (7).

$$B\,(V_1) = 1077\left(\frac{548.87\,V_1 - 1270.7}{1077}\right) - 2653.3 \tag{7}$$

Equation (7) indicates a value approach magnetic field measured by the sensor 2 will be based on the value of the voltage that is read by the sensor 1 measuring range 0-900 gauss. Equation (7) is the result of the transformation equation thus replacing the equation (1).

Just like the range of zero to 900 gauss, the determination of the equation range from 900 to 1400 gauss is done by substituting the equation (3) and (4) to the equation (5). This result in a voltage transformation equations equation (8).

$$V_2(V_1) = \frac{102920\,V_1^{\,2} - 104091\,V_1 + 2130291}{1053.9} \tag{8}$$

While equation to interpolate the value of the magnetic field is obtained by substituting the equation (8) into the equation (4). The result is obtained as equation (9).

$$B(V_1) = 1053.9\left(\frac{102920\,V_1^{\,2} - 104091\,V1 + 2130291}{1053.9}\right)\text{-}2581.1 \tag{9}$$

Equation (9) indicates a value approach magnetic field measured by the sensor 2 will be based on the value of the voltage that is read by the sensor 1 on the measurement range 900-1400 gauss. Equation (9) is the result of the transformation equation thus replacing the equation (3).

3.5. Validation of the transformation results

In this section, the testing is done by moving closer and away from the source of the Hall Effect sensors magnetic field. Sources of magnetic fields used are neodymium magnet rods with characteristics as in part A. By varying the distance between the permanent magnet rods with a layout position of a Hall Effect sensor, the test results for the magnetic field ranges from 0 to 900 gauss is as shown in Table 1. As for the range of 900 gauss to 1400 is as shown in Table 2. In Table 1, the validation is done by comparing the results of the transformation equation is B (V_1) (equation (7)) to the value of the magnetic field measured on a Hall Effect sensor reference in that equation (2). Whereas in Table 2, the validation is done by comparing the results of the transformation equation is B (V_1) (equation (9)) to the value of the magnetic field measured on a Hall Effect sensor reference in that equation (4).

Table 1. Results of Testing Transformation sensor 1 to sensor 2 range 0-900 Gauss

distance (cm)	V1 (Volt)	V2 (Volt)	V2 (V1) (Volt)	B (V1) (Gauss)	B (V2) (Gauss)	Error
0.7	3.83	3.11	3.23	881.60	855.62	2.95
0.8	3.56	2.99	3.11	703.72	694.48	1.31
0.9	3.37	2.90	3.01	577.13	569.26	1.36
1	3.23	2.84	2.93	486.60	454.58	6.58
1.1	3.13	2.78	2.87	394.88	400.61	1.45
1.2	3.04	2.73	2.82	335.20	350.91	4.69
1.3	2.96	2.68	2.77	287.82	330.50	14.83
1.4	2.90	2.65	2.74	257.98	288.37	11.78
1.5	2.85	2.63	2.69	227.10	222.57	1.99
Average of Error						**5.22**

Table 2. Results of Testing Transformation sensor 1 to sensor 2 range of 900-1400 Gauss

distance (cm)	V1 (Volt)	V2 (Volt)	V2 (V1) (Volt)	B (V1) (Gauss)	B (V2) (Gauss)	Error
0.4	4.27	3.92	3.72	1472.64	1454.26	1.25
0.5	4.24	3.71	3.70	1235.08	1226.36	0.71
0.6	4.15	3.52	3.34	1072.24	1059.08	1.23
Average of Error						**2.45**

IOP Conf. Series: Journal of Physics: Conf. Series **853** (2017) 012018 doi:10.1088/1742-6596/853/1/012018

From Tables 1 and 2 above, the increase in the working area sensor 1 is obtained with an average relative error of 3.83%. The error occurs because there are several factors which influence it, such as the distance between the sources of the magnet with a Hall Effect sensor, which is relatively close to that of Hall Effect sensors, are too sensitive. Another factor is the influence of the magnetic field around and error or tolerance measurements. The value of B (V_1) and B (V_2) is plotted for the entire range of the obtained graph such as shown in Figure 11.

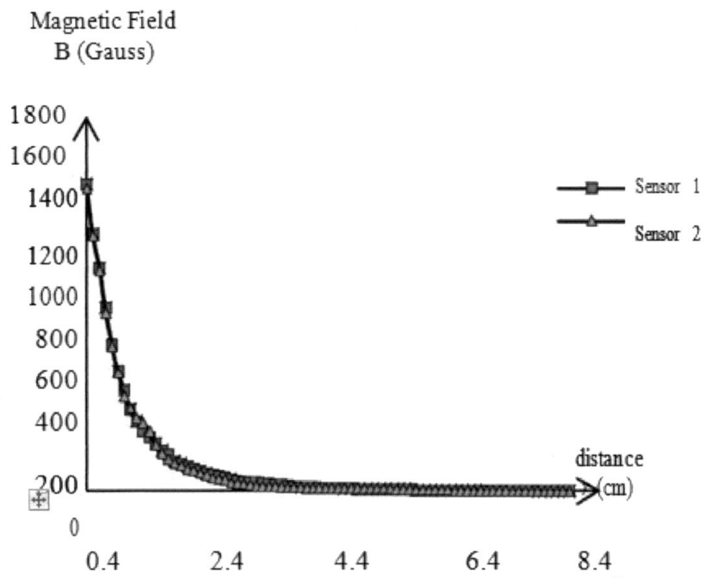

Figure 11. Graph comparing the results of measuring sensor 1 and sensor 2

The graph in Figure 11 shows that the results of the transformation curve equation sensor 1 as a whole has followed the pattern of the magnetic field measurement results which is done directly by the sensor 2. This proves that sensor 1 is already behaving like sensor 2 and can work over a range of magnetic field measurements the same as work area sensor 2.

4. Conclusion

At the sensors work area ranges that intersect, i.e. 0-900 gauss, the characteristic curve of Hall Effect sensors are not coincident with it so that the transformation should be performed in all ranges of the working area. In the span of an intersecting area of work, the process of transformation can be termed as the sensor calibration of the sensor reference.

Transformation equation characteristic curve to increase the area of the sensor work began to happen to the value of 900 gauss magnetic field. Improved working area can be made within the particular work area of the sensor reference.

References

[1] James Sedgwick, William R and Reinhold Ludwiq 1998 *Design of a Digital Gauss Meter for Precision Magnetic Field Measurements* Department of Electrical and Computer Engineering Worcester Polytechnic Institute Worcester USA.

[2] Mitra Djamal, Ramli, Rahadi Wirawan and Edi Sanjaya Sensor Magnetic GMR *Teknologi dan Aplikasinya, dalam Prosiding Pertemuan Ilmiah* XXV HFI, Jateng & DIY.

[3] Arya Krishna S and Lizy Abraham 2014 Analysis of Different Hall Effect Current Sensors for Space Applications *International Journal of Innovative Science* Engineering &Technology **1** Issue 5.

IOP Conf. Series: Journal of Physics: Conf. Series **853** (2017) 012018 doi:10.1088/1742-6596/853/1/012018

[4] Bachtera Indarto, Melania Suweni Muntini, and Darminto 2009 *Pembuatan Magnetometer ber-tranduser Efek Hall* Jurusan Fisika FMIPA ITS, Surabaya.
[5] Jaromír Jezný, Miloslav Čurilla, 2013 Position Measurement with Hall Effect Sensors *American Journal of Mechanical Engineering* **1** 7 231-235.
[6] Johan Wahyudi, Gurum Ahmad Pauzi and Warsito 2013 *Desain dan Karakteristik Pengunggunaan Sensor Efek Hall UGN3503 untuk Mengukur Arus Listrik pada Kumparan Leybold P6271 secara Non Destruktif* Jurusan Fisika FMIPA Universitas Lampung Bandar Lampung.

IOP Conf. Series: Journal of Physics: Conf. Series **853** (2017) 012019 doi:10.1088/1742-6596/853/1/012019

Design of vein finder with multi tuning wavelength using RGB LED

Franky Chandra[1*], Aries Wahyudianto[1], and M Yasin[2]

[1]Biomedical Engineering, Faculty of Science and Technology, Airlangga University, Surabaya (60115), Indonesia
[2]Physics, Faculty of Science and Technology, Airlangga University, Surabaya (60115), Indonesia

*Corresponding email address: frankysatria@gmail.com

Abstract. Detection of intra vena is very important technique in the medical clinic applications. For intravenous detection, some nurses usually have a mistake which can cause a pain or injured to the patient. When the nurses are headed with this problem, it becomes dangerous for the patient. To solve the problem, in this paper, vein finder with multi-tuning wavelength for intra vena detection is proposed and investigated. Vein finder is tested to various skin colour and body mass. The results show that vein finder was successfully designed with controllable wavelength in the range of 600- 696 nm using RGB LED.

1. Introduction

Detection of intra vena is very important technique in the medical clinic applications. For intravenous detection, some nurses usually have a mistake which can cause a pain or injure to the patient. When the nurses face this problem, it becomes dangerous for the patient. Taking of blood sample at the veins is performed by nurses. Veins are not only used to take sample of blood, but also as a media of drugs delivery. Installation of a needle on a vein is required to match correctly and to provide comfort to the patient [1]. Installation of the needle fitting that is appropriate to the blood vessels is influenced by several factors including experience of the nurse and the patient's skin thickness to find the veins easily. Patients who are difficult to trace the vein called the Difficult Venous Access (DVA)[1]. Patient with DVA can be caused by fatness or young body likes baby and toddler.

The study conducted by The Infusion Nurses Society (INS) in the United States that the installation of the needle on the blood vessels is twice the maximum one nurse for every pair, if it is twice wrong it must call an experienced nurse to pair. The tools used to assist the performance of nurses is called vein finder. Vein finder prepared on lights of Light Emitting Diode (LED) with red colour that has a wavelength of 620 nm [2]. The working principle used by vein finder is by firing red light rays of LED with a wavelength of 620 nm on a patient's arm and then if there are blood vessels, the blood vessels will have a dark red colour or tending to black, while for other networks will remain red like colour LED. Research conducted by the vein finder [3], which focuses on the detection of the baby's blood vessels ranging in age from one month to two years. Resulted in a prototype tool using four pairs of super bright red LEDs with a wavelength of 620 nm and using the battery as a voltage source. However, the tool in this study is only able to detect the veins in patients aged one month to two years. Ade Pajar Pirdioanto (2013) developed a detection tool using LED type super flux red colour that is believed to be brighter than LED super bright and fuzzy logic programming given in order to conserve battery power. This method is used in toddler. Based on several weaknesses in previous researches, the study of the vein finder needs to be developed further.

Therefore, researchers design the vein finder to detect the veins which can be set according to the wavelength of the patient's condition. Veins detector with controllable wavelengths that can be set in accordance with the patient's condition will be very useful. Wavelength used by the colour of the LED output is generated. Vein Finder that is designed is also expected to be able to be used for all ages of patients. So it would be a useful tool used by medical personnel in hospitals, clinics or health centres in order to facilitate the installation of a needle on a vein, and also to be an alternative choice and technology development of vein finder.

2. Theory

2.1. The light characteristic on skin

The skin can be penetrated by light at specific wavelengths. The light is divided into three criteria in the breakout. Three criteria in the light of which light with a wavelength of 200-400 nm can only penetrate the epidermis skin layer, light with a wavelength of 400-600 nm can penetrate the dermis layer of the skin, whereas light with a wavelengths of 600-700 nm can penetrate the skin subcutaneous tissue (Figure 1). Properties of the skin fired with the light reflect, scatter and absorb it.

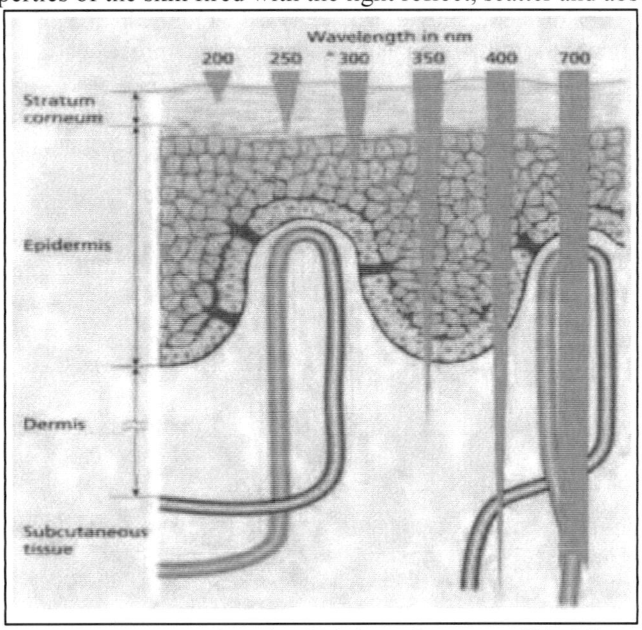

Figure 1. Wavelength of Light to Penetrate Skin [4]

When the wavelength of visible light is 400-700 nm fired onto skin then some light will be absorbed, reflected and diffused into other parts of the skin. If the light wavelength range of 400-600 nm value then the light is merely distributed on the dermis only, whereas if the light wavelength of 600-700 nm, the light will spread and reach the subcutaneous layer of the skin in which there are blood vessels. Absorption, reflection and light transmission were fired can be seen in Figure 2. Colours absorbed by the skin will be forwarded to the subcutaneous layer where a vein is located. Vena will look black on the surface of the skin when exposed to light at these wavelengths. This is because haemoglobin in the blood absorbs the light so that the veins will be looks dark.

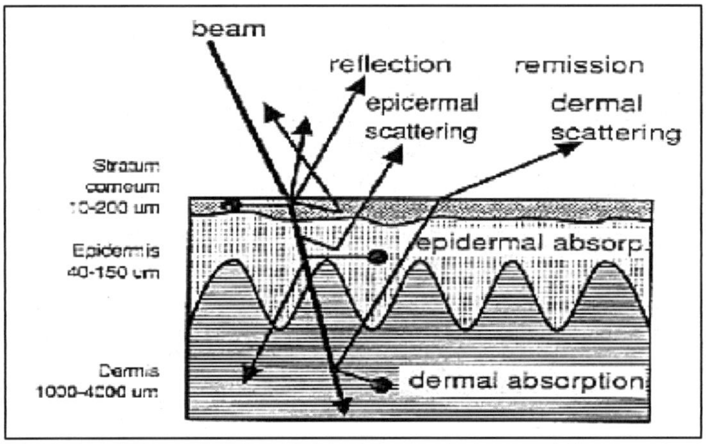

Figure 2. The Light Characteristic On Skin

2.2. Light absorption rate on haemoglobin

In humans, there are approximately 4-6 liters of blood in which the blood is drawn up by 52-62% and the rest of the blood plasma of blood cells. Blood cells have their respective functions. Red blood cell as the largest content in blood cell has a function to carry oxygen to all parts of the body. To do this, the red blood cells have a component that is able to bind oxygen is haemoglobin. When the blood flows from the heart to the lungs, the haemoglobin binds oxygen going to be continued throughout the body (HbO2), while the return from the whole body to the heart, the haemoglobin is without oxygen (Hb).

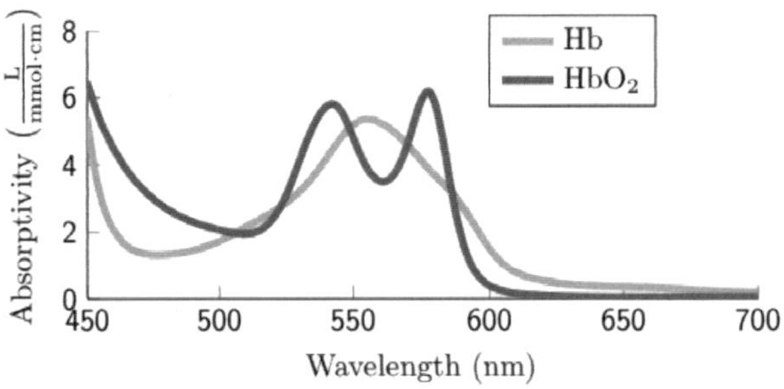

Figure 3. The absorption spectra of light in Hb and HbO2 [5]

In Figure 3, it can be seen the light spectrum absorption rate of haemoglobin carried by Zijlstra (2000) [14]. The picture show that the peak absorption of Hb and HbO_2 is at wavelengths of 530 and 590 nanometers in a range of colours of green and yellow. These colours will be tested on heart rate calculation process to find the colour that best represents the heart rate.

2.3. Light Emitting Diode (LED)

LEDs can also be arranged for the colour output. These LEDs is called LED RGB (Red, Green, and Blue). RGB LED is designed to regulate the emergence of LED colour liking and is useful to use it because the LEDs can bring one of three colours, RGB LEDs have 3 basic colours used as a benchmark universal colour (primary colours). RGB LED is shown Figure 4.

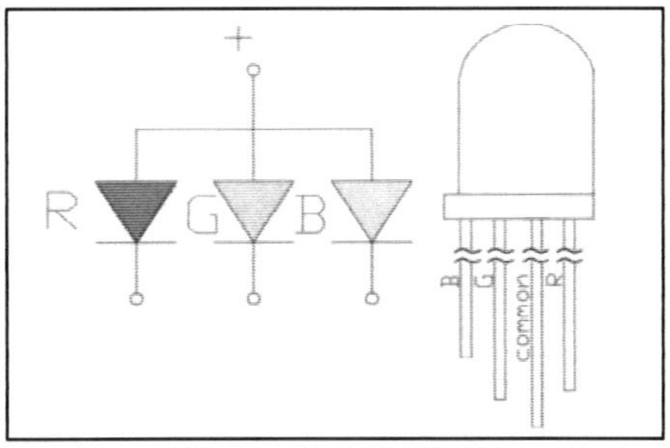

Figure 4. RGB LED

The colours on the RGB LED can be changed based on the code numbers that are universal throughout the world. When one component has a stronger intensity, colour hue is close to primary (red, green, or blue), and when the two strongest components have the same intensity, the colour is the colour tone of secondary (cyan, magenta or yellow). Secondary colour formed by the sum of two primary colours has an equal intensity: cyan is green + blue, magenta red + blue, yellow and red + green. Each secondary colour is the complement of the primary colours. The combination of the three colours red, green, and blue in the right intensity will produce white. Their three component colours in RGB LED's as well as a wide variety of colour combination, allowing us to try a wide variety of colours, wavelength, and power in the vein finder that is designed.

2.4. RGB colour model

RGB colour model is based on the concept of adding strong primary light that is Red, Green and Blue. No signal light waves are absorbed by our eyes or RGB (0, 0, 0). RGB colour wheels is shown Figure 5.

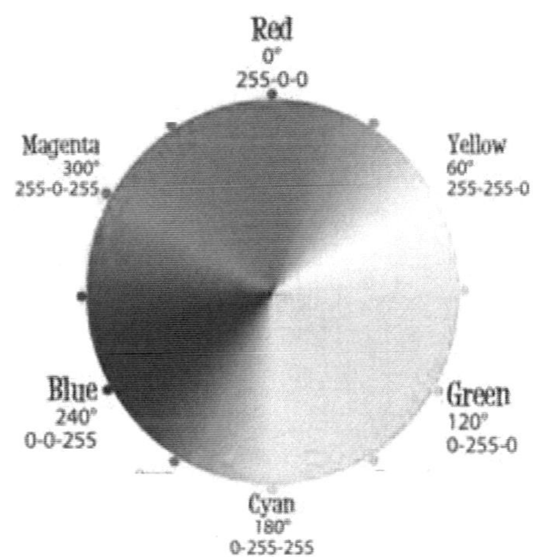

Figure 5. RGB Colour wheels[6]

If we add the red light on the room, then the room will turn red for example RGB (255, 0, 0), all the objects in the room can only be seen as red. Enlightenment and the influence of light intensity in the RGB colour space can be seen in Figure 6.

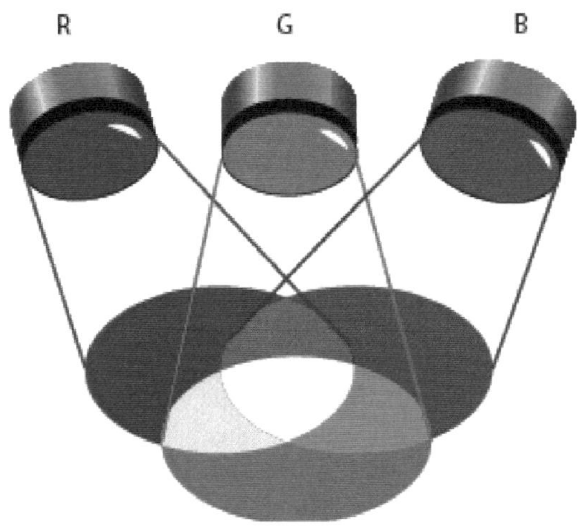

Figure 6.RGBLightness[6]

2.5. Arduino
Arduino downloader software version 1.5, used in this study is an open source electronic kits designed specifically for enabling everyone to learn to develop electronic devices that can interact with a variety of sensors and controllers. The language used in Arduino programming language is C / C ++. This software is used to program the LED lights that are in accordance with the results of the configuration specified by the user.

3. Methods

3.1. Equipment design
This research was performed by the combination of the two devices, namely hardware and software. Hardware is all electronic components used in the system. Designing appropriately for each of these components is very supportive of the design performance tool in this research. The design tool is preceded by making the hardware first. Consisting of LED driver circuit, LCD driver circuit and power supply. LED driver circuit is used as the activation function of the LED because this design uses RGB LED strip. RGB LED stands for Light Emitting Diode containing an array of colours Red, Green, and Blue. Power supply used is the battery. Circuit of RGB LED driver is shown on Figure 7.

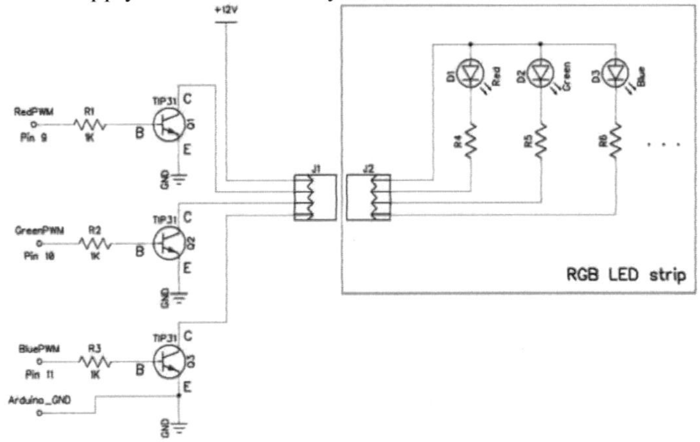

Figure 7. Circuit of RGB LED Driver

Driver circuit using three transistors that act as an amplifier of each LED is represented by three feet of the RGB LED strip. Each leg on the foot base transistor is given by 1 kΩ resistor and used as input to the PWM pin on Arduino. Designing software is used as a regulator of the RGB LED colour that will be issued. The software used is 1.5 arduino IDE. IDE 1.5 is used as a writing arduino program in order to execute commands as well as the settings on the LCD display in accordance with the desired wavelengths. LED colour settings combined between each value of PWM. Then to make a bright red colour is required to adjust the other colours so that the red colour became dominant because the LED RGB LEDs have three primary colours of red, green and blue. Programming was written in IDE 1.5 software, then the program was downloaded into the microcontroller on the Arduino. Flowchart of program is shown in Figure 8.

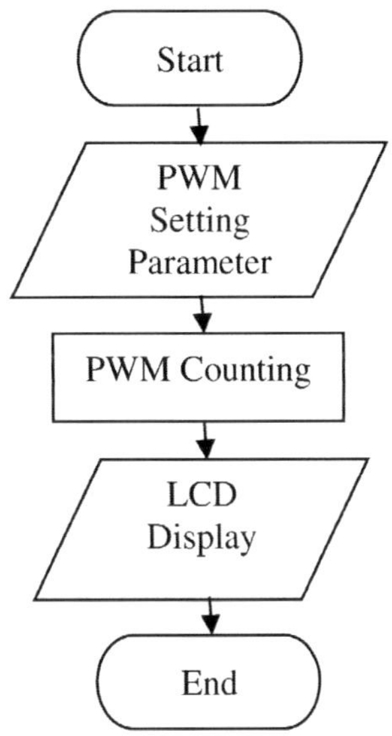

Figure 8. Flowchart

3.2. Testing

The testing phase is used as a benchmark of the performance of a series designed hardware running appropriately or not appropriately. This test includes testing driver of RGB LED. Testing the value of the wavelength produced by RGB LED according to the reach of the needs was done by using a spectrometer. Testing of Wavelength of RGB LED will be varied based on the colour on each of its components measured in wavelengths. In this test the width of the grid is set at 0.02 mm. This wavelength will be calculated using the equation (1).

$$d \sin \theta = n\lambda \qquad (1)$$

In testing of LED power phase, powermeter used in calculating the output power LED. Lit LED was to be measured in advance and placed parallel to powermeter. LED would then be moved away from the position powermeter with the displacement range of 0.5 cm. Data acquisition phase is used as an assessment of whether the design of the tool run well or not. Data retrieval was used in two circumstances, namely:

1. Colour LED wavelength was varied and then measured using a spectrometer.
2. LED output power was varied based on distance using powermeter.
3. Patients were varied then design tool used to detect a vein in the arm.

Data analysis was carried out in a visual way that is viewed by the eye sighting of the difference results from the design data retrieval tool designed. Patient used varied based on age, BMI scale, as well as skin colour.

4. Results

4.1. Wavelength testing
This research has successfully designed the vein finder wavelength that can be set as a detection tool of intravenous (IV). This device is designed to use multiple components and circuits including the power supply circuit, potentiometer circuit, LED circuit, LCD shield and Arduino Uno microcontroller.

Table 1. Wavelength Testing

No.	Colour	LED Configuration	Measured Wavelength (nm)
1	Red 1	(255,0,0)	696
2	Red 2	(245,0,0)	693
3	Red 3	(235,0,0)	682
4	Red 4	(225,0,0)	660
5.	Red 5	(215,0,0)	663
6	Red 6	(205,0,0)	647
7	Green	(0,255,0)	516
8	Blue	(0,0,255)	490
9	Orange	(255,30,0)	604
10	Orange 1	(255,60,0)	569
11	Orange 2	(255,90,0)	599
12	Orange 3	(255,120,0)	606
13	Yellow 1	(255,210,0)	591
14	Yellow 2	(255,255,0)	567
15	Magenta	(255,0,255)	524

Wavelength measurements were performed based on each order of colours seen in the telescope. As an example in Table 1, Red 1 configuration (255, 0, 0) wavelength measurements carried out as in general. But in colour measurement configuration with double or triple like orange with the configuration (255, 30, 0), measured wavelength was obtained from the average of each wavelength corresponding colour components.

According to Hossein (2014) in his article about the visible light spectroscopy, there is a range at a wavelength of a specific colour that is presented on the colour wheel. According to this colour wheel, red colour is contained in the wavelength range 630-750 nm. This is according to a test conducted on six types of red colour with a value between 205-255 bit configuration in which at least 205 configuration values have a range of wavelengths at 647 nm, while most configuration values of 255 have a wavelength of 696 nm. The other primary colours that were tested are green or blue. On the colour wheel the green wavelength is in the range of 480-560 nm, while the blue is in the range 430-480 nm. On spectrometry measurements , it was obtained that green colour wavelength had the range at 516 nm. This is in accordance with existing literature. As for the blue colour, it had a range of wavelengths at 490 nm. This is slightly different from the existing literature in which the blue colour is in the range 430-480 nm. This error can occur due to several factors including the vision is not focused during the measurement. It can be caused by human error, so that both the calculation of the angle and the light beam on the telescope sight errors may occur.

At the secondary colour measurements or merger of two primary colours. Measurement was done in the wavelength of the orange, yellow, and magenta. Orange colour is derived from mixing red and green. Orange colour in the colour wheel was contained in the wavelength range of 590-630 nm. This is in accordance with the measurements made. Where in orange 3 with the highest level of green colour on the configuration (255, 120, 0) got a wavelength in the range of 606 nm whereas the orange with the lowest level of the green colour on the configuration (255, 30, 0) got a wavelength in the range of 604 nm. In the yellow colour is also obtained from mixing red and green, a colour wheel according to the wavelengths are in the range of 560-590 nm. This is in accordance with the measurements made in this research.

4.2. Results of Output Power LED Tested

In observing the LED circuit, then the measurement of the output power of the LED is done by using a power meter. These measurements were varied by LEDs on the power meter distance. The data obtained are presented in Table 2.

Table 2. Comparison of Power to the Distance to Detect Wavelength

No	Colour	Wavelength (nm)	Distance (cm)	Power (mW)
			0.5	2.145
			1	1.217
1	Red	696	1.5	0.813
			2	0.542
			2.5	0.466
			0.5	2.101
			1	1.243
2	Green	516	1.5	0.873
			2	0.613
			2.5	0.531
			0.5	2.168
			1	1.278
3	Blue	490	1.5	0.840
			2	0.554
			2.5	0.509
			0.5	1.662
			1	1.051
4	Red 2	693	1.5	0.750
			2	0.566
			2.5	0.485
			0.5	1.993
			1	1.117
5	Red 3	682	1.5	0.786
			2	0.540
			2.5	0.496
			0.5	1.963
			1	1.153
6	Red 4	660	1.5	0.749
			2	0.499
			2.5	0.462
			0.5	1.867
			1	1.105
7	Red 5	663	1.5	0.736
			2	0.484
			2.5	0.441
			0.5	1.813
			1	1.082
8	Red 6	647	1.5	0.717
			2	0.469

IOP Conf. Series: Journal of Physics: Conf. Series **853** (2017) 012019 doi:10.1088/1742-6596/853/1/012019

No	Colour	Wavelength (nm)	Distance (cm)	Power (mW)
			2.5	0.431
9	Orange	604	0.5	2.545
			1	1.506
			1.5	1.005
			2	0.694
			2.5	0.598
10	Orange 1	569	0.5	2.8
			1	1.669
			1.5	1.102
			2	0.775
			2.5	0.663
11	Orange 2	599	0.5	3.568
			1	1.667
			1.5	1.207
			2	0.784
			2.5	0.668
12	Orange 3	606	0.5	3.336
			1	2.004
			1.5	1.327
			2	0.929
			2.5	0.792
13	Yellow 1	591	0.5	3.599
			1	2.071
			1.5	1.380
			2	0.878
			2.5	0.817
14	Yellow 2	567	0.5	3.153
			1	1.853
			1.5	1.233
			2	0.861
			2.5	0.736
15	Magenta	524	0.5	4.458
			1	2.566
			1.5	1.668
			2	1.154
			2.5	0.988

From these measurements, it can be seen that the greater the distance from the LED power meter, the smaller the power also arrested. For example, the red one with a distance of 0.5 cm, LEDs was capable of emitting a power of 2,145 mW, while at a distance of 2.5 cm, the LED was only capable of emitting power of 0.466 mW.

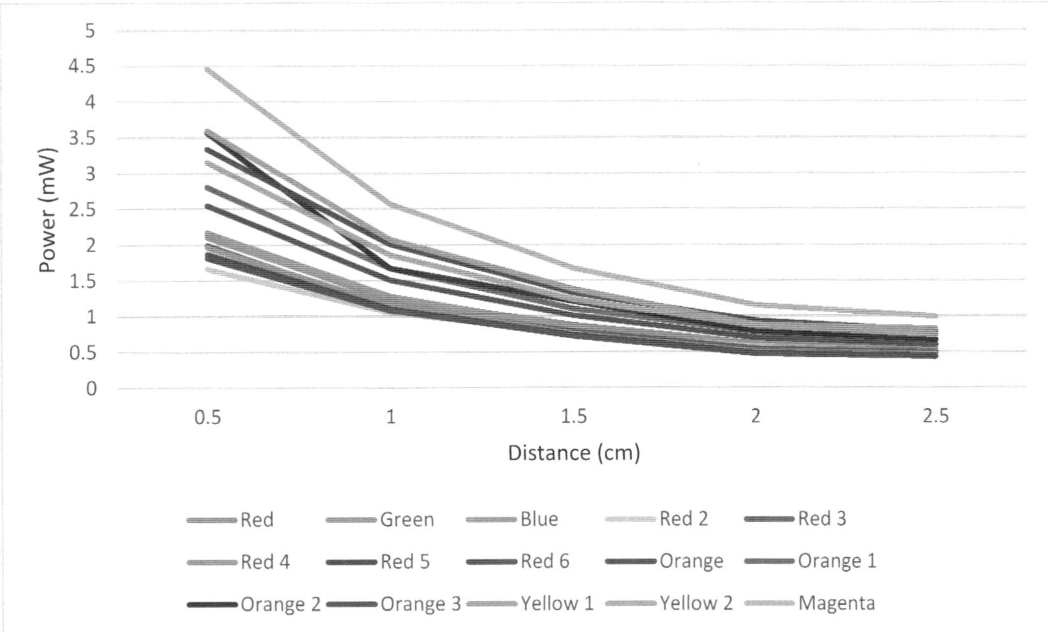

Figure 9. Graph of Relationship between Distance and Power to Detect Wavelength

As can be seen in Figure 9 of all colours have been tested, the power will be smaller if given the greater distance. This is possible because LEDs used have the properties of light spread, so if a given distance was farther then the power meter could only capture most of the power of the LED. In addition to distance, power measured is also influenced by the number of colours components. For example, the colour magenta, the power obtained at a distance of 0.5 cm is 4.458 mW. This value is much greater than the value of the power produced by the red and blue colour that is at 2.614 mW and 2.168 mW. This is because the magenta is an arrangement of red and blue colour components so as to form a magenta colour needs to be lit red and blue colour components in RGB LEDs. This is what causes the output power is greater than the magenta colour red or blue colour on the RGB LEDs.

4.3. Data analysis

The overall colours and configurations were tested first on one patient. This is done to look at the configuration of LEDs and a range of wavelengths that can penetrate the skin. In the configuration of LEDs (0, 0, 255) in blue and a wavelength of 490 nm, the light emitted by the LED is obtained not able to penetrate the skin of the patient. In the configuration of LEDs (0, 255, 0) with a green colour and a wavelength of 516 nm, it has been obtained image that may represent veins. Arising shadow is also found in each secondary colour configuration with a combination of red and blue as well as red-green. As an example, the configuration of the LED (255, 255, 0) is on yellow 2 and a wavelength of 567. In this configuration, the shadow is much evident when compared with the colour green. On the surface of the skin, red colour looks more dominating than the green colour. In red colour with LED sequential configuration of (205, 0, 0) to (255,0,0) with a wavelength of 647-696nm. The blood vessels will be more clearly seen as increasing the size of the wavelength. In the configuration of LEDs (205, 0, 0), Red 6 still looks a little bit shadow that does not represent the veins. While the configuration of the LED (255, 0, 0), the veins can be seen clearly.

The level of depth of epidermis of the skin in adults is in the range of 0-3 mm, where the veins are in the lowest layer of the epidermis of the skin [7]. This is consistent with the research that has been done. In blue colour with a wavelength of 490 nm, the light can not penetrate through the veins. While the red colour with a wavelength of 647-696 nm, the light that is able to reach the veins with the best result was at a wavelength of 696 nm.

In the combined results of a secondary colour between red and blue as well as red-green, the LED light can penetrate the skin to the surface of the vein. This is possible because on any combination of colours, the configuration is done on the red always at the maximum level of colour (255). Only with the addition of the other colours on the configuration, the appearance of the veins will be slightly reduced or interrupted with other skin components that were detected. After knowing the level of translucency of colour on the skin, vein finder was designed to be tested on 10 people patient with BMI levels and skin colour is different. List of patients tested along with BMI and skin colour respectively is shown on Table 3.

Table 3. Result of Vein Finder Tested on 10 Patients

No	Patient number	Age (years old)	Weight (kg)	Height (cm)	BMI	BMI Information	Skin Colour	Result
1.	1	24	80	164	29.74	Fat	Brown black	Visible
2.	2	25	77	175	25.14	Fat	Yellow brown	Visible
3.	3	23	69	167	24.74	Normal	Brown	Visible
4.	4	23	79	173	26.4	Fat	Black brown	Visible
5.	5	24	75	155	31.22	Level 1 Obesity	Yellow brown	Visible
6.	6	23	95	178	29.98	Fat	White red	Visible
7.	7	26	71	165	26.08	Fat	Brown	Visible
8.	8	4	18	75	32	Level 1 Obesity	Brown	Visible
9.	9	63	75	161	28.93	Fat	Brown	Visible
10.	10	40	65	168	23.03	Normal	White	Visible

From the results of tests on 10 patients, it is obtained that the whole patient visible blood vessels veins with BMI levels were different from each patient or a different skin colour. As an example, patient number 4, 23 years old with a BMI level of 26.4 is classified as fat. Figure 10(a) shows the results of the current documentation of measurement. Figure 10(b) is the results of the binary image processing software. The binary image here is intended to clarify the results of the appearance of the veins on documentation.

a b

Figure 10. Appearance of Veins of Patient Number 4 in (a) The Colour Image (b) Binary Image

5. Conclusions

5.1. Conclusions
Based on these results we can conclude that:
the vein finder was successfully designed with controllable wavelength in the range of 600- 696 nm using RGB LED. Vein finder was able to be used to detect veins of 10 patients with different age, IMT and skin colour.

5.2. Suggestions
In designing of vein finder, LED circuit should add a lens to focusing the light. A stable and longer lifetime power supply is also a need for further research.

References
[1] Luminetx 2008 Innovation in Vascular Access Near-Infrared Technology Revolutionizes Visualization of Blood vessels Luminetx Technologies Corporation
[2] Jameco 2008 Vein finder part no.2190080 www.jameco.com
[3] Harianto F 2011 Rancang Bangun Alat Deteksi Pembuluh Darah pada Bayi UMY Universitas Muhammadiyah Yogyakarta
[4] Moreira C 2014 Application of High Brightness LEDs in the Human Tissue and Its Therapeutic Response Federal University of Santa Maria UFSM Brazil.
[5] Jonus N J, etc 2014 Camera Based Heart Rate Monitoring. Department of Applied Mathematics and Computer Science Technical University of Denmark
[6] Adrian F 1998 Colour Space Conversions University of Westminster United Kingdom.
[7] Ganesh S 2007 Depth and size limits for the visibility of veins using the vein viewer imaging system
[8] Bashkatov A N 2005 Optical Properties of Human Skin, Subcutaneous and Mucous Tissues in the Wavelength Range from 400 to 2000 nm *Institute of Optics and Biophotonics* Saratov State University Russia.
[9] Derrickson, Tortora 2009 *Principles of anatomy and physcology 9th edition*
[10] Djuanda, Adhi, dkk 2007 *Ilmu Penyakit Kulit dan Kelamin* Jakarta Balai Penerbit FKUI.
[11] Hossein M, et. al 2014 *Visible and Near Infrared Absorption Spectra of Human and Animal Haemoglobin Determination and Application* Taylor & Francis.
[12] Pirdianto, P Ade 2013 Rancang Bangun Alat Deteksi Pembuluh Darah Vena pada Balita dengan Smart Energy Berbasis Fuzzy Logic *Jurnal UMY* Universitas Muhammadiyah Yogyakarta Yogyakarta
[13] Varma, Ravi N 2014 *Infrared VeinViewer* Nandurkar college of engineering and Technology Yavatmal India
[14] Zijlstra W G, et al 2000 *Visible and Near Infrared Absorption Spectra of Human and Animal Haemoglobin Determination and Application* Taylor & Francis.

IOP Conf. Series: Journal of Physics: Conf. Series **853** (2017) 012020 doi:10.1088/1742-6596/853/1/012020

Optimization image of magnetic resonance imaging (MRI) T2 fast spin echo (FSE) with variation echo train length (ETL) on the rupture tendon achilles case

Akhmad Muzamil[1, 3,*] and Achmad Haries Firmansyah[2]

[1] Department of Biomedical Engineering Post Graduate School, Airlangga University
[2]Department of Physics, Faculty of Science and Technology, Airlangga University
[3]Haji Hospital of Surabaya

[*]Corresponding email address: muzamilakhmad@gmail.com

Abstract. The research was done the optimization image of Magnetic Resonance Imaging (MRI) T2 Fast Spin Echo (FSE) with variation Echo Train Length (ETL) on the Rupture Tendon Achilles case. This study aims to find the variations Echo Train Length (ETL) from the results of ankle's MRI image and find out how the value of Echo Train Length (ETL) works on the MRI ankle to produce optimal image. In this research, the used ETL variations were 12 and 20 with the interval 2 on weighting T2 FSE sagittal. The study obtained the influence of Echo Train Length (ETL) on the quality of ankle MRI image sagittal using T2 FSE weighting and analyzed in 25 images of five patients. The data analysis has done quantitatively with the Region of Interest (ROI) directly on computer MRI image planes which conducted statistical tests Signal to Noise Ratio (SNR) and Contras to Noise Ratio (CNR). The Signal to Noise Ratio (SNR) was the highest finding on fat tissue, while the Contras to Noise Ratio (CNR) on the Tendon-Fat tissue with ETL 12 found in two patients. The statistics test showed the significant SNR value of the 0.007 ($p<0.05$) of Tendon tissue, 0.364 ($p>0.05$) of the Fat, 0.912 ($p>0.05$) of the Fibula, and 0.436 ($p>0.05$) of the Heel Bone. For the contrast to noise ratio (CNR) of the Tendon-FAT tissue was about 0.041 ($p>0.05$). The results of the study showed that ETL variation with T2 FSE sagittal weighting had difference at Tendon tissue and Tendon-Fat tissue for MRI imaging quality. SNR and CNR were an important aspect on imaging optimization process to give the diagnose information.

1. Introduction

Radio diagnostic, a radiological branch, aims to diagnose an abnormality experienced by the patient. The radio diagnostic equipment has a wide range of advanced tools for instance CT-Scan, Magnetic Resonance Imaging (MRI), Mammography, Fluoroscopy, Angiography, and X-ray. MRI is the main imaging modality in evaluating the soft tissue and central nervous system because of its ability to multilane. Besides, it also provides detailed description of the human body with good contrast and differences in anatomical imaging

In order to obtain the quality image of MRI, it has to consider many things related to imaging techniques of MRI. One of them in the calculation the sequences and contrast to suit the purpose of inspection are the determination of the parameters Time Repetition (TR), Time Echo (TE), Echo Train Length (ETL), Number of Excitation (NEX), slice thickness and the other parameters that can affect the results of MRI images [1].

Magnetic resonance imaging (MRI) is a non-ionizing scanning tool [2], which is capable to imaging the organs were examined based on the principle of magnetic resonance of atomic nuclei of hydrogen by using a radio frequency signal (RF) and magnetic field at a certain scale [3]

Fast Spin Echo (FSE) is one of the sequence spin echo (SE) with fast scan time. FSE awards pulses with one pulse of 90° followed by a multiple pulse of 180° rephasing within one repetition time (TR). Application of multiple pulses per TR will produce the sequence of echo called Echo Train Length (ETL). Each echo has differences depending phase encode each of the TR and full fill different lines in k-space which equal to the image formation [4].

Determining the value of ETL considerable influences the results of MRI image for base of the patient (cooperative and non-cooperative). Echo train length on weighted T2 FSE can cause a blur image affecting the value of the Signal to Noise Ratio (SNR) and Contras to Noise Ratio (CNR) which considered as one of the weaknesses of the FSE [4].

K-space is a space of propagation frequency where the signal of frequency derived from the patient. K-space will be filled quickly so that the scanning time will be shorter because of ETL effective which consider as the excess of the FSE and it creates a strong signal to liquid and imaged nerve in detail [5].

T2 weighting is very important in showing the image of the structure of the Achilles tendon, especially in sagittal slices compared to conventional SE technique [6-9]

Signal to Noise Ratio (SNR) and Contrast to Noise Ratio (CNR) obtained under the provisions of Bryan, [10] is:

$$SNR = \frac{S_1}{N} \tag{1}$$

Contrast to Noise Ratio (CNR) is:

$$CNR = \frac{S_1 - S_2}{N} \tag{2}$$

Where:

S1 is the first sign that will be measured its region of interests value. It has matched with the anatomy information which will be evaluated.

S2 is the second sign that will be measured its region of interests value. It has matched with the information which will be evaluated.

N is stand for noise that is signal distraction in the ROI measurement on background

2. Method

The research subject was the result of MRI imaging as many as 20 images on each variation Echo Train Length (ETL) on T2 weighted Fast Spin Echo (FSE) sagittal from some patients (at least 4 patients) which performed in the radiology Haji Hospital of Surabaya. Spin Echo T2 weighted image in MRI ankle obtained by several procedures prepared by patients for instance equipment preparation materials, patient positioning, scanning, image acquisition, determining ROI, evaluating the SNR and CNR. Scanning began with the protocol, then the variation ETL assumed as other constant parameters.

Determining Region of Interest (ROI) made directly on the computer of MRI with direct ankle ROI method. Computer MRI signal value determined on Tendon Tissue, Fat, Fibula,

Heelbone and outside the tissue or noise (noise). Selection of these tissues connected to the composition of the most tissue, then obtained the evaluation of SNR and CNR values as well as the time of the examination. Example of determining ROI is presented in Figure 1.

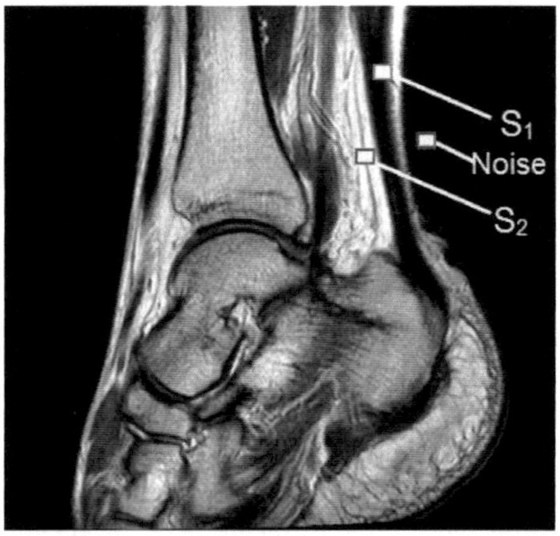

Figure 1. Determination of ROI

3. Analysis

To obtain the average value of the signal that has been processed by Fast Fourier Transform (FFT), in the form of the mean signal of direct ROI on tissue were 1 (tendons), the tissue 2 (FAT), the tissue 3 (fibula), and 4 (heelbone), and noise on the background tissue. In addition, the sending acquired scanning inspection time of each variation of Echo Train Length (ETL). The obtained data was analyzed to determine the value of Echo Train Length (ETL) optimal results and obtain the best image of the optimal scan time. Both of them can be used as the standard in MRI examination diagnosis of Achilles tendon rupture in the hospital who had Radiology.

The quantitative data analysis on ROI method was used to determine the influence of ETL in k-space to image-quality value SNR and CNR significantly on any determined tissue to do ANOVA (Analysis of Variance) [4].

4. Result and discussion

After acquiring the sagittal T2 Fast Spin Echo ankle MRI images of five patients on any variation of ETL, there were more than 20 results matched with the pictures. In scanning process, it was selected a suitable picture project from all pictures. Thus, it showed the identified tissue areas namely tendon tissue, fat, fibula,heelbone (Calcaneus) and background and collected the signal from each tissue to be identified directly with the technique of ROI (Region of Interest) on the computer MRI of each image to analyze quantitatively the value of signal to Noise Ratio (SNR) and the Contras to Noise Ratio (CNR). From the results of the calculations of normality test data and statistical significance test to ANOVA (Analysis of Variance) test, there was a meaningful difference in every variation given to each patient.

4.1. Analysis signal to noise ratio (SNR)

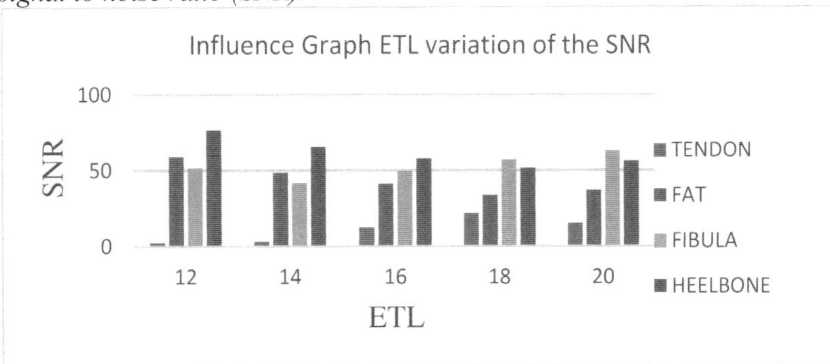

Figure 2. Influence Graph ETL variation of the SNR

From the results of research data showed that the effect of variations in the value of SNR to ETL was unstable or linear as follow variation given produce certain SNR. The decrease of SNR in each tissue can be influenced by the molecular structure of the tissue. If it has the composition of the hydrogen atoms, it will produce a lot of high SNR. The life time factor also affects the results of MRI image due to the composition of the hydrogen atom in each tissue decreases with life time.

In this study, the main focus was fatty tissue for the MRI imaging ankle with weighting T2 Fast Spin Echo. FAT with effective factors ETL sensitive imaging in tissues that contain a lot of liquid. Although the most sensitive was the water, but in this case, ankle MRI water had the lower percentage than the Fat to determine the value of SNR.

The application of ETL which too long caused any fat and hyperintens fluid. It was occurred due to the application of re-phasing pulse Radio Frequency (RF) 180° which reduced the interaction of spin in the FAT. Thus, T2 decay increased and fat would be gray in colour (J-coupling) which is a limitation of sequences of Fast Spin Echo (FSE) [8] (Westbrook, 2002). ETL excessive application caused the amplitude of the signal decreases because the application ETL is quite large with Time Echo (TE) which led increasingly to the phase encode for filling the faster K-Space [9]. As a result, it can reduce the occurred signal amplification but the scanning time will be more fast.

4.2. Analysis contrast to noise ratio (CNR)

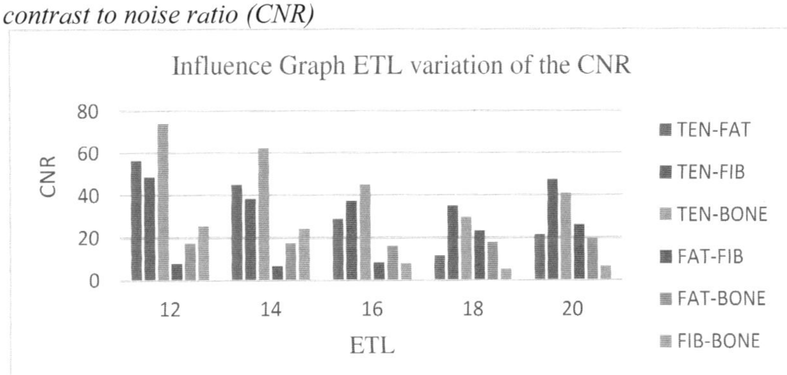

Figure 3. Influence Graph ETL variation of the CNR

The result of this study showed that the effect of variations in the value of CNR ETL was unstable or linear as any variation given to produce the CNR value. Due to distinguish contrast between one tissue to the other tissues, the hydrogen role atom composition affects significantly to the image hyperintens and hypointens. It led the contrast seen clearly. The main focus of the study was the tissue of tendons and fat because these two tissues have a different structure. The tendon has a clay structure with a large of density and it is susceptible to resonant radio frequency as the contribution to image hypointense. Fat structure remained liquid although in small amount so they were sensitive to radio frequency resonance and contributed slightly to the image of the comparison hyperintense.In addition, both tissues can be used for making CNR good value. Application ETL excessive FSE sequences as well as affect the signal and the contrast occurred between other tissues are getting lower

4.3. The relation effect of ETL to SNR and CNR

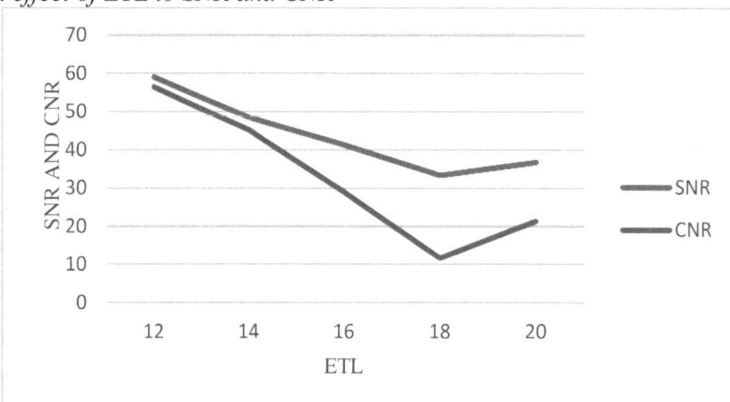

Figure 4. ETL variations influence the of SNR and CNR

Based on Figure 4, it could be concluded that when the value of the variation ETL 12 generates a graph of the increasing in value of SNR and CNR while the variation ETL 14 generates a graph of the decreasing between SNR and CNR values, the graphs can rise back to the ETL 20. This variation can be used as consideration for the technologist to determine the value of ETL variations that will be used for MRI examinations ankle resulting SNR and CNR values were high.

4.4. Effect of variations ETL to the scanning time

The result showed that the scanning process performed on five patients in each variation ETL scanning time were produced relatively the same in a variation ETL lower scanning process of 2:42 minutes. The resulted ETL was quicker scanning but the obtained signal was reduced. The analysis of SNR and CNR was decrease. In accordance with the equations to calculate the time scanning of ETL factors as follows:

$$time\ Scanning = \frac{T_R \times PE \times NEX}{ETL} \quad (3)$$

Example of the calculation of time scanning with TR of 5450 ms, PE is 320, NEX 1 and ETL 12-20 as follows.

IOP Conf. Series: Journal of Physics: Conf. Series **853** (2017) 012020 doi:10.1088/1742-6596/853/1/012020

Table 1.The Effect Of Variation To Scanning Time

ETL Value	*Scanning Time*	Figure
12	2.42 minute	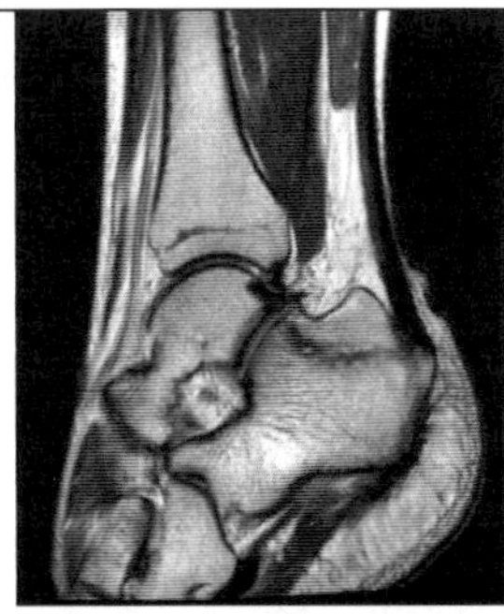
16	1.81 minute	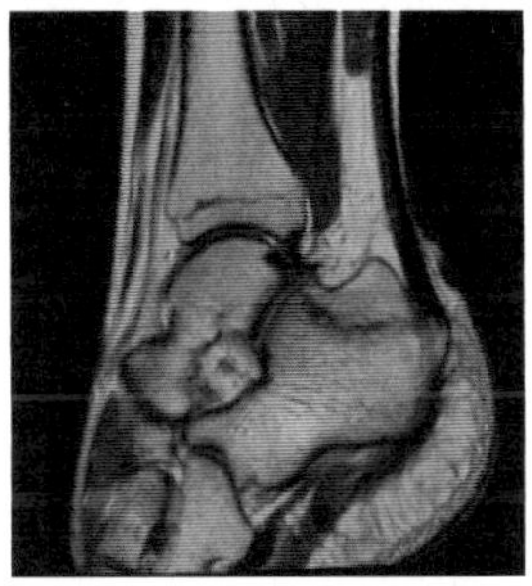
20	1.45 minute	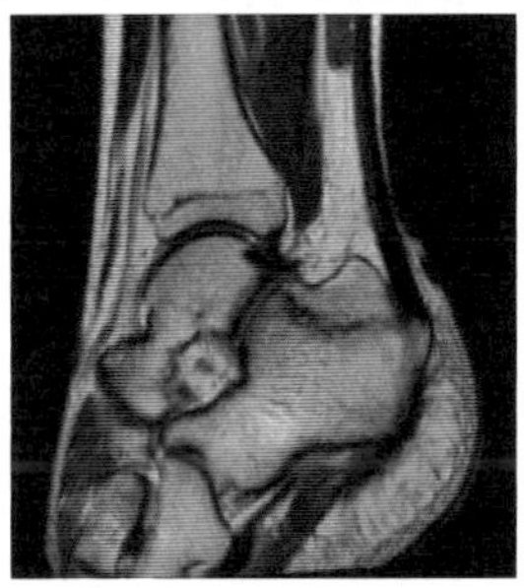
14	2.07 minute	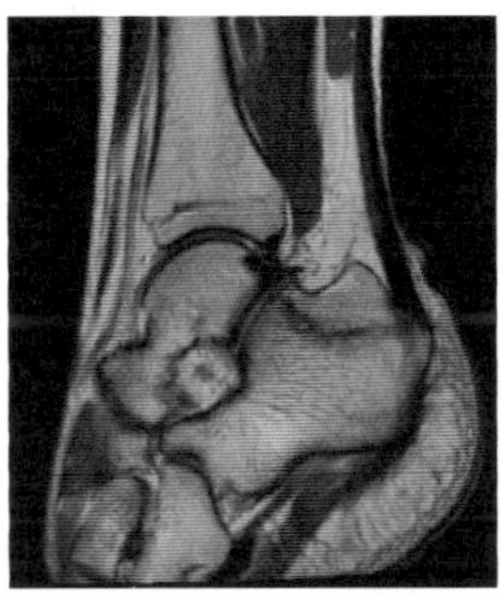
18	1.61 minute	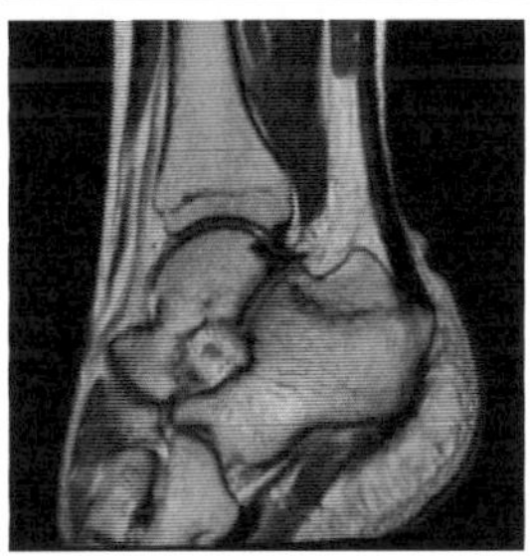

From the analysis, the calculation of scanning time was reduced if the obtained results using ETL large variation value. This is because the process of encoding the K-space would be faster, causing the scanning time would be faster as well, then it can be used as a consideration in the determination of the value of the variation ETL to get the value of the signal and the contrast is quite high.

4.5. Determination of optimal value ETL variations on image quality

The determination of the optimal ETL variations obtained from a graph showing the value of high SNR and CNR. Values optimal SNR and CNR were determined not only by the high or maximum value but also the intersection graph. The results required a proper selection on variations ETL that produce the good value of SNR and CNR. The graphs influence on the value of variations ETL SNR and CNR for five patients are shown in Figure 5 to 9.

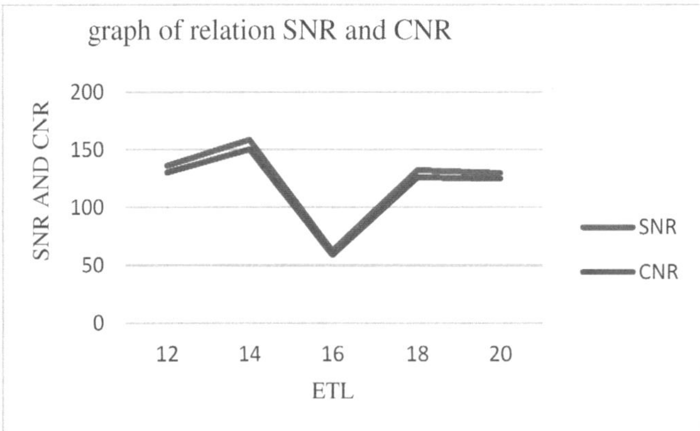

Figure 5. Influence Graph ETL variation of the SNR and CNR Patient 1

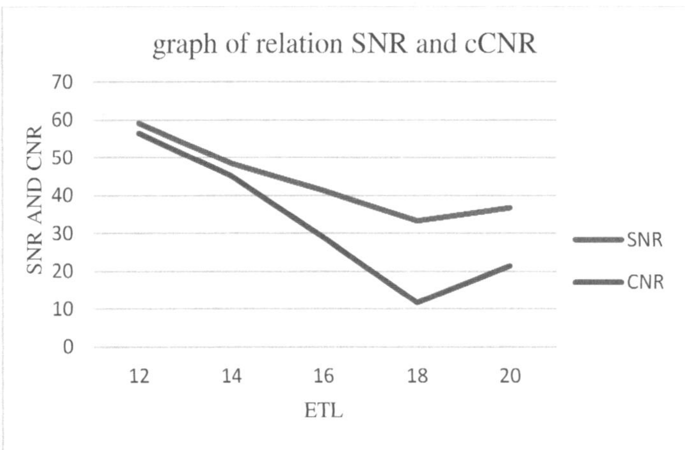

Figure 6. Influence Graph ETL variation of the SNR and CNR Patient 2

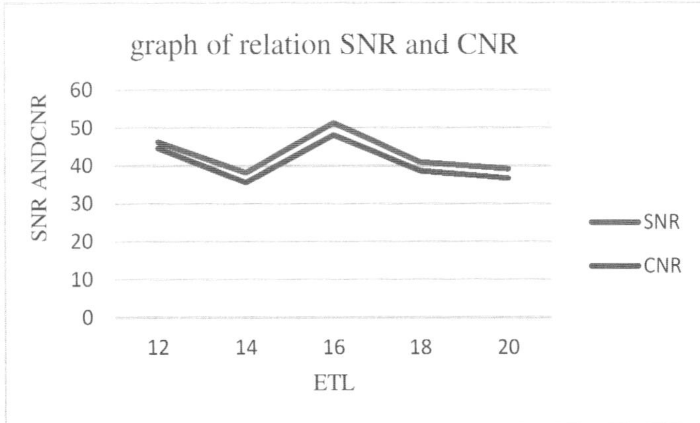

Figure 7. Influence Graph ETL variation of the SNR and CNR Patient 3

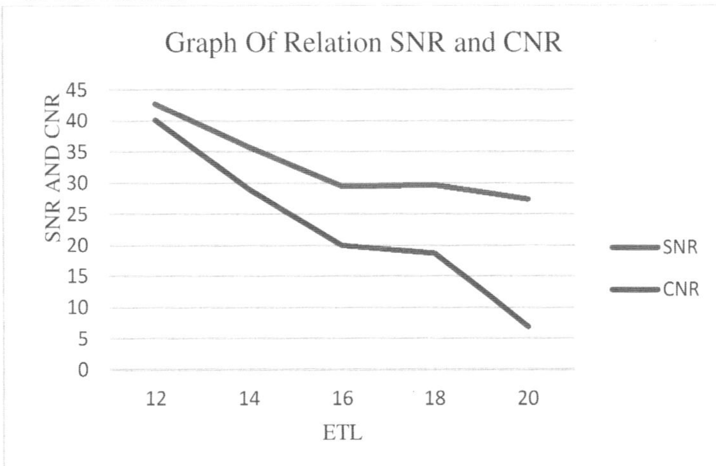

Figure 8. Influence Graph ETL variation of the SNR and CNR Patient 4

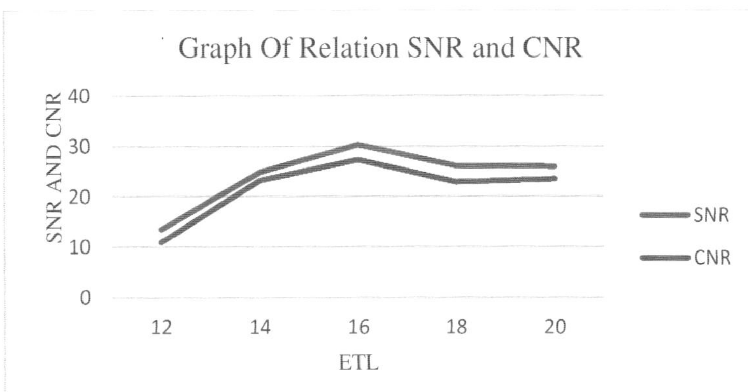

Figure 9. Influence Graph ETL variation of the SNR and CNR Patient 5

From the varies connection ETL to SNR, CNR and the scanning time, it was concluded that the value of ETL 12 units generated the high value of SNR and CNR. It can be used by the technologist to determine the value of ETL to generate the value of SNR and CNR optimal time scanning. In addition, to require analysis of the value of SNR, CNR, and the scanning time, the image processing required as a supporting influence the selection of optimal ETL variation values of the MRI image.

5. Conclusion

Based on the research, it can be concluded as follows:

1. Variation of ETL affected to the results of an MRI examination on Ruptures of Achilles Tendon.
2. Setting of ETL equal to 12 will produce high SNR and high CNR for two patients by scanning time 2:45 minutes.

Acknowledgement

This final assignment has been resolved because of the support of RSU Haji Surabaya, the first Lecturer Advisor, the second Lecture Advisor, the first Testers, the second Testers and all support from classmates. Physics 2012

References

[1] Notosiswono M and Susy S 2004 Utilization of Magnetic Resonance Imaging (MRI) as a means of diagnosis of patients *Media Research Health* **XIV** 3.

[2] Forshult and Stig E 2007 *Magnetic Resonance Imaging*- MRI- An Overview Karlstad university Swedia

[3] Bushbreg J T, J A Siebert, E M Leidholdt and J M Boone 2002 *The Essensia Physic of medical Imaging second edition* second edition Lippincott Williama and Wilkins Philadelphia USA 377-462

[4] Josepa ND Simanjuntak, Muhammad Nur and Eko Hidayanto 2014 *Studi Analisis Echo Train Lenght dalam K-Space serta Pengaruh Terhadap Kualitas Citra Pembobotan T2 FSE pada MRI* 1.5 T. **17** 1.

[5] Westbrook and Catherine 2002 *Handbook of MRI Technique* DP photo setting Lesbury Bucks Printed and Bound in Great Britain, Cambridge.

[6] Carmichael D W, Thomas D L, and Ordidge R J, 2009 Reducing ghosting due to k-space discontinuities in fast spin echo (FSE) imaging by a new combination of k-space ordering and parallel imaging *Journal Magnetic Resonance*, 200,119-125

[7] Maksymowych and Walter P 2007 Magnetic Resonance Imaging for Spondyloarthiritis Avoiding the Minefield *The Journal of Rheumatology* Publishing Company Canada.

[8] Goethem V J W M, Parizel P M and Jinkins J R 2002 *Review Article: MRI of the Postoperative Lumbar Spine,* **44**,723-739.

[9] Houghton and Victor 2006 *Imaging Intervertebral Disc Degeneratio Journal of Bone and Joint Surgery* **88-A**, Supplement 2

[10] Bryan R N 2010 *Introduction to the science of medical Imaging* 67-171 University Press Cambridge New York

Fabrication of prototype for measuring the exhaled breath temperature (EBT) to support detection of asthma

Harnawan A.A [1,*], Mariati , Fahrudin A. , Assegaf A. [2]

[1] Physics Study Program, Universitas Lambung Mangkurat, Achmad Yani Street Km. 35,8, Banjarbaru South Kalimantan, Indonesia
[2] Regional General Hospital Ulin, Jl. Ahmad Yani Km. 1, Banjarmasin South Kalimantan, Indonesia

*corresponding author : adeagungharnawan@unlam.ac.id

Abstract. Recently, EBT has been proven as the marker of airways inflammation like asthma and proposed as the non-invasive tool. Although EBT device has already been made but this device is rarely used for patients in Indonesia. The aim of this study is to develop a prototype which accurately measures EBT and is comfortably used by patients including children. This prototype was made using SHT11 as a sensor of EBT which is integrated on a thermal flask 0.5L. This flask filled up patient breath and the temperature of air breath was measured. The EBT of twelve healthy samples and seven samples with asthma was examined using this prototype, the measurement was done within three minutes for all of them. The test results of EBT on healthy samples obtained the median is 33.9°C within of 33.0°C - 34.7°C and EBT on asthma samples obtained median is 35.0°C within the range 34.9°C - 36.0°C.

1. Introduction

Health is one of the most important parts of life. The level of a person's health can be determined from several aspects, one of which is on the performance of a respirator. Asthma is a chronic disease which attacks the respiratory tract bronchial on our lungs, where there is inflammation bronchial cavity wall resulting in constriction of the airways that ultimately leads a person experience shortness of breath. Paredi et al, in their study found that patients with asthma had a rise in EBT that were faster than normal people[1]. Garcia et al, in their research also found the same thing regarding the relationship between EBT and asthma. The research found that healthy people EBT was 33.2 ± 0.2°C compared to patients with uncontrolled asthma that was 33.7 ± 0.8°C and patients with uncontrolled asthma that was 34.9 ± 0.8°C[2]. Popov et al shows that EBT can distinguish both samples with and without asthma. In their research, the obtained EBT asthmatics are in the range from 34.12 to 36.09 and from 32.29 to 35.84 °C for healthy people [3],[4],[5].

To fulfill the purposes of measurement of EBT, it needs to be made an instrument that is capable of measuring EBT precise and easy to use. This article discusses the prototype of the measuring instrument EBT using SHT11 sensor-based microcontroller ATmega16 as an instrument for supporting the treatment of asthma.

2. Materials and methods

Steps to prepare the prototype for measuring EBT to support detection of asthma can be shown in Figure 1. All steps of fabrication of the system were done in the Laboratory Instrumentation of Physics on Faculty of Mathematics and Natural Sciences, Universitas Lambung Mangkurat. Then, the step testing system to support asthma detection was done in Regional General Hospital Ulinon Banjarmasin, South Kalimantan.

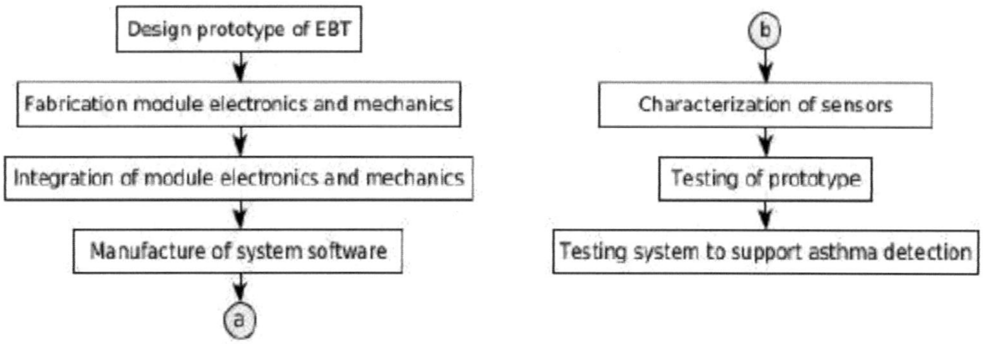

Figure 1 The steps to prepare the prototype..

2.1. Preparation tools and materials

Tools and materials used were electronics equipment, multimeter, personal computer, power supply, microcontroller module minimum system ATMega16 [6], RTC module, the SHT11 sensor with i2c interface [7], thermal flask, RS232 module, LCD 16x2 module, BASCOM, Personal Computer and several of cables.

2.2. Design prototype of EBT

A design of the system can be shown in Figure 2. The figure shows that the EBT is collected on the thermal flask then the EBT is measured with the SHT11 sensor and the data of EBT are displayed by LCD and saved by a personal computer.

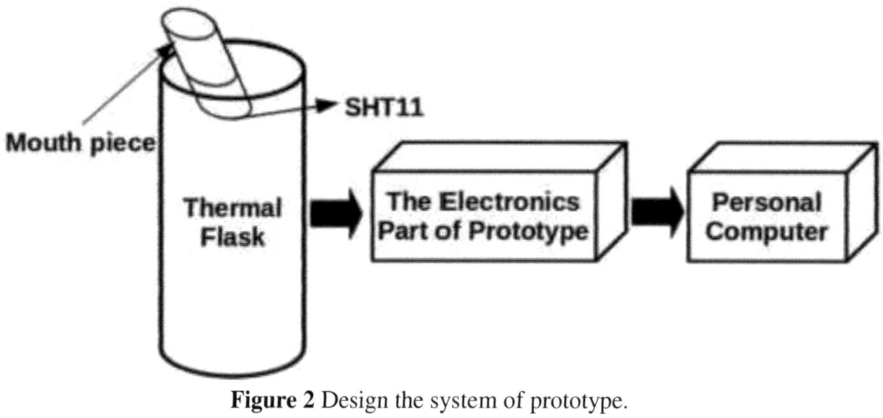

Figure 2 Design the system of prototype.

2.3. Fabrication module electronics and mechanics

The electronics part of prototype consists of sensor SHT11 placed in the thermal flask, module RTC (Real Time Clock), module LCD 2x16 and power supply. The power supply and RTC use the circuit that with microcontroller ATMega16 module. The interface of LCD 2x16 module with microcontroller module using configuration 4 bit. All the modules electronics are assembled and laid on the box.Prototype electronic parts have the following specifications power supply with the output voltage of 5V, 12V and 12V. Power supply 5 V is required to turn SHT11 and minimum system. Minimum system ATmega16 uses 11.059200 MHz crystal.

The form of mechanics module is just the modification of thermal flask used to keep warm the liquid in it. It is equipped with the mouthpiece to blow the air from breathing results.

2.4. Integration of electronics and mechanics module

The integration between the two modules uses a cable that connects the sensor module SHT11 with ATmega16 microcontroller i2c input. The connection of data from the electronics part of a prototype to a personal computer use the RS232, type of digital communication. It was selected to prevent lost data on the cable connected to them because of the length of a cable. The prototype has integrated shown in Figure 3.

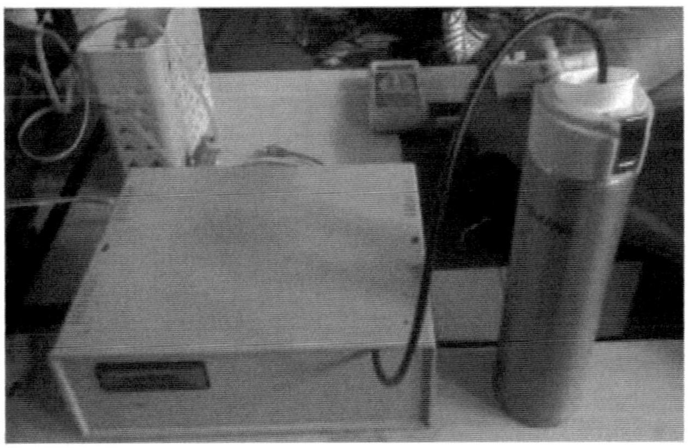

Figure 3 The prototype for measure EBT

2.5. Manufacture of system software

The algorithm embedded with Bascom to the microcontroller ATMega16 is shown in Figure 4. Microcontroller reads time from RTC module, then reads the EBT from sensor SHT11, then displays the value of sensors that reads the temperature of air in the thermal flask at LCD 2x16 module, then sends it to a personal computer.

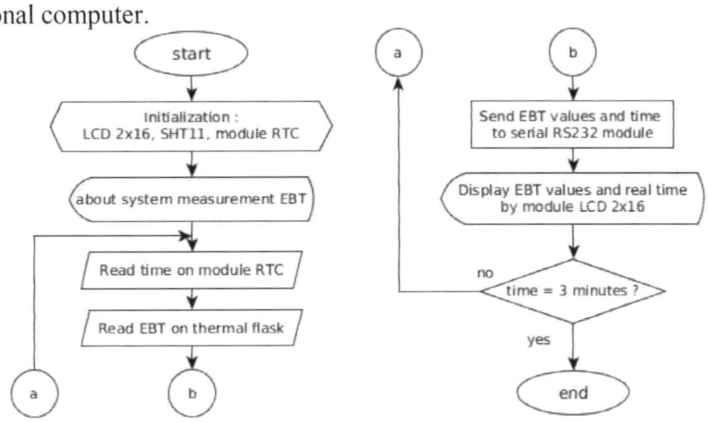

Figure 4 Flowchart embedded to microcontroler ATMega16.

2.6. Characterization of sensor SHT11

The characterization of SHT11 sensor aims to look at the performance of the sensor compared with a standard digital thermometergenerating a correction factor from the prototype to the exhaled breath temperature measured.

Characterization is done by placing SHT11 and the thermometer in the same position in the thermal flask. The hot air is passed through a small tube into a thermal flask. Measurements were performed within a temperature range of 20 - 60°C with a temperature recording every increase of 0.1°C. The result in Figure 5 (a) shows the differences of SHT11 and thermometer readings and calculation. It isobtained that the temperature correction value was 3.5°C. based on the average of the difference between the temperature value by digital thermometer with a temperature value measured by the prototype. This correction added values of SHT11 and the result has been shown in Figure 5 (b).

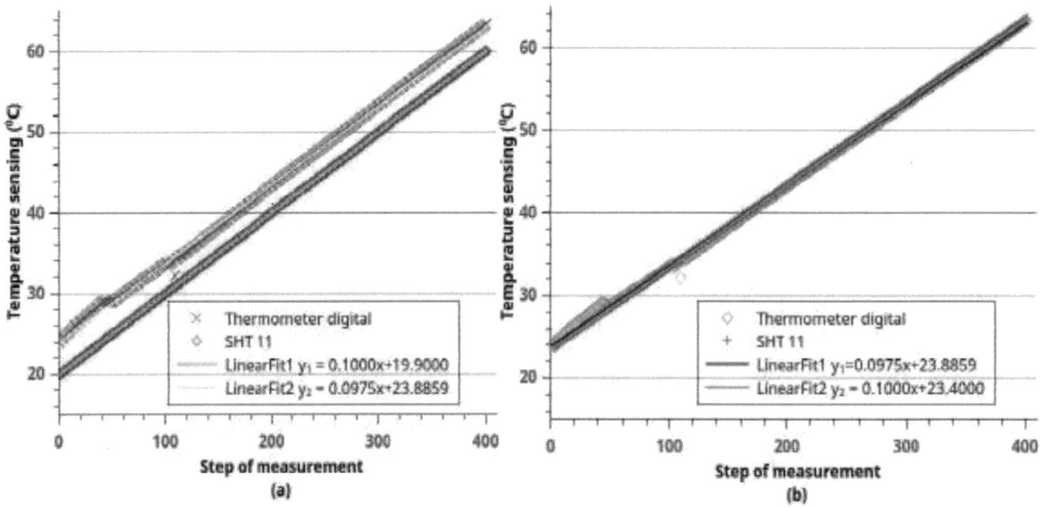

Figure 5 Characterization SHT11

On the results of correction as shown in Figure 5 (b) it was performed the paired t-test to determine whether the presence or absence of the difference in temperature readings by prototypes and digital thermometer. Based on the calculation, t_{value} was -2.02 than from the $t_{distribution\ table}$ for $df = 400$ t for 99% confidence interval of 2.5882, the value of $t < t_{distribution\ table}$ so Ho is accepted. It means that there is no difference between a prototype and a digital thermometer.

2.7. Test of prototype

Tests were performed on 10 samples with repetition of data collection three times to determine the repeatibility of measurement. The test results are shown in Table 1.

Table 1 The performance test of prototype to measure EBT

No	Samples	EBT_1 (°C)	EBT_2 (°C)	EBT_3 (°C)	Average EBT (°C)	δ deviation
1	A	29.4	28.8	29.5	29.23	±0.379
2	B	33.2	33.7	33.4	33.43	±0.252
3	C	32.5	34.5	34.6	33.80	±1.185
4	D	33.6	31.7	32.3	32.53	±0.971
5	E	30.8	33.3	33.2	32.43	±1.415
6	F	33.5	34.8	34.0	34.10	±0.656

7	G	35.0	34.8	35.1	34.96	±0.153
8	H	34.5	35.3	34.4	34.73	±0.493
9	I	34.0	33.8	33.8	33.86	±0.115
10	J	30.3	29.7	28.6	29.53	±0.862

3. Data analysis

The next test of a prototype was conducted on 12 healthy samples and samples of 7 people with asthma. The highest temperature measured was taken as the measured value of EBT. The test results in healthy samples are shown in Table 2 and the results of tests on healthy samples are shown in Table 3.

Table 2. The result of EBT on health samples

No	Samples	EBT (°C)
1	A	34.5
2	B	33.3
3	C	34.7
4	D	33.0
5	E	33.9
6	F	34.7
7	G	33.5
8	H	34.1
9	I	34.0
10	J	31.1
11	K	30.3
12	L	34.3

Table 3. The result of EBT on asthma samples

No	Samples	EBT (°C)
1	M	34.9
2	N	34.9
3	O	36.3
4	P	35.4
5	Q	35.0
6	R	36.0
7	S	35.5

The value of EBT comparison charts of one sample of healthy and asthmatic is shown in Figure 6, starting at the second to 0-180s.

4. Result and Discussion

Table 1 shows that the performance measurement repeatability deviation of EBT was in the range 0.115-1.415, meaning the prototype can measure EBT correctly on samples.

Table 2 and table 3 shows that results of prototype testing on samples of healthy EBT values obtained a range from 30.3°C to 34.7°C and the sample asthmatics EBT values obtained a range from 34.9°C to 36.3°C.

Figure 6 shows a graph EBT difference between normal samples and samples with asthma. EBT for the sample chart with asthma is increasing faster than the normal sample from 0 to 180 seconds. EBT value at 180 seconds is considered a sample value EBT that was recorded.

A thermal flask in this research was used withour the feature to clear the air after use, so they need a few minutes to equalize the initial conditions for the next EBT measurement.

The performance of the prototype needs to be improved to enhance EBT measurement and store data without using a personal computer, so that the necessary storage modules on the prototype so that data can be read by a computer for further analysis. Then it should also be used for the LCD graphic to display data EBT measured directly on the prototype.

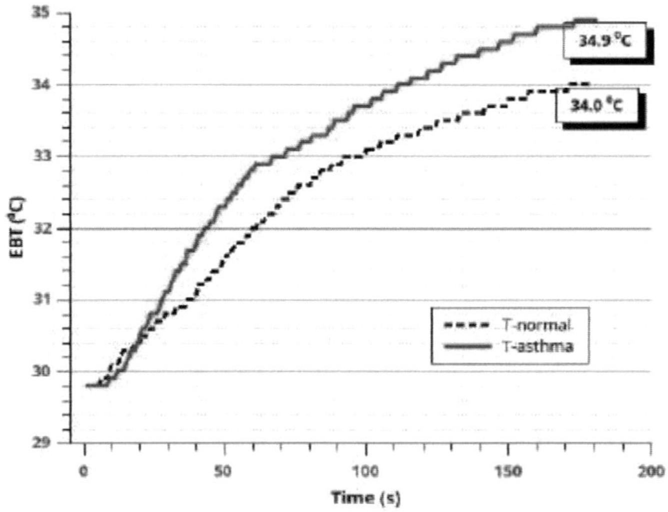

Figure 6 The difference graphics between EBT asthma and normal

5. Conclusion

Based on the results and discussion we concluded that the prototype can measure samples EBT in healthy and asthmatic. Results of prototype testing on samples of healthy EBT values obtained ranges from 30.3 to 34.7°C and the sample asthmatics EBT values obtained ranges from 34.9 to 36.3°C.

6. Acknowledgements

Researchers thank to the head of Instrumentation Laboratory of Physics Faculty of Mathematics and Natural Sciences University Mangkurat as well as all parties involved in this research.

References

[1] Paredi, Sergei A Kharitonov, and Peter J Barnes 2002 Faster rise of exhaled breath temperature in asthma. *American Journal of Respiratory and Critical care Medicine*. **165** : 181 - 184

[2] Garcia,M, M Bergna, E Uribe, A Yahez, and JB Soriano. 2013. *Increased exhaled breath temperature in subjects with uncontrolled asthma* Int J Tuberc Lung Dis 17(7) : 969 – 972

[3] Popov, T A , Dunev S, Kralimarkova T Z, Kraeva S and DuBuske L M. 2007. *Evaluation of a simple, potentionally individual device for exhaled breath temperature measurement*. Respir Med. 101 : 2044 – 2050

[4] Popov, T A T Z Kralimarkova, C T Tzachev, S S Dunev, V D Dimitrov, and J Gill 2010. Development of an individual device for exhaled breath temperature measurement. *IEEE Sensors Journal*. **10** :110 – 113

[5] Popov, T A, T Z Kralimarvo and T D Dimitrov 2012. Measurement of exhaled breath temperature in science and clinical practice. *Breathe*, **Vol 8** no 3, March 2012

[6] Atmel 2010 8-bit Microcontroller with 16K Bytes in System Programmable Flash, ATMega 16. San Jose: Atmel Corporation.

[7] Sensirion 2011 Datasheet SHT1x (SHT10, SHT11, SHT15) Humidity and Temperature sensor IC Switzerland : Sensirion AG

IOP Conf. Series: Journal of Physics: Conf. Series **853** (2017) 012022 doi:10.1088/1742-6596/853/1/012022

Characterization of transimpedance amplifier as optical to electrical converter on designing optical instrumentation

D Hanto* and R K Ula

Research Center for Physics, Indonesian Institute of Sciences, South Tangerang, Indonesia

*Corresponding email address: dwi.hanto@lipi.go.id

Abstract. Optical to electrical converter is the main components for designing of the optical instrumentations. In addition, this component is also used as signal conditioning. This component usually consists of a photo detector and amplifier. In this paper, characteristics of commercial amplifiers from Thorlabs PDA50B-EC has been observed. The experiment was conducted by diode laser with power of -5 dBm and wavelength 1310 nm; the optical attenuator to vary optical power from 0 to 60 dB, optical to electrical converter from Thorlabs Amplifier PDA50B-EC; multimode optical fiber to guide the laser; and digital voltmeter to measure the output of converter. The results of the characterization indicate that each channel amplification has a non-linear correlation between optical and electrical parameter; optical conversion measurement range of 20-23 dB to full scale; and different measurement coverage area. If this converter will be used as a part component of optical instrumentation so it should be adjusted suitably with the optical power source. Then, because of the correlation equation is not linear so calculation to determine the interpretation also should be considered in addition to the transfer function of the optical sensor.

1. Introduction

Recently, developments in Optoelectronics have been growing very rapidly. At the end of this decade, Optoelectronics was not only dominated by optical fiber for telecommunications, but it was also found in other fields such as health equipment[1–3], material processing[4], and measurement instrumentation. As mentioned at the last, i.e. measurement instrumentations field was found by using optical fiber as sensors for weight of vehicles [5–8], structural health monitoring[9], [10], geotechnical monitoring [11], [12], oil and gas monitoring[13], and other strategic necessity. The use of optical fiber as sensors has several advantages such as resistance to electromagnetic interferences, small size and light weight, small power, high sensitivity, wide bandwidth transmission as well as more resistant to the dangerous area.

Besides optical fiber as a well-known chosen sensors in instrumentation-based on optoelectronics, one of the important component in these fields is an optical-electrical converters. These converters can use various kinds that have developed in the marketplace or research. One kind which is well-known in optoelectronics is transimpedance amplifier[14], usually also called a high-impedance amplifier.

This type of converter is quite good in optical instrumentation, especially in terms of sensor readings that is based on the amplitude because it has a small noise and has been proportionally changing [15].

In this paper, measurements using amplifier from PDA50B-EC Amplifier from Thorlabs has been conducted in the laboratory. This amplifier is a commercial amplifier that is available in the marketplace. The purpose of this paper is to determine for characteristics of the amplifier in order to this amplifier that will be used for electrical optic converters in the design of optical instrumentation. On the other hand, these characteristics also can be used for the development of further amplifiers in order to have better characteristics because of rapid development and complex requirement in the field of optoelectronics.

2. Experiment

The main important component of the optical to electrical converter is a photodiode. The photodiode is a semiconductor element that is sensitive to light energy so as to change it in the form of an electric current. When the energy of the light is greater than the energy band gap in the photodiode (ie., hν> = Eg), it excites the electrons at the valence band to the conduction band. This vacancy of electron creates hole resulting in flow of electricity that is called the photocurrent. The magnitude of the photocurrent can be expressed by the following equation:

$$I_p = \frac{P_{in}\,(1-R)e}{h\upsilon}[1 - \exp(-\alpha d)] \tag{1}$$

Where is the photocurrent produced by photodiode, Pin = total optical power entering photodiode, R = fresnell reflection coefficient at the air-semiconductor, e is electron charge, h = Plank's constant, υ = frequency of the incoming light, α = absorption coefficient of the semiconductor at the incident wavelength, and d is the width of the absorption region[16].

A photodiode can be operated in one of two modes: photovoltaic (zero-bias) or photoconductive (reverse bias). The photovoltaic method can be seen at Figure 1. The characteristic of a current source is that its voltage must be determined by other elements in the circuit. Thus, as shown in Figure 1, if the photodiode is connected to a load resistance of RL, the output voltage of the circuit is given by:

$$V = P\rho R_L \tag{2}$$

Where V is the output voltage, ρ is responsitivity of photodiode in the selected wavelength of optical source, and RL is resistance of load in this circuit. This equation is valid as long as V < VS. Figure 1 shows the simplest way to bias a PIN diode to convert the optical signal into a voltage signal.

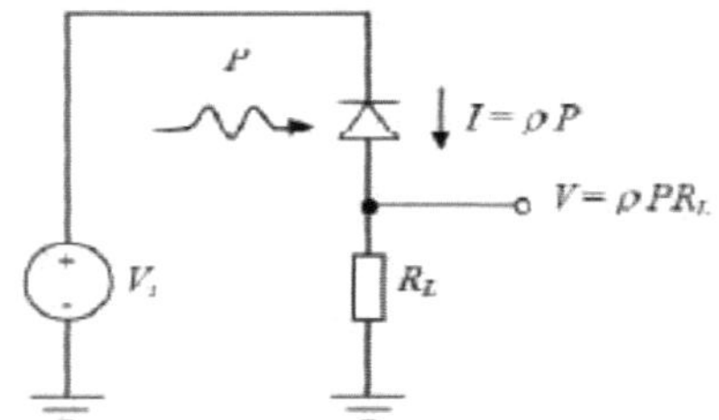

Figure 1. Photovoltaic mode of optical to electrical converter [17]

The other one is photoconductive mode using operational amplifier for the optical to electrical converting like the Figure 2. The measured output current is linearly proportional to the input optical power. The photodiode is reversed biased, thus improving the bandwidth while lowering the junction capacitance. The gain of the detector is dependent on the feedback element (Rf)[17][18].

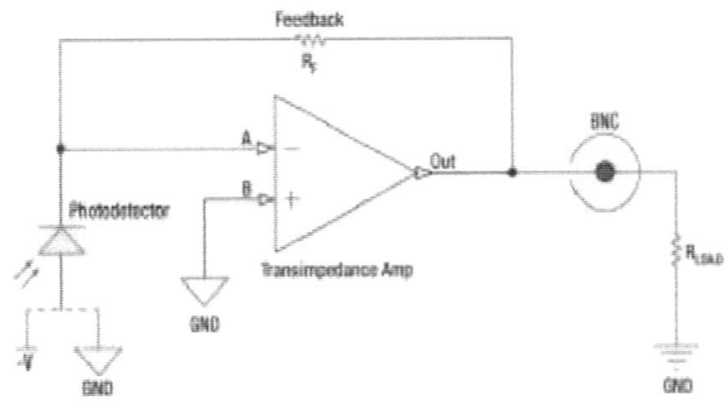

Figure 2. Photoconductive mode of optical to electrical converter [18]

Experiments have been conducted using a laser diode, optical attenuator from Attenuator HP 8156A, Thorlabs Amplifier PDA50B-EC [19], and digital multimeter. Some modules are arranged as shown in Figure 3. The diode laser used to have a wavelength of 1310 nm and an optical power of -5 dBm. The beam of the laser was connected to the multimode optical fiber cable. The variation of the optical power was obtain by pressing to attenuation button from 0 to 60 that has been provided by the this attenuator. Then, the output from attenuator was connected with Thorlabs Amplifier at the photodiode with optical fiber cable too. Measurement results in electrical voltage were measured through a digital multimeter.

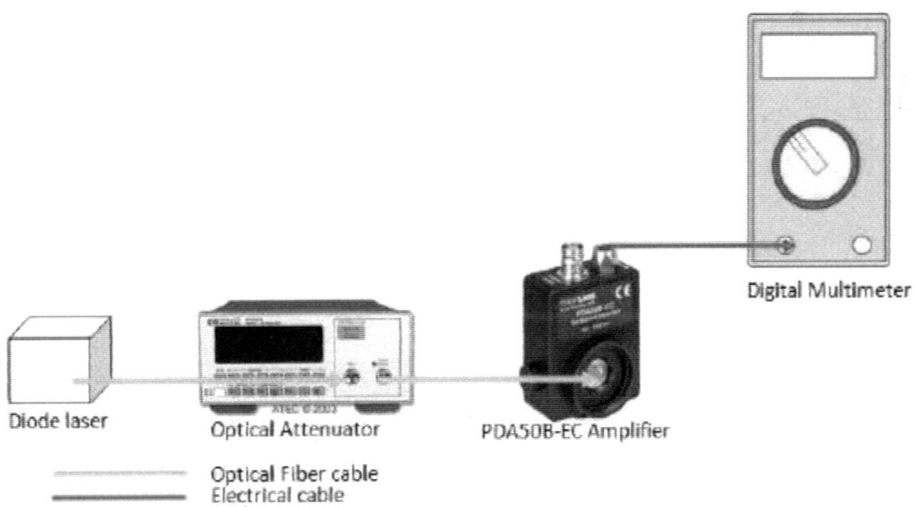

Figure 3. Set up experiment

Thorlabs Amplifier used in this experiment has eight kinds of gain or amplification as shown at the Table 1. These experiments also have used 12 V from DC power supply. Experiments on each gain channel from Amplifier have been done by providing optical power

IOP Conf. Series: Journal of Physics: Conf. Series **853** (2017) 012022 doi:10.1088/1742-6596/853/1/012022

attenuation from 0 to 60 dB. The results of the measurement show in the graph to know characterization of Thorlabs amplifier.

Table 1. Addressing of gain channel for Thorlabs Amplifier PDA50B-EC

Channel	Gain/Amplification (dB)
1	0
2	10
3	20
4	30
5	40
6	50
7	60
8	70

3. Discussion

The experimental results are shown in Figure 4. According to this figure, there are some performance indications of each channel gain/amplification. Generally, all of the correlation between optical and electrical showed non-linearly and approached exponentially. Regarding to Equation 1, relation between optical power and current also produced exponentially. On the other hand, the voltage have linier correlation with photocurrent on the transimpedance amplifier circuit. Thus, it should be a non linear relationship between optical power and voltage like the Figure 4.

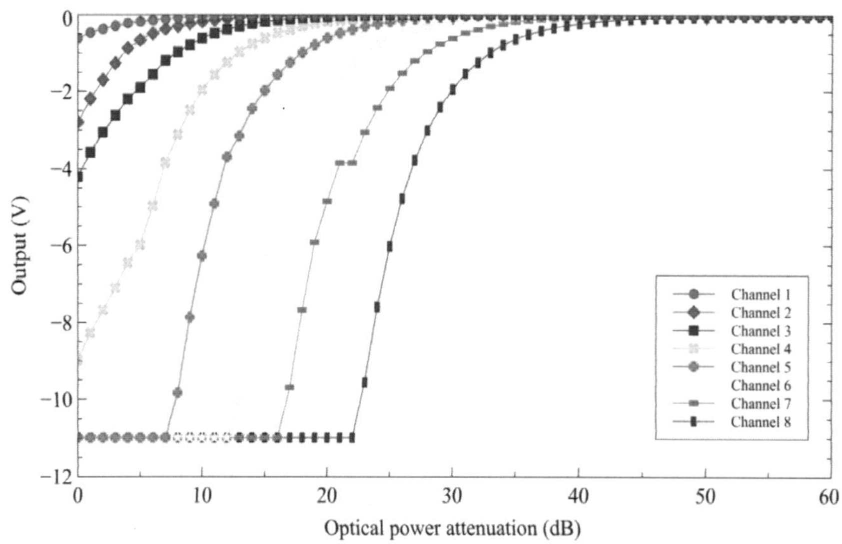

Figure 4. Result of Thorlabs Amplifier PDA50B-EC characterization

Channels 5 to 8 have saturation area, that is closer to the power supply voltage. Consequently, this area causes no conversion from optical power into electrical because of no output changing. At channel 5, saturation occurs when the optical power was attenuated less than 9 dB. Because of the experiments using the diode laser of -5 dBm so it's saturation was obtained around -14 dBm. In addition, when the optical power is greater than -14 dBm , saturation will occur for this channel. In the same way, it will be determined for other channel.

For instance, channel 6 reached out saturation when optical power is higher than -17 dB; channel 7 when reached out higher than -21 dB , and then channel 8 is higher than -27 dB.

Whereas, output of amplifier has reached out around 0 at any conditions. This conditions also cannot be used in the optical instrumentation because it also causes no conversion from optical into electrical parameter. Each channel from Thorlabs Amplifier PDA50B-EC has this area as shown the Table 2.

Table 2. Channel of Thorlabs Amplifier PDA50B-EC causes low output

Channel	Optical power made too low output (dBm)
1	< -13
2	< -18
3	< -23
4	< -27
5	< -33
6	< -37
7	< -43
8	< -50

For optical instrumentation purposes, characteristics of amplifier are very important to be selected. It caused disadvantages including a short-range, sensitivity is too small, and even too big sensitivity if the characteristic is obeyed. In terms of sensitivity and dynamic range, the amplifier has been chosen in area neither saturation nor close to zero output. Thus, if Thorlabs Amplifier PDA50B-EC is used, areas should be noticed as shown as Table 3. In addition, the values shown in the Table 3 are also coverage of readout that can be converted properly. Generally, each channel has a dynamic range of 20 to 23 dB. The channels 1-3 are less than 20 dB because of the limitations of the optical power used in the experiment. Basically, if it is considered 20 dB dynamic range so channel 1 can be used up to 7 dBm, channel 2 up to 2 dBm, and channel 3 can be used up to -2 dBm.

Table 3. Coverage area of Thorlabs Amplifier PDA50B-EC

Channel	Optical power (dBm)	Dynamic range (dB)
1	-13 to -5	8
2	-18 to -5	13
3	-23 to -5	18
4	-27 to -5	22
5	-33 to -12	21
6	-37 to -17	20
7	-43 to -21	22
8	-50 to -27	23

4. Conclusion

In optical instrumentation, characteristic of the amplifier is needed to be used for appropriate purposes. The range output from the optical-based sensors must be adapted to the characteristics of the amplifier as described above. The right choice of amplifier characteristics and output sensor can contribute to the measurement accuracy, especially in the field of optics so it should be adjusted suitably with the optical power source. Then, because the correlation equation is not linear so calculation to determine the interpretation also should be noticed beside of the transfer function of the optical sensor.

Acknowledgments

The work has been supported by Riset Unggulan LIPI 2016. The authors also thank to Lilyanti, O. Jaya, Bryan Saputro, and Nurul from State University of Jakarta for helping this experiments.

References

[1] D Dei, G Grazzini, G Luzi, M Pieraccini, C Atzeni, S Boncinelli, G Camiciottoli, W Castellani, M Marsili, and J Lo Dico 2009 *Sensors* **9** 4 pp. 2574–2585
[2] A Bozkurt, A Rosen, H Rosen, and B Onaral, Jan. 2005 *Biomed Eng.* Online **4** p. 29.
[3] D Tsiokos, G T Kanellos, G Papaioannou, and S Pissadakis 2012 Biomedical Engineering- *Technical Applications in Medicine* pp. 245–268.
[4] T Talako, A Ilyuschenko, and A Letsko, 2009 *KONA Powder Part* J **27** 27 pp. 55–72
[5] L Y Mimbela, J Pate, and S Copeland 2003 *Report: New Mexico State University*
[6] R B Malla, A Sen, and N W Garrick 2008 *Sensors* **8** pp. 2551–2568.
[7] M Bin and Z Xinguo, 2010 Int. Conf. Intell. Comput. Technol. Autom., **1** pp. 458–461
[8] M Lydon, S E Taylor, D Robinson, P Callender, C Doherty, S K T Grattan and E J Obrien 2014 *IEEE Sens J* **14** 12 pp. 4284–4296
[9] A Setiono, R K Ula, D Hanto, B Widiyatmoko and R W Purnamaningsih (2016) 2017 AIP Conference Proceedings 1711 030003 pp. 1–6
[10] W Li, J Liu, J Pan, J Pang and X Lu 2014 *Photonic Sensors* **4** 3 pp. 269–273
[11] T Liu, C Wang, Y Wei, Y Zhao, D Huo, Y Shang, Z Wang, and Y Ning 2009 20th Int. Conf. Opt. Fibre Sensors. Proc. SPIE, **7503**, p. 75030H–75030H–5
[12] Y Yin, W Hongde, Y Gao and X Li May 2010 *Landslides*, **7** pp. 339–349
[13] FFT 2014 *High performance long distance pipeline protection.*
[14] S Yin, P B Ruffin and F T S Yu 2008 *Fiber Optic Sensor* 2nd New York : CRC Press,
[15] R S Quimby 2006 *Photonics and Lasers An Introduction*, 1st ed. New jersey: John Wiley & Sons, Inc.
[16] R P Khare 2004 *Fiber Optics and Optoelectronics*, Oxford University Press.
[17] M Azadeh 2009 Fiber Optics Engineering Springer.
[18] Thorlabs Inc 2016*Photodiode Tutorial* Thorlabs.
[19] Thorlabs Inc 2016 *Ge Transimpedance Amplified Photodetectors,* Thorlabs.

IOP Conf. Series: Journal of Physics: Conf. Series **853** (2017) 012023 doi:10.1088/1742-6596/853/1/012023

Density characterization of radiochromic film through source axis distance (SAD) technique in linac with slab phantom for radiotherapy applications

Yousida Hariani[1*] and Bambang Haris[2]

[1]Physics Department, Science and Technology Faculty, Airlangga University
[2]Radiology Division of Dr. Soetomo Hospital, Surabaya, Indonesia

[*]Corresponding email address: yousida-hariani@fst12.unair.ac.id

Abstract. Characterization of radiochromic film density is accomplished through Source Axis Distance (SAD) technique in a slab phantom Linac with various depths and breadths of field. Type of the film used is gafchromic RTQA2. The dose of radiation exposure of the film may cause changes in the film density. This research aims to determine the relation between the density and the dose depth through the characteristic of curves to identify the depth of the dose and particular breadth of the field as a reference for the dose of radiotherapy patients. The result shows that the higher the dose is absorbed, the darker the film will be, yet the lower the density is obtained. The dose depth is determined by measuring the amount of dose received at various depths and breadths of field using film that is placed on the slab phantom with 6 MV linac radiation and dose of 300 cGy. The variation of the depth at 1.5 cm; 4 cm; 6 cm; 8 cm; 10 cm, the field size at 4x4 cm^2, and the dose depth at 359.7 cGy; 315.3 cGy; 281.4 cGy; 241.2 cGy; 220.5 cGy were settled. The field size 6x6 cm^2 takes the dose depth 354.6 cGy; 314.1 cGy; 282.6 cGy; 244.5 cGy; 224.7 cGy. The field size 8x8 cm^2 takes the dose depth 351.6 cGy; 313 cGy; 283.8 cGy; 247.2 cGy; 228 cGy. The field size 10x10 cm^2 takes the dose depth 348.9 cGy; 342.6 cGy; 248.4 cGy; 249.6 cGy; 231 cGy.

1. Introduction

2014's World Cancer Report (WCR) recorded 12.4 million deaths caused by cancer [1]. Cancer is a disease caused by the abnormal growth of body tissue cells (growing extremely fast and irrepressible; infiltrates and suppresses body tissues thus affecting organs [2]). Cancer is frequently found as cervical cancer, breast cancer, nasopharyngeal cancer, and skin cancer [3]. During this time, the treatments for cancer include surgery, chemotherapy, and radiotherapy. Radiotherapy is a treatment using ionizing radiation, such as *X-rays, gamma rays*, or high-energy electrons [4]. The use of ionizing radiation and radionuclides in cancer treatment is common among the societies. According to *Brazzilian Journal of Physics* in 2009, 50% of cancer treatments used radiotherapy [5]. Treatments with radiotherapy can be divided into two types: *Teletherapy* (External Radiotherapy) and *Brachytherapy*. External Radiotherapy is performed by giving radiation on the surface of the patient body at a certain distance with the best modalities of Linear Accelerator (Linac) with a certain dose. While *Brachytherapy* is performed by inserting the radiation source into the body [6].

Linac can produce photon beam which is used in irradiating cancer [5]. It also produces dose rates that varies in units called Monitor Units (MU) in its applications. This MU is equivalent to one cGy. Determination on the variation of the dose rate should be concerned as it affects biological tissues [6]. The radiation dose given should meet the purposes of radiotherapy which include curative radiation and palliative radiation. Excessive dosage will cause damage to normal tissues that can lead to tissue necrosis. While insufficient dosage in killing the malignant cells will lead to relapse or recurrent [7]. In a previous study [8], by Linac superficial dose variation through Source Surface Distance (SSD) technique, radiation dose is obtained with high accuracy values at the result of the irradiation characteristic curve with 6 MV photon energies above 150 cGy. In another study [8], obtained characteristic of gafchromic RTQA2 film gradient curve tends to decrease after the administration of radiation dose at 150 cGy.

Based on this background, a research must be conducted to measure the dose accuracy of the characteristic curve which can be used as a reference in determining the dose depth of radiation therapy by using a film through Source Axis Distance (SAD) technique in Linac which varies in the depth and field size. Researchers choose the gafchromic RTQA2 film since it is fluorescenceat high energies and resistant to temperatures up to 70^0 C [9]. Average selection of SAD technique is reinforced by research [10]. Department of Radiation Oncology, Virginia Commonwealth University Health System, USA, declares that the technique Source Axis Distance (SAD) which gives IMRT (Intensity Modulated Radiation Therapy) effect in Linac increases the equal distribution of the dose in the target.

Characteristic curve is a graph showing the relationship between the number of exposition and the results of the film density, with exposure doses given by Linac at 300 cGy and irradiation angle at 0 ° with a Slab Phantom, which makes the level of darkness or density of the resulting film visible. The dose absorbed in the film can be read through the level of darkness or film density by using a densitometer. The photon energy used in this study is 6 MV. This number was chosen as the photon energy commonly used in radiotherapy for patients lies in this range. By knowing the depth of the dose and the density of the film, we are able to find the characteristic curve to obtain the accurate dose depth to be used as a reference in radiotherapy.

2. Materials and methods

Materials and tools used in this experiment are:

2.1. *Linear accelerator (linac):*

The specifications of Linac involve date of install April 2013, Model/ tipe 2300 Ix,Serial number 5578, Varian Medical System USA,SSD 100 cm,MLC 120Photon, energy 6 MV, Field size of maximum radiation 4x4 cm^2, 6x6 cm^2, 8x8 cm^2, 10x10 cm^2 on Isocenter position, and depth of target 1,5 cm, 4 cm, 6cm, 8cm, 10cm [1].

2.2. *Densitometer*:

As the film density gauges the absorbed dose in the film; specifications include Model/tipe 10000 XL, Seri number 028872, Scanner Epson and made of Japan in 2013 [2].

2.3. *Gafchromic film*:

RTQA2 Model, International Specialty Products (ISP), Temperature resistance reaches 70 , Field size 14 cm x 17 cm, high spatial resolution intensity of light, can be used in ambient light [2,6].

2.4. *Slab phantom*:

Consists of Water equivalent white polystyrene “RW3” Material for high-energy photon and electron energy, Energy field Photons: Co-60 – 25 MV Electrons: 4 – 25 MeV, Material composition of Polystyrene (C8H8) with an admixture of 2.1% ± 0.2% TiO2, Plate dimension Surface area: 300 x 300 mm ± 0.1 mm Thickness: 1 mm ± 0.05 mm 2 mm ± 0.05 mm 5 mm ± 0.05 mm 10 mm ± 0.05 mm

Adapter plates: 20 mm ± 0.1 mm Mass density: 1.045 g/cm3 (Z/A)r value: 0.536 Electron density (e/g): 3.386 x 1023 Electron concentration (e/cm3) 3.539 x 1023[10].

In this research, the first step is the preparation of Linear Accelerator (LINAC) equipment to organize K-fil therapy consul; Turn-on on a computer consul and wait for time delay. Prepare MLC (Multi-leaf Collimator) with jaw to set the field size with variations of 4x4 cm^2, 6x6 cm^2, 8x8 cm^2 and 10x10 cm^2. Adjust the depth on a slab phantom with variations of 1.5 cm, 4 cm, 6 cm, 8 cm, 10 cm. Warm-up with the selection of energy at 6 MV. And as the last step, on a plane linac, set-up a film source gafchromic RTQA2 on the slab phantom with distance at 90 cm.

On the preparation of RTQA2 gafchromic film, put the film between slab phantom and depth variation with 90 cm's distance above the flat table surface of slab phantom with radiation sources. Slab phantom is placed just below the radiation source with irradiation angle of 0 °. Determination of the film density is a measurement of air dose Linac output absorbed in the film. The film is irradiated with the plane linac. After the film is exposed to radiation, the film color will change. This change in color indicates the magnitude of the dose absorbed by the film. The dose absorbed in the film is measured by using a densitometer. This treatment is implemented with a variation of field size and depth on energy 6 MV of Linear Accelerator. The next step is to take the TMR (Tissue Maximum Ratio) data of TPS (Treatment Planning System) computer produced by isodoses curve obtained from CT-Scan. Characteristic curve is obtained from the calculated ratio of density and dose depth. The characteristic curve is a curve graph showing the relation between the number of exposition and the results of the film density to get a reference picture of dose in radiotherapy.

3. Results and discussion

3.1. Radiation of gafchromic RTQA2 film on slab phantom

Irradiating gafchromic RTQA2 film is performed by modulating energy 6 MV linac. The energy selection is based on the standard commonly used in radiotherapy. Linac has two kinds of beam; photon beam and electron beam. In medical field, Linac used for cancer radiotherapy can be divided into three tiers based on the value of energy and the type of particle accelerated. Those are low-energy photons and electrons (4-6 MV), medium energy photons and electrons (6-15 MV), and high energy photons and electrons (15-25 MV) [11]. Multi energy beam of electrons and photons in third edition of The Physics of Radiation Therapy for the purposes mentioned, electron energy ranges around 4-22 MV and photon energy around 6-18 MV - 6 MV photon energy included [6].

One principle of radiotherapy is the optimization of the dose given to cancerous tissue which seeking for protection from a healthy organ tissue around the cancer. Therefore, the linac output dose used in this study is 300 cGy. The previous research states that the value of high accuracy in determining dose is better done by using a higher dose around 150-200 cGy [9]. However, results confirm that the value of high accuracy in determining the dose is better at the dose higher than 200 cGy [7], as well as at the dose of 300 cGy which perceives maximum output in irradiation. Additionally, irradiation angle used in linac is 0^0. It means that the linac gantry position is perpendicular to the film that will receive radiation exposure. Angle 0^0 is selected thus the distribution of the dose received by the film as well as to produce constant exposure of the film perpendicular to standard deviation slightly [10]. Distance Determination from the radiation source to the gafchromic RTQA 2 film uses SAD (Source Axis Distance) technique. SAD is a technique setting the field and the depth of target, with distance between the source of radiation from Linac and the film surface is about 90 cm and 10 cm D-max. Selection on the distance is based on the best irradiation efficiency to deliver radiation with the best modalities linac. Following is the description of SAD techique on the film data retrieval with various depths and breadths of field.

Gafchromic RTQA2 film with various breadths of field is placed isocentric above the slab phantom. Phantom is made from a material that resembles the human body. Slab phantom is designed for medium and high energy radiation therapy dosimetry with photon energy around 4mV MV-25 and electron energy around 4-25 MeV. Phantom consists of 33 RW3 plate with dimension of 30cm x 30cm

IOP Conf. Series: Journal of Physics: Conf. Series **853** (2017) 012023 doi:10.1088/1742-6596/853/1/012023

[1]. During this measurement, dose of radiation obtaining a picture as a reference dose should be received by the patient radiotherapy using medium water phantom, since the human body is composed of 80% water [11]. Aside to suit the medium and high energy, slab phantom can perform an absolute quality control and relative dosing. Other advantages are able to vary the depth measuring up to 30 cm increments of 1mm made of water equivalent (Goettingen White Water) with thickness tolerance ± 0.05 mm. Plate adapter for various types of detectors can be inserted without an air gap [1].

The breadth variation of the field of gafchromic RTQA2 film are 4x4 cm^2, 6x6 cm^2,8x8 cm^2, and 10x10 cm^2. Selection of the breadth variation is done by a width measure of the smallest field less than 15x15 cm^2. The variation can be applied to patients of cancer therapy with relatively small size of cancer. MLC (Multi Leaf Collimator) on air Linac is comprehensively set to a large field thus it produces a square shape with a breadth variation of the required field. MLC is a mechanical component integrated with the linac gantry that can be controlled automatically by using a computer and independent to each other thus produce a field or fields in various forms [6]. In addition to the breadth variation, in this research, fields are also varied in the depth of gafchromic RTQA2 film which get the linac radiation exposure. The depth variation selected includes 1.5 cm, 4 cm, 6 cm, 8 cm and 10 cm. Selection based on the depth of the cancer position in the body that is often encountered.

3.2. The effect of field size and depth variation of the density values

Varying field size (breadth) and depth on the characterization of radiochromic film density through Source Axis Distance (SAD) technique in a LINAC with slab phantom for radiotherapy applications aims to determine the influence of the breadth and depth of the density of the resulting film. The variety of field size and depth in radiation therapy is a parameter that should be taken into account with the value of high accuracy as it relates to the principle of radiotherapy in seeking the protection of a healthy organ tissue around the cancer [7]. Gafchromic RTQA2 film density measurement results with various depths and field sizes are presented on table 1.

Table 1. The results of gafchromic film RTQA2 density

Field Size (cm)	Depth (cm)	Distance Irradiation (cm)	Radiation Angle	Density
4x4	1,5	100	0^0	60
	4	100	0^0	67
	6	100	0^0	68
	8	100	0^0	76
	10	100	0^0	79
6x6	1,5	100	0^0	71
	4	100	0^0	71
	6	100	0^0	77
	8	100	0^0	80
	10	100	0^0	85
8x8	1,5	100	0^0	69
	4	100	0^0	74
	6	100	0^0	80
	8	100	0^0	81
	10	100	0^0	80
10x10	1,5	100	0^0	63
	4	100	0^0	71
	6	100	0^0	71
	8	100	0^0	77
	10	100	0^0	81

IOP Conf. Series: Journal of Physics: Conf. Series **853** (2017) 012023 doi:10.1088/1742-6596/853/1/012023

From the density of Gafchromic RTQA2 film measurement results with various depths and field sizes presented above, the relation chart between the depth variation and density in various field sizes can be obtained as follows:

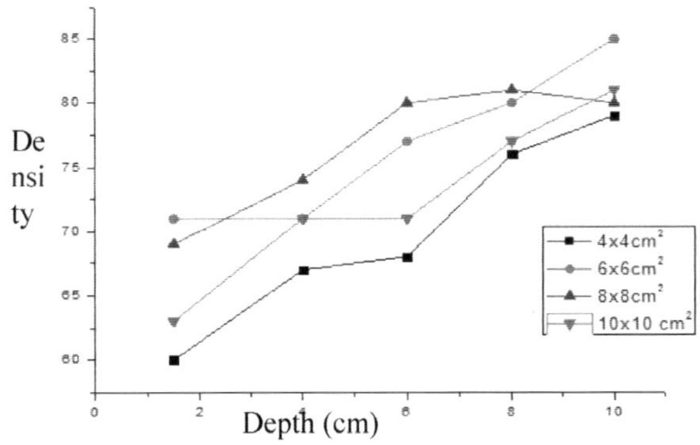

Figure 1. The Graphic relation between the depth and density variation in various field sizes

The graphic relation between the depth variation and density in various field sizes at figure 1 shows that the greater the depth value used, the higher the density obtained, and vice versa. While the density is inversely proportional to the level of darkness in the film after irradiation. The higher the level of darkness in the film, the smaller the density value is, whereas, the dose absorbed by the film increases. . It means that the higher the dose absorbed, the darker the film will be, yet the lower the density obtained. This is consistent with the theory of absorbed dose to the equation:

$$D = \frac{dE}{dm} \tag{1}$$

with D: Dose Absorption,
dE: Energy average, and
dm: Massa.

The amount of the dose absorbed will be higher in the tissues close to the surface rather than deep within the body. Absorbed dose may also be defined through the absorbed energy material per unit mass. Mathematical Analysis of the relation between density and depth is calculated in equation:

$$OD = \mu x \tag{2}$$

with μ: attenuation coefficient and
x: the absorption coefficient of the passed material.

The formulation indicates the relation between density and depth is OD = μx. Hence, the graph above shows that the higher the density level, the greater the depth value. Below is the film before and after getting radiation.

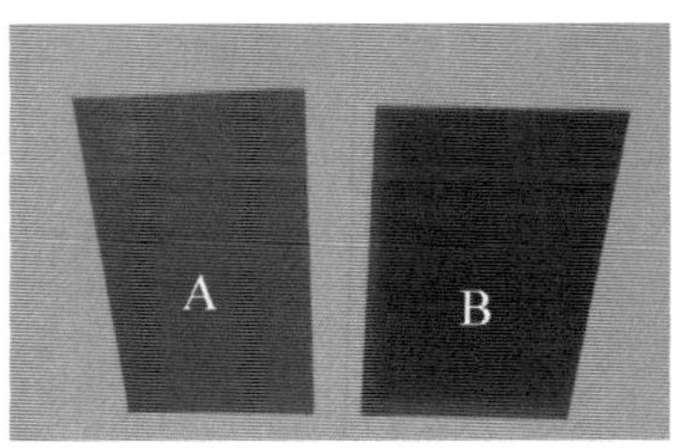

Figure 2.the film (A) before and (B) after getting radiation.

3.3. Dose figures resulting from the accuracy value of the characteristic curve between density and gafchromic RTQA2 film dosimetry results

The characteristic curve is a curve graph showing the relation between the number of exposition and the results of the film density. This curve was first discovered by Hurteen and Drifield in 1890. In this research, the characteristic curve describes the relation between the dose and the film density varied in the depth and field size. The denoted dose value % of TMR% is obtained from TPS (Treatment Planning System) presented in Table 2, to calculate the dose depth values. Dose Depth is the dose at a certain depth value received by phantom or patient. The dose values analyzed can be used as a reference picture of doses at various depths and spacious grounds with energy of 6 MV linac. The dose depth can be found using equation:

$$MU = \frac{D}{TMR\,\%\,x\,OF} \tag{3}$$

(Monitor Units) is a linac output dose, (Depth Dose), (Tissue Maximum Ratio), and (Output Factor) of linac. For the field size 4x4 cm^2, 6x6 cm^2, 8x8 cm^2, 10x10 cm^2 is 1.00. So we get the equation :

$$\text{Dose Depth (cGy)} = \text{TMR\% x MU} \tag{4}$$

The dose 300 cGy as the percentage of the dose at the maximum depth is always normalized to 100%, with the linac output dose used in this research 300 MU, which equivalent to 300 cGy. All fields obtained by calculating the dose depth (cGy) are presented in Table 2:

Table 2. The calculation result of the depth-dose of fields variation

Field Size(cm)	*Depth(cm)*	*Density*	TMR (%)	*Depth-Dose* (cGy)
4x4	1,5	79	119,9	359,7
	4	76	105,1	315,3
	6	68	93,8	281,4
	8	67	80,4	241,2
	10	60	73,5	220,5
6x6	1,5	71	118,2	354,6
	4	71	104,7	314,1
	6	77	94,2	282,6
	8	80	81,5	244,5
	10	85	74,9	224,7
8x8	1,5	69	117,2	351,6
	4	74	104,5	313,5
	6	80	94,6	283,8
	8	81	82,4	247,2
	10	80	76	228
10x10	1,5	63	116,3	
	4	71	114,2	342,6
	6	71	94,8	248,4
	8	77	83,2	249,6
	10	81	77	231

Based on the calculation results of depth-dose value above, characteristic curve showing the relation between the density with depth-dose is obtained as follow:

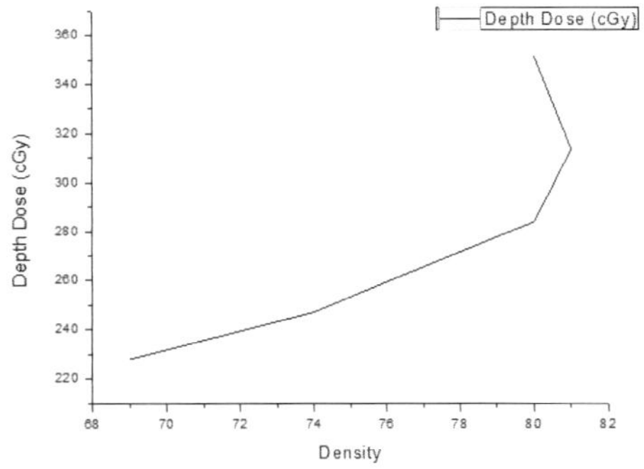

Figure 3. The characteristic curves of relation between gafchromic RTQA2 film density with depth-dose on 4x4 cm^2 field size

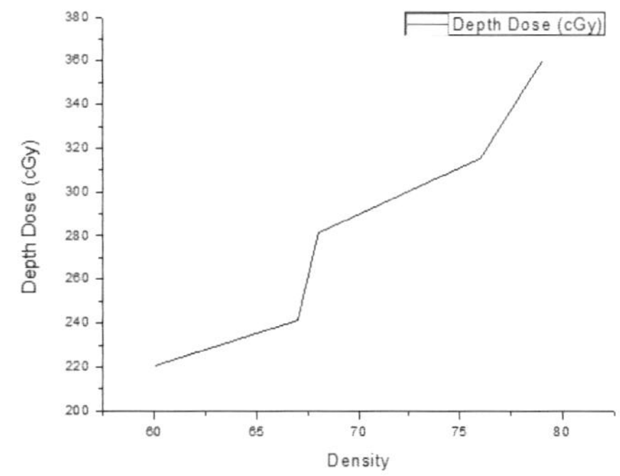

Figure 5. The characteristic curves of relation between gafchromic RTQA2 film density with depth-dose on 8x8 cm^2 field size

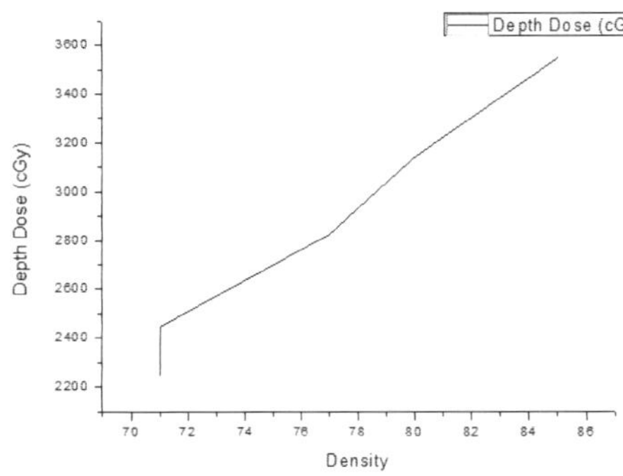

Figure 4. The characteristic curves of relation between gafchromic RTQA2 film density with depth-dose on 6x6 cm^2 field size

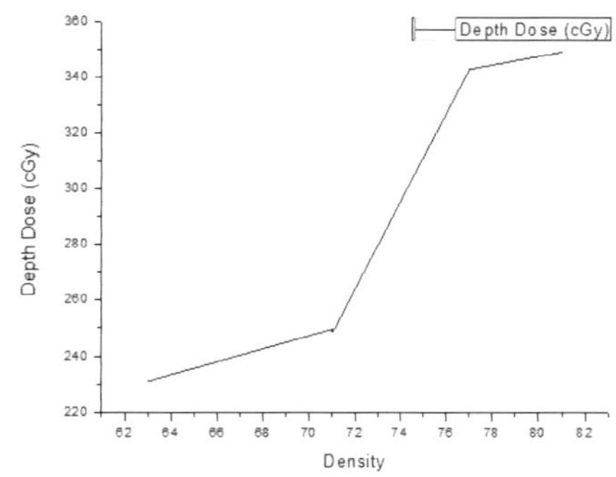

Figure 6. The characteristic curves of relation between gafchromic RTQA2 film density with depth-dose on 10x10 cm^2 field size

Analysis of depth-dose relationship with the density can be calculated by the equation:

$$\dot{X} = \frac{\Gamma A}{X^2} = D = f\frac{\Gamma A}{X^2} t \tag{5}$$

to obtain the linear equation:

$$D = \frac{a}{OD^2} = D = aOD^{-2} \tag{6}$$

Where is a constant which can be obtained from the graph of with depth dose which is the value of the gradient chart, is the conversion factor of the rate of exposure to the rate of absorbed dose (rad / R), is irradiation (R), is depth of target, is Absorbed dose (rad), is Optical density, is the power source (a specific constants) and is Activitation (Ci).

In the all of fields obtained by calculating the depth-dose (cGy) shows that the higher the density, the further depth-dose value increase. From the calculation, magnitude of specific field size can be seen and can be used as guidelines for the calculation of depth dose as a reference depth dose received by patients for the purpose of verification of the dose received by the patient using a dosimeter of gafchromic RTQA2 films.

Where is a constant which can be obtained from the graph of with depth dose which is the value of the gradient chart, is the conversion factor of the rate of exposure to the rate of absorbed dose (rad / R), is irradiation (R), is depth of target, is Absorbed dose (rad), is Optical density, is the power source (a specific constants) and is Activitation (Ci).

In the all of fields obtained by calculating the depth-dose (cGy) shows that the higher the density, the further depth-dose value increase. From the calculation, magnitude of specific field size can be seen and can be used as guidelines for the calculation of depth dose as a reference depth dose received by patients for the purpose of verification of the dose received by the patient using a dosimeter of gafchromic RTQA2 films.

4. Conclusion

Based on the background and research that has been done, researchers concluded that:

1. Density film is even more valuable when the value of the current film position depth variation has greater exposure. Conversely, the lower the depth value, the density value of the film decreases. The density is inversely proportional to the level of darkness in the film after irradiation, the higher the level of darkness in the film, the smaller the density values, whereas the dose absorbed by the film decreases. This means that the higher the dose absorbed, the higher level of darkness in the film, but the density decreases.
2. The characteristic curves obtained can calculate the magnitude of the dose at a certain depth and specific field size. So, it can be used as a guide of a reference dose calculations and dose the patient received a picture for purposes of verification of the dose received by patients radiotherapy.

References

[1] Ministry of Health RI.2014. The rate of cancer growth in Indonesia. www.depkes.go.id Accessed on 20 September 2015 at 14:50 pm
[2] Khan, Faiz M. 2010. *The Physics of Radiation Therapy*. Third Edition. Lippincott Williams and Baltimore
[3] El Barouky J. Fournier-Bidoz N, Mazal A, Fares G, Rosenwald JC. Partical use of Gafchromic EBT films in Electron Beams for in-Phantom Distribution Measurements an Monitor Units Verification. Phys Med. 2011;27:81-8
[4] Chair, Azam Niroomand-Rad. 1998. *Radiochromic Film Dosimetry*. Recommendations of AAPM Radiation Therapy Committee task Group No. 55. American Association of Physicsts in Medicine
[5] Keall,P. 2001. *The effect of source-axis distance on integral dose: implications for IMRT.* Department ofRadiation Oncology, Virginia Commonwealth University Health System. Virginia: USA
[6] Khan, Faiz M. 2010. *The Physics of Radiation Therapy*. Third Edition. Lippincott Williams and Baltimore
[7] Cember, Herman. 1983. *Pengantar Fisika Kesehatan.* Semarang: Pergamon Press
[8] Darmawati. 2012. Implementasi Linear Accelerator dalam Penanganan Kasus Kanker. Pascasarjana Fisika-UGM. Yogyakarta
[9] Butson, Martin J. Yu, Peter K.N. Cheung, Tsang, dan Metcalfe, Peter. *Radichromic Film for Medical Radiation Dosimetry*. Material Science and Engineering R (2003), 41: 61-120
[10] Butson, M. J., Cheung, T., dan Yu, P.K.N. *Absorption Spectra Variations EBT Radiochromic Film from Radiation Exposure.* Phys. Med. Biol. 50 (2006). N135-N140
[11] *International Specialty Products*, 2010. Struktur lapisan gafchromic RTQA2 film
[12] Podgorsak E B, 2005. *Review of Radiation Oncology Physics: A Handbook for Teachers and Students.* Education Report Series. International Atomic Energi Agency. Vienna. Austria

Permanent magnetic field, direct electric field, and infrared to reduce blood glucose level and hepatic function in *mus musculus* with diabetic mellitus

Suhariningsih[1,*], Hari Basuki Notobroto[2], Dwi Winarni[3], Saikhu Achmad Hussein[3] and Tri Anggono Prijo[1]

[1]Department of Physics, Faculty of Science and Technology, Airlangga University, Surabaya, Indonesia
[2]Faculty of Public Health, Airlangga University Surabaya, Indonesia
[3]Department of Biology, Faculty of Science and Technology, Airlangga University, Surabaya, Indonesia

*corresponden email address: suhariningsih@fst.unair.ac.id

Abstract. Blood contains several electrolytes with positive (cation) and negative (anion) ion load. Both electrolytes deliver impulse synergistically adjusting body needs. Those electrolytes give specific effect to external disturbance such as electric, magnetic, even infrared field. A study has been conducted to reduce blood glucose level and liver function, in type 2 Diabetes Mellitus patients, using Biophysics concept which uses combination therapy of permanent magnetic field, electric field, and infrared. This study used 48 healthy mice (*mus musculus*), male, age 3-4 weeks, with approximately 25-30 g in weight. Mice was fed with lard as high fat diet orally, before Streptozotocin (STZ) induction become diabetic mice. Therapy was conducted by putting mice in a chamber that emits the combination of permanent magnetic field, electric field, and infrared, every day for 1 hour for 28 days. There were 4 combinations of therapy/treatment, namely: (1) permanent magnetic field, direct electric field, and infrared; (2) permanent magnetic field, direct electric field, without infrared; (3) permanent magnetic field, alternating electric field, and infrared; and (4) permanent magnetic field, alternating electric field, without infrared. The results of therapy show that every combination is able to reduce blood glucose level, AST, and ALT. However, the best result is by using combination of permanent magnetic field, direct electric field, and infrared.

1. Introduction

The aim of this research is to determine the best combination of permanent magnetic field with an electric field and infrared for lowering blood sugar levels as well as ALT and AST in mice with type I diabetes mellitus.

1.1. Interaction of magnetic field, electric field, and infrared in the body

Blood in the body contains a variety of electrolytes, such as sodium (Na^+), potassium (K^+), calcium (Ca^{2+}), Magnesium (Mg^{2+}), chloride (Cl), HCO^{3-}, SO^{4-}, and so on. Both types of these electrolytes (cations and anions) work together to deliver impulses as required by the body. Electrolytes will have certain effects when external disruption of electric and magnetic field is given.

Exposure of external electric field to the body will cause the rearrangement position of free ions in the blood vessels. Negative ions will tend to move in the opposite direction to the direction of the external electric field, while the positive ions will move in the direction of an external electric field that allows collision with negative ions. The interaction of positive ions and negative ions to move due to a disturbance of external electric fields on the body causes electronic polarization be $P = \varepsilon_o \chi S$, so there will be a potential difference of polarization V (t). Potential polarization that appears will be countered by the potential immune L (t) of the body, so that in healthy individuals, the balance of potential is G (t) = 0 (see figure 1). On the contrary, if the body is sick, the potential immune of the body cannot restore or is against the potential polarization formed by the external electric field, so that it results in G (t) ≠ 0.

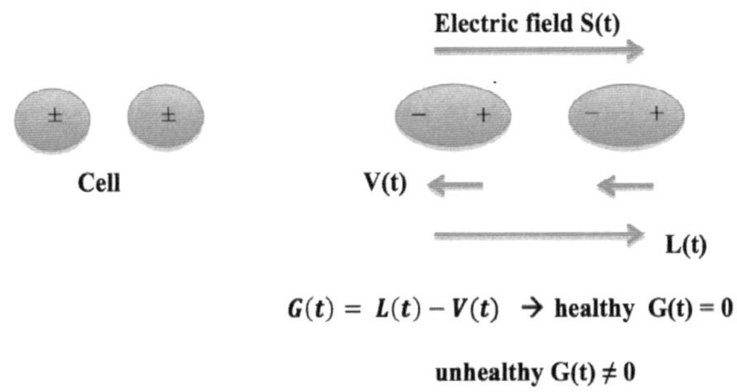

Figure 1. External electric field that goes inside the body

Electronic polarization will cause an electric shift with vector density of $D = \varepsilon o\ (1 + \chi)\ S$. Electric shift that occurred resulted in changes in the refractive index of the cell that satisfies the equation $n_e(S) = n_e - ½\ n_e^3 r_{33} S$. The changes of refractive index that occur will increase the permeability of the cell plasma membrane, resulting in the increased flow of ions and water through the plasma membrane. The flow of ions and water would lead to changes in the relative conductivity of the plasma membrane.

The relationship between electric field, refractive index, and permeability is also expressed by Saleh and Theich [1], that is the exposure of electric field on the body will reduce or affect refractive index, that is, increasing the permeability of the plasma membrane particularly the lipid group.

Blood (containing iron ferromagnetic) is also regarded as magnetic materials which has paramagnetic properties. When given the disruption in the form of an external magnetic field, the molecules in the blood cells will align with the direction of the external magnetic field is given as shown in Figure 2, although only temporarily because the orientation of the molecules that is subject to interference will return to its original position when external magnetic field is eliminated.

Figure 2. Magnetization in ferromagnetic materials after being given an external magnetic field

One of the molecules that play a role in this case is the haemoglobin molecule. Haemoglobin molecule is considered to have an important role because it constitutes Fe atoms which are bonded with oxygen molecules (as in Figure 3). When haemoglobin molecules are in aligned position due to an external magnetic field magnetization, this will facilitate the molecular oxygen to be separated from the bond and oxidize damaged cells due to the same symmetrical position.

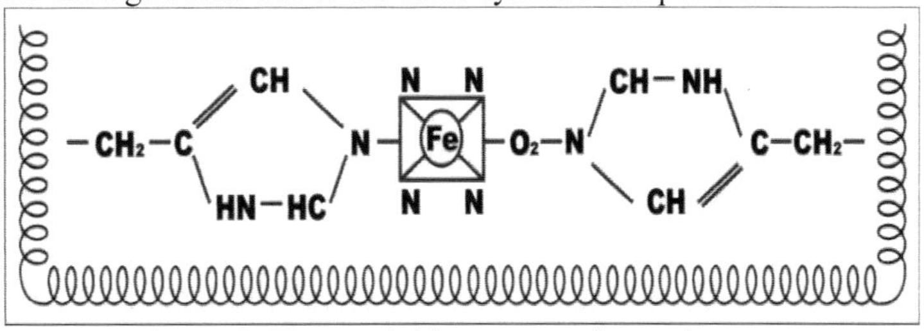

Figure 3. Haemoglobin structure in erythrocytes

Magnetic field can vibrate the ferromagnetic electric charge in the blood vessels, this vibration will cause a movement of energy and heat. Motion and heat energy generated is useful to dilate the blood vessels (vasodilation). Therefore, the blood flow that has previously been carried out by an electric field, will become more fluent in the presence of a magnetic field.

Infrared wavelengths can penetrate into the skin tissue a few centimetres deep. In this process, the water molecules in the network play a very important role, because the infrared light that is absorbed by water molecules causes the water molecules to vibrate. This increases the vibration energy of the water molecules. Because the energy increases, the temperature rises and the body that is exposed to infrared light will feel warm. Other effects are the blood vessels become wider and the blood flows more smoothly. Thermal effects caused by infrared light is used for therapy.

1.2. Diabetes Mellitus

Type 2 diabetes mellitus (DM 2), formerly referred to as non-insulin dependent diabetes (NIDDM) or adult-onset diabetes, is 90-95% of the cases of DM. DM 2 patients experience insulin resistance and relative insulin deficiency, therefore, at the beginning or even along the lives of patients with type 2 diabetes do not require additional insulin. In general, not all people with type 2 diabetes mellitus have overweight or obese. Being overweight lead to various degrees of insulin resistance. Patients with type 2 diabetes who are not overweight or obese are likely to experience an increase in body fat percentage mainly distributed in the abdominal [2].

The bond between insulin to the insulin receptor (IR) causes IR auto phosphorylation, which will trigger various signalling pathways, including phosphorylation of tyrosine residues of insulin receptor substrate (IRS). IRS 1 phosphorylation activates the signalling pathways for the translocation of vesicles containing glucose transporter (GLUT 4) from the cytoplasm to the cell membrane and activates anabolic pathway in muscle and fat cells (adipocytes), while the anabolic effect to the liver is mediated by the phosphorylation of IRS 2. Translocation of GLUT 4 to the cell surface allows the uptake of glucose by cells. Insulin-stimulated signalling pathways can be attenuated or tuned down by various regulators. IRS auto phosphorylation can be restored to its original state by Tyr phosphatase-1B (PTB1B). IR tyrosine kinase activity is inhibited by protein suppressor of cytokine signalling 1 (SOCS1) and SOCS2. Tyrosine phosphorylation and IRS activation is inactivated by serine phosphorylation caused by the excess of nutrients or activation of stress pathways.

Insulin resistance is a condition failure signalling by insulin, which is commonly associated with chronic over nutrition occurs in conditions of obesity (Muoio & Newgard, 2008). Ann Louise (2012) mentioned that insulin resistance is a condition where the cells are not able to respond to the normal action of insulin. The result of insulin resistance is the uptake of glucose by cells that are sensitive to insulin cannot occur. The fat cells (adipocytes) plays a role in insulin resistance because adipocytes can produce adipokines (group of hormones and cytokines) and are capable of storing excess lipids in obesity conditions. The excess amount of lipid can lead to a redistribution of lipid abnormalities to the other organs and tissues. Adipokines produced by adipocytes, among others, are the leptin hormone, adiponectin, retinol-binding protein and resistin, and proinflammatory cytokines such as interleukin

(IL-6) and tumor necrosis factor-α (TNF α). Leptin and adiponectin are categorized as anti-diabetogenic, because it can decrease the synthesis of triglycerides (TG), stimulate β-oxidation of lipids and improve insulin action both in the skeletal muscle and liver. In obesity case, the leptin level is increased but adiponectin is decreased, indicating that in the obesity case, the resistance to leptin occured (Muoio & Newgard, 2008).

Since the inception of obesity or high-fat diet (high-fat diets), metabolites derived from lipids (lipid-derived metabolits) accumulates outside adipocytes (including skeletal muscle, heart and liver). The accumulation in the liver resulted in the change of track long- chain acyl Coas (LC-Coas) to the endoplasmic reticulum and the cytosolic lipid such as diacylglycerol (dags), ceramide and triglycerides. This change is mainly governed by the presence of elevated levels of malonyl CoA induced by glucose. Malonyl CoA acts as both an intermediate precursor of de novo lipogenesis and carnitinpalmitoyltransferase allosteric inhibitor-1 (cpt1), which acts as a rate-limiting enzyme for carrying LC-Coas into the mitochondria for β–oksidasi (Muoio & Newgard, 2008). Insulin resistance lowers the effect of insulin inhibition in peripheral lipolysis, increases the level of free fatty acids that can be taken up by the liver, and increases the risk of fatty liver (Krauss, 2004).

Lipids are transported from the liver in the form of very low density lipoprotein (VLDL), which is formed by the microsomal triglyceride transfer protein (MTP) that inserts triglycerides in apolipoprotein B (apo B). The decrease of the activity of MTP and the synthesis and secretion of apo B will reduce the liver ability to produce lipid and cause accumulation of triglycerides in the liver (Paschos & Paletas, 2009) (Figure 4). Prolonged insulin resistance will increase triglycerides, decrease level of high density lipoprotein (HDL) (Goyal et al. (2014) and increase small dense low-density lipoprotein (LDL), despite the plasma levels of LDL is normal (Bitzur et al., 2009).

Disruption or damage in liver function may increase levels of aspartate transaminase (AST / SGOT) and alanine transaminase (ALT / SGPT) (Oh & Hustead, 2011). Goyal et al. (2014) states that the increase of SGPT levels reached 20% of total patients with diabetes, especially in poorly controlled diabetes. Compared to AST, ALT which plays a role in gluconeogenesis is more correlated with the onset of lipid accumulation in the liver than AST.

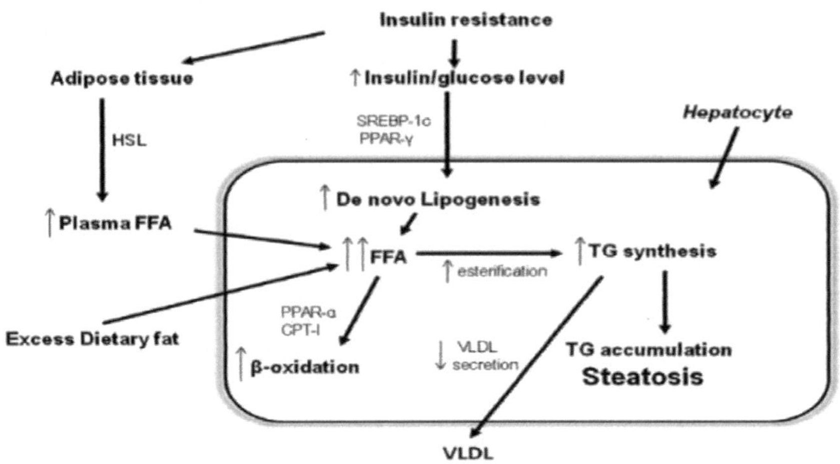

Figure 4. Diagram showing the relation between insulin resistance and metabolic changes in liver

Chronic excessive nutrients induce dysfunction and pancreatic β cell death. Insulin resistance condition and pancreatic endocrine function declining are the causes of type 2 diabetes mellitus.

Increased blood glucose level is correlated with the increased blood viscosity, even though it is to a non-diabetic individual [3]. Tamariez et al. (2008) states that the increase of blood viscosity correlates with the onset of insulin resistance and type 2 diabetes are caused by problems with the arrival of insulin and glucose in tissues which is sensitive to insulin. Increased blood viscosity in patients with diabetes also affects the increasing workload of the heart in pumping blood to the vascular bed resulting in disruption in microcirculation, and the dense blood can damage blood vessels [3]. Disturbances in the microcirculation, according to Yagihashi et al. (2002) in Kostova et al [4] will result in the reduction of oxygen supplies and nutrients to the capillaries. This condition which was mentioned by Peduzzi et al. (1984) in Kostova et al [4] is a major cause of retinopathy in DM patients. Hyperlipidemic condition that often occurs in people with diabetes can also increase the blood viscosity [3].

Popel et al. (2005) in Kostova et al [4] mentions that regarding to the increasing of blood viscosity, erythrocyte (red blood cells) in DM patients are more prone to aggregate compared with non-diabetic individuals. The increase of erythrocyte aggregation was also observed in several pathological conditions. The increase of erythrocyte aggregation will also decrease the rate of blood flow and blood ability to carry oxygen to various peripheral tissues. (Foresto et al., 2000, [4])

2. Method

This study was conducted in two stages:

2.1. First stage

Instrument design that generates electric field, magnetic field, and infrared rays.

2.1.1. Design PSI equipment

The designed instrument is a combination of a permanent/constant magnetic field source of 100 mT (P), direct electric field (S) of 600V/m, and infrared (I) of 1 mW/cm^2, which would be presented to the group of experimental animals (mice). Each group of experimental animals are put in cages of a plastic tray (open top surface).

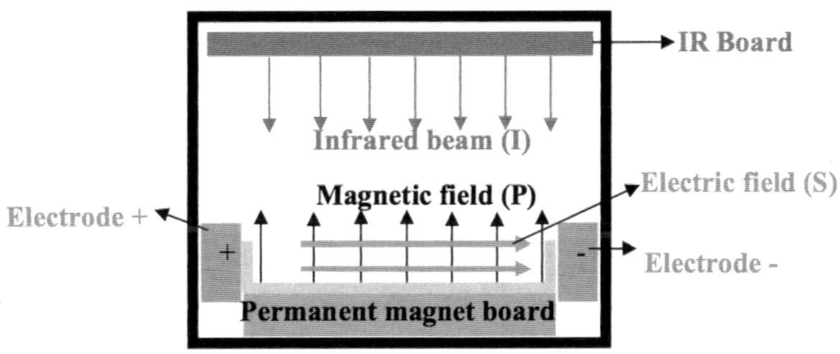

Figure 5: Front-view PSI instrument, including infrared beam direction, constant magnetic field, and direct electric field

2.1.2. PBI instrument design

This instrument is a combination of a permanent magnetic field source (P) 100 mT, an alternating electric field (B) with the frequency of 16 kHz, and infrared (I) of 1 mW/cm^2, which would be presented to the group of experimental animals (mice). Each group of experimental animals are put in cages of a plastic tray (open top surface). Front-view of Figure 8 of the treated animals sample box shows the direction of the light beam infrared, magnetic and electric field which was exposed to the experimental animals.

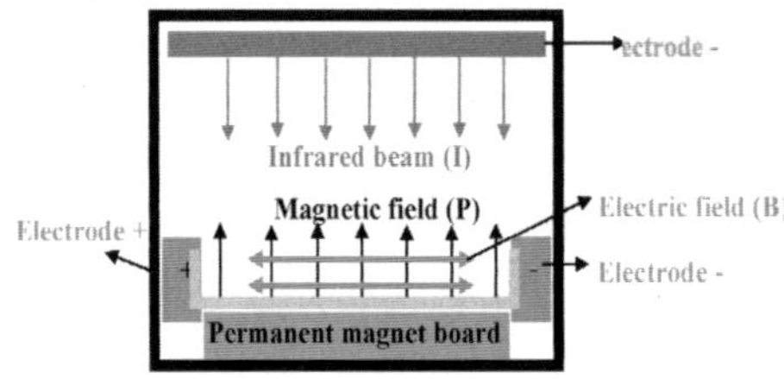

Figure 6: Front-view of PBI instrument, including infrared beam direction, constant magnetic field, and alternating electric field

2.2. Second stage.

To test the instrument from the first stage to know the combined effect of electric fields, magnetic and infrared rays to the mice with Diabetes mellitus

Animals/sample used in this research are healthy mice (*mus musculus*), male, 3-4 weeks old with a weight range of 25-35 g as many as 48 animals. Mice were induced into diabetic mice by the injection of streptozotocin (STZ). The procedure of mice induction based on the research by Novelli et al [6]. in which the mice were induced by STZ dissolved in citrate buffer pH 4.5 at a dose of 20 mg/kg intrapertonially in mice for 5 consecutive days (multiple low dose). On day 7 and 14 after the induction, blood glucose levels are evaluated in mice. Mice with glucose levels above 200 mg/dL will be classified as diabetic mice.

Subsequently the samples were divided into groups of healthy and groups of diabetic mice. Samples of diabetic group are divided into eight subgroups. Each group consisted of 4 mice. Each group treated with panel generating a permanent magnetic field, electric field, and the infrared except for the normal control group and diabetic control.

3. Result and discussion

The magnetic field affects the ferromagnetic substance- Erythrocyte aggregate conformation arrangement that cause the targeted cell membrane depolarization, decreasing the Blood viscosity so that Insulin resistance will decrease.

External electric field which is exposed to mice will lead to electronic polarization $\mathbf{P} = \varepsilon_o\chi\mathbf{S}$, electronic polarization will cause an electric shift with vector density of $\mathbf{D} = \varepsilon_o(1 + \chi)\mathbf{E}$. Electric shift that occurred resulted in changes in refractive index of the cell that satisfies the equation $n_e(E) = n_e - \frac{1}{2} n_e^3 r_{33}E$. The changes of refractive index that occur will increase the permeability of the cell plasma membrane, resulting in the increased flow of ions and water through plasma membrane. The flow of ions and water would lead to an increase in the conductivity of the plasma membrane that induce vasodilation then cause increasing blood flow and decreasing insulin resistance.

Infrared rays which is absorbed by water molecules cause the water molecules vibrate. Activation of the water at speeds increasing the ionization of water: H^+, OH^- in microcirculation increase then produce heat that rise the body temperature, induce chemical reactions, affecting cell membrane permeability that increases Cytosolic free Ca^{2+} then causes a rise of Insulin receptor activity and decreases of Insulin resistance.

The experimental results (Figure 7 to 10) proved that the treatment by combining the type of permanent magnet, the type of electric current (direct current and alternating), and the use of infrared light (with and without IR) to diabetic mice showed an improvement in fasting blood glucose levels.

These improvements demonstrated a significant reduction in blood sugar levels after 28 days of treatment in all combination groups (PSI +, psychotic, PBI +, PBI-) compared to diabetic control group (KD) and no significant difference with normal control group (KN). Tao and Huang [5] states that the provision of separate magnetic treatment can alter the aggregate red blood cells caused by hyperglycemia in diabetes condition into long chains. This change will reduce the viscosity of the blood and allows erythrocytes into the capillaries so the oxygen and nutrients transport to the tissues becomes better. Electric currents resulting in vasodilation (widening of blood vessels) (Sullivan & Davison, 2001) and the use of infrared thermal effects make the treatment more effective. The decrease of blood viscosity, vasodilation and the effect of heat allows the reactivation of insulin signalling pathway, thereby signalling function returns to normal. Normally, the signalling function of insulin lowers blood sugar levels to normal blood sugar levels.

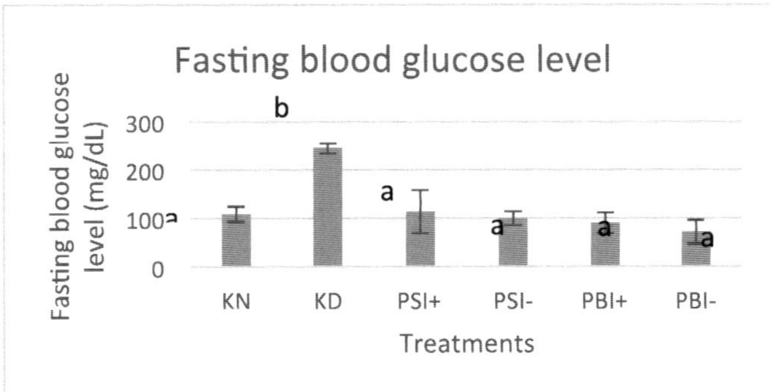

Figure 7. Profile of fasting blood glucose level to the treatments

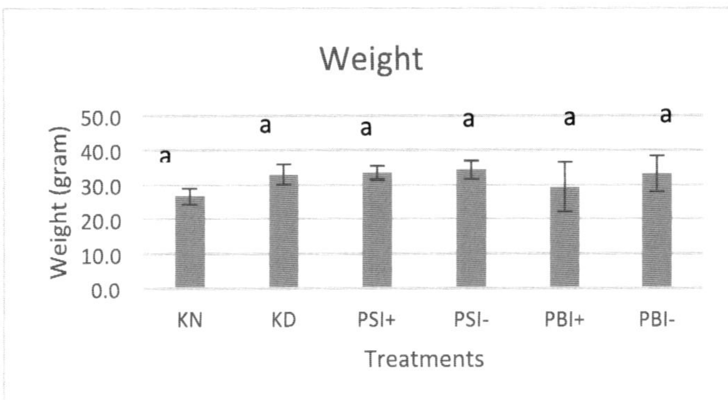

Figure 8. Profile of weight to the treatments

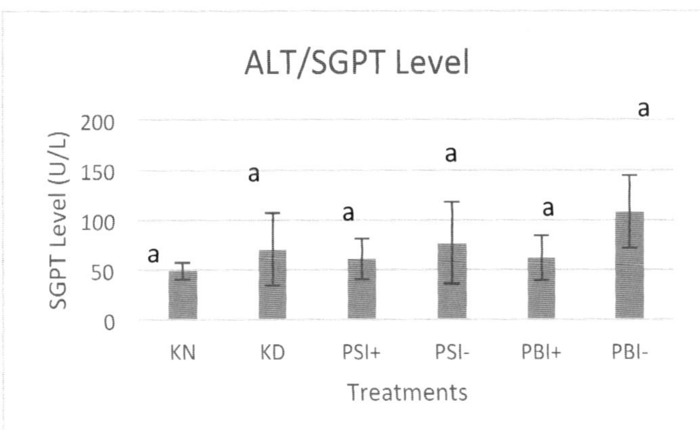

Figure 9. Profile of ALT level to the treatments

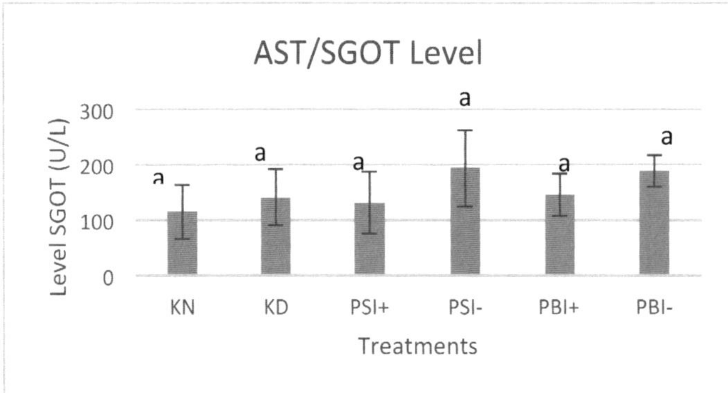

Figure 10. Profile of AST level to the treatments

4. Conclusion

From the results of this study, it can be concluded that:

1. For Diabetes Mellitus case, all combinations of permanent magnetic field, electric field, and the infrared rays (PSI+, PSI-, PBI+, and PBI-) can lower blood sugar levels, but the decline was contained in the optimum combination of PSI+
2. For Weight Loss case, all combinations of permanent magnetic field, direct electric field, and infrared light (PSI+, PSI-) and combination of a permanent magnetic field, alternating electric field, and infrared light (PBI+, and PBI-) can increase body weight.
3. For liver function (ALT and AST), the combination of PSI+ lowers ALT and AST significantly.
4. This study shows the potential of combination therapy instrument of permanent magnetic field, direct electric field, and infrared rays, which is effective and safe to improve blood glucose levels and hepatic function in diabetic mice.

References

[1] Saleh B E A, Teich MC 2007 *Fundamentals of Photonics* John Wiley & Sons Inc Hoboken New Jersey Canada

[2] American Diabetes Association 2011 *Diagnosis And Classification Of Diabetes Mellitus* Diabetes Care 34:S62-9

[3] Irace C, Carailo C, Scavelli F, *de Franceschi MS, Esposito* M, Gnasso A 2014 *Blood Viscosity in Subjects With Normoglycemia and Prediabetes* Diabetes Care 37 P 488-492.

[4] Kostova V, Antonova N, Velcheva I and Ivanov I 2012 *Comparative Analysis of The Rheological Properties of Blood Patients With Type 2 Diabetes* Series on Biomechanics 3-4 P80-85.

[5] Tao R and Huang K 2011 *Reducing blood viscosity with magnetic fields* Physical Review 00 001900 doi: 10.1103/PhysRevE.00.001900 http://ec.europa.eu/health/ph_risk/risk_en.htm accessed June 2016.

[6] Novelli MB, Bonamassa M, Masini M, Funel N, Canistro D, De Tata V, Martano M, Soleti A, Campani D, Paolini M and Masiello P 2010 *Persistent correction of Hypergycemia in streptozotocin-Nicotinamide-Induced Diabetic Mice by A non-Conventional Radical Scavenger* Naunyn-Schimied Archive Pharmacology 382:127-137.

Hydroxyapatite coating on cobalt alloys using electrophoretic deposition method for bone implant application

Aminatun[1*], Shovita M[1], Chintya K I[1], Dyah H[1] and Dwi W[2]

[1]Department of Physics- Faculty of Science and Technology- Airlangga University, Mulyorejo, Campus C Unair Surabaya, 60115
[2]Department of Biology- Faculty of Science and Technology- Airlangga University, Mulyorejo, Campus C Unair Surabaya, 60115

*Corresponding email address: aminatun@fst.unair.ac.id

Abstract. Damage on bone due to osteoporosis and cancer triggered high demand for bone implant prosthesis which is a permanent implant. Thus, a prosthesis coated with hydroxyapatite (HA) is required because it is osteoconductive that can trigger the growth of osteoblast cells. The purpose of this study is to determine the optimum concentration of HA suspension in terms of the surface morphology, coating thickness, adhesion strength and corrosion rate resulting in the HA coating with the best characteristics for bone implant. Coating using electrophoretic deposition (EPD) method with concentrations of 0.02M, 0.04M, 0.06M, 0.08M, and 0.1M was performed on the voltage and time of 120V and 30 minutes respectively. The process was followed by sintering at the temperature of 900 ^{0}C for 10 minutes. The results showed that the concentration of HA suspension influences the thickness and the adhesion of layer of HA. The higher the concentration of HA-ethanol suspension the thicker the layer of HA, but its coating adhesion strength values became lower. The concentration of HA suspension of 0.04 M is the best concentration, with characteristics that meet the standards of the bone implant prosthesis. The characteristics are HA coating thickness of 199.93 ± 4.85 μm, the corrosion rate of 0.0018 mmpy and adhesion strength of 4.175 ± 0.716 MPa.

1. Introduction

Damage on bone due to osteoporosis and cancer triggered high demand for bone implant prosthesis which is a permanent implant. Until now, the fulfilment of the needs for bone implant prosthesis in Indonesia still relies on the imported products. The imported bone implant prosthesis often used today is the cemented type that its application would lead to aseptic loosening. Aseptic loosening is the most common failure mechanisms in replacement joints that cause pain to the patient and eventually will need the repeated surgery. To avoid aseptic loosening, prosthesis coated HA (HA) is required because HA is an osteoconductive which can stimulate the growth of osteoblast cells. Therefore, recent researches tend to focus on the coated implant, which generally coats the metal alloy implants with HA.

Various methods of coating HA to the substrate alloy Ti have been done, including the method of pulsed laser deposition (PLD) [1-5] and RF magnetron sputtered [6]. Both two methods are high-technology and expensive. In addition, the entire coating uses titanium alloy as a substrate, whereas titanium alloys are very expensive which will generally be a problem for all people. The alternative metal besides titanium alloy is a cobalt alloy. Its mechanical properties are equivalent to titanium alloy, its biocompatibility is slightly below the titanium alloy, but the price is cheaper than titanium alloys [7-9].

A research was done by Aminatun [10] which conducted HA coating on cobalt alloy and titanium alloy with dip coating method. This study resulted in uneven HA coating on the surface of the cobalt alloy substrate. Therefore, it is necessary to find another simple and unexpansive coating method. That method is electrophoretic deposition (EPD). EPD is a method of suspension deposition on the metal surface due to the influence of external electric fields. Factors that affect the quality of the coating on the coating process by the method of EPD are deposition time, the applied voltage, and the concentration of the suspension. Deposition rate is proportional to the deposition time and voltage used. Meanwhile, the concentration of the suspension HA affects the rate of deposition HA particles on the surface of the metal alloy substrate [11]. Until now, there is no precise information on the composition of the suspension concentration of HA-ethanol used in the coating of cobalt alloy. Therefore, it is necessary to do a more in-depth study of the HA coating of cuttlefish bone to cobalt alloy using the EPD method especially in terms of optimization of the suspension concentration of HA-ethanol.

2. Materials and methods

This research was conducted in 4 stages: a) Preparation of a cobalt alloy substrate and HA of cuttlefish bone; b) the coating process by using the EPD method with variation of the concentration of HA suspension; c) the sintering at a temperature of 900°C for 10 minutes; and d) characterization of samples.

2.1. Preparation of a cobalt alloy substrate and HA of cuttlefish bone

Cobalt alloy substrate was synthesized through fusion method using tri arc melting furnace on the current of 200 A with a composition of 62.25% Co-31.5% Cr-5% Mo-0.5% Mn-0.5% Si-0.25%N which was adapted to ASTM F75 (American Society for Testing and Materials F75) and was then continued with homogenization process at a temperature of 1300°C for 2 hours. The materials that were used are powder chromium [Cr (99+) - Merck], powder cobalt [Co (99+) - Sigma-Aldrich], molybdenum powder [(Mo (99+) - Merck], powder manganese [Mn (99+) - Merck], silicon powder (Si - Merck) and powder Cr2N (Nilaco). After that, the alloy was cut using wire cut with varying sizes of 0.5 mm x 0.8 mm and thickness of 2-3 mm, rubbed with coarse 80-grit silicon, washed with distilled water and then soaked in ethanol [12]. HA of cuttlefish bone was made by hydrothermal method. First of all, extraction of $CaCO_3$ from the lamella cuttlefish bone, then it was heated at a temperature of 350°C for 3 hours. Subsequently, the mixture suspension $CaCO_3$ 1M and $NH_4H_2PO_4$ 0.6M was heated at the temperature of 200°C for 12 hours in a closed vessel. The mixture powder from the heating was washed with distilled water until pH neutral, followed by methanol. The final process was sintering at a temperature of 900°C for 1 hour [8].

2.2. The coating process using the EPD method

HA suspension was made in ethanol with various concentrations of 0.02 M, 0.04 M 0.06 M, 0.08 M and 0.10 M. After that, the coating process was done by using the EPD method with cobalt alloy served as the cathode and platinum as the anode. The coating process was done at a constant voltage and coating time which was at a voltage of 120 V for 30 minutes. Samples of the coating were dried at room temperature.

2.3. The sintering process

The sintering process was done on dried samples layer a tube furnace STF & TZF Models at a temperature of 900°C with 10 minutes of holding time, the increasing rate of 10°/min and the rate of decline 5°/min, under vacuum conditions flowed with argon and with the flow rate of 31 / min [12].

IOP Conf. Series: Journal of Physics: Conf. Series **853** (2017) 012025 doi:10.1088/1742-6596/853/1/012025

2.4. Characterization of sample

The crystal structures of HA coating samples on cobalt alloy were characterized using X-ray diffractometer PAN alyticalX'Pert PRO and its surface morphology was characterized using Scanning Electron Microscopy (SEM- Fei Inspect S50). XRD test was characterized using a source with a wavelength of CuK_α = 1.54060Å, at the angle = 5°-60°, 0.0170° step size, scan step time 135.2550s, operated at 40 kV and 30 mA. SEM test was performed at a voltage of 20 kV with WD = 15.2 to 15.7 mm. Adhesion strength test was characterized using Shimadzu autograph AG-10TE and corrosion test used the potentiostatmeter. Adhesion strength test and corrosion test have been repeated three times. Adhesion strength used Equation 1 and the value of corrosion rate used Equation 2.

$$\sigma = \frac{F}{A} \qquad (1)$$

With σ is adhesion (MPa), F is the force applied (N) and A is the surface area of the HA layer [13].

$$corrosion\ rate(mpy) = \frac{0{,}13 \cdot I_{corr} \cdot EW}{\rho} \qquad (2)$$

0.13: metric conversion factors and time, I_{corr}: current density when E_{corr} (uA/cm^2),
ρ : density (g /cm^3), EW: equivalent weight [14]

3. Results and discussion

3.1. Morphology and thickness of HA layer

HA layers on cobalt alloy from various concentrations of HA suspensions characterized by SEM on surface morphology and a cross-sectional, are presented in Fig. 1 and Fig. 2.

a b c

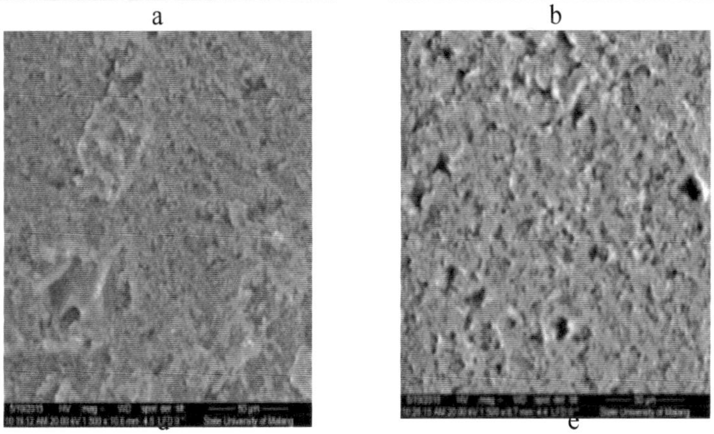

d e

Fig. 1. Morphology Surface of Samples with magnification of 1500x: (a) 0.02M, (b) 0.04M, (c) 0.06M (d) 0.08M and (e) 0.10M

a
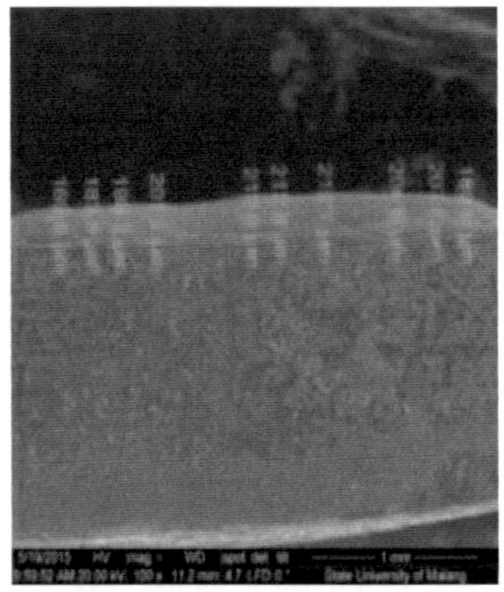
b
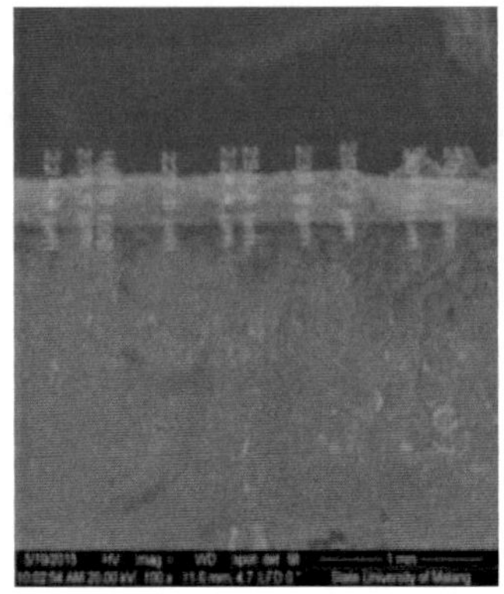
c
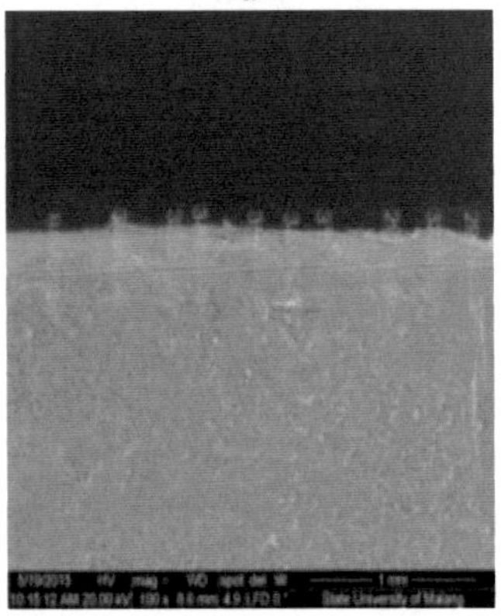
d
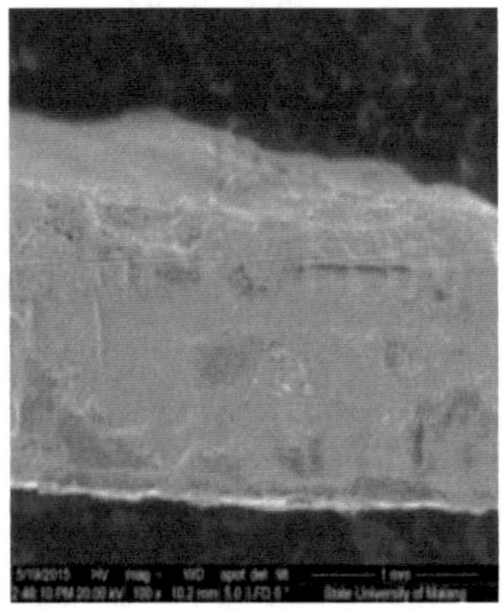
e

Fig. 2. Cross sectional samples of (a) 0.02M, (b) 0.04M, (c) 0.06M (d) 0.08M and (e) 0.10M

Based on the results of SEM magnification of 1500x in Fig.1, it appears that the HA is covering cobalt alloy substrate. However, it seems that the coating formed is uneven and there are some which coagulates as in Fig. 1a and 1b. SEM test was also done on a cross section sample presented in Fig. 2. It was intended to determine the thickness of the HA coating on the surface of the substrate in each sample. The average thickness of a layer of HA obtained from SEM measurements is presented in Fig. 3.

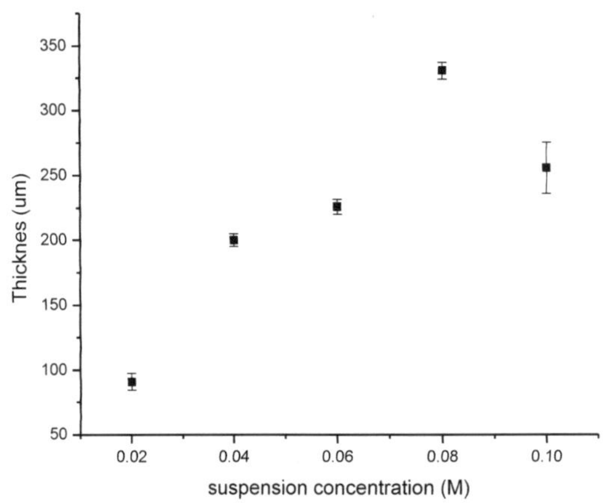

Fig. 3. The thickness of HA layer variation suspension concentrations of HA

Based on Fig. 3, it can be seen that the higher the suspension concentration of HA-ethanol of the concentration of 0.02M up to 0.08M, the greater the value of the HA layer thickness. However, it is not the case at a concentration of 0.10 M, HA coating thickness is lower than in samples of 0.08 M. It is presumed that at a suspension concentration of HA 0.10 M, HA coating that has been coating the surface of the cobalt alloy substrate was not strong enough to hold the HA particles continuously onto the surface of the substrate due to the influence of high power voltage of 120 V so that the outside of the HA layer falls back into suspension, causing the HA layer formed become thinner. Best terms of HA layer thickness as the medical applications standard which do not cause any clinical problems is 50-200μm, seen from the aspect of solubility of the SBF [15]. Based on Fig. 3, it shows that the samples which meet the medical implant applications standard are suspension concentration of HA 0.02 M and 0.04 M respectively in a row with an average thickness of 90.92 ± 6.57 μm and 199.93 ± 4.85 μm.

3.2. Adhesion strength HA layer

Adhesion strength of HA layer on the cobalt alloy substrate was done by conducting shear strength test to determine the strength of HA layer attached to the cobalt alloy substrate. The adhesion strength of HA layer is presented in Fig. 4.

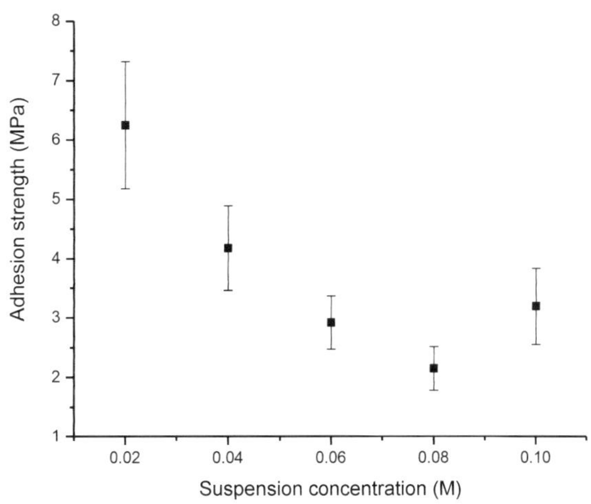

Fig. 4. Adhesion strength vs. suspension concentration of HA

Based on Fig. 4, the highest adhesion strength of HA layer is at 0.02 M of 6.248 MPa. The sample of 0.02 M has the lowest HA layer thickness among other samples. The low strength of adhesion is because the hold time is too short in the sintering process at the temperature of 900^{0}C, so that the HA coating is still not strong enough to retain its position attached to the substrate cobalt alloy.

This shows that the thicker the layer of HA, the lower the adhesion strength of HA coating. According to Fig.4, it also reveals that the higher the concentration of the suspension of HA-ethanol used in the coating process, the lower the adhesion strength of HA coating. The adhesion strength values are still far from the results from the research conducted by Bunyamin [16] which results in adhesion strength of HA coating layer in SS316L with the thickness of 20-25 μm by 20-25 MPa, whereas HA coating on Ti 6Al 4V with the thickness of 25-40 μm has the adhesion strength of 10-14 MPa. The low of adhesion strength in this study is because the resulted layer is very thicker than the expected.

3.3. The corrosion rate in the SBF solution

In this study, the corrosion test was done to all samples in a solution of *simulated body fluid* (SBF). The result of corrosion tests to all samples' concentrations is presented in Fig. 5. According to Fig. 5, it appears that the corrosion rate of all samples coated with HA in SBF solution is smaller than the corrosion rate control of cobalt alloy without coating. This means that the HA layer covering the surface of the cobalt alloy is able to reduce the corrosion rate of cobalt alloy. Cobalt alloy without HA coating has the corrosion rate of 0.288mpy = 0.0074mmpy [10], it is still below the standard tolerated by body which is 0.457 mpy = 0.012 mmpy [17].Thus, the use of cobalt alloys as the implant material will not harm the body because the corrosion rate is still far below European standards. The HA coating further reduce the corrosion rate of cobalt alloy, therefore it is qualified as the material of implant prosthesis.

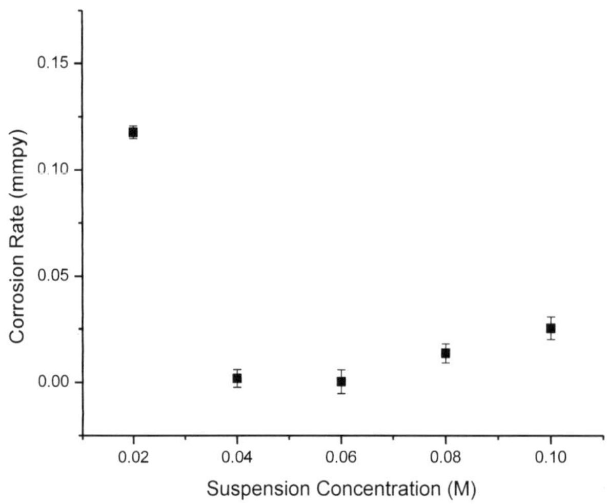

Fig. 5. The corrosion rate variation suspension concentration of HA

Based on Fig. 5, it can be seen that the corrosion rate decreases with the increase of HA suspension, especially at a concentration of 0.02 M – 0.08 M. This is due to the greater concentration the thicker layer resulted (Fig. 3). The presence of HA layer on the cobalt alloy inhibits the corrosion of cobalt alloy. Samples with suspension concentration HA-ethanol of 0.02M has the highest corrosion rate value compared to other samples, which is 0.1177 mmpy. This is because the sample of 0.02 M, HA which coated substrate was not distributed evenly. In the sample of the suspension concentration HA-ethanol of 0.04 M and 0.06 M, they have relatively low corrosion rate values compared to most other samples, which are 0.0018 and 0.0003 mmpy. In both samples, the HA layer covering the cobalt alloy substrate surface was distributed evenly. Therefore, the HA coating can protect the surface of the substrate, resulting in the low value of corrosion rate.

4. Conclusion

Suspension concentration of HA-ethanol 0.02 M to 0.1 M on cobalt alloy coating by using EPD method has effects on the thickness and adhesion strength of HA layer. The higher the suspension concentration of HA-ethanol, the thicker the layer of HA, but the value of the adhesion strength layer becomes lower. Suspension concentration of HA 0.04 M is the best concentration, with characteristics that qualified the standards of the bone implant prosthesis which are HA coating thickness of 199.93 ± 4.85 μm, the corrosion rate of 0.0018 mmpy and adhesion strength of 4.175 ± 0.716 MPa.

Acknowledgement
The researchers would like to thank to DIKTI giving the grant of BOPTN Airlangga University for this research.

References
[1] Fernandez-Pradas *et al.* 2001 Influence of the Interface Layer on the Adhesion of Pulsed Laser Deposition HA Coating on Titanium Alloy *Applied Surface Science* pp 31-37
[2] Blind O, Lorena H, Klein, Bruno D, Laurence J 2005 Characterization of HA Films Obtained by Pulsed Laser Deposition on Ti and Ti-6Al-4V Substrates *Dental Material* pp 1017-1024
[3] Quanhe B, Chuanzhong C, Diangang W 2005 Pulsed Laser Deposition and Its Current Research Status In Preparing HA Thin Films *Applied Surface Science* **252** pp 1538-1544
[4] Quanhe B, Chuanzhong C, Diangang W, Junming L 2008 The Influences of Target Properties And Deposition Times on Pulsed Laser Deposited HA Films *Applied Surface Science* **255** pp 619-621
[5] Dinda G P, Shin J, Mazumder J 2009 Pulsed Laser Deposition on HA thin Film on Ti-6Al-4V, Effect of Heat Treatment on Structure and Properties *Acta Biomaterial* pp 1821-1830
[6] Samuel 2008 The Deposition, Characterisation and Biocompatibility of HA and Silicon Doped HA Thin Film Coating for Orhopedic Applications *Thesis submitted to the University of Nootingham for the Degree of Doctor of Philosophy*
[7] Aminatun, Siswanto, Fitri W 2011 The Effect of Co:Cr Variation to Mechanical Properties and Cytotoxicity Co-Cr-HA Composite *Proc. of 3rd International Conference and Workshop on Basic and Applied Science, Universitas Airlangga*
[8] Aminatun, Siswanto, Penga Y, Istifarah, Apsari R 2013 The Effect Of Sintering Process On The Characteristics Of HA From Cuttlefish Bone (*Sepia Sp.*), *Research J. of Pharmaceutical, Biological and Chemical Sciences* **4** (4) pp 1431-1442
[9] Shamsul A Z, Nur H, Ruzaldi C 2007 Fabrication and Propertis of Cobalt-Chromium HAP Composite *Internasional J. Science.* **3** (1) pp 35-31
[10] Aminatun, Apsari R, Yusuf Y, Suhariningsih 2015 Synthesis And Characterization of HA Layer on Cobalt Alloys Through Dip Coating Method as a Prosthetic Bone Implant Candidate *J. Of Optoelectronic And Biomedical Materials* **7** (1) pp 11-18
[11] Besra, L and Liu M 2006 A Review of Fundamentals and Applications of Electrophoretic Deposition (EPD) *Progress in Materials Science* **52** 1-61
[12] Aminatun, Efinda P, Arista I, Umi H, Dyah H, Suhariningsih 2014 Synthesis of Cobalt Alloy Through Smelting Method and Its Characterization as Bone Implant Prosthesis *AIP Proc.* pp 137-143
[13] Van V L H 1989 *Element of Materials Science and Engineering*. 6th ed. USA
[14] Butarbutar, Sofia L, Febrianto 2009 Pengujian Mesin Edaq Untuk Mengukur Laju Korosi, Pusat Teknologi Reaktor dan Keselamatan Nuklir – BATAN **13** (2) Sigma Epsilon ISSN 0853-9103
[15] Dumbleton J and Michael T M 2004 Hydroxiapatite –Coated Prostheses in Total Hip and Knee Arthroplasty *J. of Bone and Joint Surgery, Inc.* pp 2526-2540
[16] Bunyamin A and Hanyaloglu C 2008 Bioceramis dip – coating on Ti-6Al-4V and 316L SS Implant Materials *J.Mater Sci: Mater Med.* **19** pp 2097-2104
[17] Prasetyo A 2010 The Effect of Silicon on The Corrosion Rate Cobalt Alloy (ASTM F 75) With Potensiodinamic Polarization *Thesis Enginering Faculty of Indonesia University*

IOP Conf. Series: Journal of Physics: Conf. Series **853** (2017) 012026 doi:10.1088/1742-6596/853/1/012026

The effectiveness of laser diode induction to *Carica Papaya L.* chlorophyll extract to be ROS generating in the photodynamic inactivation mechanisms for *C.albicans* biofilms

S Dewi Astuty[1,3,*] and A Baktir[2]

[1] Student of Doctoral Program, Faculty of Science and Technology, Universitas Airlangga, Surabaya, East Java, Indonesia
[2]Department of Chemistry, Universitas Airlangga, Surabaya, East Java, Indonesia
[3]Department of Physics,Faculty of Science and Technology, Universitas Hasanuddin, Makassar, South Sulawesi, Indonesia

*Corresponding email address: tuty_ilyas@yahoo.com

Abstract. Research on the effectiveness of photo inactivation of *C.albicans* biofilms led by a-PDT system mediated by chlorophyll-diode-laser-induced was done. This research was done using *in vitro* technique in order to effectively determine chlorophyll extract of ROS-generated *Carica Papaya L.* using *in situ* technique. This technique induced laser diode on different dose and *C. albicans* with reduced degree. This research is a preliminary study in efforts to find anew sensitizer agent candidate made of chlorophyll extract and antifungal of *Carica Papaya L.* The effectiveness of eradication has been tested with MDA's content and OD of biomass biofilms as well as analyzed using ANOVA and Tukey Test (α=0.05).The characteristic of chlorophyll extract of *Carica Papaya L.* has maximum absorptions on blue areas (λ_{max}= 420 nm) and red areas (λ_{max}= 670 nm). The MIC value of *Carica Papaya L.*'s*chlorophyll* extract against *C. albicans* planktonic and biofilms cell is 63.8 μM and 31.9 μM respectively. The result shows that treatment using laser which was combined with chlorophyll extract is more effective than that with laser only or chlorophyll extract only. The treatment using laser combined with chlorophyll extract obtained more than 65% (α=0.05) (more than that of negative control) for P_2L_1 group with OD_{595} 0.915. The MDA's content showed that group of laser which was mediated with chlorophyll extract had larger values than group of laser or chlorophyll extract only.

1. Introduction

Photo inactivation is part of photodynamic therapy (PDT), known as a-PDT (antimicrobial photodynamic therapy) [1,2]. Photo inactivation aims to inhibit the growth of pathogenic microbes which cause infections in human body. Mechanisms of photo inactivation based on interaction between photons of light with molecules of the sensitizer produce some kinds of ROS (reactive oxygen singlet). ROS is reactive compounds produced from the reaction of the radical compound substrate triplet sensitizer to molecular oxygen, whereas singlet of oxygen molecules formed from the reaction of triplet sensitizer to molecular oxygen. Both types of reactive compounds contribute to oxidation biomolecule by hydrolysis cell mechanisms (Figure 1.).

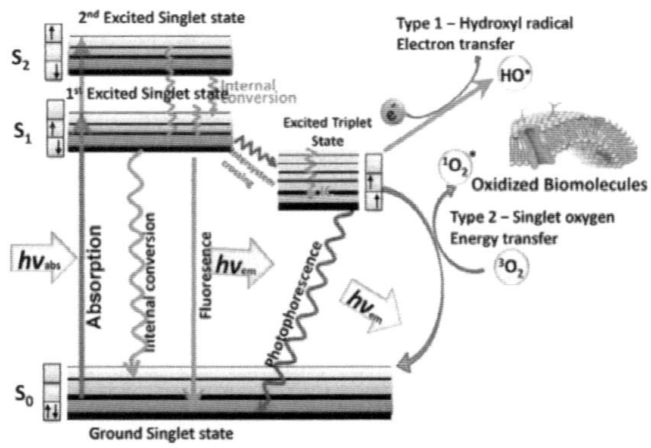

Figure 1. Photoinactivation mechanisms [3]

While the amount of ROS accumulated in a tissue such as the cell membrane or cell wall of microbes, hydrolysis, and ion exchange out and into the cells causes an imbalance of cellular metabolism and DNA chain termination, it also causes cell death by necrosis or apoptosis.

Photons of light absorbed by a molecule or tissue cause multiple interactions depending on the power density of photons of light: photochemical interaction (0.01 to 50 W/cm^2), photothermal (10^0-10^6 W/cm^2), photoablation (10^7-10^8 W/cm^2), plasma-induced ablation (10^{11}-10^{18} W/cm^2), photodisruption (10^{16} W/cm^2) [4]. The interaction of photo inactivation for *C.albicans* is expected to occur through the transfer of photochemical interaction outer electrons in the molecule guanine group so that the electrons become unstable and tend to form free radicals which in turn will induce cell death [4]. In order to avoid excessive thermal effects, the power density of 10-100 mW/cm^{2}is used [5].

C.albicans is microbial pathogens with a high level of virulence, known as triggers infection called candidiasis [6]. Candidiasis can occur in the mouth, gastrointestinal tract, and respiratory and genital mucosa [7] called systemic candidiasis. Systemic candidiasis can be associated with opportunistic nature of *C.albicans* as likely to be severe if patients experience pathologicalimmune-compromise such as HIV-AIDS patients. Pathogenicity degree of *C.albicans* is caused by several factors: the process of adhesion and invasion, toxin secretion of enzymes and metabolites, dimorphism properties, and the phenotype of switching from white to opaque and biofilm formation [8]. Biofilms are considered as a major mediator of infection, with an estimated 80% incidence of infections associated with biofilm formation [9,10].

Morphology of *C.albicans* cell includes the type of eukaryotic organisms with complex structures and has a cell membrane with a more rigid and thicker cell structure. The existence of the extracellular matrix in the form of a layer biofilm β-1.3 glucan improves the rigidity and thickness of *C.albicans* [11,12]. Eradication of pathogenic microbe photo inactivation *C.albicans* through a mechanical consistent with the wavelength of red light was conducted in order to get deeper tissue penetration.

Chlorophyll and its derivatives are potential new photosensitizers in cancer therapy [13,14] because they have similar structure of Heme blood that they can easily accumulate in the body, select only tissue cancer for its hydrophilic properties, and is rapidly cleared from normal tissue because of hydrolytic properties. The main properties of chlorophyll in photosynthesis is a high energy absorption therefore it is also potential as a photosensitizer in photodynamic therapy. The important characteristic of chlorophyll for photodynamic mechanisms is the lifetime of triplet state of a-PDT which is one of photo dynamics mechanisms for inhibition growth of microbial pathogen.

One of the plants which contain high levels of chlorophyll content is medicinal plant [15] with antifungal substances such of flavonoids, saponins, tannins, and papain. Chlorophyll extract of *Carica Papaya L.* has a molecular structure similar to the porphyrin, and it has two characteristic wavelength of maximum absorption in the region: blue and red areas.

This study aims to analyze characterization and effectiveness chlorophyll extract of *Carica Papaya L.,* related to reduction of OD values and MDA's values on photo inactivation mechanisms against *C. albicans* biofilms growths.

2. Method

2.1. Isolation of chlorophyll from Carica Papaya L..
The mixtures for isolation of chlorophyll by maceration methods were made of methanol and petroleum ether in a 2:1 ratio, and they were also re-extracted with petroleum ether and diethyl ether in 1:1 ratio. The chlorophyll fraction used column chromatography step gradient polarity methods with silica gel with the adsorbent and *n*-hexane/acetone mixture as the eluent with ratio from 1:0 to final 1:1 [16].

2.2. Characterization of extract chlorophyll of Carica Papaya L..
This step included several tests such as stability test absorbs spectrum in different temperature of storage (4^0C, 27^0C, 60^0C) and pH of solutions(pH 4.5 – 7.5) [17], phytochemical test on the content of flavonoids, saponins, and tannins, the MIC and MFC test to determine the optimal concentration in photo inactivation treatment. The concentration series was made of 4.3, 8.5, 21.3, 31.9, 42.5, 63.8, and 170 μM (mixture of chlorophyll extract 2% and 20% distilled water).Purity test was done using TLC (*thin layer chromatogram*) compared to *standard chlorophyll TLC 6%chl-a* [16].

2.3. Culture and biofilms preparation.
C. albicans was reactivated by cultivation in PDA media incubation for one night with 37^0C temperature before experiment. One dose of a culture was inoculated into 10 mL of BHI-B and was grown aerobically at 37^0C for one night and shaken at 160 rpm. The aliquot was harvested after centrifugation at10000 rpm and 4^0C for 15 min, washed twice with PBS (sterile pH 7.4), and re suspended in PBS for concentration of 0.5 McFarland (10^7 CFU/mL) with Densichek Calibration Standard. 100 μL of aliquot was next transferred into 96-well microtiter plate (the well number filled aliquot was three times of the number of all treatment groups) and incubated for 1.5 hour for adhesion phase at 37^0C and shaken at 160 rpm. After adhesion phase, the supernatant was removed from the well plate and gently washed twice with 150 μL of PBS to remove non adherent cells. Next, 200 μL of BHI-B, containing 8% glucose, was transferred to each well and incubated and shaken at 160 rpm and 37^0C for 48 hours. Then, biofilms were grown on well plate and washed twice with PBS to remove nonadherent cells [18,19]. For treatment group of chlorophyll extract alone, 100 μL of PBS for both concentrates prepared was transferred into well filled with biofilms and was incubated at room temperature for 2 hours before irradiation with time of illuminating difference. It is the same procedure for positive control group using fluconazole. After all treatment groups had been finished (including illumination), all of biofilms on each well were washed twice with PBS then staining was conducted with 200 μL of crystal violet0.1%. It was next incubated at room temperature, washed twice again with PBS, and added 200 μL of ethanol 98%. 100 μL of aliquot which had been through staining phase was then transferred into new 96-well microtiter plate and measured for its turbidity using Elisa readerλ_{595}.

All of treatment groups should be processed separately to avoid difference of start treatment, especially on time of biofilm incubation. It can also be overcome by creating a biofilm in 96-well separately for each treatment group such as the photosensitizer group alone and the combination of laser-sensitizer group, which was added to chlorophyll extract and incubated simultaneously before the irradiation can be done in parallel. Then other treatment groups (laser alone, positive and negative controls) were stored at refrigerator to slow down the growth of biofilms before all of treatment groups proceeded through phases of staining and reading of their OD.

2.4. Parameters of light and photo inactivation methods

The distance of illuminating of 450 nm of Blue Laser diode with 53.880 mW of output power was 3 cm, whereas its irradiation durations were 60, 30, and 15 seconds (marked as treatment L_1, L_2, and L_3 respectively). Concentration of chlorophyll extract based on MIC biofilms values (P_2 = 31.8 μM)was then compared to another concentration (P_1 = 42.5 μM). The treatment group as much as 13 each in triple namely: (P_1-P_2) photosensitizer only (2 groups), (L_1-L_3) laser only (3 groups), (P_1L_1-P_2L_3) laser combine photosensitizer (6 groups), (F+) positive control groups with antifungal only, and (B-) negative control groups with no biofilms.

2.5. Measurement levels of ROS formation,

In order to measure the levels of ROS formation, this study used MDA-TBA (Malondialdehyda-2Thiobarbituric acid) method. MDA is the number of radicals formed in the sample due to a particular treatment. MDA test aims to detect how much free radical compound accumulated in animal tissues after extreme treatments whereas in the system photo inactivation radical compounds are formed through a photochemical process after the photosensitizer absorbs the light in photophysics process. Radicals formed from a chain of chemical reactions can be assumed as a result of free radicals oxidative stress that it can be measured by MDA method. Chlorophyll extract was made in several rows of isolated and irradiated concentration with a laser diode which has been characterized previously by the length of irradiation time(irradiation dose determination). 0.5 μL of sterile PBS was mixed with 0.05 gr of test samples vortex and centrifuged drying (20 minutes, 10000 rpm, 4^0C) obtained filtrate (f_1). 250 μL of f_1 was mixed with 4.75 mL of sterile PBS vortex and centrifuged (15 minutes, 3000 rpm)obtained f_2. 4 mL of f_2 was mixed with 1 mL TCA vortex and centrifuged (15 minutes, 3000 rpm) obtained f_3. Finally, 4 mL of f_3 was mixed with 1 mL TBA vortex, incubated twice (80^0C for 15 minutes than room temperature for 60 minutes), and centrifuged (15 minutes, 3000 rpm) obtained $f_{4.}$ The last filtrate determined the absorbance with UV-Vis spectroscopy whose absorbance value was detected as MDA's content on test sample.

2.6. The analysis

The data treatment was analyzed with ANOVA test to significantly determine the difference among treatments and with Tukey test whose degree of confidence was above 90%.

3. Results

The isolation of *Carica Papaya L.*'schlorophyll extract showed that chl-*a*, chl-*b*, and chl-total with Argon formula(1949) were 4.431 mg/L, 4.631 mg/L, and 14.058 mg/L. The characteristic of chlorophyll is similar to porphyrin as a dye with an absorbance spectrum with two major absorption bands in the visible range, namely Q band (red area) and Soret band (blue area) [20]. The maximum of absorption obtained two peaks at blue and red areas which could be viewed from the wavelength. The absorbance value at blue areas on 414 nm of wavelength was 0.233 au and in red areas on 670 nm of wavelength was 0.123 au. These were higher than the chlorophyll standard which are 420 nm, A= 0.123 au (blue areas) and 670 nm, A=0.093 au (red areas) (Figure 2.). The literature shows that maximum absorption wavelength for acetone solution were Chl-a with 662, 430, Chl-b with 645.5, and 456.9 nm [16,21].

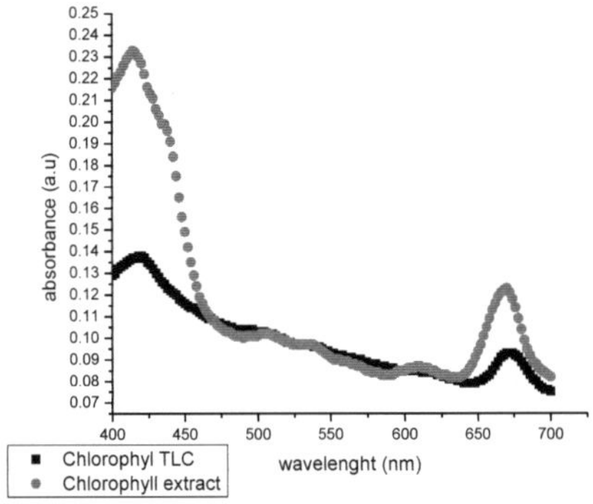

Figure 2. Chlorophyll Absorption Spectrum

HPLC analysis, when starting the retention time of *Carica Papaya L.*'schlorophyll extract is 2.8 minutes (Area=186.9 mAU.s) compared to the standard which is 3.0 minutes (Area=388.5 mAU.s) (flow rate 1 mL/minutes, pressure 35 bar, λ_{ref} 220 nm). In a study was mentioned for chl-*a* and chl-*b* had 13.6 minutes and 9 minutes of retention time [16].

The stability properties of chlorophyll extract of *Carica Papaya L.* have been tested with different storage temperature and pH conditions against maximum absorption wavelength. Acid of conditions had nosignificant effect to changes of maximum absorption spectrum. However, absorbing values relatively decreased because the acid condition made more dilute solution. Despite storage temperature stability, the solutions of chlorophyll extract was not stable for thermal storage above 60^0C because the solution changing color became clear, and it meant there was degradation. Even so, maximum absorption wavelength did not change. The absorption values of *Carica Papaya L.*'s chlorophyll extract storage against temperature of 4^0C, 27^0C, and 60^0C are 0.717, 0.697, and 0.731 (for λ_m=668 nm), and 2.824, 2.523, and 2.340 (for λ_m=415 nm) respectively. A study by Araujo(1995) mentioned the value of molar absorption coefficient against room temperature and refrigerator storage with the dark condition after 84 days with the value are 82.29 L.mmol^{-1}.cm^{-1} (refrigerator) and 80.73 L.mmol^{-1}.cm^{-1} (room temperature) at λ_{665} did not change significantly [22].

The MIC values were needed in determining optimum concentration of *Carica Papaya L.*'s chlorophyll extract (sensitizer) for a-PDT. They were needed to streamline the mechanism of photo inactivation in a-PDT that when applied as a sensitizer, they only act as alight absorber which then produced ROS compound. The result of MIC test showed that the MIC values against *C. albicans* planktonic and biofilms cell are 63,8μM and 31.9 μM respectively. There was reduction about 50% of biofilms concentration compared to that of planktonic cell which was applied to a-PDT mechanisms in which MIC value of biofilms was (31.9 μM) combining with another concentration.

Photo inactivation mechanisms which were applied against biofilms' growth showed effective combination between blue light of laser diode and *Carica Papaya L.*'s chlorophyll extract.

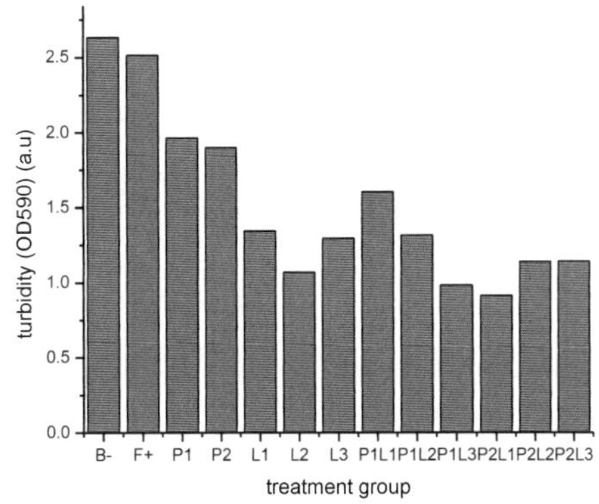

Figure 3. a-PDT mechanisms compared to control

The low turbidity means treatment was done more effectively than in high turbidity. The value of higher OD_{595} showed much biomass which was detected using crystal-violet staining. In this study, the turbidity was impacted by surviving cell in biofilms. The turbidity of negative control (B- treatment group) was 2.632 au, for positive control with antifungal conventional by fluconazole (F+ treatment group)which was 2.515 au (4.46% reduction), for treatment chlorophyll extract only treatment group P_2 (31.8 μM) was 1.899 au (48.97% reduction), for treatment lower than P_1 (42.5 μM) was 1.965 au (27.86% reduction), for treatment laser only treatment group L_2 (30 s) was 1.070 au (59.35% reduction), for treatment lower than L_1 (60 s)was 1.343 au (48.97% reduction), and for treatment group L_3 (15 s)was 1.294 au (50.84 % reduction). For combination treatment of laser and chlorophyll extract (a-PDT mechanisms), the turbidity for P_2L_1 treatment group with 0.915 (65.22 % reduction) was lower than other treatment groups on combination laser with exogenous photosensitizer.

The MDA's value had not shown variability which was significant among groups. The MDA's value of photosensitizer only and laser only groups showed similar value of 6.4906 nmol/mL. However, for laser-mediated chlorophyll group showed higher value than that of photosensitizer only or laser only groups which was about 7.2134 nmol/mL. Based on MDA's value, in this study was noted that radical compound can be detected using MDA's content test, and it can also be proven that there was an interaction between light and a sensitizer in order to produce a radical compound.

4. Discussion

The clinical application of photodynamic therapy with *in vitro* technique on wound infection was carried out. In some microorganisms, it has been proven effective in reducing gram-positive and gram-negative bacteria using a blue light with or without the addition of exogenous photosensitizer [20,23]. In contrast to bacteria in general, microbial *C.albicans* with more complex and thick structures is more effectively inactivated by red light [5] because it has a strong penetrating power. MB dye induced by 660 nm of diode laser for 98 seconds could more effectively reduce *C.albicans* biofilm formed in association with other microbes compared to TBO dye [24].Something similar was also reported that 630 nm of LED with a dose of 30 J/cm^{2}was no less effective photosensitizers to kill planktonic cells of *C.albicans* [25]. MB dye with 660 nm of Laser diode capable of producing the type of ROS sodium dodecyl sulphate, caffeine, and hydrogen peroxide [26] and 660 nm of the low-power laser 100 mW are more reactive against gram-positive bacteria rather than against yeast and gram-negative bacteria [27].

The next stage of further research, that the relation with one of the characteristics of the biofilm is limited zone of oxygen in the layer [28], which is the resistance of photodynamic therapy such as photo inactivation [29]. Limitations of oxygen in suspension and biofilm have been investigated inrelevance to the viability reduction of cells in applications of photo inactivation [20,30] which was allegedly affected by the lack of effective ROS, where the availability of sufficient oxygen into research opportunities especially for applications against *C.albicans* biofilm.

5. Conclusion

Carica Papaya L.'s chlorophyll extract has been isolated with antifungal active substance content, and absorbent molecules with specific characteristics of the two intensities are Soret band and Q band which area potential photosensitizers in photo inactivation specifically for eradication against *C.albicans* biofilm. Furthermore, combining 450 nm of laser diode L_1 27.84 J/cm^2 (60 s) with *Carica Papaya L.*'s chlorophyll extract P_2 (31.875 μM) is more effective in inhibiting 65% growth of *C. albicans* biofilms.

Acknowledgments

We would like to give thanks to Suhariningsih as my promotor and to my co-promotor, Suryani Dyah and Afaf Baktir for all motivation, support, and advice in finishing this paper. This research is part of my great doctoral research and its preliminary study. We would also like to give thanks to DIKTI who has given financial support for Disertation Doctoral Research (PDD) in 2017. Finally, we would like to give thanks to all colleagues at RSKI and Biochemical Laboratory, of whom I cannot mention one by one, for support for instruments used in this research.

References

[1] Ruiz-González R, Agut M, Reddi E, Nonell S 2015 Article A Comparative Study on Two Cationic Porphycenes Photophysical and Antimicrobial Photoinactivation Evaluation *International Journal of Molecular Science* **16** pp 27072–27086
[2] Dai T, Huang Y, Hamblin 2009 Review : Photodynamic Therapy for Localized Infections - State of The Art Photodiagnosis and Photodynamic Therapy **6** pp 170-188
[3] Silva Z S, Sandra K, Kristianne P, Ying-Ying H, Michael R 2015 Animal models for photodynamic therapy (PDT) *Bioscience Reports* **13** 35 (6)
[4] Niemz M H 2007 *Laser-Tissue Interaction, Fundamentals and Applications* Third enlarged edition Springer-Verlag Berlin.
[5] Calzavara-Pinton P G, Rossi M T, Ortel B 2011 Photodynamic Inactivation of Microbial Pathogen : Medical and Environmental Applications, Chapter 14 : Photodynamic Antifungal Chemotherapy *European Society for Photobiology* pp 361-375
[6] McGinnis M R and Tyring S K 1998 *General Concepts of Mycology* Medical Microbiology 4th ed.
[7] Samaranayake Y H *et al.* 2005 In Vitro Method To Study Antifungal Perfusion In Candida Biofilms *J. Clin. Microbiol.* **43** (2) pp 818-825.
[8] Mayer F, Wilson D, and Hube B 2013 Candida Albicans Pathogenicity Mechanisms *Virulence Landes Bioscience* **4**:2 pp 119–128
[9] Archer N K *et al.* 2011 Staphylococcus aureus biofilms Properties, regulation and roles in human disease Virulence *Landes Bioscience Special Focus* **2**:5 pp 445-459
[10] Melo *et al.* 2013 Photodynamic inactivation of biofilm: taking a lightly colored approach to stubborn infection *NIH Public Access Expert Rev Anti Infect Ther.* **11**(7) pp 669–693
[11] Stubss H J *et al.* 1999 Hydrolase and transferase activities of the ß-1, 3-exoglucanase of Candida albicans *Eur. J. Biochem.* **263** pp 889-895
[12] Baktir A *et al.* 2012 Novel Materials for Eradicationof Biofilm Extracell Matrix of pathogenic candida *J. of Materials science and Engineering B2* **12** pp 650-658
[13] Indrawati R *et al.* 2010 Perkembangan Sensitizer pada Terapi Fotodinamik Tumor dan kanker Hingga Penuntunan Nanopartikel (Nanoparticulate Targeting) Dengan Antibodi Monoklonal *Indonesian J. of Cancer* **4**(3) pp 101-110
[14] Brandis A S *et al.* 2006 *Chlorophylls and Bacteriochlorophylls : Biochemistry, Biophysics, Functions and Applications,* pp. 461-483, Springer, Printed in The Netherlands.

[15] Setiari N and Nurchayati Y 2009 Eksplorasi Kandungan Klorofil pada beberapa Sayuran Hijau sebagai Alternatif Bahan Dasar Food Supplement J. BIOMA **11** (1) pp 6-10
[16] Milenkovic S M *et al.* 2012 The Identification of Chlorophyll and Its Derivatives in he Pigment Mixtures : HPLC-Chromtography, Visible and Mass Spectroscopy Studies *Advanced Technology* **1**(1) pp 16-24
[17] Budiyanto AW, Notosudarmo, Limantara L 2008 Pengaruh Pengasaman terhadap Fotodegradasi Klorofil a *J. Matematika dan Sains* **13** (3) pp 66-75
[18] Dovigo L N, J.C. Carmello, C.A. de Souza Costa, C.E. Vergani, I.L. Burnetti, V.S. Bagnato, A.C. Pavarina 2013 Curcumin-Mediated Photodynamic inactivation of Candida albicans in a murine model of oral candidiasis *Medical Mycology J.* **51** pp 243-251
[19] Andrade A.P.D. Ribeiro, L.N. Dovigo, I.L. Brunetti, E.T. Giampaolo, V.S. Bagnato, A.C. Pavarina, 2013, Effect of different pre-irradiation times on curcumin mediated photodynamic therapy against planktonic cultures and biofilms of Candida spp, archives of oral biology, **58** pp. 200-210.
[20] Astuti S D 2011 *Potensi* Light Emitting Diode (LED) Biru Untuk Fotoinaktivasi Bakteri Staphylococcus Aureus dengan Porfirin Endogen *Disertasi, Program Doktor MIPA, Program Pascasarjana Univeritas Airlangga*
[21] Jeffrey S W *et al.* 1996 *Phytoplankton Pigments in Ocenography : Guidelines to Modern Methods,* UNESCO Publishing, Paris, p.1.
[22] Araujo P W and Brereton R G 1995 Stability Studies of Chlorophyll a Using Different Storage systems *Analytical Proc. Including Analytical Communications* **32** pp 415-417
[23] Dai T *et al.* 2012 *Concepts and Principles of Photodynamic Therapy As An Alternative Antifungal Discovery Platform* Front. Microbiol.
[24] Pereira C A *et al.* 2011 *Original* Article : Susceptibility of Candida albicans, Staphylococcus aureus, and Streptococcus mutans Biofilms to Photodynamic Inactivation : an in vitro Study *Laser Medical Science* **26** pp 341-348.
[25] Matthes D F P *et al.* 2012 Photosensitization of In Vitro Biofilms Formed on Denture Base Resin *The J. of Prosthetic Dentistry* **112** (3) pp 632-637
[26] Kato L T *et al.* 2013 Antimicrobial Photodynamic Inactivation Inhibits C.albicans Virulence factors and Reduces In Vitro Pathogenicity *Antimicrobial Agents and chemotherapy* **57** (1) pp 445-451
[27] DeOliveira B P *et al.* 2014 Original Article : In Vitro Antimicrobial Photoinactivation With Methylene Blue in Different Microorganisms *Brazilian J. Oral Sciences* **13** (1) pp 53-57
[28] Walters M C *et al.* 2003 Contributions of antibiotic penetration, oxygen limitation, and low metabolic activity to tolerance of *Pseudomonas aeruginosa* biofilms to ciprofloxacin and tobramycin. *Antimicrob. Agents Chemother.* **47** pp 317–323
[29] Taraszkiewicz A *et al.* 2013 Review Article Innovative strategies to overcome biofilm resistance *BioMed Research International*
[30] Maclean M *et al.* 2008 The Role of Oxygen in The Visible-Light Inactivation of Staphylococcus Aureus *J. of Photochemistry and Photobiology B : Biology* **92** pp 180-184
[31] Grossweiner L I 2005. *The Science of Phototherapy: An Introduction* Springer USA.
[32] Prasad P N 2003 *Introduction to Biophotonic* Wiley Interscience John Wiley and Sons, Inc, Pub.

IOP Conf. Series: Journal of Physics: Conf. Series **853** (2017) 012027 doi:10.1088/1742-6596/853/1/012027

Preparation of creating active carbon from cigarette filter waste using microwave-induced KOH activation

Yanuar Hamzah* and Lazuardi Umar

Department of Physics, Universitas Riau, Kampus Bina Widya Sp. Baru, Pekan Baru 28293, Indonesia.

*Corresponding email address : yanuar.hamzah@lecturer.unri.ac.id

Abstract. For the first time, cigarette filter waste, which is an environmental hazardous material, is used as basic material prepared for creating activated carbon (AC) via KOH chemical activation using a microwave input power of 630 W and irradiation time of 20 minutes. Active carbon was characterized by TGA, x-ray diffraction, scanning electron microscopy, energy dispersive x-ray, nitrogen adsorption-desorption, and absorption of methylene blue (MB). The results of x-ray diffraction showed that active carbon has a semi-crystalline structure with peaks of 2θ of 22.87° and 43.70°. Active carbon microstructure analysis showed that the layer height (L_c) is inversely proportional to the width of the layer (L_a), and the distance between the two layers is d_{002} and d_{100}, which depends significantly on the ratio of AC: KOH. It was found that the optimum BET surface area and adsorption capacity for MB were 328.13 m^2 / g and 88.76 m^2 / g, respectively. The results revealed the potential to prepare activated carbon from cigarette filter waste using microwave irradiation.

1. Introduction

There are 5.6 trillions of cigarette filter butts which have been produced by tobacco industries worldwide, and it is now the biggest challenge for global environment protection every year [1]. Cigarette filter butts (CF) are made of nearly 95% non-biodegradable cellulose acetate (plastic) which looks like cotton. CF is disposal, toxic, and hazardous material waste thrown to the environment after usage despite its extremely slow degradation process [2]. It has a microscopic-sized fibrous matter characterized with Y shape made of cellulose acetate. A plasticizer, glycerol triacetate, is applied to bond the fibers [3]. Cellulose acetate is a type of polymer made of paper and similar to cotton of fiber composites in thousands of fiber. For this research, the writer could not find any similar research using cellulose acetate precursor of cigarette butts waste as an active carbon.

The CF waste has the potential to be used as raw material for active carbon. It has been reported that active carbon materials containing a meso-/micropore structure can be produced using carbonization process with 262 m^2/g of its surface area [4]. The porous carbon of CF has also been studied using pre-treatment sulfuric acid (H_2SO_4) which is carbonized at 500°C temperature. Using the Braumauer Emet and Teller (BET) method, surface area was at 117.2 m^2 / g and the adsorption-desorption isotherm surface area decreased by around 72% [5]. These active carbons are especially suited to oil-water separation, methylene blue (MB) adsorption, and super capacitors [6-8]. The use of active carbon as alternative adsorbent material is because it has large surface area and structural diversity.

In this present work, we report for the first time the preparation of active carbon from CF butts waste by microwave induced KOH activation and a way of making use of a waste material. The CF butt which was a hazardous waste material was selected as promising alternative due to its abundance, thus reducing it as a disposal problem. Microwave heating method has been applied to prepare active carbons because it rapidly and uniformly heats [9]. In addition, it is an energy-efficient method compared to conventional methods by efficiently improving the quality of final products. Effect of the activation ratio of KOH to enlarge the absorption capacity and MB absorptions of the CF active carbon products were then investigated to determine the optimum preparation condition.

2. Materials and method

2.1 Experiment

A set up experiment has been done using microwave induce active carbon from biomass with KOH activation method as reported in [9]. In this research, the cigarette filters (CF) from Marlboro Red (Philip Morris, Indonesia) were collected and used for this purpose. The CBF were then washed with hot water and unwrapped from the cigarette paper. They were then washed again with 96% ethanol for three times, dried in open air for 72 hours, and placed into a vacuum oven for 3 hours at 230°C for heat treatment. The CF samples were then crushed into a powder size of about 100 μm. The CF powder was activated with potassium hydroxide (KOH) solution at ratio carbon-to-alkali from 1, 2 and 3 labeled as CF-1, CF-2, and CF-3, respectively. The activation step was induced in a tubular glass reactor with a microwave input power of 630 W at 2.45 GHz frequency and irradiation time for 20 minutes. The samples were washed with hot distilled water (85°C) until the pH of the filtrating solutions was neutral (pH = 7.0). Finally, the cleaned samples were immediately dried in a vacuum oven at 100°C for 60 hours to eliminate moisture.

2.2 Characterization Technique

Thermal decomposition analysis of the CF was determined using Thermal Gravimetric Analysis (TGA) of Shimadzu TGA-50 technique at the temperature range from 30°C up to 600°C at the heating rate 1°C/min in liquid nitrogen flow. Crystallographic properties of the prepared active carbon of the sample were characterized using x-ray diffractometer (XRD) of Siemens D5000m, with nickel-filtered Cu K K-α target (λ = 1,5418 Å) at the scan rate 0.5°/min in the 2θ range from 10-60°. The morphology and microstructural investigation of the prepared samples were then performed using Scanning Electron Microscopy (SEM) and Energy Dispersive X-Ray (EDX) (JEOL-JSM 6360LA) technique. The adsorption-desorption isotherm of surface area was analyzed using Quantachrome Instruments with liquid nitrogen flowing at 77K. Finally, all of the samples were studied on MB adsorption.

3. Results and discussion

The TGA analysis has shown the obtained thermal decomposition of CF in the first stage of heating. A mass loss was observed in 250°C to 400°C temperature. This stage was due to the dehydration of water content in the CF samples. The highest mass loss (74%) trend was observed from second stage starting at 260°C and was completed 350°C. At higher temperatures, the sample mass decreased slightly up to 600°C. The results of weight loss showed a CF carbon yield of 16% of the final mass, indicating the decomposition of cellulose which was a structure with higher stability.

The effect of pyrolysis of cellulosic acetate with a mass loss in the second stage, reported in [4], yielded 12% of final weight loss and started at 280°C to 370°C temperature. In [5], cellulose acetate filter with CF carbon yield of 10% to 15% using carbonization process was conducted at 280°C to 360°C temperature, and the result had good agreement with our results.

IOP Conf. Series: Journal of Physics: Conf. Series **853** (2017) 012027 doi:10.1088/1742-6596/853/1/012027

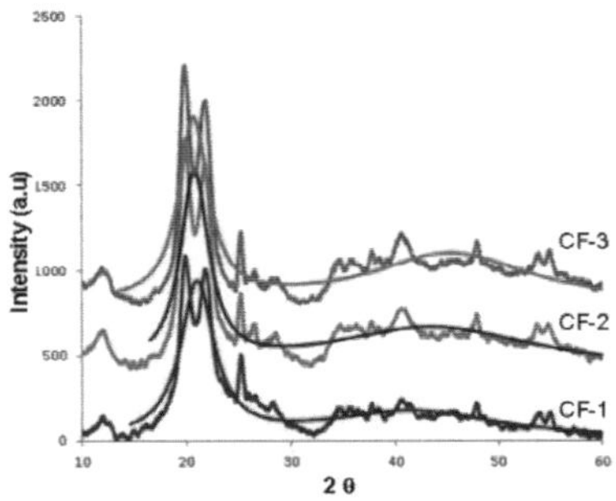

(a)

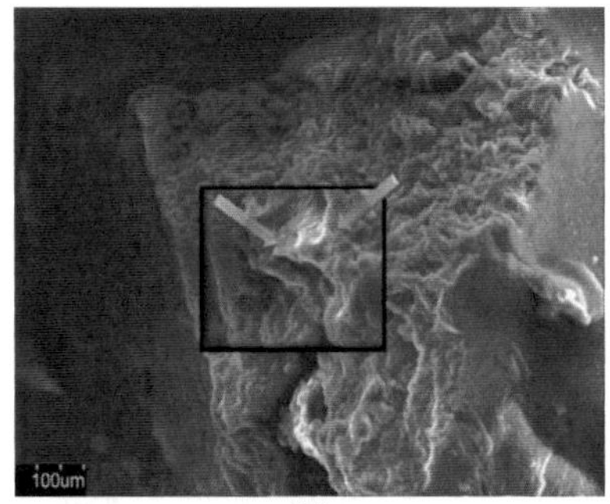

(b)

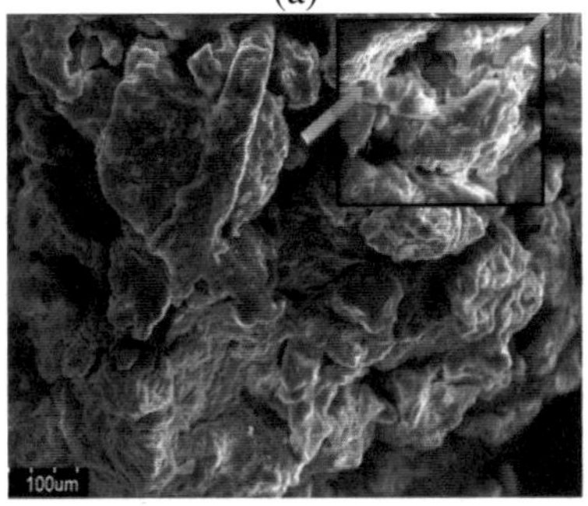

(c)

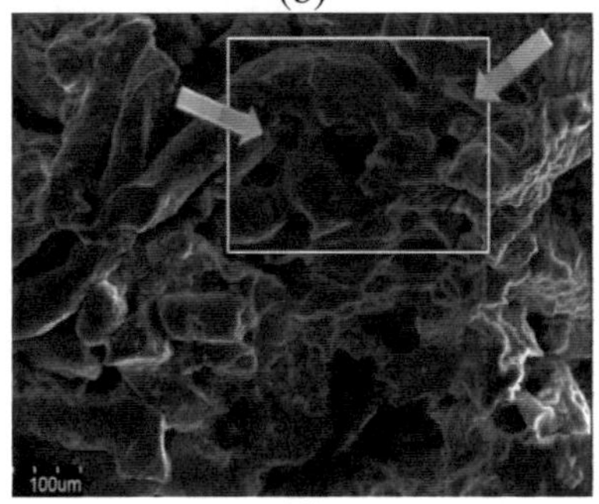

(d)

Figure 1. (a) XRD pattern and (b), (c), (d) SEM images of the active carbon CFs.

Figure 1 (a) shows the results of XRD pattern of all samples indicating two major reflections at the position 22.87° (002) and 43.70° (100). The XRD analysis shows a shift angle of reflection for each peak due to KOH activation ratio. The relative crystallinity all the sample (2θ) was quantified range from 10° to 60° and fit using the Origin software (Microcal Inc., Version 8). From the determination of the interlayer interferences, the average layer height L_c was 3.53 nm with the average width of the layer L_a was 0.59 nm for the intensity of peaks (002) and (100). The interlayer spacing was calculated for sample CF-3 and resulted in d_{002} (0.38 nm) and d_{100} (0.21 nm). These reflections were similar to non-graphitic carbons with significant degree of rotational and translational disorder in the stacking of the graphenes, resulting in better layer alignment [4]. Active carbon CF microstructure analysis showed that the layer's height (L_c) was inversely proportional to the width of the layer (L_a), and the distance between the two layers of d_{002} and d_{100} depends significantly on the ratio of CF and KOH.

Figure 1 (b), (c), and (d) show SEM image of samples CF-1, CF-2, and CF-3, respectively. The porous active carbon of all samples is meso-/micro porous-composed with the diameter of ~ 50 μm and the length of ~ 10 μm. Figure 1 (d) shows a magnified images depicting numerous micro porous structure on the surface of the active carbon CF. As comparison, the SEM reported that CF carbonized at 800°C for 7 hours has the same morphology with our result of CF-3 [5]. EDX analysis of all samples shows the presence of C and O elements with the following composition: 62.23% C, 37.77% O (CF-1); 65.50% C, 34.50% O (CF-2); and 68.68% C, 31.32% O respectively. The result shows that there is no other element observed which was composed in our active carbon CF.

To study the effect of active carbon CFs porosity after KOH activation, nitrogen adsorption-desorption isotherms measurement was then calculated using BET method, and it resulted specific surface area of 303.49 (m^2/g) with pore diameter of 3.01 nm for CF-1 and 328.13 (m^2/g) with pore diameter of 3.04 nm for CF-3 compared to 262 m^2/g for porous carbon from cellulose acetate fibers [5].

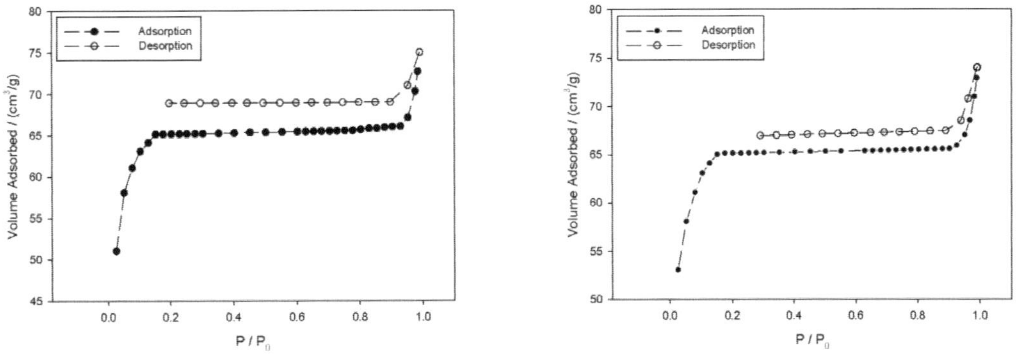

Figure 2. Nitrogen adsorption-desorption isotherms of active carbon CFs (a) CF-1 and (b) CF-3

The ability of active carbon to absorb the MB is a quality standard to absorb dye. The active carbon CFs prepared in this work showed relatively high MB adsorption capacity of 88.77 mg/g for CF-3, as compared to some previous works reported in the literature [10].

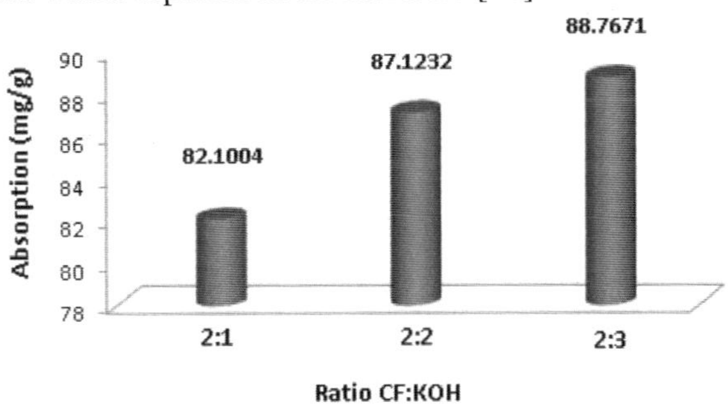

Figure 3. Absorption of methylene blue Ratio of CF:KOH

The observed result of MB active carbon is verified with the analysis of the microcrystalline dimension, the average layer height L_c, the average width of the layer L_a, as well as surface area of active carbon produced. The microcrystalline dimension of active carbon CFs L_c increased inversely proportional and showed a similar trend to that of the L_a micro crystalline dimension (ratios-to-alkali) ranging from 1, 2 and 3 due to enlarged surface area for absorbed MB.

4. Conclusion

The result shows the adaptability of microwave-induced in activating the KOH/cigarette filters waste with simple activation time for 20 minutes and 630 Watt at 2.45 GHz frequency. The two broads peak at (002) and (100), and shifted diffraction peaks in XRD curve due to variation on the mass ratio of CF/KOH. The BET surface area and pore diameter were 328.13 m^2 / g and 3.04 nm respectively. This indicates the optimum adsorption capacity of active carbon for MB of 88.77 mg / g, verified by the morphological and nitrogen adsorption-desorption isotherms studies.

Acknowledgement

The authors would like to thank Ms. Nanda Nurbaiti for preparing the samples, Mr. Valendry Harvenda for manuscript preparation, and also Mr. Awitdrus for result analysis and discussion of this research.

References

[1] Healton C G, Cummings K M, O'Connor R J and Novotny T E 2011 *Tob. Control* 20 i1.
[2] Smith E A and Novotny T E 2011 *Tob. J.* 20 2-9.

[3] Barnes R 2011 *Tob. Control* 20 45-8.
[4] Polarz S, Smarsly B and Schattka J H 2002 *Chem. Mater.* 7 2490-5).
[5] Soltani S M and Yazdi S K, 2012 2nd Intl. Conf. on Environmental and Industrial Innovation (Singapore) vol 35 (Singapore : IACSIT Press Singapore) p11.
[6] Liu C, Chen B and Li C 2015 J. Adhesion Sci. and Tech. 29 2399-407.
[7] Foo K Y and Hameed B H 2011 Chem. Eng. J. 166 792–5.
[8] Lee M, Kim G P, Song H D, Park S and Yi J 2015 Nanotech. 25 345601-68.
[9] Ji B Y, Li T, Zhu L, Wang X and Lin Q 2007 Appl. Surf. Sci. 254 506-12.
[10] Kopac T, Dogu G and Dogu T 2008 Chem. Eng. J. 144 400-06.

Synthesis and characterization of breast-phantom-based gelatine-glutaraldehyde-TiO_2 as a test material for the application of breast cancer diagnosis

Nuril Ukhrowiyah*, Novi Setyaningsih, Dyah Hikmawati, and Moh Yasin

Department of Physics, Faculty of Science and Technology, Airlangga University, Surabaya 60115, Indonesia.

*Corresponding email address: nurilukhrowiyah@yahoo.co.id

Abstract. Synthesis of breast-phantom-based on gelatine-glutaraldehyde-TiO_2 as testing material of breast cancer diagnosis using Near Infrared-Diffuse Optical Tomography (NIR-DOT) is presented. Glutaraldehyde (GA) is added to obtain optimum breast phantom which has same elasticity modulus with mammae. First, synthesis is conducted by mixing gelatine with various amounts of 1 g, 2 g and 3 g with saline solution on 40^0 C temperature for 30 minutes until they become homogenous. Next, GA with concentration of 0.5 and 1.0% is added. The characterization includes FTIR test, physical test, and mechanical test used to identify group of gelatine's functions. Elasticity modulus of breast phantom of gelatine composition 2 g and 0.5% GA is obtained at 53.46 kPA which is the approximation of mammae culture elasticity. This composition is chosen to synthesise the next step. In the second step, TiO_2 is added with variation of 0.01 g, 0.015 g, 0.02 g, 0.025 g, and 0,03 g. With this variation, it is aimed to get a breast phantom providing image with optimum absorption. The test of this material uses Differential Scanning Calorimetry (DSC), homogeneity test, and analysis of coefficient absorption. The result shows the sample has a good thermal property in the range of 40 – 70^0 C with a good homogeneity and absorption coefficient of 0.4 mm^{-1}.

1. Introduction

Cancer is a disease caused by abnormal cell / tissue which is malignant and grows fast, whereas breast cancer is a type of cancer mostly found in women, and the highest number of breast cancer case can be found in Indonesia [1] Plaint or early symptoms of this disease are often unrecognized by sufferers, therefore, it needs a diagnosis method which is fast, safe, inexpensive, effective, and noninvasive. One technique for early detection of Breast cancer which is easily accessible by public is the near infrared-diffuse optical tomography (NIR-DOT). The advantage of NIR-DOT is that the radiation source used is in the form of near infrared (NIR) which is non-invasive and non-ionizing .Unfortunately, direct exposure to visible light or NIR on mammary tissue often causes concern in most women. Therefore, in an effort to build early detection equipment with NIR-DOT requires replacement material to act as tissue simulations to ensure the presence of Carcinoma mammae. For that purpose, breast phantom is used.

Phantom is defined as an object which is able to act as a tissue simulation to replicate the characteristics of human's or animal's tissue [2] This requires test material to have elastic characteristics with the corresponding value of the absorption and scattering similar to those of breast. One of the materials which meet these properties is gelatine [3]

Gelatine is a natural polymer extracted from bones and skins of animals which are used extensively both in pharmaceutical and industrial fields. Gelatine swells when it absorbs water. It is capable of absorbing water 5-10 times its weight, forming a gel at 35-40° C temperature. Furthermore, it is soluble in hot water and can be reversibly changed from sol to gel [4]. In making this gelatine-based breast phantom, saline solution is used as a solvent, Copy of isotonic fluid which represents the human body fluids [5] However, gelatine has the disadvantage that it decays quickly that it is necessary to add GA as a crosslinking agent (crosslink) to keep it long lasting. Crosslink itself is a bond which connects one polymer chain with another polymer chain which can increase resistance to heat, chemical resistance, and mechanical properties such as tensile strength It can also increase the thermal stability [6]. Breast phantom GA is easily obtained from a lower toxicity than from formaldehyde Inelastic, durable breast phantom must also have optical properties whose absorption and scattering properties can be detected by the means of optical tomography. It also needs extra TiO_2 as scattering agents since it is nontoxic, stable, and environmentally friendly non-corrosive [7].

Based on the background the above, this research aims to create gelatine-glutaraldehyde-TiO_2 synthesized breast phantom glutaraldehyde with saline solvent. Characterization applied in this research is FTIR test, and physical characterization includes ease of samples when removed from the mould and durability against environmental sample, press test, DSC test, homogeneity test, and analysis of the absorption coefficient.

2. Research methods

Materials needed in this research were gelatine, glutaraldehyde, TiO_2, NaCl, and distilled water. Equipment used included digital balance, beakers, thermometers, pipettes, stirrers, magnetic stirrer, stirrer bar, and plastic pots. The tools used for sample characterization included FTIR, DSC set of test equipment, tensile strength equipment, and a set of optical tomography.

Synthesis of breast phantom began with a saline solution made of 0.9 g of NaCl dissolved in 100 ml of distilled water and 0.5% of glutaraldehyde solution preparation and 1%. of Breast phantom was synthesized with ratio of saline and gelatine used was 1:10 [8] The next step was making variation of gelatine of 1 g, 2 g, and 3 g, or making gelatine solutions of 10% (w / v), 20% (w / v), and 30% (w / v). The materials used in the first stage before varied by TiO_2 are presented in Table 1.

Table 1. Materials for First Stage

Sample	Gelatine (g)	Glutaraldehyde (%)
A	1	0.5
B		1
C	2	0.5
D		1
E	3	0.5
F		1

Gelatine was mixed with saline solution at 40° C for 30 minutes until becoming homogeneous. Stirring was done with a minimum speed to reduce bubbles in the sample. After 30 minutes, the gelatine solution and glutaraldehyde were added into the sample at 1:2 ratio [6] Then, the sample was poured into a plastic pot and cooled in the refrigerator for 24 hours. Samples with the best variation that were printable and durable and had similar elasticity of breast tissue characterization based on the first stage next were varied by TiO_2. TiO_2 variation of 0.01 g, 0,015 g,; 0.02 g, 0.025 g, and 0.03 g [9] Next, TiO_2 was added to a mixture of gelatine and saline solution for 1 hour until it became homogeneous, then glutaraldehyde was added. The solution was then poured into the mould and stored in the refrigerator for 24 hours.

Characterization of the samples was done through FTIR test, physical characterization test, tap test, DSC test, and materials test, and material absorption coefficient (μ_a) homogeneity test. FTIR test was performed to determine the functional groups of each ingredient. The physical characterization was performed to determine the ease of sample when removed from the mould and durability so that samples can be used optimally. Pressure test was performed to analyse the sample's modulus of elasticity. DSC test was carried out on the effect of temperature on test material to ensure that during the tomographic test, the sample was not degraded due to the heat radiation of the laser. The next test was homogeneity and optical properties test which was performed by measuring the absorption coefficient value (μ_a) samples using optical tomography.

3. Results and discussion

Result of the sample obtained shortly after synthesis process and after stored in the refrigerator for 24 hours are shown in Figure 1. The sample was gelatinous, rubbery, and transparent.

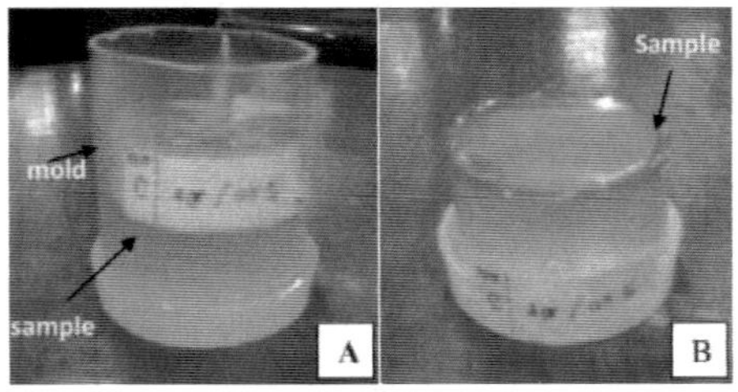

Figure 1. (A) The sample after synthesis; (B) samples after being stored in the refrigerator for 24 hours

Once the samples were synthesised, it was then tested using the FTIR, physical, and compression test. Physical tests included ease of samples when removed from the mould and test on the sample's durability. The best sample based on physical tests and pressure tests were next varied with TiO_2. Variations of TiO_2 used were 0.01 g; 0.015 g; 0.02 g; 0.025 g, and 0.03 g. Various results of the sample after TiO_2 was added were shown in Figure 2. In Figure 2 (a) were shown differences in the samples before and after TiO_2 was added. Samples after synthesis before TiO_2 was added were transparent brown, whereas samples became murky brown after TiO_2 was added. The greater the composition of TiO_2 added, the murkier the samples' colour were. As for the absorption coefficient test sample preparation, samples were printed cylindrically flat with 2.7 cm in diameter, and they had various thickness of 0.3 cm, 1.5 cm, and 0.3 cm as shown in Figure 2 (b).

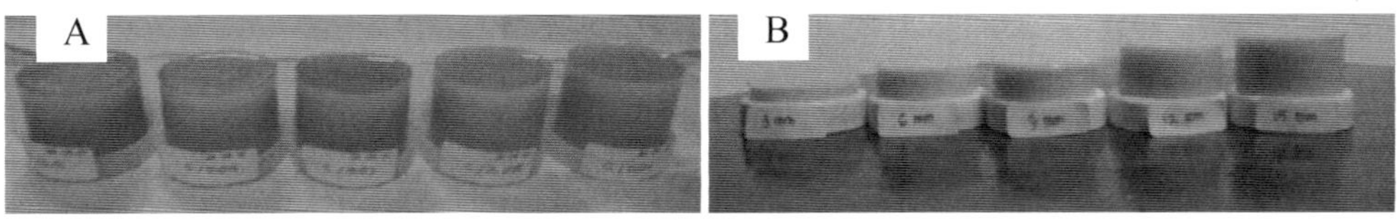

Figure 2. (A) The sample after TiO_2 varied, (B) The test sample material absorption coefficient

3.1. *FTIR test*

FTIR characterization test result performed on samples of gelatine, gelatine composite-GA, and GA was shown in Figure 3. FTIR analysis was performed by matching the wave absorption unique area to each functional group. Results of functional groups along with local uptake obtained by FTIR spectra were shown in Table 2.

Table 2. Results of FTIR Spectrum Analysis

Function Group	Wave Number (cm^{-1})		
	Materials		Sample
	Gelatine	GA	Gelatine- GA
OH	3398.69	3492.55	3439.19
NH *Bending*	1656.91		1541.18
CH *Stretching*	2881.75 2943.47	2874.03 2951.19	
COO^-	1244.13 1336.71		1338.64 1396.51
C=O		1633.76	
C=N			1649.19

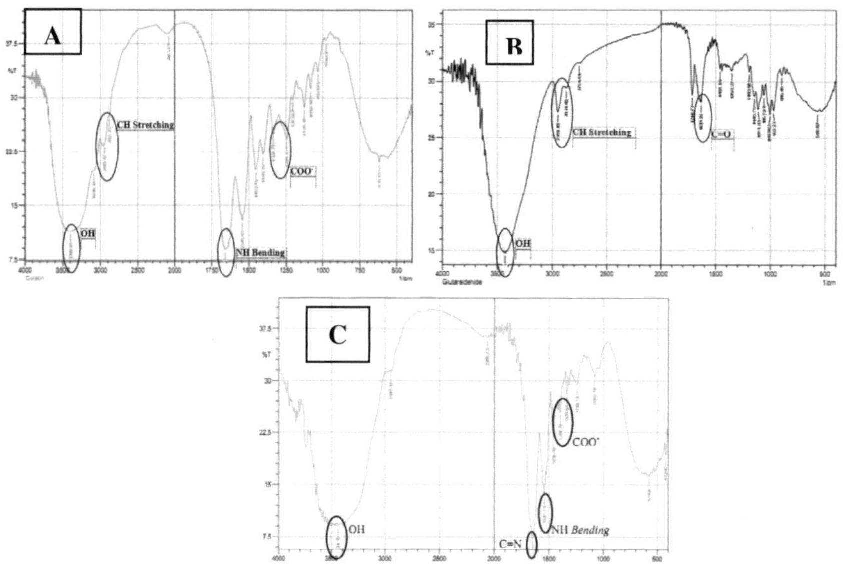

Figure 3. FTIR spectra of (A) gelatine, (B) GA, and (C) gelatine-GA

Based on the FTIR test results, each group had a specific and different absorption. Gelatine had a fairly long chain with several groups contained therein such as amines, carbonyl, and hydroxyl. Bending NH amine groups were present in the wave absorption 1656.91 cm^{-1}. Carbonyl groups shown in the FTIR spectrum were a group of COO derived from the proline in the gelatine whose absorption wavelength was 1244.13; 1336.71 cm^{-1} hydroxy 1. The OH group of gelatine was contained in uptake 3398.69 cm^{-1}. CH Stretching was on the uptake of 2881,75 and 2943.47 cm^{-1}. The FTIR spectrum of GA included two functional groups based on the chemical structure of the GA. The cluster functionality in the GA was CH and C = O stretching. CH Stretching on GA detected in absorption wavelength was 2874.03 and 2951.19 cm^{-1}. Carbonyl group C = O was contained in the uptake of 1633.76 cm^{-1}. Composite gelatine-GA that has been characterized by FTIR was known as the functional groups that represent each ingredient. FTIR spectra representing gelatine was a group of NH Bending at 1541.18 cm^{-1} absorption. OH groups were at 3439.19 cm^{-1}, and catchment area group C = N as the identity formation of crosslink and amide group C = O gelatine was at a similar wave, respectively - each was 1690-1640 cm^{-1} and 1680-1630 cm^{-1}. The composite spectrum gelatine-GA was detected in absorption of 1649, 19 cm^{-1}. This led to the formation of crosslink to the composite which could not be identified through the test FTIR.

Physical tests were conducted to determine the ease of sample when removed from the mould and samples' durability or resistance to the environment when the sample was placed at room temperature. Observations on convenience samples when removed from the mould was done by releasing the samples that have been stored in the refrigerator for 24 hours from the mould, as shown in Figure 4. Based on test results convenient samples when removed from the mould, the samples A and B were hardly removed compared to samples C, D, E, and F. This was because the gelatine concentration in the samples A and B were lower than that in other samples. 1 g of gelatine could make the sample's texture less dense because of the amount of crosslink between gelatine and GA on samples A and B was fewer than that of other samples. When samples A and B were removed from the mould, the bottom of the samples kept sticking to the mould board so that the samples were damaged in the bottom part. In contrast, the C, D, E, and F samples were easier to remove that the bottom did not stick to the mould or damaged. Samples were solid when removed from the mould as shown in Figure 4 (a) and the sample did not stick when released from the mould as shown in Figure 4 (b).

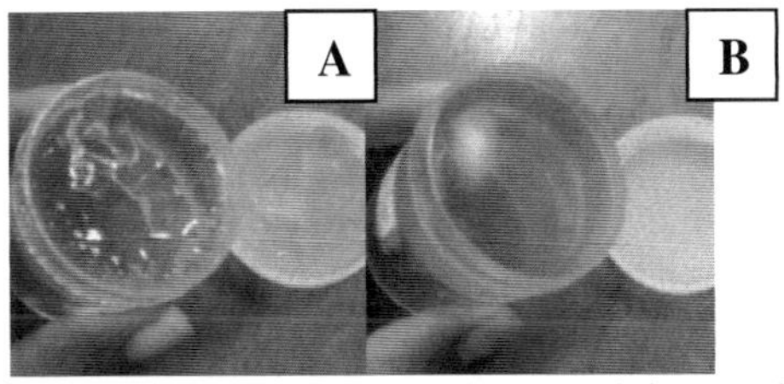

Figure 4. The samples when removed from the mould (A) stick; (B) does not stick

Observation on the samples' resistance or durability to environmental was done by observing the samples which were placed at room temperature once every 24 hours until the samples were damaged or mouldy. Based on the observation result, it was found that samples A, C, and E with 0.5% of glutaraldehyde concentration lasted for 3 days, while samples B, D, and F with 1% of glutaraldehyde lasted for 4 days. The result of the samples' lasting endurance test and physical test results for all samples are presented in Table 3.

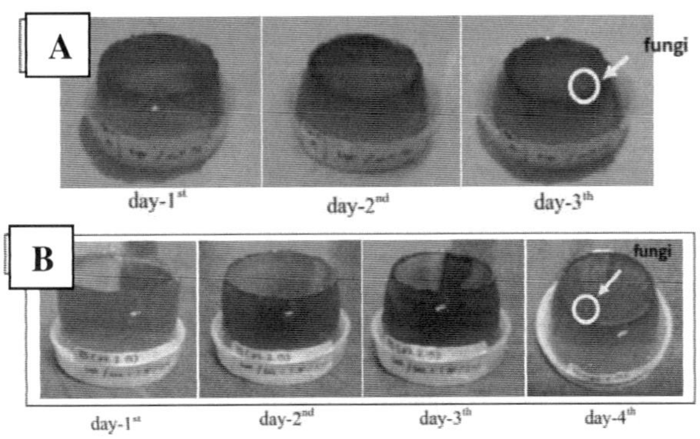

Figure 5. (A) Sample A on days 1 to 3, (B) Sample D on days 1 to 4

Sample A on day 1 to 3 is shown in Figure 5 (a), and sample D on day 1 to 4 is shown in Figure 5 (b). Sample A on the first day had not been damaged. However, it began to shrink in the upper part on the second day. This occurred because the sample was dehydrated, in other words, the water in the sample evaporated into the air which made the top part of the sample contract. On the third day, sample A was overgrown by fungus on its surface which was indicated by the white arrow in Figure 5 (a). Meanwhile, on the third day, sample D had not been invaded by fungi, but its upper part started to shrink. On the fourth day, sample D started to be overgrown by fungi as indicated by white arrow shown in Figure 5 (b). Sample C and E experienced similar process since they had same concentration of GA which was 0.5%. Meanwhile, sample B and F, which had similar concentration to that of sample D, experienced similar process as sample D did. Examination of the physical tests showed that samples with higher concentration of glutaraldehyde has stronger resistance to the environment, in other words, they are more durable. This indicates that the crosslink agent glutaraldehyde makes samples more durable because of its degree of crosslink among gelatine and higher GA. That also depends on the degree of density crosslinking agent

Table 3. Physical sample test results

Sample	Gelatine	GA (%)	Sample removable	sample durability in room
A	1	0.5	difficult	3
B		1	difficult	4
C	2	0.5	easy	3
D		1	easy	4
E	3	0.5	easy	3
F		1	easy	4

3.2. Press test

Characterization of compressive strength to produce data in the form of the force (F) required to press the sample until the sample is damaged / injured. The sample used for the pressure test is a cylinder with 3.6 cm in diameter and 2 cm in height. The process was done by pressing up to a depth of 1 cm or 50% of depth on all samples. Indenter which is a cylinder with a diameter of 1.2 cm was used. The elasticity of each sample is presented in Table 4. Meanwhile, the relation between elasticity modulus and GA concentration of 0.5% and 1% is presented in Figure 6.

Table 4. Result of compressive strength test analysis

Sample	Gelatine (g)	GA (%)	F(N)	$E(kPa)$
A	1	0.5	7.170 ± 0.005	15.04 ± 1.18
B		1	9.680 ± 0.005	20.30 ± 1.59
C	2	0.5	25.490 ± 0.005	53.46 ± 4.17
D		1	40.070 ± 0.005	84.03 ± 6.55
E	3	0.5	31.30 ± 0.005	65.77 ± 5.13
F		1	47.870 ± 0.005	100.39 ± 7.82

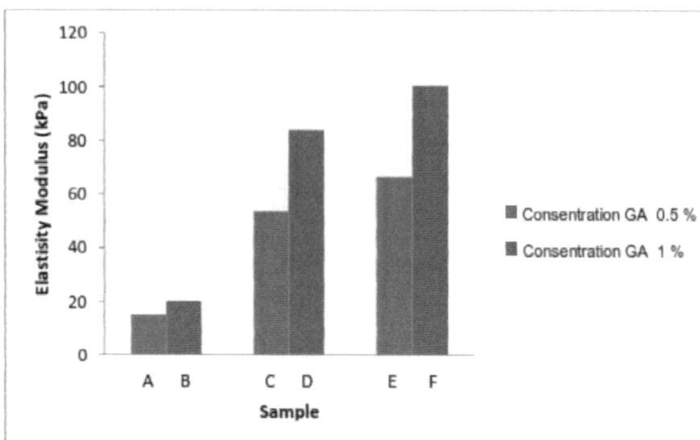

Figure 6. Graph on relationship between variation of the modulus of gelatine elasticity and GA concentration of 0.5% and 1%

Based on the graph in Figure 6, there is an increase modulus of elasticity along with increasing concentrations of either gelatine. This is because the resulting mechanical properties depended on density of crosslinking gen. Variations of GA gelatine at a concentration of 1% had greater elasticity modulus than that of GA 0.5%. This was because higher concentration of glutaraldehyde which affected crosslink the gelatine also increased. In other words, the higher the crosslink is, the higher the elasticity modulus becomes [6] Based on the literature, it is known that the value of the elasticity modulus of breast tissue is 0, 167-29 kPa [10], and that means the elasticity modulus of samples A and B were included in the range of elasticity modulus of breast tissue. However, it was difficult to form samples A and B that sample C (gelatine 2 g / GA 0.5%) with a elasticity modulus of 53.46 ± 4.17 kPa was the sample to be varied by TiO_2 in the homogeneity test, and it was determined for its material absorption coefficient in order to make its elasticity similar to that of breast tissue.

3.3. *Differential scanning calorimetric test (DSC)*

DSC test in this study aims to determine an acceptable temperature samples and ensure the samples were not degraded when irradiated by the laser. In this DSC test, 13.897 mg of sample was used and heated at 40-70°C temperature, and the temperature was raised 5° C per minute. Temperatures used in DSC test was based on the characterization of laser in which laser output generates temperatures of 20°C every 10 minutes and the test process is performed for ± 10 minutes for one sample. DSC thermogram of synthesized gelatine-GA-TiO_2 is shown in Figure 7.

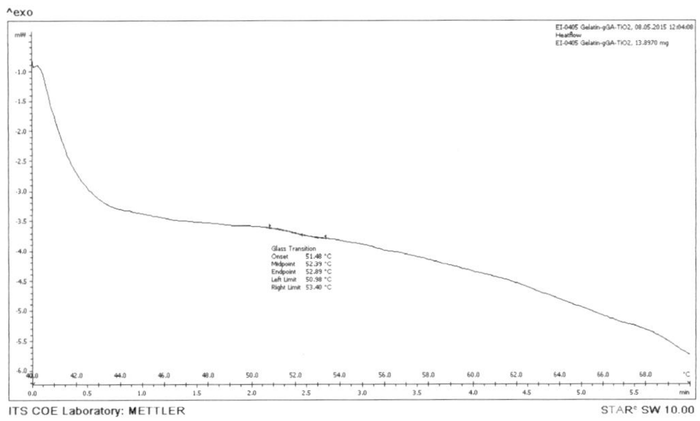

Figure 7. DSC thermogram gelatine composite-GA-TiO_2

Based on the thermogram above, it can be seen that in the temperature range test neither endothermic nor exothermic occurred which illustrates that the sample did not emit or absorb heat (stable). The glass transition temperature (Tg) occurred at of 51,40- 53,40°C temperature, and it shows a sample on balance. The result of DSC test analysis above shows that the sample was quite stable and did not undergo degradation in the temperature range 40-70°C in the optical tomography test.

3.4. *Homogeneity test*

In optical tomography test, one of the parameters observed in the sample is the scattering coefficient [11]. Therefore, the samples varied with TiO_2 as light scattering agent that samples scatter light beam was given and could be detected by all detectors. Samples were already varied with TiO_2 when irradiated with laser as shown in Figure 8 (b).

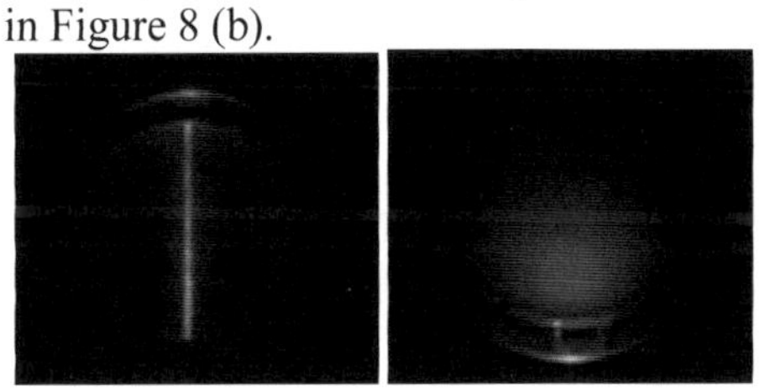

Figure 8. The sample when irradiated with laser (a) before the varied with TiO_2, (b) after varied with TiO_2

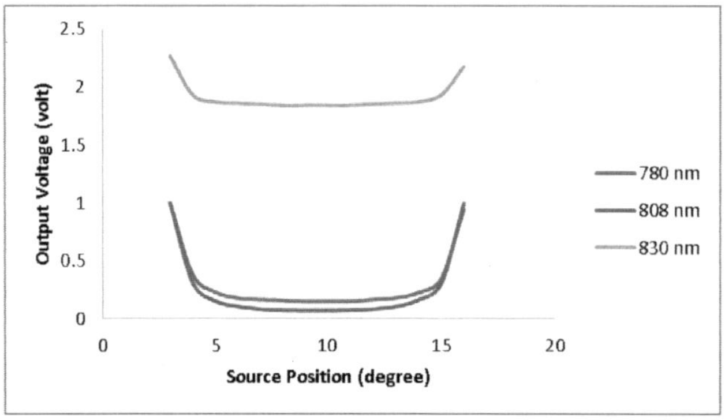

Figure 9. Graph of the relations between the output voltage at the source position on each wavelength (sample C / 0.01 g TiO_2)

In this homogeneity test used samples C with varied TiO_2 namely 0.01; 0,015; 0.02; 0,025; and 0.03 g. The laser with 3 variations NIR wavelength that is 780 nm, 808 nm and 830 nm are launched to this sample. Based on sample homogeneity test results obtained by the graph the relationship between the output voltage at the source position number for each of the various wavelengths. Results for samples C with a concentration of 0.01 g TiO_2 is presented in Figure 9.

In this study uses only one position detector which detector A. If the sample has a good homogeneity, then the position of sources that have the same distance to the detector A to produce the same output voltage. Under these conditions, the graph of the output voltage generated by the source position has had a trendline or a U-shaped pattern in Figure 9 can be seen that for all variations of wavelengths have a trendline or a similar U-shaped pattern.

NIR laser with a wavelength of 830 nm produces the highest intensity that can be detected NIR laser detector compared with the wavelength of 780 and 808 nm, so that the fund would be used 830 nm laser absorption material to the analysis phase. The material to be used for the absorption of the test will be selected based on the absorption of the material at each variation of TiO_2 (in the NIR laser wavelength of 830 nm). Determination is done by creating a linear graph for the lowest voltage at each sample variation of TiO_2 based on sample homogeneity test. The lowest voltage present in 9th place and 10th which is the farthest position from the position detector, so that made a graph of the output voltage with a variation of TiO_2 in a position to source 10 shown in Figure 10.

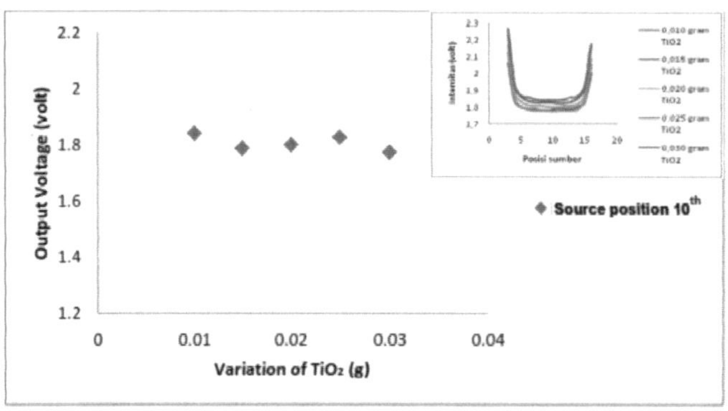

Figure 10. Graph of the relationship between the output voltage with a variation of TiO_2 in the 10th position of the source (λ = 830 nm) .Insert figure (Graph relationship with the intensity variations of TiO_2)

Based on the graph in Figure 10, it can be seen that the samples with a variation of 0.03 TiO_2 have the highest power absorption or low scattering power. On the concept of optical tomography, scattering power is a parameter to be considered in recording the sample shape based on the observed sample's scattering power [11]. Samples with the lowest scattering power are selected because if the low scattering power can be detected by a detector, then the larger power must also be detected. In this study, sample at 0.03 g TiO_2 variations was selected to be used for the analysis of absorption material.

3.5. Results of value analysis absorption coefficient materials
Based on the sample analysis and homogeneity test, it was shown that laser with a wavelength of 830 nm which generated the highest output voltage could be detected by detector, and sample C with a variation of 0.03 g TiO_2 at a wavelength of 830 nm (the source position to 10th) was a sample which has the power maximum absorption. In the analysis of the absorption coefficient of this material was used sample C with a variation of 0.03 g TiO_2, and an infrared laser with a wavelength of 830 nm was also used as the laser source. Absorption coefficient test results are shown in Table 5.

Table 5. Results of the test material absorption coefficient

Sample Thickness (mm)	Io(Volt)	I(Volt)
3		8.56
6		1.18
9	10.86	0.5
12		0.14
15		0.06

Based on Table 5 was made the logarithmic between graph intensity after getting through source material (I) and sample thickness as shown in Figure 11.

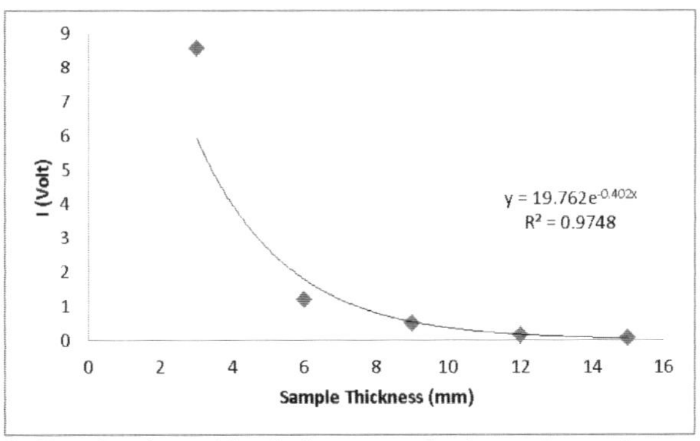

Figure 11. Graph o logarithmic between the intensity and sample thickness

The graph in Figure 11 shows the thicker the sample, the smaller the intensity detected by the detector. This shows the absorption of the sample events which occurred as a result of an interaction of light with matter. Based on this graph, sample thickness variations can be used to determine the absorption coefficient of the sample. Based on absorption test results which is indicated by the graph in Figure 11 and obtained y = 19,76e-0.40x based on the Lambert-Beer law, it can be seen that the value of the absorption coefficient (μ_a) of 0.40 mm-1. Absorption range coefficient used in optical tomography process ranges from 0.01 to 1 [12]. Therefore, it can be concluded that absorption material value of 0.40 mm^{-1} included in the range can be used in the process of optical tomography.

4. Conclusion
The composition of breast phantom which can result the best physical characteristics and elasticity similar to breast tissue is that of sample C. Sample C lasts for 3 days, and it is easily removed from the mould. Furthermore, its elasticity approaching is similar to that of breast tissue which is equal to 53.46

± 4.17 kPa). The homogeneity of the breast phantom is pretty good with a maximum intensity at a wavelength of 830 nm. Sample C with a variation of 0.03 g TiO_2 produces maximum absorbance value in the farthest position (position 9) based on the homogeneity which is equal to 0.178 volts. Value of the absorption coefficient (μ_a) of breast phantom test based absorption coefficient is 0.401 mm^{-1}.

References

[1] Novianti F A and S W Purnami 2012 Analisis Diagnosis Pasien Kanker Payudara Menggunakan Regresi Logistik dan Support Vector Machine (SVM) Berdasarkan Hasil Mamografi *Jurnal Sains dan Seni ITS* **Vol1** Surabaya.

[2] Pogue, Brian W and Michael S Patterson 2006 Review of tissue simulating phantoms for optical spectroscopy imaging and dosimetry *Journal of Biomedical Optics* **11** (4) 041102.

[3] Siddiqi A K 2009 *Development of Tissue-Equivalent Heat-Sensitive Gel for The Experimental Verification of Near Infrared (NIR) Laser-Mediated Cancer Detection and Therapy* Thesis Georgia Institute of Technology.

[4] Maddu A K Modjahidin, SarSardy, dan H Zain 2006 Pengaruh Kelembaban Terhadap Sifat Optik Film Gelatine *Makara Sains* **Vol 10** No1.

[5] Nurjanah N 2011 *Studi Komparasi Efektivitas Kompres Normal Salin, Air Hangat, dan Alkohol Terhadap Derajat Flebitis pada Anak yang Dilakukan Pemasangan Infus di RSUP Dr.HasanSadikin Bandung* Thesis Universitas Indonesia.

[6] Bigi A G Cojazzi, S. Panzavolta, K Rubini and N Roveri 2001 Mechanical and Thermal Properties of Gelatine Film at Different Degrees of Glurtaraldehide Crisslinking Biomaterial **22** 763-768.

[7] Wijaya K I Tahir and N Haryanti 2005 Synthesis of Fe_2O_3-Montmorillonite and Its Aplication as a Photocatalyst for Degradation of Congo Red Dye *Journal FMIPA* Universitas Airlangga.

[8] Wang J 2009 *Broadband Near-Infrared Tomography for Breast Cancer Imaging* Tesis Department of Physics and Astronomy Darmouth College.

[9] Akarcay, H Gunhan, S Preisser, M Frens and J Ricka 2012 Determining the Optical Properties of a Gelatine-TiO_2 Phantom at 780 nm *Biomedical Optics Express* **Vol 3.**

[10] McKee C. T 2011 Indentation Versus Tensile Measurements of Young's Modulus for Soft Biological Tissues *Tissue Engineering*, Part B **Volume 17**.

[11] Choe R. 2005. *Diffuse Optical Tomography and Spectroscopy of Breast Cancer and Fetal Brain* Disertation University of Pennsylvania.

[12] Heilscher, AH AY Bluestone, GS Abdoulaev, AD Klose, J Lasker, M Stewart, U Netz, dan J Beuthan 2002 Near-Infrared Diffuse Optical Tomography, *Disease Markers*, **18** (313-337).

Collagen-chitosan- glycerol bio-composite as artificial tympanic membrane for ruptured inner ear organ

Prihartini Widiyanti [1,2*], Rara Setya Angtika, Brillyana Githanadi, Ditya Hanif Kharisma[1], Tarikh Omar Asyraf[1], Adita Wardani

[1]Biomedical Engineering, Faculty of Science and Technology, Airlangga University, Surabaya
[2]Institute of Tropical Disease, Airlangga University, Surabaya

*Corresponding author: pwidiyanti@fst.unair.ac.id

Abstract. WHO data in 2012 shows that 5.3% of world population highly suffers from hearing loss and deafness. One of the deafness causes is rupture of tympanic membrane. Tympanic membrane damage which occurs often is perforated tympanic membrane, and it is also commonly known in medical term as tympanic membrane perforation. The causes, for instance, are high frequency of using earphones, traumatic accidents, noise, bacteria, viruses, and infectious microorganism. Tympanoplasty becomes the only treatment that can be widely accepted despite of deficiencies in postoperative complications. Therefore, this research aims to create artificial tympanic membrane made of natural materials such as type I collagen composited with chitosan and made of addition of glycerol to improve its mechanical strength and biodegradability. The method included the process of dissolving acetic acid in distilled water and mixation with chitosan. The solution is next added with glycerol and stirred to be homogeneous. After that, it was minted in petri dish and aerated before characterized. The sample characterization included tensile strength of which tensile test results showed that the value of the elasticity modulus tended to decrease with an increase in collagen concentration. The elasticity modulus values in a row for the variations of 7: 3, 8: 2, and 9: 1 were 35.10 MPa, 54,52 MPa, and 47,45 MPa respectively. The morphological test with 1000x, 2500x, and 5000x magnification showed their interaction in the formation of pores. Cytotoxicity results, moreover, showed that those samples were non-toxic and safe for the body due to the percentage of living cells. The sound absorption coefficient was between 1000 Hz - 2000 Hz which means that it could use as sound absorbing material. The antibacterial test results showed that all the sample variations were anti-bacterial due to the diameter of the clear zone. In conclusion, collagen and chitosan composite with addition of glycerol could be used for potential artificial tympanic for to its characterization

1. Introduction

WHO data on the number of hearing loss and deafness is shocking. In 2000, there were 250 million (4.2%) of the world population suffering from hearing loss, and approximately half (75-140 million) were in Southeast Asia where the prevalence of hearing loss and deafness was high at 4.6%, including Indonesia, which was steadily rising [1]. Tympanic membrane, which is an organ of the inner ear, carries forward impulse sound from the outer ear to the ossicles in the middle ear, then reinforces it before subsequently forwards it to the auditory nerve. The tympanic membrane damage that often occurs is perforated tympanic membrane which is also commonly known, in medical term, as perforation of the tympanic membrane. The causes are inter alia, trauma from an accident, high frequency of using earphones, noise, infectious microorganisms, viruses, and bacteria. The impact of deafness is profound which means that if someone cannot hear, then the individual is inhibited to talk and communicate.

Furthermore, the patient will have difficulty to study and become a retarded citizen meaning that he is a human with low quality and eventually a burden to the family, the community, and the country. Based on previous research, the use of polymer and natural polymers continues to be developed to find the best alloy in the manufacture of artificial tympanic membrane which meets the physical and mechanical properties. Natural material derived from collagen or its derivatives promises better properties in biocompatibility. Collagen is the most important structure in the inner lining of the tympanic membrane which helps to regenerate tissue including cells [2]. Additionally, Collagen in the manufacture of the membrane should be composited with other polymers such as chitosan to be potentially used as a *patch* in order to form the complex structure and interaction of aggregate matrix through covalent and ionic *cross-linking* [3]. Membrane is flexible to be a marker of mechanical quality improvement together with the addition of a *plasticizer* agent, i.e. glycerol, as amino acid-based *polyols* that bridges the polymerization and vegetable *cross-linking* bonds [4].

Therefore, in order to improve the quality of mechanical properties of the acoustic layer of biodegradable membrane and permanently restore hearing ability with sensitivity to power and velocity of sound in the material layer membrane system outside the tissue, functional group analysis test must be conducted via *Fourier Transform Infrared* (FTIR), tensile strength and elasticity modulus test, sound absorption coefficient test, and cytotoxicity assay test. Biomaterials are assumed to be made of chitosan-collagen composite with the addition of glycerol as potential artificial tympanic membrane in ruptured inner ear organ.

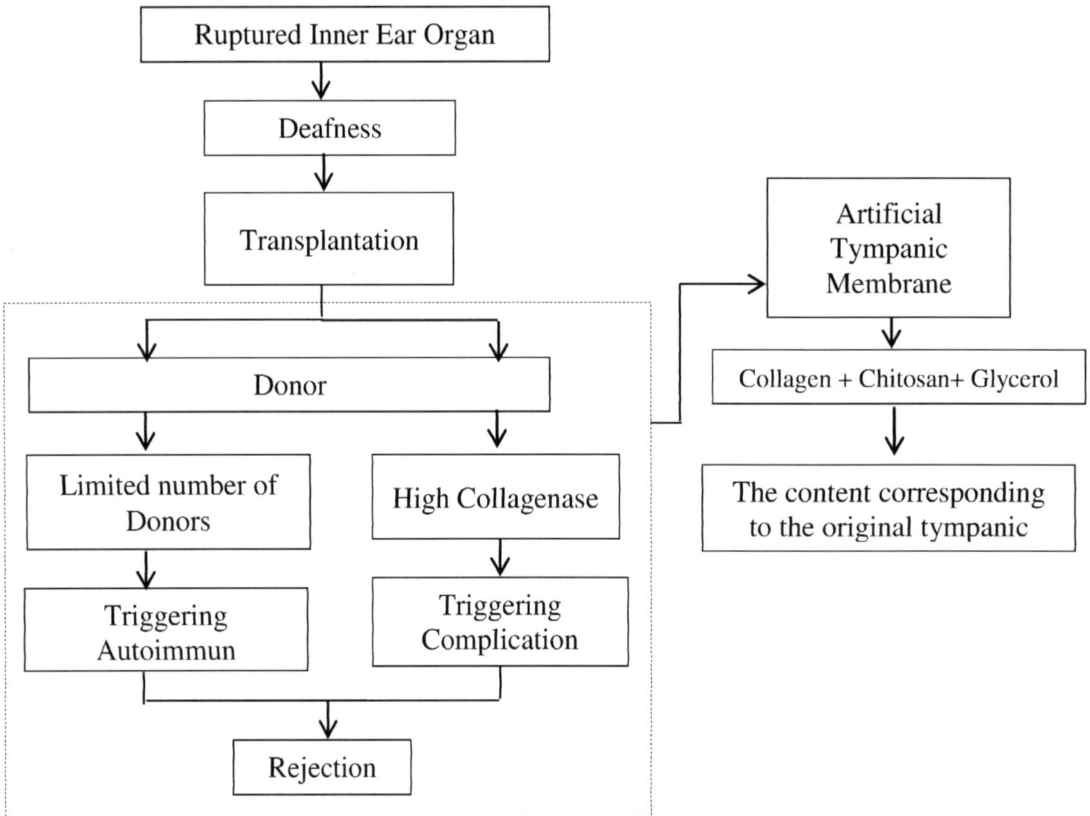

Figure 1. Scheme of research conceptual framework

2. Materials and method

2.1. Materials
Tools used in this research were digital Mettler Toledo scales, magnetic Yellow MAG HS 7 stirrer, Fourier Transform Infrared (FTIR) of Perkin Elmer, IMADA autograph, and Elisa reader. Meanwhile, materials used were collagen type I from red snapper produced by BATAN, chitosan, glycerol, distilled water, and acetic acid

2.2. Methods

2.2.1. Synthesis of Collagen and Chitosan Composite with Glycerol Addition.
First, 2.85 ml of 0.5 M acetic acid was dissolved in 97.2 ml of distilled water. Second, a composition of various solutions was made of a solution of collagen mixed with chitosan in acetic acid solution in order to be homogeneous. Next, the homogeneous solution resulted was added with 0.5 ml glycerol and stirred to be homogeneous. Finally, printing was performed on a petri dish and then aerated before characterized.

2.2.2. Fourier Transform Infrared (FTIR).
FTIR was used to determine the functional groups of the materials used as well as a new functional group obtained from the synthesis conducted on the sample. Next, samples were then added to taste KBr powder which was then compressed using a hydraulic clamp then placed on the specimen and irradiated with infrared wave number 4000-450 cm^{-1} [5].

2.2.3. Tensile Test and Elasticity Modulus Test.
Tensile test was used to determine the elastic properties of the materials when implanted and the mechanical response when the material interacted with the body's tissue samples formed into the shape of a dog bone with 63.5 mm in length and 10 mm and 5 mm in width on each sample variation in accordance with the American Society for Testing Materials (ASTM D 1822 L). Next, the samples were added with a 50 N load at the speed of 10 mm/min. The results obtained were the maximum voltage value (MPa) and the amount of strain and elongation [6]. The formula used to describe the correlation of stress with strain was:

$$E = \sigma / \varepsilon \quad (1)$$

2.2.4. Sound Absorption Coefficient Test.
In this method there are two techniques used to determine the normal sound absorption coefficient. They are standing wave tube method and two microphone tube method. The experiment was set up according to ASTM E1050/ISO 10534-2 test Method for Impedance and Absorption of acoustical Materials by Two Microphone Impedance Tube. This test method is similar to C384 test method which used an impedance tube with a sound source connected to one end, and the test sample was mounted to the other end. However, the measurements techniques for these two methods were different. ASTM E1050/ISO 10534-2 standard was used as a research screening tool, useful for manufactures and researchers in evaluating the absorption of materials. This method has been used in order to determine sound absorption coefficient of absorptive materials at normal incident that is 0 degree by using the reflector of a surface on material. The reflector of a surface on material can be used to determine the absorption coefficient of a surface. The absorption coefficient of a surface is a very useful parameter as it states the fraction of that energy that is absorbed when sound is incident to a surface. [7]

2.2.5. Cytotoxicity Test.
Cytotoxicity test aims to find the toxic effects of a material on cell cultures applied as biomaterial which has met the standards of medical applications [8] through MTT Assay test method. The samples

were sterilized, incubated, and then added with MTT reagent in the 3-4 hour incubation before analyzed using *Elisa* reader. Data obtained from the ELISA reader, such as absorbance readings (OD) of each pitting, were then converted into a percentage of the cell.

3. Results and Discussion

The result of this research was artificial tympanic membrane in which the samples on all variations were made with a thickness of 30-90μm in accordance with the original of tympanic membrane (Ruah *et al.* 2000). Next, the artificial tympanic membrane generated was tested to conform to the parameters required to be applied through tensile test, morphology test, cytotoxicity test, and absorption coefficient test.

3.1. Fourier Transform Infra-Red (FTIR),

It was found that cross-links occur in group C = O stretch which is an ester formation at 1647,43 cm^{-1}.

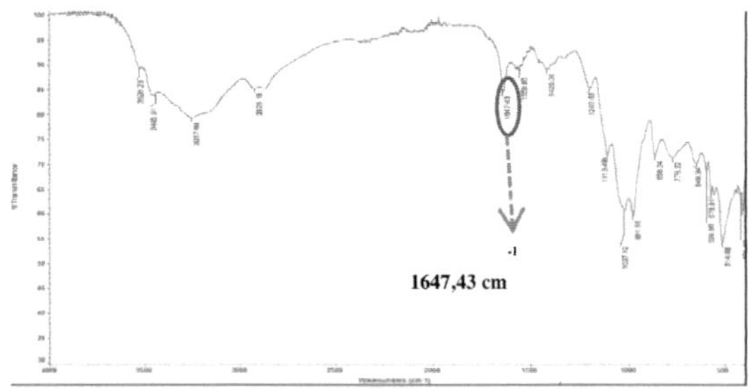

Figure 2. Fourier Transform Infrared (FTIR) Result

3.2. Tensile Strength Test and Elasticity Modulus Test

Tensile strength test was then performed to determine the elasticity of the materials. The test results showed the elasticity modulus values tended to decline with increased concentrations of collagen.

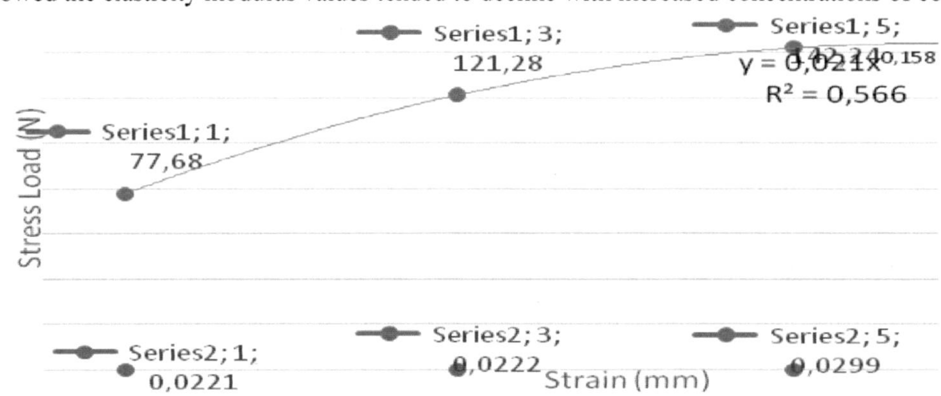

Figure 2. Composite Tensile Test Results

The values of elasticity modulus in a row for the variations of 7: 3, 8: 2, and 9: 1 were 35.10 MPa, 54.52 MPa, and 47.45 MPa respectively. In general, the values of the modulus elasticity decreased as the collagen concentration increased because collagen could conduct brittleness properties in the composite. In accordance with a literature [9], the modulus of elasticity of the human tympanic membrane of 20-60 MPa indicated that the entire sample variations were still in accordance with the functional standard range of human tympanic membrane.

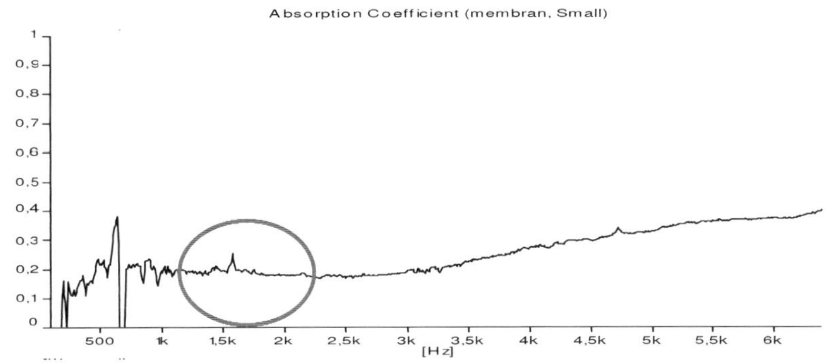

Figure 3. Membrane Absorption Coefficient Spectrum Tympanic Membrane

3.3. Sound Absorption Coefficient Test
The frequency range of human hearing is between 20- 20000 Hz [10]. According to when the coefficient is large (> 0.2), the material is referred to as sound absorbent material, otherwise if <0.2 is called reflective material. In addition to the results graph of FFT spectrum Analyzer in the range of Pure Tone Average / PTA (the threshold of human hearing) is capable of receiving the intensity of the incoming sound up to 100 dB and forward 85 dB into the acceptable range of intensities at normal tympanic membrane in normal hearing. [11]

3.4. Cytotoxicity Test result

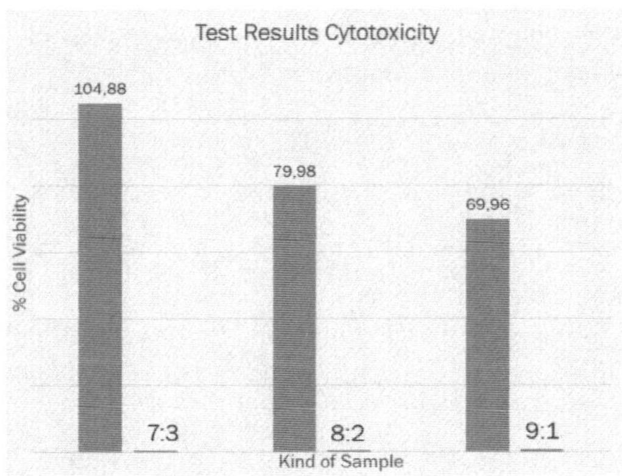

Figure 4. Cytotoxicity test result

The results showed that the cytotoxicity assays (%) of the cell viability in the variations of 7: 3.8: 2, and 9: 1 were 104.88%, 79.98%, and 69.96% respectively. The results obtained indicated that the viability (%) of all samples was more than 50% which means non-toxic [8].

4. Conclusion
Synthesis of Colagen-Kitosan Composite with the addition of glycerol could be obtained by mixing collagen and chitosan with glycerol and printing it in accordance with the dimension of the human tympanic membrane.

The test results showed that the samples had a tensile modulus of elasticity values in the range of 30-60 MPa. The result of Sound Absorption Coefficient Test was showed capable of receiving intensity of the incoming sound up to 100 dB and forwarding 85 dB into the acceptable range of intensities at normal tympanic membrane in normal hearing. Cytotoxicity test results showed that the cell viability (%) was more than 50% which indicates that the samples were not toxic.

Acknowledgment
The author would like to thank the Ministry of Research and Technology of Indonesia for the funding of this research

References

[1] WHO, 2007 SEA-Deafness-10 Distribution: General Situation Review and Update on Deafness Hearing Loss and Intervention Programmes

[2] Feenstra L, Kohn FE and Feyen J 1984 The concept to fanartificial tympanic membrane Clinical Otolaryngology 9 (1):215-220

[3] Begum AA, Radhakrishnan R and Nazeer KP 2011 Study ofstructure-property relationship on sulfuric acid crosslinked chitosan membranes *Malaysian Polymer Journal* 6 (1):27-38

[4] Mao L, Imam S, Gordon S, Cinelli P and Chiellni E 2002 Extruded cornstarch glycerol polyvinyl alcohol blends mechanical properties, morphology, and biodegradability *Journal of Polymers and the Environtment* Vol 8 (4): 205-211

[5] Smith and Brian C 2011 Fundamentals of Fourier Transform Infrared Spectroscopy Washington: CRC Press

[6] Purwanti A 2010 Analisis kuat tarik dan elongasi plastik kitosan terplastisasi sorbitol *Jurnal Teknologi* Volume 3 Nomor 2 Desember 2010 99-106

[7] Robert Oldfield 2006 Measuring Absorption Coefficient: "*Improved Membrane Absorbers*" University of Salford Salford: UK 41-56

[8] ATCC 2011 MTT Cell Proliferation Assay ATCC® 30 -1010K

[9] Fay J, Puria S, Decraemer WF and Steele C 2005 Three approaches forestimating the elastic modulus of the tympanic membrane *Journal of Biomechanics 38*: 1807–1815

[10] Caldarelli, David D and Ruth S Campanella Ear World Book Online Americas Edition 26 May 2003.

[11] UIUC Physics 406 Acoustical Physics of Music 2016

IOP Conf. Series: Journal of Physics: Conf. Series **853** (2017) 012030 doi:10.1088/1742-6596/853/1/012030

Effects of hyperbaric oxygen therapy in enhancing expressions of e-NOS, TNF-α and VEGF in wound healing

Imam Susilo[1*], Anita Devi[2], Azham Purwandhono[3] and Sunaryo Hadi Warsito[4]

[1]Department of Pathology, Faculty of Medicine, Airlangga University, Surabaya, Indonesia
[2]Hyperbaric Division of Sanglah Hospital, Denpasar, Bali, Indonesia
[3]Department of Pathology, Faculty of Medicine, Jember University, Jember, Indonesia
[4]Department of Animal Husbandry, Faculty of Veterinary Medicine, Airlangga University, Surabaya, Indonesia

*Corresponding email address : imam_susilo@fk.unair.ac.id

Abstract. Wound healing is a physiological process that occurs progressively through overlapping phases. Tissue oxygenation is an important part of the complex regulation for wound healing. Hyperbaric Oxygen (HBO) therapy is a method of increasing oxygen delivery to tissues. The therapy improves tissue oxygenation and stimulates the formation of H2O2 as a secondary messenger for Tumour Necrosis Factor alpha (TNF α), e-NOS, VEGF and Nuclear Factor Kappa Beta phosphorylation (NF-Kb) which play an important role in the rapid transcription of a wide variety of genes in response to extracellular stimuli. This study aims to determine the effects of Hyperbaric Oxygen therapy in enhancing the expressions of e-NOS, TNF-α, VEGF and wound healing. This study is an animal study with a 'randomized control group of pre-test and post test design' on 28 Wistar rats. Randomly, the rats were divided into 4 groups with 7 rats in each group. The HBO treatment group 1 received 5 sessions of HBO 2.4 ATA in 3x30 minutes; the HBO treatment group 2 received 10 sessions of HBO 2.4 ATA in 3x30 minutes; and each of the control groups were without HBO. Each of the 28 male rats were given a full thickness excisional wound of 1x1cm. Examinations of e-NOS, TNF-α, VEGF expressions and wound healing were performed on day-0 (pre-HBO) and day-5 HBO or on day-0 (pre-HBO) and day-10 HBO.The resultsshowthat the Hyperbaric Oxygen therapy can improve e-NOS ($p=0.02$), TNF-α ($p= 0.02$), VEGF expression ($p=0.02$) and wound healing ($p=0.002$) significantly in the provision of HBO 2.4 ATA for 3x30 minutes in 5 sessions over 5 consecutive days. While the 10 sessions of HBO 2.4 ATA for 3x30 minutes over 10 consecutive days only increase e-NOS ($p=0.02$), TNF-α ($p=0.04$), VEGF expression significantly ($p=0.03$) but do not improve wound healing significantly ($p=0.3$) compared with no HBO. The study concludes that HBO can improve the expressions of e-NOS, TNF-α, VEGF and wound healing in the provision of HBO 2.4 ATA for 3x30 minutes in 5 sessions, while the 10 sessions of HBO 2.4 ATA for 3x30 minutes only increase e-NOS, TNF-α, VEGF expression but do not improve wound healing.

1. Introduction

Wound healing is a normal process through a highly complex mechanism. The process occurs progressively through overlapping phases; haemostasis, inflammatory, proliferation, contraction and tissue remodelling [1]. The healing process can be ended by fibrosis or scarring even in optimal conditions, resulting in multiple disabilities [2] Acute injuries can become chronic wounds. The problem of chronic wounds is now a challenge for modern society resulting economically and socially significant medical impact. A Chronic wound is one of the most frequent causes of morbidity in developing countries [3].

An estimated 1.5% incident of the population suffered injuries at any points in time. Laceration or wound often encountered daily (estimated at 20 million cases each year) as a result of cut. A post operation wound is the biggest cause of skin trauma. There are more than 110 million surgical incisions every year, some which can't recover completely due to various causes. The wound developed into chronic wounds .The large number of the population and the burden of costs require serious attention on early detection, prevention, diagnosis and treatment of chronic wounds [2].

The Hyperbaric Oxygen therapy, as an adjuvant therapy in wound healing, results an increase in RNS mainly in NO through iNOS and e-NOS, as well as an increase in ROS, especially H2O2. Excessive doses of RNS and ROS are a threat to the survival of the cells, while the therapeutic doses (2.4 ATA) of HBO provide positive benefits for wound healing [4]. In clinical use, the dose of HBO is restricted from 2-2.5 ATA to a maximum of 3 ATA for 60-120 minutes. The HBO therapy varies between 5-30 times [5]. The provision of 100% oxygen at a pressure of 2 ATA improves tissue oxygenation of 30-40 mmHg in normal circumstances up to 250-300 mmHg, while the pressure of 3 ATA allows the improved tissue oxygenation up to 10-15 times [6].The Hyperbaric Oxygentherapy increases tissue oxygenation and stimulates the formation of H2O2 as a secondary messenger for the phosphorylation of Nuclear Factor Kappa Beta (NF-kB). The Nuclear Factor Kappa Beta/Rel transcription proteins play an important and quick role on a wide variety of genes in response to extracellular stimuli. Many aspects of the inflammatory response are also regulated by NF-kB, including the production factors of complement cascade and the induction of proinflammatory cytokines such as IL1 and TNF α [7]

The Tumour Necrosis Factor (TNF) group consists of 19 cytokines which have an important role in regulating the development and functions of the immune system. Cytokines from the group of Tumour Necrosis Factor (TNF) regulates the inflammatory and immune processes. The lack of TNF can trigger cell deaths. Although TNF-α cytotoxicity affects some tumour cells, TNF α will only trigger apoptosis if the NFkB signalling is inhibited. TNF α is an activator ofNFkB, proinflammatory transcription factors that help cells survival through activation of pro-survival genes and suppression of pro-apoptotic genes.

Neovascularization holds a central function in wound healing. In vitro studies indicate that VEGF, the most potent angiogenic protein, was induced by NO. Barriers will disrupt NO synthesis of VEGF expression and angiogenesis, thus slowing down the process of wound healing.

The Hyperbaric Oxygen therapy suppresses the inflammatory response to prevent chronic wounds [8] Supplementary oxygen through the blood stream accelerate sangiogenesis in ischemic wounds. Therefore, accelerates the healing time [6]. Numerous studies have proven the benefits of ATA 2.4 HBO five sessions and ten sessions [9]

Our previous studies result that HBO increases the expression of iNOS ($p = 0.001$), while the wound healing ($p = 0.002$) significant at 2.4 ATA HBO administration for 3x30 minutes in 5 sessions. Furthermore, the provision of 2.4 ATA HBO for 3x340 minutes in 10 sessions only enhance the expression of iNOS significantly ($p = 0.002$) but does not improve wound healing significantly ($p = 0.3$) compared with no HBO administration [10].

This study aims to determine the influence of the Hyperbaric Oxygen therapy (HBO) to increase the expression of TNF-α and its relationship with the increased expressions of e-NOS, TNF-α and VEGF in wound healing. This study tests the influence of hyperbaric oxygen therapy (HBO) to the increase of eNOS, TNFα and VEGF expression by the immune histochemical examination and the macroscopic wound healing process by measuring the area of wound closure on Rattusnorvegicus Wistargallus. The results are expected to be a reference for drafting standard operational procedures (SOPs) of the use of hyperbaric oxygen therapy (HBO) for wound healing in health care centres.

2. Methods

This study was randomized by a controlled group of the pre test-post test designed in 28 Wistar rats. The samples were divided into 4 groups by random allocation, each consists of 7 rats. Treatment group 1 was given HBO 2.4 ATA for 3x30 minutes in 5 sessions; the treatment group 2was given HBO 2.4 ATA for 3x30 minutes in 10 sessions, and each of the control groups were without HBO. Allof the 28 rats were given a 1x1cm full-thickness excisional wound. An immunohistochemical examination of the expressions of eNOS, TNFα, VEGF and the wound size measurement were conducted to determine the wound healing. The immunohistochemical examination and the wound size measurement to the treatment group 1 and the control group 1 were conducted on day 0 (pre-HBO) and day 5 of the HBO therapy. The immunohistochemical examination and the wound size measurement on the treatment group 2 and the control group 2 were conducted on day 0 and day 10 of the HBO.

3. Results and discussion

The study was conducted after the experimental animals had been acclimatized for 7 days. Skin tissue sampling was done by making a full-thickness of 1x1 cm excisional wound on the dorsal part of the experimental animals. The skin tissue sampling on day 0 was performed on all animals (a total of 28 samples) in the treatment group and the control group. The treatment group was given Hyperbaric Oxygen treatment in 5 sessions and 10 sessions. Comprehensive measurement of a wound closure and tissue sampling were done after the treatment of Hyperbaric Oxygen on day 5 (a total of 14 samples) and on day 10 (a total of 14 samples). The slides was made and immunohistochemical staining was conducted.

Immunohistochemical examination for e-NOS, TNF-α and VEGF expressions was conducted under a microscope. This examination was intended to count the number of immunoreactive cells against e-NOS, TNF-α and VEGF. The quantitative data obtained by counting the number of cells in 10 different fields of view at 40x magnification.

3.1. Day 0 (Pre HBO) condition of variables of expressions of e-NOS, TNF-α, VEGF, and wound healing

The examination of e-NOS, TNF-α, VEGF expressions and wound healing done in groups with and without HBO while the extent of pre HBO wound receives the same treatment, i.e. full thickness of1x1cm (100 mm^2)excisional wound size. On day 0 the expressions of e-NOS, TNF-α and VEGF in inflammatory cells with low intensity were observed.

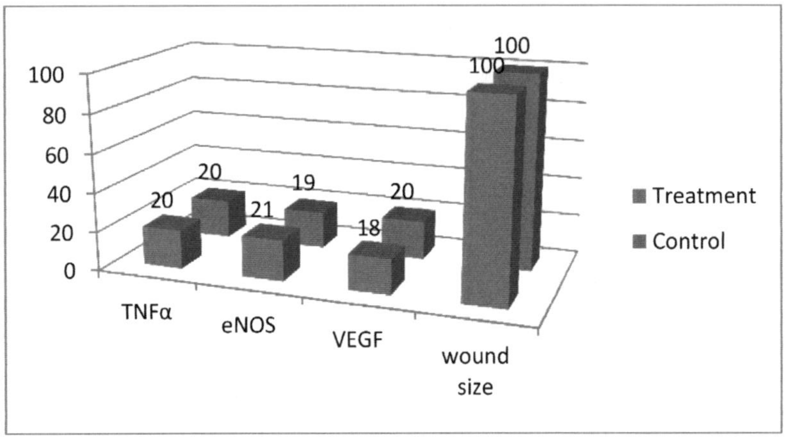

Figure 1. Average HBO on Day 0; Variables of expressions of eNOS, TNFα, VEGF and wound size

3.2. HBO condition of variables of expressions of e-NOS, TNF-α, VEGF and wound healing on day 5

The administration of HBO 2.4 ATA for 3x30 minutes in 5 sessions conducted in the treatment group, followed by an examination of e-NOS, TNF-α, VEGF expressions and the wound size to the treatment group and the control group. Figure 2 shows the average results of the variable expressions of e-NOS, TNFα, VEGF and the wound size. On day 5 there are increased of e-NOS, TNF-α, VEGF expressions with high intensity in the treatment group and the control group. The increased expressions of e-NOS, TNF-α, VEGF in the treatment group are higher than those in the control group. Meanwhile, the wound healing in the HBO treatment group is higher than that in the control group (smaller wound size and better healing).

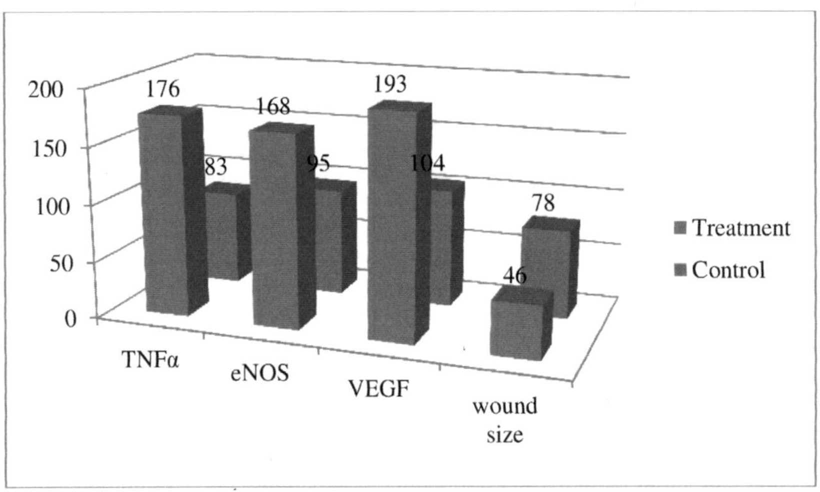

Figure 2 Average HBO on day 5; Variables of Expressions of e-NOS, TNF-α, VEGF, and wound size

3.3. HBO condition on day 10; variables of expressions of e-NOS, TNF-α, VEGF, and wound size

The HBO administration of 2.4 ATA in 3x30 minute each session was conducted 10 times in the treatment group and followed by an examination of the expressions of e-NOS, TNF-α, VEGF and the wound size to the treatment group and the control group of HBO administration on day 10. Figure 3 shows that the intensity of the expressions of e-NOS, TNFα, VEGF were still high but lower than that of on day 5. There is no significant difference of the wound healing variable between the treatment group and the control group.

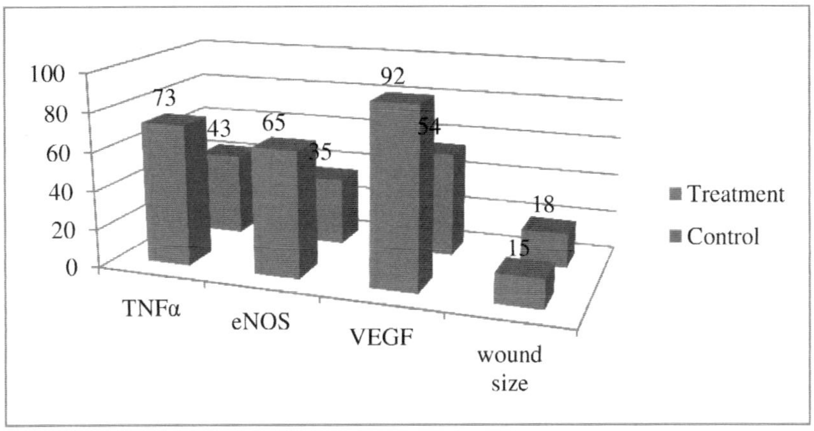

Figure 3. Average HBO on Day 10; e-NOS, TNF-α, VEGF Expressions and Wound size

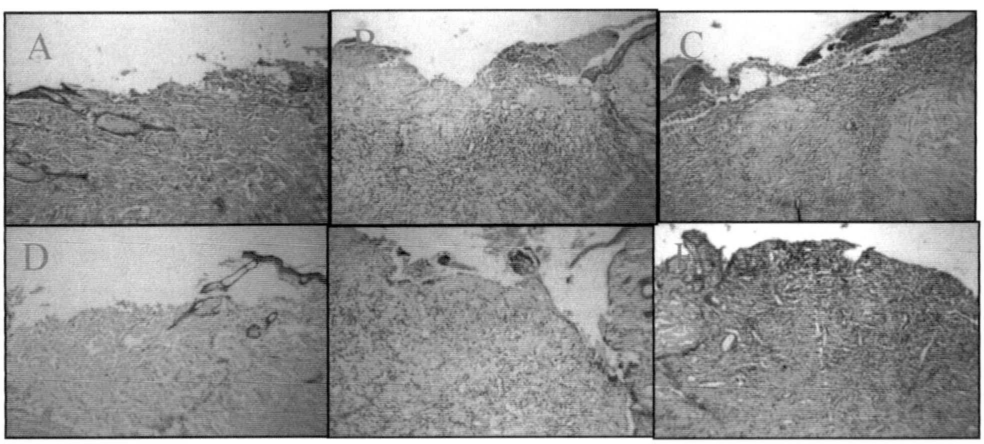

Figure 4. TNFα Expressions using a light microscope Olympus CX31 objective magnification 40x. A: control group day 0, B: control group day-5, C: control group day 10, D: treatment group day 0, E: treatment group day-5, F: treatment group day 10. The intensity of expression of TNF-α is still low in the control group day 0 but the day-5 intensity increases the expression of TNF-α, then the 10th day intensity of the expression of TNFα declines again.

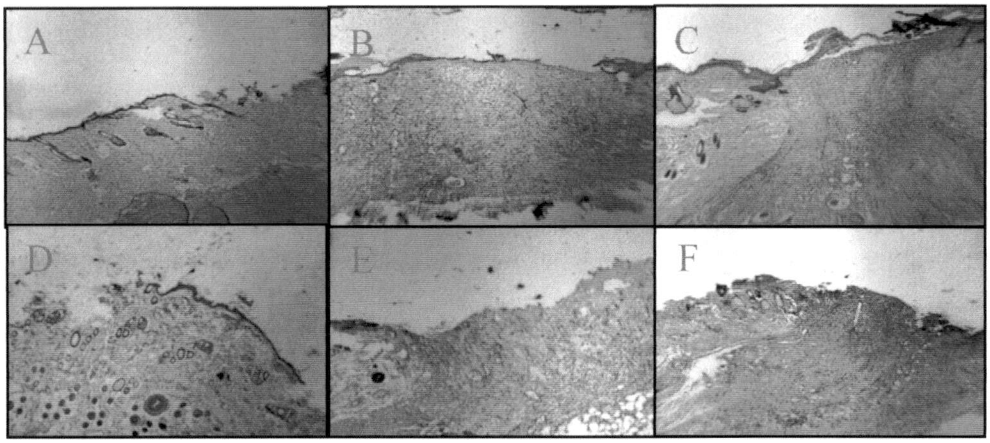

Figure 5. Expressions of e-NOS under a light microscope Olympus CX31 objective magnification 40x. A: control group day 0, B: control group day5, C: control group day 10, D: treatment group day0, E: treatment group day5, F: treatment group day 10. In the control group day 0, the e-NOS expression remains in low intensity; The intensity of e-NOS expression increases on day 5 but declines again on the 10th day.

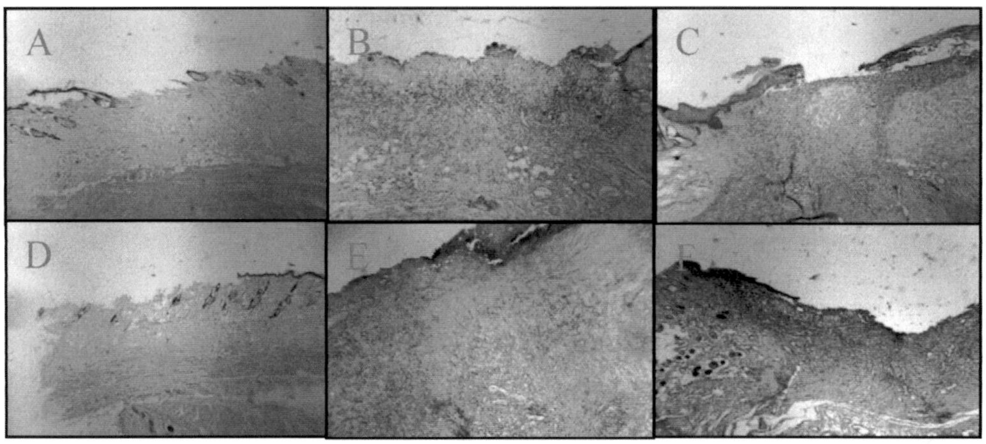

Figure 6. VEGF Expressions under a light microscope Olympus CX31objective magnification 40x. A: control group day 0, B: control group day5, C: control group day 10, D: treatment group day 0, E: treatment group day5, F: treatment group day 10. The intensity of VEGF expression is low in the control group day 0; The VEGF expression intensity increases on the 5th day but decreases again on the 10th day.

The results of examination on the expressions of e-NOS, TNFα and VEGF in the treatment group and the control groups on day0 show that the expressions of e-NOS, TNF-α and VEGF intensity are low (1-1). After the administration of HBO 2.4 ATA in five sessions, on the 5th day there seems to be increased expressions of e-NOS, TNF-α and VEGF (2-2) along with the increased number of inflammatory cells, in which the increased expressions of e-NOS, TNF-α and VEGF in the treatment group is significantly higher than those of the control groups. On the 10th day the intensity of the expressions of e-NOS, TNF-α, VEGF is still high. However, the number of cells that express them reduces and the number of inflammatory cells remains high (2-1).

TNFα is a cytokine regulator and major effectors, mainly secreted by monocytes and macrophages that have been stimulated [11]. HBO administration increases the production of e-NOS in cultured endothelial cells [12]. Several biological processes and growth factors are stimulated and increased with Hypoxia and HBO. Such processes include angiogenesis, collagen synthesis, the activity of osteoclasts, and the release of VEGF and TNF-α. Oxygen stimulates biological processes under conditions of hypoxia and hyperoxia. It is called Oxygen Paradox. One mechanism for forming collagen fibroblasts is stimulated through peroxide. This occurs in HBO hypoxia wounds and during therapy. Peroxide formed in the HBO therapy resembles the stimulus that occurs during hypoxia. Another mechanism is cytokine stimulation which begins with hypoxia, and increases in cytokine regulation which is supported by the presence of hyperoxia occurring in the HBO therapy. VEGF, TNF-α TGF-β and PDGFβ are released at wound hypoxia as an initial stimulus, but its activity increases in hyperoxia conditions, especially in the presence of lactic acid [13].

The evidence shows that HBO can accelerate wound healing through mechanisms expressions of e-NOS, TNFα and VEGF in the inflammatory phase and the proliferative phase. The administration of HBO can potentially be used for a post-surgical adjuvant therapy for certain cases, thus it is expected to reduce the duration of treatment (Length of Stay / LOS) in the hospital. Indeed, it is expected to reduce the cost and recovery time for patients after surgery so that patients can be immediately productive and return to their normal activities. Therefore, we need to do more research on the effects of hyperbaric oxygen therapy against various types of wounds, including wounds that are conventionally difficult to cure, post-trauma wounds and postoperative wounds.

The evidence also shows that the administration of HBO continuously for a long time does not provide significant benefits for wound healing. Therefore, time off (recovery time) needs to be given after administration of certain HBO in 5 sessions before continuing to the next session. Further research is needed to determine whether or not there are effects of the recovery period and the range of speeds of the wound healing/recovery time interval before the next continuing HBO therapy session. Further research with more dose variations (between 1-5 sessions and between 5-10 sessions) is needed to obtain the optimal dose of the Hyperbaric Oxygen therapy for acute wounds without infections. Further research needs to be conducted to determine the repetition time interval after the administration of the Hyperbaric Oxygen therapy in5 sessions (recovery period) as well as the use of HBO as an adjuvant therapy for various types of injuries.

4. Conclusion

This study concludes that applying 2.4 ATA HBO for 3x30 minutes in 5 sessions increases the expressions of e-NOS, TNF-α, VEGF and accelerates wound healing, while giving HBO 2.4 ATA for 3x30 minutes in 10 sessions enhances the expression of e-NOS, TNF-α and VEGF but does not accelerate wound healing.

Furthermore, the evidence shows that giving HBO can accelerate wound healing through mechanisms of e NOS, TNF-α and, VEGF expression. The HBO administration can be potentially used for a post-surgical adjuvant therapy and specific cases as the administration is expected to shorten the duration of treatment in the hospitals, and reduce the cost and recovery time for patients after surgery so that they can return to normal activities.

References

[1] Finochietto P V, Franco M C, Holod S, Gonzalez A S, Converso D P, Arciuch VGA, Maria P, Poderoso J J and Carrearas MC 2009 Mitochondrial Nitric Oxide Synthase: A Masterpiece of Metabolic Adaptation, Cell Growth, Transformation, and Death *The Society for Experimental Biology and Medicine* **234** (9) 1020-8

[2] Jiang Y, Huang S, Fu X, Liu H, Ran X, Lu S, Hu D, Li Q, Zhang H, Li Y, Wang R, Han C and Chen H 2011 Epidemiology of Chronic Cutaneus Wounds in China *Wound Repair Regeneration* **19** 181-188.

[3] Sander A L, Henrich D, Muth C M, Marzi I, Barker J H and Frank J M 2006 In Vivo Effect of Hyperbaric Oxygen on Wound Angiogenesis and Epithelization, Wound Repair and Regeneration *Wound Healing Society* **17** 179-184.

[4] Schreml S, Szeimies R M, Prantl L, Karrer S, Landthaler M and Babilas P 2010 Oxygen in Acute and Chronic Wound Healing, *British Journal of Dermatology, British Association of Dermatology* **163** (2):257-68.

[5] Kumar S and Wong P F Leaper D J 2004 What is New in Wound Healing ? *Perspective in Medical Science*, Turk J Med Sci **34** 147-160

[6] İşçimen A and Küçüktaş M 2008 Wound Healing and Hyperbaric Oxygen treatment, *Journal of Turkish Academy of Dermatology*.

[7] Ghosh S 2007 *Handbook of Transcription Factor NF-kappaB*, CRC Press Taylor & Francis Group

[8] Thom SR, 2010 Hyperbaric Oxygen: its Mechanisms and Efficacy *Wound Healing Plastic and Reconstructive Surgery* **127** 131-138

[9] Lahat N, Bitterman H, Yanif N, Kinnarty A, and Bitterman N 1995 Exposure to hyperbaric oxygen induces tumour necrosis factor-alpha (TNF-alpha) secretion from rat macrophages,Clinical and Experimental Immunology *Clinical Sciences* Ltd **102** 655-659

[10] Xu X, Wang Z, Li Q, Xiao X, Lian Q, Xu W, Sun X, Tao H,and Li R 2009 Endothelial nitric oxide synthase expression is progressively increased in primary cerebral microvascular endothelial cells during hyperbaric oxygen exposure.

The first-principle study on the stability of trans-HCOH in various solvents

Rizka Nur Fadilla[1], Nufida Dwi Aisyah[1], Hermawan K Dipojono[2] and Febdian Rusydi[1,*]

[1]Theoretical Physics Research Group, Department of Physics, Faculty of Science and Technology, Airlangga University, Kampus C, Jl. Mulyorejo, Surabaya, 60111, Indonesia
[2]Laboratory of Computational Material Design, Research Group of Engineering Physics, Bandung Institute of Technology, Jl. Ganesha 10, Bandung, 40132, Indonesia

*corresponding email address: rusydi@fst.unair.ac.id

Abstract. We attempt to study about the solvent effects of the stability of trans-HCOH molecules using the density-functional theory. Experimentally, trans-HCOH rearranges to H_2CO with half-life of two hours [1] which we theoretically proved that it occurs through quantum tunneling. [2] In this work, we calculate the rearrangement rate of the molecules in various solvents. The solvents are selected based on their dielectric constant values, from lower to higher ones; they are benzene, dichloroethane, benzaldehyde, acetone, methanol, ethanediol, dimethylsulfoxide, formic acid, water, and formamide. We use polarizable continuum to model the solvents (PCM). We begin from determining the reaction path from trans-HCOH to H_2CO and its corresponding energy barrier using intrinsic reaction coordinate calculation with PCM. Then, we use Wentzel-Kramers-Brillouin (WKB) approximation to calculate the rearrangement rates. The calculation results showed a general trend in which there were arrangement rate was decreasing inversely proportional to dielectric constant value.

1. Introduction

The existence of the trans-HCOH had been theoretically predicted since1974, but it was isolated for the first time in 2008 by Schreineretal. He found that trans-hydroxymethylene rearranges naturally to formaldehyde (H2CO) through a mechanism called as quantum tunneling with two hours half-life [1]. Our two previous studies have proved that the rearrangement of the molecules occurs through quantum tunnelling. The first study was based on the value of the molecule's half-life [2], and the second study was based on the monodeuteriation effects on the molecule [3].

This study reports our continuous study on the stability of trans-HCOH. The aim of this study is to see whether or not the solvents could alter the rearrangement rate of trans-HCOH to H2CO. Our theoretical investigation is conducted based on the rearrangement path of the molecules. We use density-functional theory and Wentzel-Kramers-Brillouin (WKB) approximation[4] to obtain the rearrangement rate of the molecules under the effect of solvents.

2. Computational details

Our previous study showed that our calculation method is proper for the rearrangement of trans-HCOH by comparing the results to the experimental data. [2] Thus, we continue using 6-31G(d,p) basic set and B3LYP exchange-correlation functional to perform our calculations for this study. [2] We perform the calculations with trans-HCOH and H_2CO as our molecule models which are shown in Fig. 1.

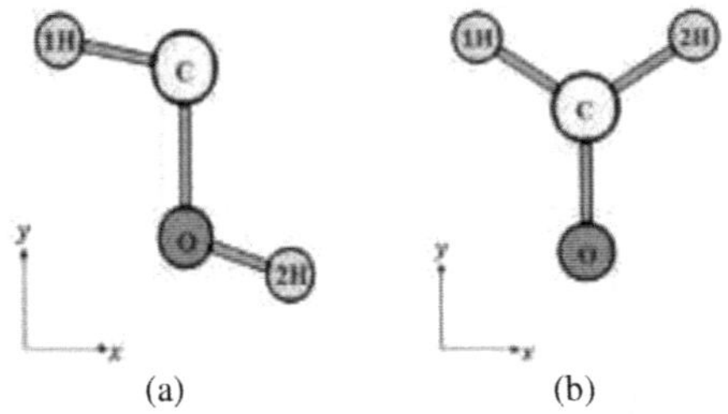

Fig. 1. Molecular geometry of (a) trans-HCOH (b) H2CO. Symbol 1H means hydrogen atom number 1, and symbol 2H means hydrogen atom number 2.

We selected ten solvents based on their dielectric constant values [5] from lower to higher ones; they arebenzene, dichloroethane, benzaldehyde, acetone, methanol, ethanediol, dimethylsulfoxide (DMSO), formic acid, water, and formamide. The solvents and their respective dielectric constant values are listed in Table 1.

Table 1. These are ten selected solvents and their respective dielectric constant values. We notate the solvents as written in the second column of this table.

solvents	notation	ε
(gas phase)	{0}	1.000
benzene	{1}	2.271
dichloroethane	{2}	10.125
benzaldehyde	{3}	18.220
acetone	{4}	20.493
methanol	{5}	32.613
ethanediol	{6}	40.245
DMSO	{7}	46.826
formic acid	{8}	51.100
water	{9}	78.355
formamide	{10}	108.940

We determined the ground spin state electronic structure of trans-HCOH and H_2CO inside those solvents using polarizable continuum model (PCM) calculation.

We predicted the transition state structure using STQN method coupled with PCM calculation.[6] We confirmed the existence of imaginary frequency through the frequency calculation.

We performed IRC calculation [7] with fifteen numbers of points and 10 bohr. $\sqrt{\text{amu}}$ of step sizes to map out the reaction path of trans-HCOH which rearranges to H_2CO under the effect of solvents. In order to ensure that we get the true minimum energy of the final structure along the reaction path, we added the calculated energy of the optimized H_2CO through optimization calculation. Then, we combined it with the IRC calculation data. We found a fitting function which best fits with our IRC calculation data. The fitting function would act as a potential barrier function. Hence, we would be able to calculate the probability of quantum tunneling which was given by WKB approximation with

$$T \cong \exp\left[-\frac{4\pi}{h}\int_{x_i}^{x_f} \sqrt{2(V_b(x)-E_{zp})dx}\right], \tag{1}$$

where h was Planck constant, and$m$ was the particle's mass which tunneled through the energy barrier, $V_b(x)$ was the energy barrier as a function of position, and E_{zp} was zero-point energy (ZPE). We performed the integration within the initial position (x_i) to the final position (x_f).

We obtained the rearrangement rate with

$$=\nu T, \quad (2)$$

In which was the rearrangement rate, ν was the vibrational frequency, and T was the probability of quantum tunneling. Finally, we would obtain the half-life ($\tau_{1/2}$) with

$$\tau_{1/2} = \frac{\ln(2)}{k}. \quad (3)$$

The DFT, frequency, PCM, and IRC calculations were performed by using Gaussian09 suite program. [8]

3. Results and discussions

3.1 The comparison between the molecules inside the solvents and those in the gas phase

Our DFT calculation coupled with PCM obtained the geometric structure parameters of trans-HCOH and H_2CO inside ten solvents which were shown in Table 2 and Table 3. The results showed that the ground state electronic structures in solvents were not significantly different. The difference was in order of 10^{-6}eV. It implied that the solvents did not affect the geometric structures.

Table 2. Geometric structure parameters of trans-HCOH in singlet spin state configuration from our DFT calculation under the effects of ten solvents.

Parameter	{0}	{1}	{2}	{3}	{4}	{5}	{6}	{7}	{8}	{9}	{10}
		solvents									
dielectric constant	1.000	2.271	10.125	18.220	20.493	32.613	40.245	46.826	51.100	78.355	108.940
∠OCH (degree)	101.3	101.6	101.9	102.0	102.0	102.0	102.0	102.0	102.0	102.0	102.0
C-O distance (Å)	1.320	1.316	1.313	1.312	1.312	1.312	1.312	1.311	1.311	1.311	1.311
C-H distance (Å)	1.124	1.124	1.125	1.125	1.125	1.126	1.126	1.126	1.126	1.126	1.126
O-H distance (Å)	0.973	0.980	0.991	0.993	0.993	0.994	0.994	0.995	0.995	0.995	0.995

Table 3. Geometric structure parameters of H2CO in singlet spin state configuration from our DFT calculation under the effects of ten solvents.

Parameter	solvents									
	{1}	{2}	{3}	{4}	{5}	{6}	{7}	{8}	{9}	{10}
dielectric constant	2.271	10.125	18.220	20.493	32.613	40.245	46.826	51.100	78.355	108.94
∠HCH (degree)	115.4	115.6	115.7	115.7	115.7	115.7	115.7	115.7	115.7	115.
C-O distance (Å)	1.209	1.212	1.212	1.212	1.212	1.212	1.212	1.212	1.212	1.21
C-H distance (Å)	1.110	1.109	1.109	1.109	1.109	1.109	1.109	1.109	1.109	1.10

Besides the molecular geometry, we obtained the vibrational frequency of trans-HCOH inside the solvents as it was shown at Fig.2. The results showed that there was no significant difference in the vibrational frequency.

We obtained six vibrational modes of the molecules. There were two out of six vibrational modes which involved atom 2H (see Fig. 1) in its vibration. Based on our previous study, we chose the vibrational mode based on its type of motion and its probability to occur. We chose the second vibrational mode of the molecule in each solvent. We used this vibrational frequency to calculate the rearrangement rate of the molecules.

3.2 Energy barrier

After we studied about trans-HCOH and H_2CO inside solvents separately, we predicted the transition state structure inside the solvents followed by the reaction path for each reaction which is shown in Fig. 3

IOP Conf. Series: Journal of Physics: Conf. Series **853** (2017) 012031 doi:10.1088/1742-6596/853/1/012031

Based on the result of IRC calculation, we obtained the molecule's energy in each coordinate reaction. The results showed that the increasing dielectric constant value affected the height of the energy barrier: the higher that the dielectric constant was, the higher that the energy barrier was.

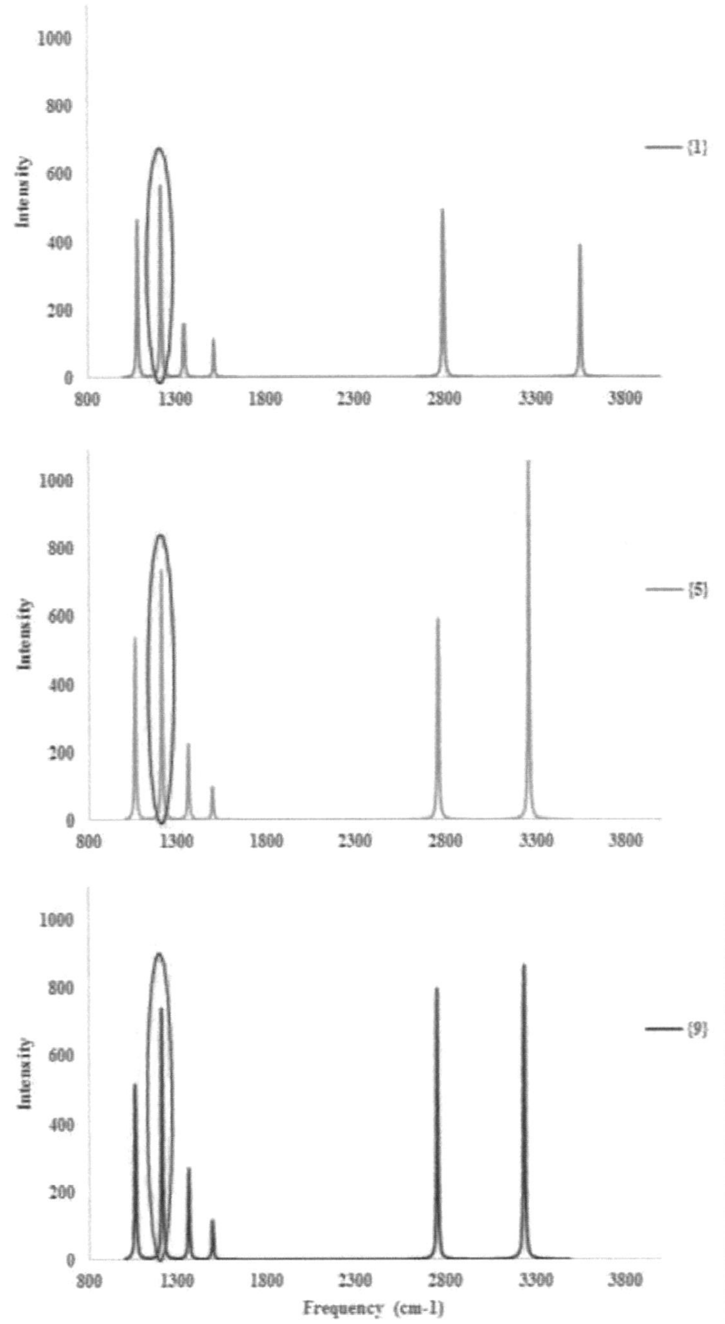

Fig.2. Vibrational frequencies of trans-HCOH normal modes inside benzene ({1}),methanol ({5}), and water ({9}) as the result of frequency calculation. The mode which is enclosed by red circle is the chosen normal mode that contributes to the rearrangement of trans-HCOH.

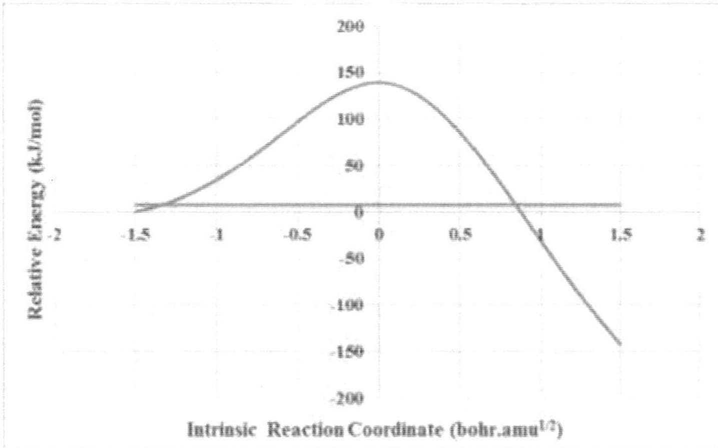

(a)

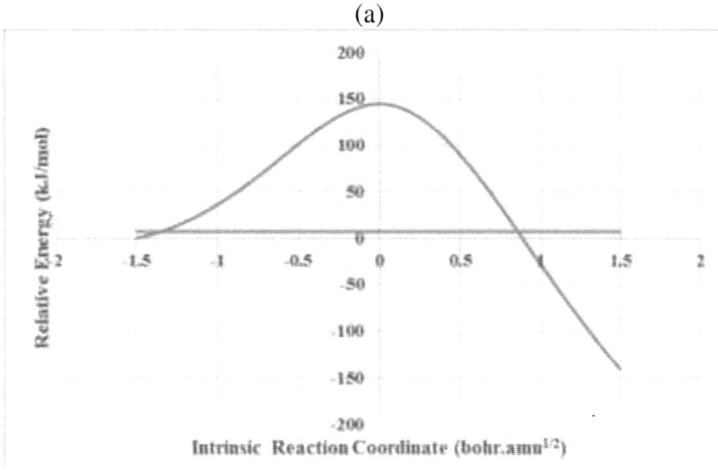

(b)

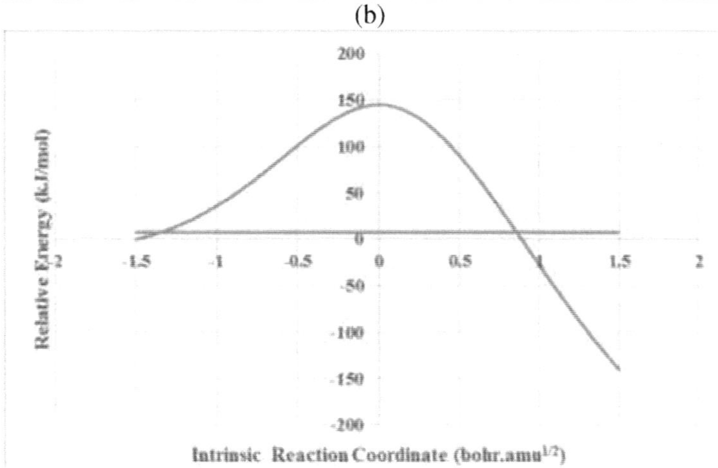

(c)

Fig.3. Potential energy curves are the reaction path of the rearrangement inside (a)benzene(b)methanol(c)water. The energy of trans-HCOH is set as a reference point. The orange line represents zero-point energy (ZPE) which corresponds to the rearrangement of trans-HCOH. The blue line is the fitted function which acts as energy barrier.

We fitted some trial functions to our IRC calculation data and we found the best fitting function was the eight-degree polynomial function. Then, the fitted function acted as the energy barrier function of the rearrangement.

The energy barrier function determined the probability of quantum tunneling. We integrated the energy barrier function from xi to xf (see equation (1)) and resulted in the probabilities of quantum tunneling which are shown in the following Table 4. The results showed that the perturbation coming from solvents lowered the probability of quantum tunneling.

Table 4. Probability of quantum tunneling in gas phase ({0}) *[2]* and in ten selected solvents ({1}-{10})

ε	Probability of Quantum Tunneling
{0}	2.100E-18
{1}	8.101E-19
{2}	3.048E-19
{3}	2.672E-19
{4}	2.621E-19
{5}	2.428E-19
{6}	2.401E-19
{7}	2.084E-19
{8}	2.332E-19
{9}	2.233E-19
{10}	2.122E-19

3.3 Rearrangement rate under the effect of solvent

The probability of quantum tunneling led us to acquire the rearrangement rate of the molecules in each solvent. Based on equation (2), we obtained the rearrangement rates of the molecules for ten solvents with mode 2 as the consideration mode which initiated the rearrangement. The results are shown in Table 5.

Table 5. Rearrangement rate in gas phase ({0}) *[2]* and in ten selected solvents ({1}-{10})

ε	Rearrangementrate
{0}	7.640E-05
{1}	2.942E-05
{2}	1.107E-05
{3}	9.704E-06
{4}	9.517E-06
{5}	8.816E-06
{6}	8.718E-06
{7}	7.567E-06
{8}	8.469E-06
{9}	8.107E-06
{10}	7.706E-06

Compared to the rearrangement rate of our previous study which resulted in $7.640 \times 10^{-5} \frac{1}{s}$ in the gas phase (see Table 4), the perturbation from solvents has decreased the rearrangement rate of the molecules: the higher that the dielectric constants value was, the lower that the rearrangement rate was. It indicated that the stability of trans-HCOH was increasing due to the existence of solvents. The trend of the decreasing rearrangement rate is shown in Fig. 4.

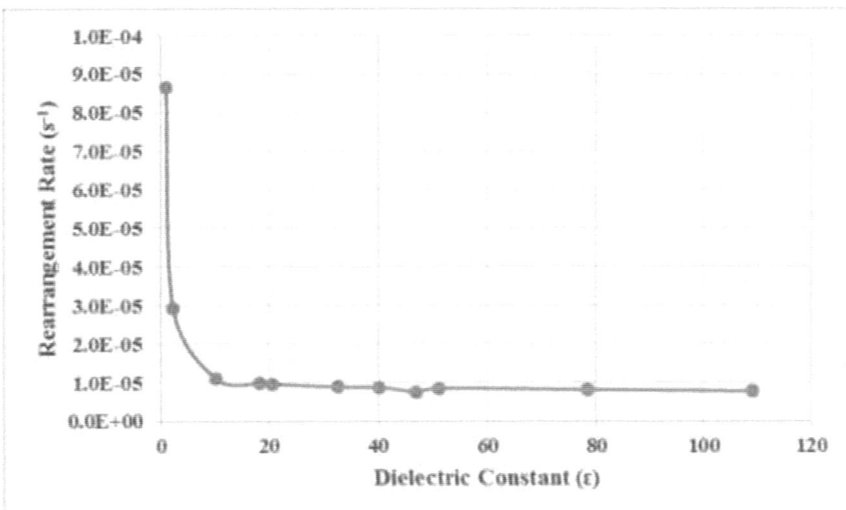

Fig. 4. The decreasing rearrangement rate of the molecules in the gas phase ($\varepsilon = 1$) and in each solvent based on dielectric constant values.

4. Conclusion

This study reports that the perturbation coming from solvents increases the energy barrier of the rearrangements. Hence, it lowers the rearrangement rates of the molecules. We have obtained the rearrangement rate of the molecules under the effect of solvents, which are lower than those in the gas phase.

Acknowledgment
All computational calculations were performed in a high performance computing system, Center of Advanced Sciences, Institut Teknologi Bandung, Indonesia.

References
[1] P R Schreiner, H P Resienauer, F C Pickard IV, A C Simmonett, W D Allen, E Matyusand and A G Csaszar 2008 *Capture of hydroxymethylene and its fast disappearance throughtunneling* Nature **453** 12 June, p. 906.
[2] R N Fadilla, N D Aisyah, H K Dipojono and F Rusydi A theoretical study of the rearranging trans-HCOH to H_2CO via quantum tunneling with DFT and WKB approximation *Procedia Engineering* 2016 (Submitted, underreviewed).
[3] N D Aisyah, R N Fadilla, H K Dipojono and F Rusydi 2016 A theoretical study of monodeuteriation effect on the rearrangement of trans-HCOH to H_2CO via quantum tunneling with DFT and WKB approximation *Procedia Engineering* (submitted, underreviewed).
[4] D J Griffiths 2005*Introduction to Quantum Mechanics* New York Pearson Prentice Hall
[5] SCRF, [Online] Available: http://dev.gaussian.com/scrf/?tabid=7. [Accessed 30 November 2016] A reference
[6] J B Foresman and A Frisch *Explorong Chemistry with Electronic Structure Methods*: A Guide to Using Gaussian, 2nd ed., Pittsburgh, PA: Gaussian, Inc.
[7] K Fukui *The Path of Chemical Reaction* - The IRC Approach, American Chemical Society **14**, pp. 363-368, 1981.
[8] Wallingford CT *Gaussian 09, Revision C.01* Gaussian Inc., Wallingford CT 2010

IOP Conf. Series: Journal of Physics: Conf. Series **853** (2017) 012032 doi:10.1088/1742-6596/853/1/012032

Analysis of thin layer optical properties of A-Si:H P-Type doping CH_4 and P-Type without CH_4 is deposited PECVD systems

Soni Prayogi[1,*], Ayunis[1], Kresna[1], Yoyok Cahyono[1], Akidah[2] and Darminto[1]

[1] Department of Physics-FMIPA, Institut Teknologi Sepuluh Nopember (ITS)
[2] LPPM Institut Teknologi Sepuluh Nopember (ITS)

*Corresponding email address : prayogi.physics@mail.its.ac.id

Abstract. The study of a thin layer growth of hydrogenated amorphous silicon (a-Si: H) using the technique of plasma enhancing chemical vapor deposition (PECVD) has been conducted. Material a-Si: H is one type of materials that is applied as solar cells. In this study, a thin layer of a-Si: H grown on glass substrates by using CH_4 and without CH_4. Most sources Si gas were used in silane gas (SiH_4) 20% dissolved in hydrogen gas (H_2). The addition of CH_4 gas greatly affects the structure of layer morphology and energy gap in thin layers. Based on the results of characterization using AFM, it was obtained a layer thickness which was added by CH_4 100 nm and layer thickness of 45 nm without CH_4. While the optical energy band gap were conducted, based on the data from characterization using UV-Vis in the wavelength range of 400-800 nm, it was obtained layer optical energy band gap added by CH_4 that was 1,95 eV and layer without CH_4 that was 1.89 eV.

1. Introduction

Since the early 1970, the study of the development of crystalline to amorphous semiconductors has been carried out by the researchers. The study is characterized by hydrogenated amorphous silicon, a-Si: H, which is semiconductive [1],[2]. As the example of application of the device a-Si: H is solar cells, photovoltaic, color sensors, thin film transistor (TFT) and others [3]. This development is followed by the development of electronic device technology itself, as a result of various studies of the basic constituent materials.

Material a-Si:Hcanis grown by using the methods of Plasma Enhanced Chemical Vapor Deposition (PECVD), with plasma as growth media. This method uses Silane gas (SiH_4) as a source gas containing 10% in hydrogen gas (H_2) and amorphous silicon material obtained by the hydrogen content of about 10-20% [4]. The effect of hydrogen content in the process of growing resulting in defect rate moved into the conduction band and valence, so that the density of states (state density) is reduced sharply, as shown in Figure 1. For the band gap energy is strongly influenced by the hydrogen content and the existence of a defect in a thin layer of a-Si: H.

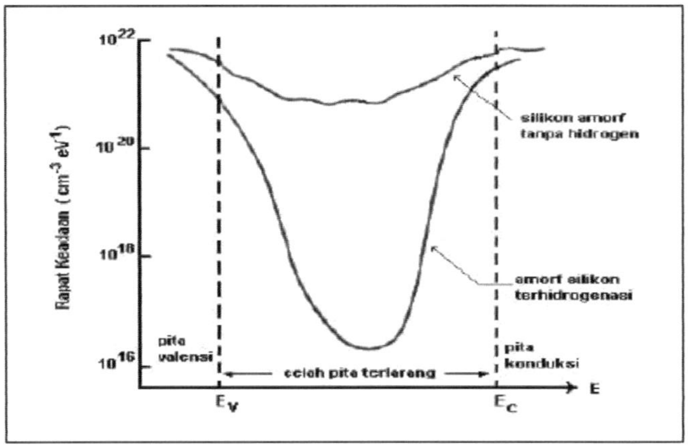

Fig.1 The effect of hydrogen content in the amorphous silicon to a state density [5].

The growth of material a-Si: H has some optimization parameters that determine the physical properties generated with the aims to improve the conversion efficiency of solar cells [6]. The optimization parameters are: The flow rate of hydrogen gas (H_2), chamber pressure, substrate temperature, RF frequency, RF power, time of deposition, dopant concentration [7], [8].

This research will be conducted on the comparison of the optical properties of a thin layer of a-Si: H p-type doping CH_4 with p-type without CH_4 is deposited by PECVD system with RF 13.56 MHz.

2. Experimental Methods

Thin layer of a-Si: H p-type was grown on the ITO substrate with PECVD system with gas flow SiH_4, B_2H_6 and added CH_4 During the process, it was given the deposition chamber pressure of 4800 mTorr, a temperature of 210 degree. RF power of 5 watts and time 15 minutes. The rate of gas fixed at 20 sccm SiH_4 and B_2H_6 gas rate fixed at 2 SCCM, while the rate on the layer used CH_4 gas at 30 SCCM.

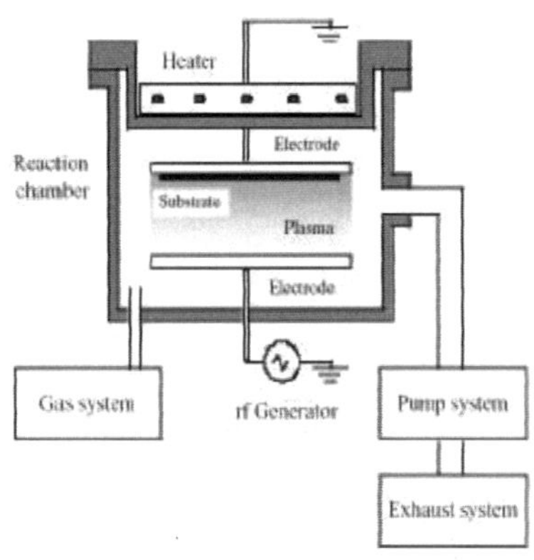

Fig. 2 Schematic RF-PECVD system [9]

Table 1.Parameters layer deposition of a-Si: H p-type

layer of a-Si:H p-type added CH4	Substrate	ITO glass
	Substrates Temperature	210 degree
	RF Frequency	13,56 MHz
	FR Power	5 watt
	Pressure Deposition	4800 mTorr
	Source gas	SiH_4 20 sccm, B_2H_6 2 sccm and CH_4 30sccm
	Time	15 Minute
layer of a-Si:H p-type without CH4	Substrate	ITO glass
	Substrates Temperature	210 degree
	RF Frequency	13,56 MHz
	FR Power	5 watt
	Pressure Deposition	4800 mTorr
	Source gas	SiH_4 20 sccm, B_2H_6 2 sccm
	Time	15 Minute

After the manufacture sample was completed, each sample sought optical properties that were the band gap by using a UV-VIS spectrometer. A thick of the sample layer was characterized by using testing tools Atomic Force Microscopy (AFM).

Microstructures, particle size and topography layer of a-Si:H were studied using AFM. For the measurement of tapping mode, AFM has been used for the purpose of investigation of surface topography of the A-Si:H. AFM Nanoscope III-A (Digital Instruments, Veeco, Metrology Group) utilized the optical beam deflection to monitor the placement of silicon cantilever having a spring constant 42 N/m and the resonance frequency of 300 kHz. It has been done with silicon probes that were above the cantilever in tapping mode. This method significantly showed improvement in the lateral resolution and resolution surface observations thin layer. Roughness of the surface was carried out by measuring the roughness parameters (RMS), which was defined as the average of the high RMS (Z) obtained from the mean data plane [10].

$$RMS = \sqrt{\frac{\sum_{i=1}^{N}(Z_i - Z_{ave})^2}{N}} \quad (1)$$

Z is a Z value at the time of measurement and N is the number of points between cursors. The method used for particle analysis can be performed with a height. This type of analysis works well on material by determining particle size that is connected to high.

3. Result and Discussion

The nature of a-Si: H p-type that has grown with PECVD system relies heavily on the particles (atoms or molecules) that is charged in the plasma and its interactions on the surface of the substrate. The particles (atoms or molecules) charged were called radical. Radicals were formed in the plasma include SiH molecules, SiH_2, SiH and atom H and atom B.

The results of the thickness measurement by Atomic Force Microscopy (AFM) are mentioned in the form of graph. Figure. 3 shows the morphology structure and thickness of thin layer a-Si: H to the rate of2 SCCM B_2H_6, 20 SCCM SiH_4 and30 SCCM CH_4 obtaining thickness of the layer of a-Si: H p-type by adding CH_4 at 100 nm. Figure 4 shows the morphology structure and thickness of thin layer a-Si: H to therate of 2 SCCM B2H6 and 20 SCCM SiH4 obtaining thickness of the layer of a-Si: H p-type without adding CH4 is 45 nm.

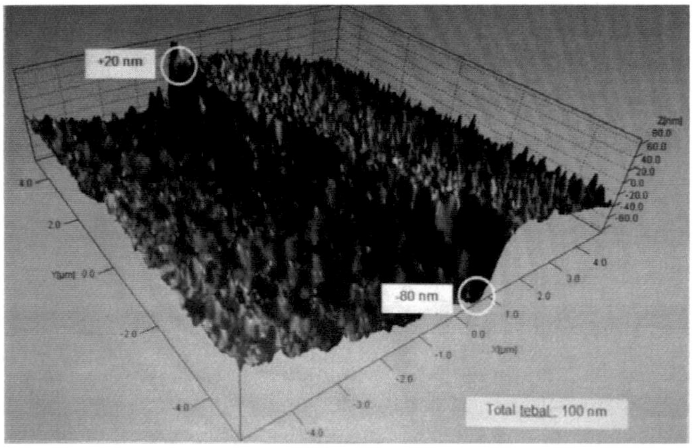

Fig. 3 The Morphology Structure and Thickness of Thin Layer a-Si: H p-type with the Addition of CH_4.

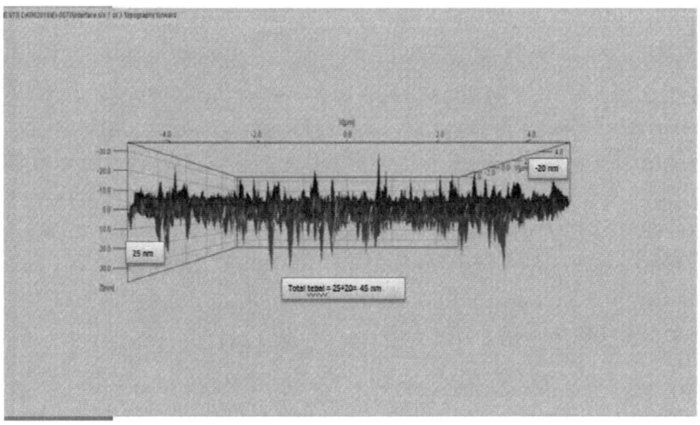

Fig. 4 The morphology Structure and thickness of thin layer a-Si: H p-type without addition of CH_4.

The calculation result of the energy gap by using a UV-VIS spectrometer was analyzed in graphic form with the relationship between the wavelength and absorption. Figure 5 shows the magnitude of the energy gap of a thin layer of a-Si: H to the rate of 2 SCCM B_2H_6, 20 SCCM SiH_4 and 30 SCCM CH_4obtaining energy gap layer of a-Si: H p-type by adding CH4 ~1.95 eV. Figure 6 shows the magnitude of the energy gap of a thin layer of a-Si: H to the rate of 2 SCCM B_2H_6 and 20 SCCM SiH_4obtain energy gap layer of a-Si: H p-type without adding CH_4~1.89 eV.

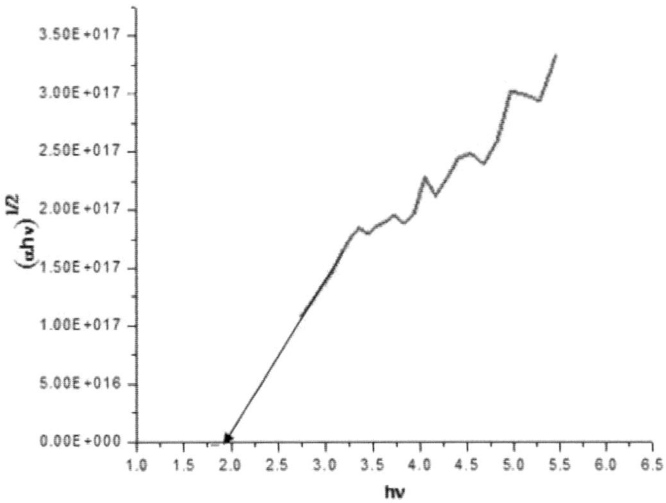

Fig. 5 Graph of the relationship between the wavelength with the absorption of a thin layer of a-Si: H p-type with the addition of CH_4

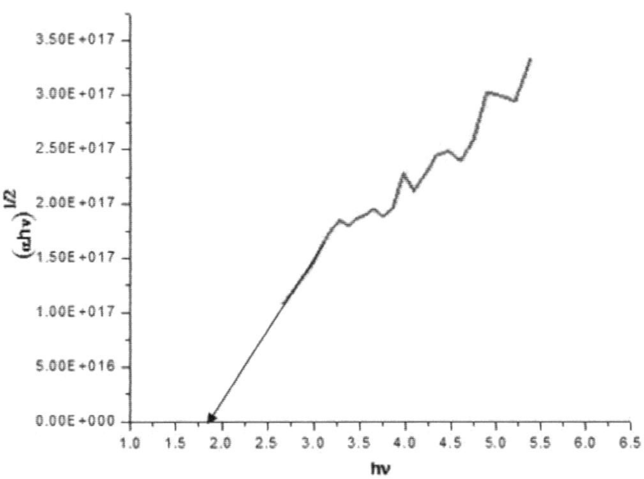

Fig. 6 Graph of the relationship between the wavelength with the absorption of a thin layer of a-Si: H p-type without the addition of CH_4.

The parameter of optimization is a key indicator of the physical properties that will be generated. However, the development to get the material a-Si: H is still based on technology that is easy, inexpensive, efficient and of good quality.

The phase of characterization became the last stage that shows the physical properties produced by optimization of growth parameters that were performed [11],[12]. Physical properties that were obtained became a benchmark for subsequent optimization stages in order to get the quality of a thin layer of a-Si: H better. The increase in the rate of hydrogen causes more irregular layer a-Si: H p-type are formed but the rising rate of Boron causes irregularity of layer formed.

4. Summary

Based on this research, it can be concluded that the layer properties of a- Si: H p-type are influenced by the rate of source gas (SiH_4, B_2H_6 and CH_4). The rate of source gas affects the number of radicals formed in the plasma. Thus, it also influences the effectiveness of interactions (bonding) between atoms on the layers of a-Si: H p-type with CH_4, its thickness 100 nm and its energy gap ~1,95 eV, whereas a layer a-Si: H p-type without CH_4, with its 45 nm thickness and its energy gap ~1.89 eV.

Acknowledgments

Author S.P. thanked the Ministry of Finance of the Republic of Indonesian, through the Institute of Education Fund Management (LPDP), which has provided financial support through the Scholarship Indonesian Masters 2015-2017. Also the author would like to thank the team of solar cell material physics research department of physics Institute Technology Sepuluh November (ITS) Surabaya.

References

[1] K Takahashi and M Konagai 1986 *Amorphous Silicon Solar Cells* North Oxford Academic Publishers Ltd London
[2] B Maennig, J Drechsel, D Gebeyehu, P Simon, F Kozlowski, A Werner, F Li, S Grundmann, S Sonntag, M Koch, K Leo, M Pfeiffer, H Hoppe, D Meissner, N S Sariciftci, I Riedel,V Dyakonov, and J Parisi Mar 2004 *Organic p-i-n solar cells Appl Phys A* **79** 1 pp. 1–14
[3] J Poortmans and V Arkhipov 2006 *Thin Film Solar Cells: Fabrication Characterization and Applications* John Wiley & Sons
[4] B Liu, L Bai, X Zhang, C Wei, Q Huang, J Sun, H Ren, G Hou, and Y Zhao Sep 2015 Fill factor improvement in PIN type hydrogenated amorphous silicon germanium thin film solar cells: Omnipotent N type μc-SiOx:H laye *Solar Energy Materials and Solar Cells* **140** pp 450–456
[5] T H Yeh and K H Khokhani1968 *Electrochemical Society Boston, May* pp. 5–9
[6] J D Malago 2002 *Pengembangan Lapisan Tipis Silikon Amorf dan Paduannya untuk Aplikasi Divais Elektronik dengan Teknik Plasma Enhanced Chemical Vapor Deposition*, Ph. D. Dissertation Institut Teknologi Bandung.
[7] J Drechsel, B Männig, F Kozlowski, M Pfeiffer, K Leo, and H Hoppe Jun 2005 Efficient organic solar cells based on a double p-i-n architecture using doped wide-gap transport layers *Applied Physics Letters*, **86** 24, p. 244102.
[8] C C Tsai and H Fritzsche 1979 Effect of annealing on the optical properties of plasma deposited amorphous hydrogenated silicon *Solar Energy Materials* **1** 1 pp. 29–42.
[9] S Nacer and A Aissat Dec. 2015 Simulation of InGaN p–i–n double heterojunction solar cells with linearly graded layers Optik *-International Journal for Light and Electron Optics* **126** . 23 pp. 3594–3597.
[10] D E Carlson and C R Wronski 2005 *IIc-1 - Amorphous silicon solar cells A2 – Markvart Tom in Solar Cell* L Castañer Ed. Oxford: Elsevier Science pp. 217–251.
[11] H Okamoto, Y Nitta, T Yamaguchi, and Y Hamakawa Apr. 1980 Device physics and design of a-Si1ITO/p-i-n heteroface solar cells *Solar Energy Materials*, **2** **3**, pp. 313–325.
[12] T Winata and I Usman 2008 The Influence of Silane Gas Flow Rate on Optoelectronic Properties of μc-Si:H Prepared by HWC-VHF-PECVD Technique *Journal of Mathematical and Fundamental Sciences* **40** **2**, pp. 130–137.

The detection of organic solvent vapor by using polymer coated chemocapacitor sensor

Apik Rusdiarna Indrapraja*, Muhammad Rivai, Achmad Arifin and Djoko Purwanto

Department of Electrical Engineering, Institut Teknologi Sepuluh Nopember (ITS), Surabaya

*Corresponding email address: lku@uad.ac.id

Abstract. A chemocapacitor consists of planar interdigital electrodes (IDE) made by two comb electrodes on a substrate. A dielectric film was applied on the electrodes in which the absorbed vapor will modify its permittivity. This study has fabricated chemocapacitor with the IDE distance of 0.5 mm, while the dielectric film was a sensitive layer consisting of a polymeric material. The deposition of the polymeric film was accomplished by drop casting. A sensor array consisting of four chemocapacitors coated with different polymers namely PEG-1540, PEG-20M, PEG-6000, and PVP was used to obtain the pattern of shift in the capacitance. The integrated circuit AD7746 was used as the capacitance to-digital converter (CDC). The organic solvents of ethanol, benzene, and aceton were used as the vapor samples in this experiment. The results showed that the change in the capacitance value increases proportionally to the concentration of vapour where sensors coated with PEG-1540 and PVP have higher sensitivity, i.e. 0.0028pF/part per thousand and 0.0027pF/part per thousand, respectively. Based on the capacitance to digital conversion capabilities, the system provides there solution of 0.4084ppm. The sensor array could produce a different pattern for each of the vapor sample. The Neural Network pattern recognition system could identify the type of vapor automatically with the root mean square error of 10^{-5}

1. Introduction

Organic solvents are widely used in industries such as paints, adhesives, cosmetics, fragrances, etc. Due to its volatile and less polar nature, the solvents will easily enter bodies through inhalation or skin absorption. It is dangerous for health because it can cause eye irritation, skin irritation, respiratory problems and cancer. In the solvent storage, leakages must always be detected to avoid pollution and fires. Therefore, it is necessary to have a type of sensor that can recognize the type of solvent vapor and measure its concentration.

The detection of organic solvent vapors using a type of lock and key can only recognize one type of vapor. To recognize a lot of organic solvent vapors, it will require a huge number of sensors. The other method is to use a sensor array consisting of sensor elements having overlapping responses. The type of sensor will produce a different pattern for each vapor.

The use of polymers as the material for the detection of organic solvents has been widely applied to various types of gas sensors because they have rapid adsorption and desorption at room temperature, low power consumption, and resistant to poisoning by other compounds [1].

IOP Conf. Series: Journal of Physics: Conf. Series **853** (2017) 012033 doi:10.1088/1742-6596/853/1/012033

Researchers have tried to create a device that can detect vapors from a variety of materials such as polymer coated Chemocapacitor [2][3][4], chemoresistor [5], Quartz Crystal Microbalance [6] and metal oxide semiconductors[7] with satisfactory results. This study investigates an organic solvent detection system using chemocapacitor sensor array coated with polymermaterial.

2. Experimental

Chemopacitor is a capacitor which has a selective absorbent material such as a polymer that serves as the dielectric material. The chemicals absorbed in the dielectric will change the capacitance of the sensor [4]. Chemocapacitor consists of planar inter digital electrodes (IDE), which was two comb electrodes on substrate.

The design of IDC is shown in Figure 1. The sensor was made of Printed Circuit Board (PCB) with a fiber substrate. The IDC has 20 electrodes, the length (L) is 10 mm, the metallization ratio (η) is 0.5 and the spatial wavelength (λ) is 2 mm. Series chemocapacitor sensor used in this study consisted of four IDC, each sensor was coated with a different polymer, i.e. PEG-1540, PEG-20M, PEG-6000, and PVP.

The hardware system in this study is shown in Figure 2. Silica gel is used to produce the reference air having low humidity. The flow rate was set at 0.1 LPM regulated by valves K1. Material samples were ethanol, acetone, and benzene. IC AD7746 was used as Capacitance to Digital Converter (CDC). The 24- bit monolithic converter has a high resolution of 0.5 fF, high linearity of ± 0.01%, I2C interface mode, and supports the expansion of the input range with an external circuit. The Chemocapacitor sensor array consisted of 4 sensors which requires analogue multiplexer to select the capacitor to be accessed. Data of each sensor was sent to the computer via serial communication.

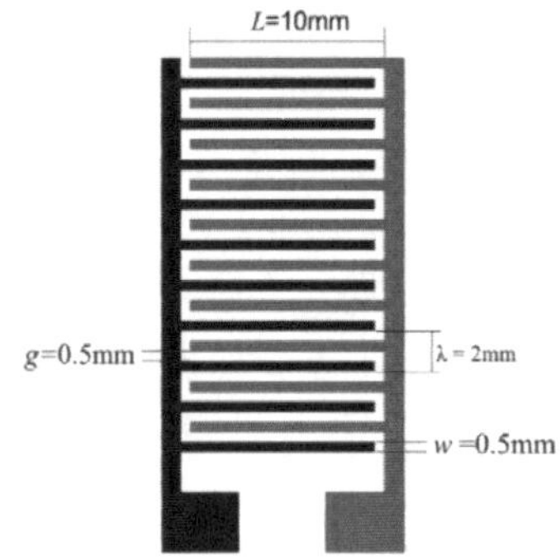

Figure 1. The layout design of IDE.

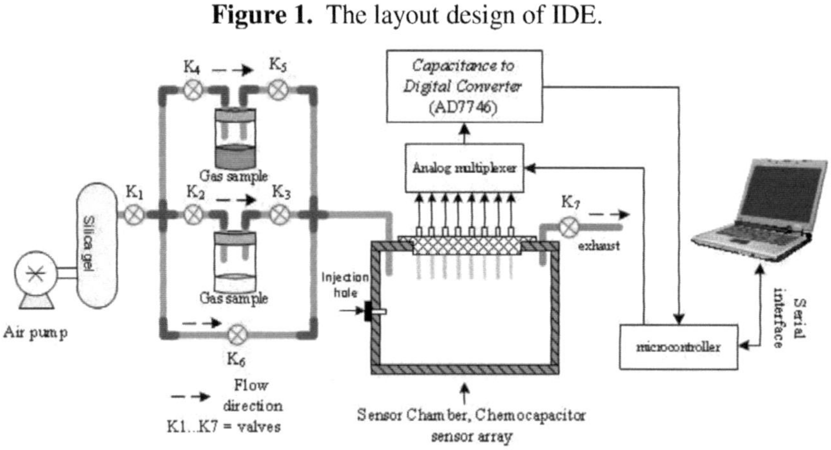

Figure 2. The hardware system used in the experiment.

3. Result and discussion

Sensor calibration was performed by observing the value of digital data to the capacitance value for each channel of the multiplexer. The relationship between the digital data to the capacitance is shown in Figure 3. This showed that thc calibration curve of digital data and the capacitance was linear and the maximum value that can be measured was 32pF. Linear regression equation was used to convert digital data to the capacitance value.

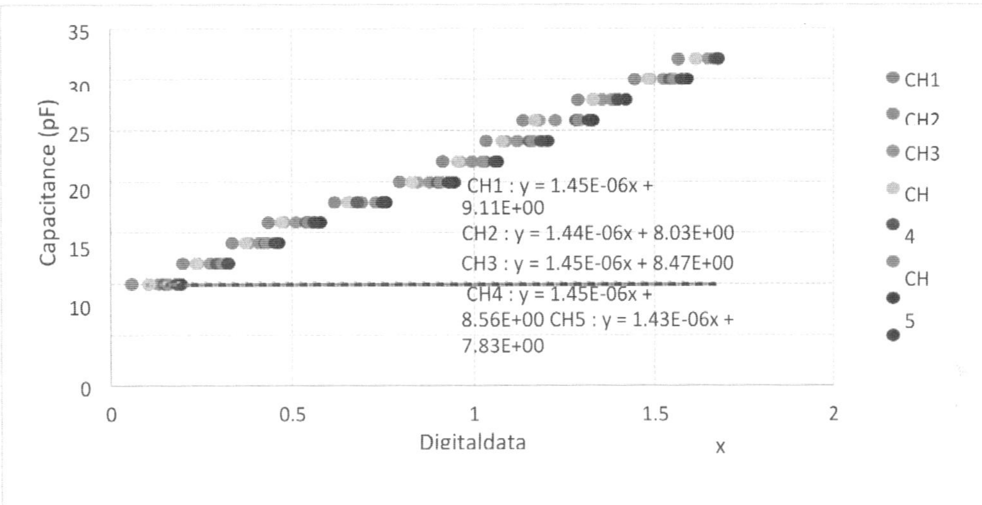

Figure 3. The relationship between the digital data and the capacitance.

The dry air was flowed into the test chamber until a stable capacitance value was reached. Then the sample gas was fed into the chamber. Each capacitance sensor changed in time domain is shown in Figure 3. When the sample gas is injected continuously, it will reach steady state and for each sensor will have a different response. The pattern of change in capacitance is normalized to the maximum value, shown in Figure 5. Each gas could produce a typical pattern for each sample.

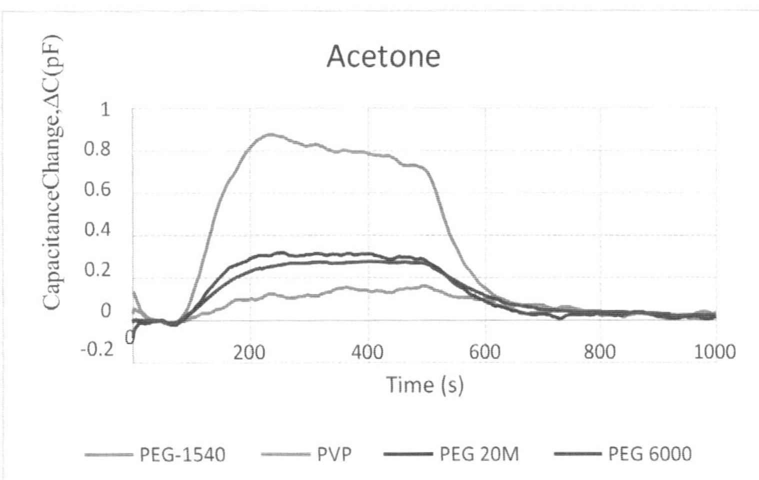

Figure 4. Capacitance changes to acetone vapor.

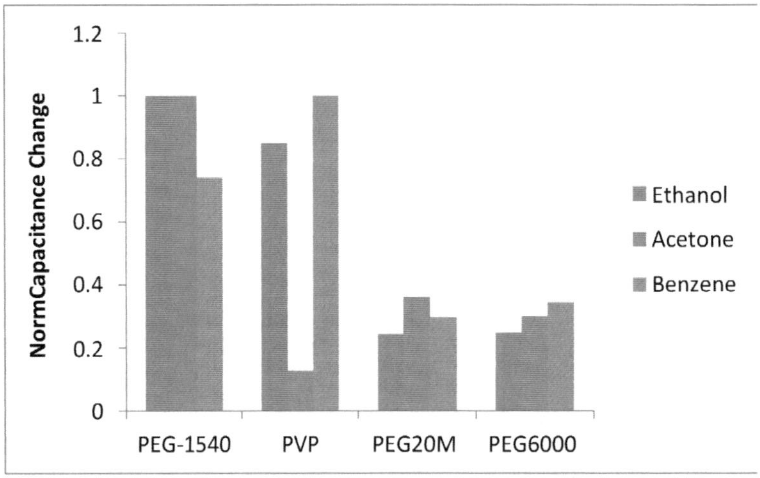

Figure 5. The pattern of change in capacitance to the vapors.

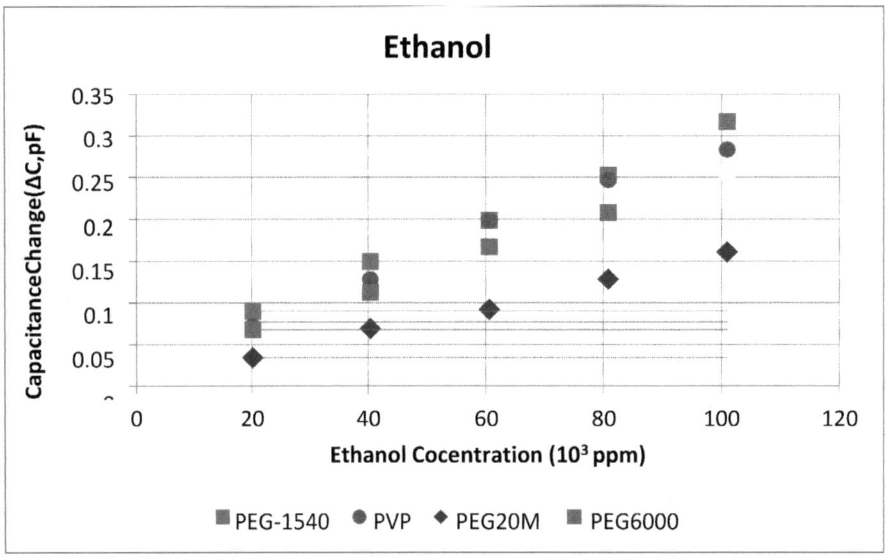

Figure 6. The relationship between ethanol vapor concentration and the capacitance change.

References

[1] Persaud K C 2005 *Polymersfor Chemical Sensing Material Today* **8** page.38-44.

[2] Oikonomou, P Raptis I and Sanopoulou M, 2014 Monitoring and Evaluation of Alcoholic Fermentation Processes Using a Chemocapacitor Sensor Array *Sensors* **14** page 16258-16273.

[3] Botsialas, A Oikonomou, P Goustouridis, D Ganetsos, Th Raptis I and Sanopoulou M, 2013 A iniaturized Chemocapacitor System for The Detection of Volatile Organic Compounds *Sensors and Actuators B* **177** p.776–784.

[4] Patel S V, Mlsna T E, Fruhberger B, Klaassen E, Cemalovic S and Baselt D R 2003 Chemocapacitive Microsensors for Volatile Organic Compound Detection *Sensor and Actuator B* **96** p.541-553.

[5] Gunawan B, Rivai M, and Juwono H, 2012 Characterization of Polymeric Chemiresistors for Gas Sensor *Telekomunika* **10** 2 p.275-280.
[6] Kimura M, Liu Y, Sakai, Sato S, 2011 Detection of Volatile Organic Compounds by Analyses of Polymer-Coated Quartz Crystal Microbalance Sensor Arrays *Sensor and Materials* **23** 7 p.359-368.
[7] Edward J, Wolfruma, Robert M, Meglen, Darren Peterson, Justin Sluiter, 2006 Metal oxide sensor arrays for the detection, differentiation, and quantification of volatile organic compounds at sub-parts- per-million concentration levels *Sensor and Actuator B* **115** p.322-329.

IOP Conf. Series: Journal of Physics: Conf. Series **853** (2017) 012034 doi:10.1088/1742-6596/853/1/012034

The dosage optimization of He-Ne laser energy as a candidate for photodynamic therapy of cancer cells with exogenous photosensitizer variations

R Apsari[1,2,*], **I Maghfiroh**[1], **H Setyawati**[3], **Winarno**[1,2], **D Arifianto**[1,2] **and A Zaidan**[1]

[1]Department of Physics, Faculty of Science and Technology, Airlangga University, Surabaya 60115, Indonesia

[2]Department of Engineering, Faculty of Vocational Studies, Airlangga University, Surabaya 60286, Indonesia

[3]Department of Chemistry, Faculty of Science and Technology, Airlangga University, Surabaya 60115, Indonesia

[*]Corresponding email address: apsari.unair@gmail.com

Abstract. The aim of this research was to find out the most effective energy dose of laser energy per unit area irradiated to the subject, used in T47D cancer cell photodynamic therapy. The experiment used three variations of energy doses and two exogenous photosensitizers. The laser used was the He-Ne laser with a wavelength of 632 nm. Laser exposure on the T47D cancer cells was conducted in vitro at a distance of 1 cm with energy doses of 30.4 J/cm^2, 18.8 J/cm^2, and 30.4 J/cm^2. The photosensitizers used in the experiment were chlorophylls isolated from sour-sop leaves and *Protoporphyrin IX* (*PpIX*). The concentration of photosensitizers which is not toxic to cells, based on the toxicity test, is 300 ppm for chlorophylls and 350 ppm for *PpIX*. At the energy dose of 30.4 J/cm^2, the percentage of cell death was 96.25% with chlorophyll addition, while with *PpIX* addition, the percentage was 81.62%. At the same energy dose, cell death in apoptosis was 95.78% with chlorophyll addition and 80.56% with *PpIX* addition. Based on the result of this research, it was revealed that the energy dose of He-Ne laser output of 30.4 J/cm^2 was the most effective energy dose between the energy dose variation that could yield T47D cell death in an apoptosis way.

1. Introduction

Cancer is a disease caused by abnormal and uncontrollable body tissue's cell growth. The prevalence of cancer in Indonesia is 1.4 per 1000 people and has become a major cause of death around the world [1]. At the moment, there are several alternative medications for cancer with their own downsides. As a consequence, alternative medication which is minimally invasive and selective against cancer cells is needed.

One type of medication that can be an alternative against cancer is photodynamic therapy [2]. Photodynamic therapy involves three main components: light, photosensitizer, and oxygen.

Chlorophylls and Protoporphyrin IX (PpIX) have chromophore and auxochrome clusters, allowing them to be made as photosensitizers. Both of them have absorption spectrums within the red wavelength [2]. In photodynamic therapy, there is a process where photosensitizers are localized on cancer tissues and when exposed with lasers the photosensitizers become active and produce Reactive Oxygen Species (ROS). ROS then induces apoptosis death. Apoptosis is an organized and programmable death, so pathologically only apoptosis death is allowed [3].

Li *et al.* (2010) have stated that the combination of C-phycocyanin and He-Ne laser irradiation is able to increase the activity effects of anti-tumor, body immune, and pro-apoptosis against cancer cells [4]. Similar research was also conducted by Al-Khafaji *et al.* (2010) who showed that the most damaging effect on cancer cells resulted from the combination of Hematoporphyrin derivative photosensitizers and He-Ne laser irradiation on low power with energy doses of 3.6 and 7.2 J/cm^2 [5]. Fadilla (2015) also conducted research using PpIX and chlorophylls with radiation from lasers with a wavelength of 650 nm, and a laser energy dose of 18.8 J/cm^2 could cause apoptosis death to cancer cells [6]. The weakness of this research is that the laser interaction effects on cells could not be achieved with an effective energy dose. Therefore, the aim of this research was to find out the most effective energy dose of lasers using non-toxic photosensitizer solutions between variation of energy doses similar to the previous studies.

2. Research Methods

The laser used in this research is the He-Ne laser LSW-10 from Jianxi Liansheng Technology Co., Ltd with a wavelength of 632 nm. The stability of the laser was characterized using an experimental setup shown in Figure 1.

Chlorophylls as photosensitizers in this research were extracted from the leaves of sour-sop. Sour-sop leaves were chosen because they can grow abundantly only from one tree which will not disturb the ecosystem by endangering the existence of sour-sop plants due to exploitation. They were also chosen because they are not a food commodity so they can be used for medical purposes as photosensitizers to cure cancer due to their chlorophyll content. The chlorophyll isolation was conducted by drying the sour-sop leaves that had been cut into small pieces in the open air but without direct sunlight. The process was then continued with maceration, evaporation, partition, and VLC (Vacuum Liquid Chromatography) to obtain 11 fractions of sour-sop leaf extract as shown in Figure 3. The green bluish fraction is the chlorophylls [8]. Then, a thin layer chromatography was conducted and the fractions were photographed with 366 nm UV light to know the fraction node patterns and to determine fractions with the most chlorophylls. The chosen fractions were then dried further in a 40°C oven to gain thick chlorophyll extracts as shown in Figure 4.

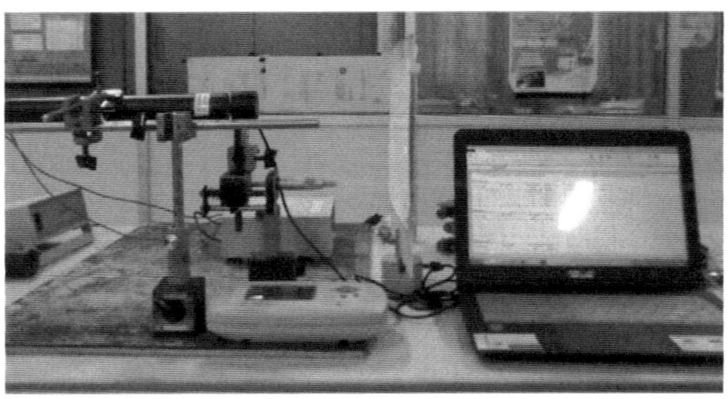

Figure 1. Checking of He-Ne Laser Power with A Digital Powermeter.

Another photosensitizer also used in this research was the commercial Protoporphyrin IX (PpIX) from Sigma-Aldrich. Absorption wavelength characterization was then conducted on both photosensitizers using a UV-Vis spectrophotometer.

Simultaneously, the T47D cancer cell culture reproduction in a petri dish using a set of cell culture tools such as a micropipette (Thermo), a centrifuge, a CO_2 incubator (Thermo), a 96-well plate, a conical tube (Biologix), a vortex, a light microscope (Nikon), ELISA Reader (Bio-Rad iMark), air flow laminar (Dalton), a fluorescence microscope (Olympus). The growth medium used in the culture reproduction of T47D cancer cells is DMEM medium with PBS (Phosphate Buffered Saline) 10%, Penicillin-streptomycin 1%, and fungison 0.5%.

Before the treatment to expose the laser to cancer cells with and without the photosensitizer, there was a photosensitizer toxicity test toward cancer cells with the MTT Assay method to find out the concentration of photosensitizers which was not toxic (safe) to cells. In the toxicity test, the photosensitizers (chlorophylls and PpIX) were prepared, with varied concentrations: 100 ppm, 150 ppm, 200 ppm, 250 ppm, 300 ppm, and 350 ppm. The photosensitizers were solved with DMSO 0.1% solvent in DMEM cancer cell growth medium. DMSO 0.1 % was chosen because it was safe to cells [7]. A safe to cell photosensitizer solvent was chosen to avoid the influence on the result of the photosensitizer toxicity test.

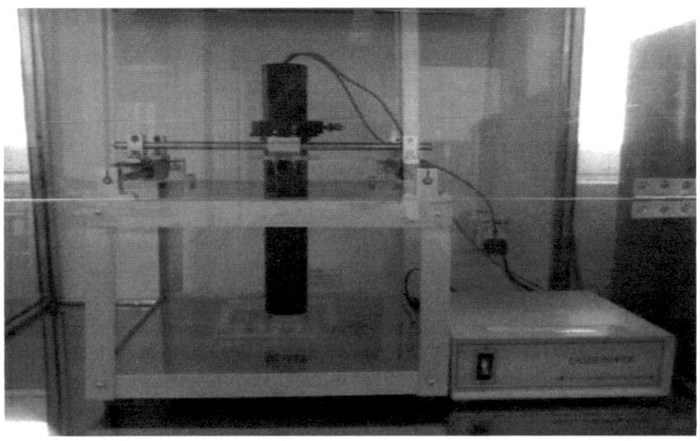

Figure 2. Setting Up the T47D Cancer Cell Exposure with Red He-Ne Laser inside Laminar Air Flow (LAF).

Figure 3. Eleven fractions produced from the VLC process.

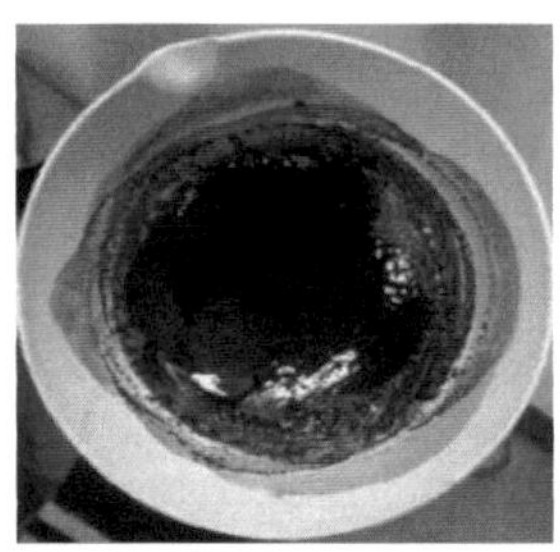

Figure 4. Thick chlorophyll extract after being in an oven.

T47D cancer cells which had been grown to reach enough population for the test, were then divided into sample groups. There were three treatment groups, namely a laser without photosensitizer addition (L), a laser with chlorophyll addition (LK), and a laser with PpIX addition (LP). Cells were exposed with lasers inside Laminar Air Flow (LAF) so that they were still in sterile condition, with mountings and a slider that had been arranged in such a way that the lasers could slide along the X and Y axis (Figure 2). Before the exposure began, the laser mounting tools were cleaned first with alcohol.

Cells that had been incubated with chlorophyll and PpIX photosensitizers were put on a sample plate. Then, cells in each well were exposed with the He-Ne laser one by one with a variation of energy doses of 7.2 J/cm^2, 18.8 J/cm^2, and 30.4 J/cm^2. These energy doses are used due to an effective dose for the goals of the therapy. Each treatment was repeated three times. After the exposure finished, the cells were observed with a fluorescent microscope with an addition of coloring reagents, Acridine Orange (AO) and Ethidium Bromide (EB), to know the number of living and dead cells, based on the different colors produced. The living cells percentage was calculated using equation (1).

$$\%\text{ living cells} = \frac{\text{OD Treatment} - \text{OD Media}}{\text{OD Cell Control} - \text{OD Media}} \times 100\% \quad (1)$$

3. Results and Discussion

3.1. *Checking of the He-Ne laser power*

Checking the power of the He-Ne laser was conducted by varying the laser distances from the samples, which were 1.0 cm up to 3.0 cm. Power measurements were conducted repeatedly up to three times, and an optimum average power of (0.603 ± 0.010) mW from the distance of 1 cm was obtained. Laser power changes due to variations of exposure distances were proven to be insignificant, which had a significant value of 0.128 .05), so the laser power was stated as stable against changes in exposure distances.

In checking the stability of the He-Ne laser power against the exposure time, it was revealed that the He-Ne laser in the distance of 1 cm started to be stable 30 minutes after turning on the laser. Besides power, the calculation of laser energy doses also considered the laser area that could be measured if the diameter was known. Laser has a Gaussian file in which the Gaussian file diameter depends on the laser exposure distance. In this research, the laser file diameter was adjusted with the data sheet of LSW-10 He-Ne laser which was 1.2 mm.

3.2. *Photosensitizer characterization (chlorophyll and protoporphyrin IX)*

Based on a photosensitizer absorbency test, it was as expected that chlorophyll had the highest absorbent peaks at the red and blue wavelengths (Figure 5), which was in line with the statement of Lichtenthaler and Claus in 2009 [9]. This result indicates that our isolation method is appropriate for this current work. As seen in Figure 6, in PpIX photosensitizer there were several absorbent peaks in the wavelengths of 462 nm, 542 nm, 598 nm, and 646 nm (red area wavelength). This is the same with research conducted by Rossetti *et al* in 2010 who found that the absorbent spectrum of PpIX showed a strong absorbency in the wavelength of 406 nm, and four weak peaks in the wavelengths of 516 nm, 554 nm, 600 nm, and 626 nm (red area wavelength) [10].The results of photosensitizer characterizations of both chlorophylls and PpIX were found to have the absorbency peaks in the red wavelength that were the same with the wavelength of the laser used in this research. The chlorophyll absorbency at the red wavelength (λ = 671 nm) is 0.129 while for the PpIX, the absorbency at the red wavelength (λ = 646 nm) is 0.088.

3.3. *Photosensitizer toxicity test*

The reading of toxicity tests was shown by Optical Density (OD) values which represent the number of living cells. Overall, with concentration variations from 100 ppm until 350 ppm, the toxicity tests showed that the percentage of living cells was more than 50%, so it could be concluded that with those concentrations, both photosensitizers (chlorophylls and PpIX) were not toxic. The concentrations of each photosensitizer with cell percentages are 300 ppm for the chlorophylls with the percentage of living cells of 97.6% and 350 ppm for PpIX with the percentage of living cells of 75.9% were chosen for the treatments with sample cells.

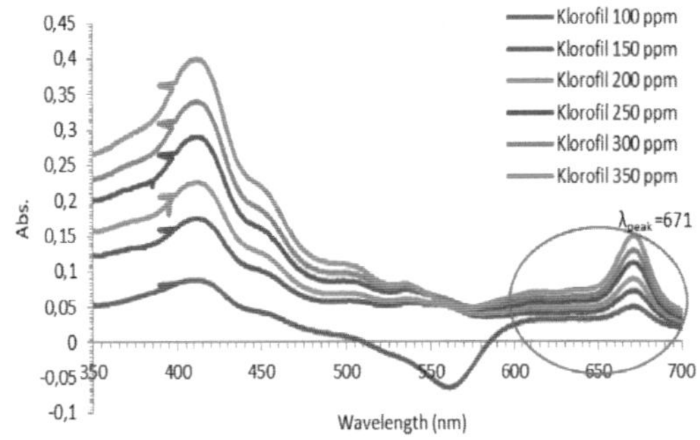

Figure 5. Chlorophyll Absorbance with Various Concentrations via UV-Vis Spectrophotometer.

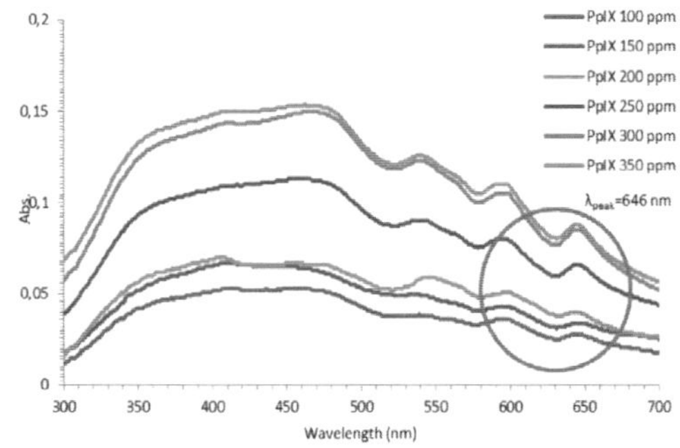

Figure 6. Absorbance of Protoporphyrin IX with Concentration Variations through UV-Vis Spectrophotometer.

3.4. *T47D cancer cell death due to the red He-Ne laser exposure with or without photosensitizer additions*

The percentages of cell death after the treatment in each group of L, LK, and LP with certain laser energy doses are presented in the form of a graph in Figure 7. As seen in the Figure 7, the highest cell death percentage occurred with the laser energy dose of 30.4 J/cm^2 and LK treatment was more effective in producing cell death compared to other groups of L and LP. It should be noted that the energy dose that became the independent variable in this research was the laser output energy dose (air dose), not the energy dose which was absorbed by the tissue of cancer cells (quantum yield).

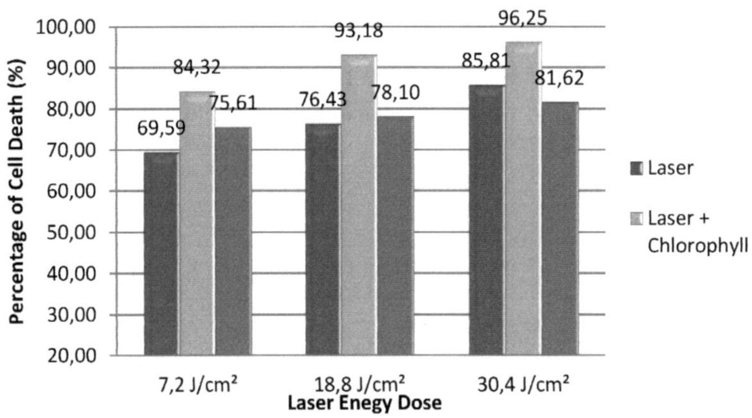

Figure 7. Graph of the Percentage of Cell Death After Treatment.

3.5. *Observation of the number of apoptosis cells on T47D cancer cells after treatment*

Coloring reagents (AO and EB) were given to differentiate the colors of the living cells and apoptosis ones. The difference in colors comes from the DNA fluorescent which is not same for each type of cells. Living cells with intact membranes have a uniform green color. When cells are in apoptosis stage and the membranes start to bleed, EB can enter into the cells and give an orange color. The picture of T47D cancer cells which was obtained from observation with a fluorescent microscope is divided into their treatment groups of control group, L, LK, and LP, respectively, as seen in Figure 8 and the percentage of apoptosis cell death due to treatments on each group is presented in the form of a graph as shown in Figure 9.

This graph shows that cancer cells exposed to the lasers with additions of photosensitizers with PpIX or chlorophylls were able to influence the cell death, and an addition of chlorophyll photosensitizer was found out to be more effective in affecting the cell death in an apoptosis way than addition of PpIX photosensitizer. This is because chlorophylls have more chromophore clusters than PpIX because chlorophylls experience more photophysical and photochemical interactions due to the more energy being absorbed from the given laser. Besides that, the peak absorption at the wavelength of 671 nm which falls into optical window range (650 nm – 850 nm) has quite high tissue penetration and enough triplet state energy to produce singlet oxygen. Meanwhile, PpIX has an absorption peak of a wavelength less than 650 nm (outside the optical window range for PDT, making it less effective in producing singlet oxygen). This is the reason why apoptosis cell death occurred more in the laser treatment group with chlorophyll addition (LK).

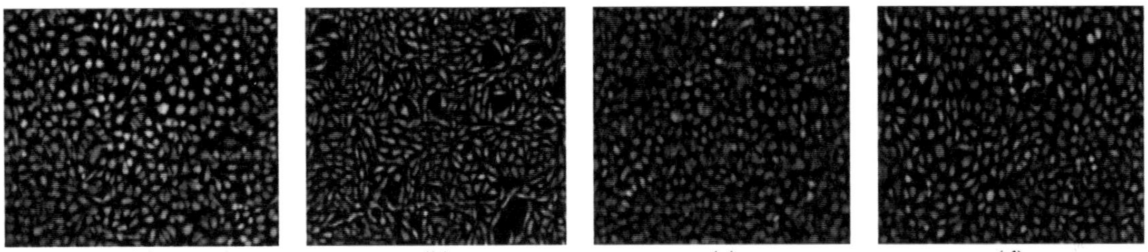

(a) (b) (c) (d)

Figure 8. T47D Cancer Cells after Treatment; (a) Control group (b) L (c) LK (d) LP.

This cell death caused by laser exposure with a photosensitizer addition is due to the interaction between the laser and the tissue. In the photodynamic therapy on the cancer cells, there is a photophysical and photochemical interaction mechanism whereas when photosensitizers are activated by light there is a production of Reactive Oxygen Species (ROS). ROS, especially singlet oxygen, can easily react with many biological substrates such as amino acids, various unsaturated lipids, and basal DNAs and RNAs. Singlet oxygen will attach with basal DNAs which will change the structure of basal DNAs of cancer cells. This structure change will disturb the metabolism process of the cancer cells making them unable to grow well [11].

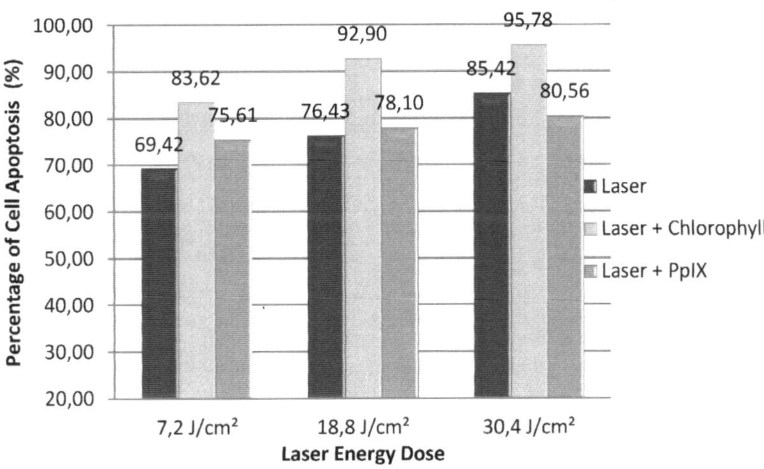

Figure 9. The Graph of the Percentage of Cell Apoptosis after Treatment.

4. Conclusions

Based on the toxicity test, the concentration of chlorophyll photosensitizers which is not toxic to cancer cells is 300 ppm with a living cell percentage of 97.6% and 350 ppm for the PpIX photosensitizers with the percentage of living cells reaching 75.9%. The percentage cell death is the biggest on the red He-Ne laser output energy dose of 30.4 J/cm^2 which is 85.81% (without a photosensitizer addition), 96.25% (with a chlorophyll addition), and 81.62% (with the Protoporphyrin IX photosensitizer addition). The optimum energy dose of the red He-Ne laser to obtain apoptosis cell death is 30.4 J/cm^2 with cell apoptosis reaching 95.78% (chlorophyll addition) and 80.56% (with the Protoporphyrin IX photosensitizer addition).

Acknowledgment
We would like to express our gratitude to the Government of Indonesia through the Directorate General of Higher Education (PUPT Research) for financial support in this research.

References

[1] Basic Health Research (Riskesdas) Jakarta Research Agency of Ministry of Health Republic of Indonesia 2013
[2] Agostinis P et al 2011 *Photodynamic Therapy of Cancer*: An Update. CA: A Cancer J. for Clinicians. **61(4)** 250-81
[3] Rode H J 2008 *Apoptosis, Cytotoxicity and Cell Proliferation Mannheim: Roche Diagnostics GmbH* p 178 (Doctoral Thesis)
[4] Li B, Xianming C, Meihua G and Wuxiu L 2010 *Apoptotic Mechanism of MCF-7 Breast Cells In Vivo and In Vitro Induced by Photodynamic Therapy with C-Phycocyanin.ActaBiochimBiophys Sin.* **42** (1) 80-9
[5] Al-Khafaji A S, Ahmed M H A and Firas S 2010 Photodynamic Action of Low Power He-Ne Laser on Photosensitized Human Hep-2 and AMN3 Cell Lines with Hematoporphyrin derivative In Vitro *Iraqi J of Canc and Med Gen* **3**(1) 54-60
[6] Fadilla N R, *The Dosage Optimization of Red Laser Energy as The Candidate for Photodynamic Therapy of Cancer Cells with Exogenous Photosensitizer* [thesis] Surabaya: Universitas Airlangga 2015
[7] Wang G H, Tzung H C, Rong J L, Jyh H S, Shih H W and Chia H L 2009 Cytotoxic Effect of the Genus Sinularia Extracts on Human SCC25 and HaCaT Cells J Toxicology. 634868 1-8
[8] Budiyanto A W, Soenarto N and Leenawaty L 2008 *Effect of Acidification against Photodegradation of Chlorophyll a. J. Math.* Sci. **13** (3) 66-75
[9] Lichtenthaler H K and Claus B 2001 Chlorophylls and Carotenoids: Measurement and Characterization by UV-VIS Spectroscopy *Current Protocols in Food Analytical Chemistry* (CPFA) Wiley, New York, F4.3.1-F4.3.8
[10] Rossetti F C, Lívia V D, Antônio C T and Maria V L B B 2010 *Fluorometric Quantification of Protoporphyrin IX in Biological Skin Samples from in Vitro Penetration/Permeation Studies* Brazilian J Pharm Sci.**46**(4) 753-60
[11] Wiseman H, Barry H 1996 *Damage to DNA by Reactive Oxygen and Nitrogen Species: Role in Inflammatory Disease and Progression to Cancer* Biochem J.**313** 17-2

IOP Conf. Series: Journal of Physics: Conf. Series **853** (2017) 012035 doi:10.1088/1742-6596/853/1/012035

The effect of TiO_2 thin film thickness on self-cleaning glass properties

Nandang Mufti*, Ifa K R Laila, Hartatiek and Abdulloh Fuad

Department of Physics, Faculty of Mathematics and Natural Sciences,
Universitas Negeri Malang (State University of Malang)
Jl. Semarang 5, Malang 65145, Indonesia

*Corresponding email address: nandang.mufti.fmipa@um.ac.id

Abstract. TiO_2 is one of semiconductor materials which are widely used as photocatalyst in the form of a thin film. The TiO_2 thin film is prepared by using the spin coating sol-gel method. The researcher prepared TiO_2 thin film with 3 coating variations and X-Ray Diffraction characterization, UV-Vis Spectrophotometer, Electron Microscopy Scanning, and examined its hydrophilic and anti-fogging properties. The result of X-Ray Diffraction showed that the phase formed is the anatase on 101crystal field. The Electron Microscopy Scanning images showed that TiO_2 thin films had a homogeneous surface with the particle sizes as big as 235 nm, 179 nm, and 137 nm. The thickness of each thin film was 2.06 μm, 3.33 μm, and 5.20 μm. The characterization of UV-Vis Spectrophotometer showed that the greatest absorption to the wavelength of visible light was in the thin film's thickness of 3 coatings with the band-gap determined by using 3.30 eV, 3.33 eV, and 3.33 eV Plot Tuoc. These results indicated that the rate of absorption would be increased by increasing the thickness of film. The increasing thickness of the thin film makes the film hydrophilic able to be used as an anti-fogging substance.

1. Introduction

The use of glass remains to be a choice for buildings in the 21st century [1]. Besides the improved knowledge on the properties of transparent materials, glass also beautifies the property and becomes an alternative to save energy [2] due to its benefits in regard with the room lighting and heating of a building. The development of glass production process leads to a more widespread use of this material in almost all parts of buildings [3]. Glass is a transparent material that also has several shortcomings, such as being contaminated by dirt if it is not cleaned gradually which will cause the reduction of lighting in a room and even can damage the surface of the material. Hence, it is necessary to find a solution to handle such problems. Various strategies have been used to modify these materials in order to assist human's work, one of which is through the surface engineering technology [4].

In recent years, Titanium dioxide (TiO_2) is the most extensive material to be investigated [5]. Titania thin film morphology plays an important role in capturing sunlight. The existence of different crystalline phases of TiO_2 will affect the effectiveness of photon energy absorption by TiO_2 thin film [6–8]. The photoactive TiO_2 crystalline phases cover the anatase and rutile phases with energy gap (Eg) respectively of 3.2 and 3.0 eV. In addition, the morphology of the TiO_2 thin film surface affects the distinctive adsorption capacity [9,10]. Different morphologies provide different and long adsorption capacities. The amount of organic pollutants adsorbed on the TiO_2 thin layer will affect the effectiveness of the interaction between h+ species or radical -OH and the persistent organic pollutants [7].

One application of titania film is as photocatalyst [10]. Photo catalysis has been widely applied in life, especially after the discovered hydrophilic nature on the surface of TiO_2 thin films. This application is then constantly evolving as anti-fogging and self-cleaning[11]. One small example of the usefulness of photocatalytic anti-fogging applications is the use of TiO_2 thin films on glass [12].

Chun et al, (2010) synthesized TiO_2 films through the use of sol-gel method by using titanium precursor isopropoxide (TTIP), with a hydrolysis process which was carried out under N_2atmosphere conditions. The obtained results acquired a percentage of transmittance as big as 90%. According toB.-H. Moon (2013),[13] sol-gel method is commonly used in the synthesis of thin film or modification of pore since its multilayer deposition can control the structure, composition, and activity of the thin film's result. F. Gherardi (2016) [4] synthesized TiO_2 films by using a TTIP precursor added by HCl and ethyl alcohol and stirred with a magnetic stirrer for 1 hour. After that, the coating on glass substrates was donevia spin coating method. The film was calcinedat the temperature varied of 250°C, 300°C and 350°C. Transmittance test results showed that the temperature of 350°C by using XRD was already known in the form of crystalline anatase with the transmittance measurement of as big as 90%.

Based on the description of data in the preliminary study and some research done by other researchers, anti-fogging materialis made of hydrophilic TiO_2 thin films on a glass substrate, affected by the variations of lap time on the spin coating. However, the effect of coating variations is lack investigated. Therefore, in this study we investigated the effect of the thickness of the TiO_2 thin films grown on glass with coating variations to structural properties, morphology and self-cleaning properties. TiO_2 thin films were synthesized by using sol-gel method. In addition, the hydrophobic properties and anti-fogging of TiO_2 thin films at different thickness will be discussed.

2. Experimental method

The preparation stage is to prepare the used material, which is a glass substrate of 2.5 x 2.5 x 0.1 cm. This substrate preparation included cleaning the substrate prior to be coated by using the material. This is very important as it would determine the bonding film of the substrate. The glass substrate was cleaned / washed in 96% alcohol for 15 minutes and acetone for 15 minutes. The substrate was then dried in an oven at 100°C for 30 minutes. The next stage is the synthesis of TiO_2 sol-gel to produce a solution of TiO_2, 99% $TiCl_4$ (precursor) and 96% ethanol (solvent) with the ratio of precursor and solvent volume as big as 1: 1. The solution was stirred by using an 800-rpm magnetic stirrer for 60 minutes at room temperature, so the solution was mixed entirely. As the solution had been mixed with salt hydrolysis reaction, the equation was:

$$TiCl_4 + 2C_2H_5OH \rightarrow TiO_2 + 4HCl + 4CH_2 \qquad (1)$$

The solution that would be formed was still in the form of sol and it would form a gel after the polyvinyl alcohol solution (PVA) was added. Both solutions were mixed and stirred by using an 800 rpm magnetic stirrer for 60 minutes so that the solution was completely mixed. The process of making the thin film was done by taking four drops of TiO_2 solution to the dripped substrate that has been placed on the spin coating apparatus. Next, the substrate which has been etched with a solution of TiO_2 was rotated at 1500 rpm rotational speed. The next step was preheating, followed by heating (annealing) which was performed by using the furnace at 500°C temperature for 1 hour. The variations in thickness were done in1x, 2x, 3x coating. Annealing process was done in order to improve the structure of the crystal.

The crystalline structure was measured by an X-ray diffractometer (Philips Brand Expert-ProType) in 2θ range from 20^0 to 80^0 by 2^0 s^{-1} steps. The UV–vis spectroscopy of the films was taken by a spectrophotometer (SP-300) from 320 nm to 800 nm wavelengths. The thickness of the films was measured by the optical interference method. The morphology and thickness of the TiO_2 thin films were characterized by using SEM *(Scanning Electron Microscopy-Electron Dispersive Analysis X-*

Ray) FEI Brand Inpect-S50 Type. The hydrophilic property of the films was evaluated by measuring the contact angle of distilled water droplet through the use of *software Image J.*

3. Results and discussions

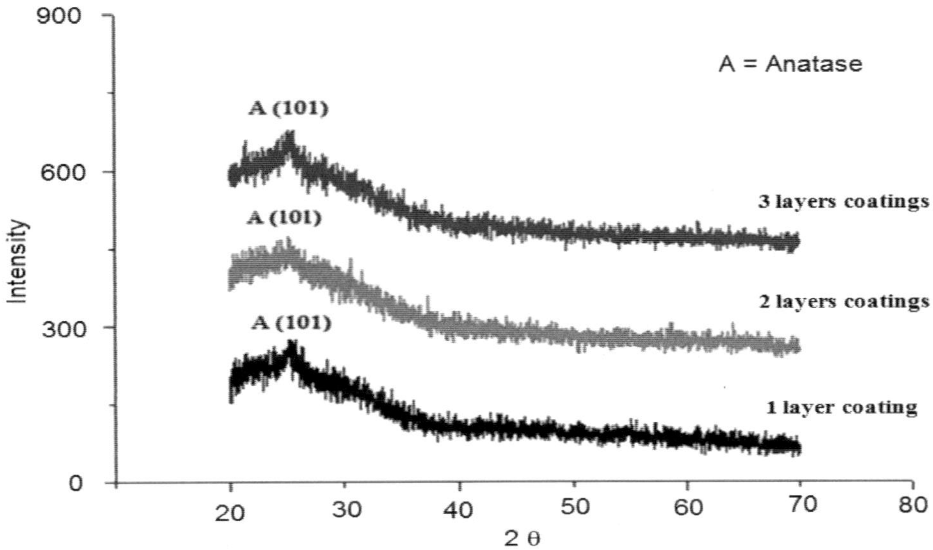

Figure 1. The diffraction pattern of TiO_2 thin film at different coating variations

Figure 1 shows the diffraction patterns of TiO_2 thin films with different coating variations. The peak was formed on TiO_2 films at scattering angle around 25degrees. This peak was related to TiO_2 anatase crystalline structure at crystal planes 101 as observed by Chu et al.(2016). Chu et al. stated that crystals grew with anatase structure at about 550°C[10]. The (101) plane peak then analyzed using Scherrer equation to calculate crystallized grain size as summarized in Table 1.

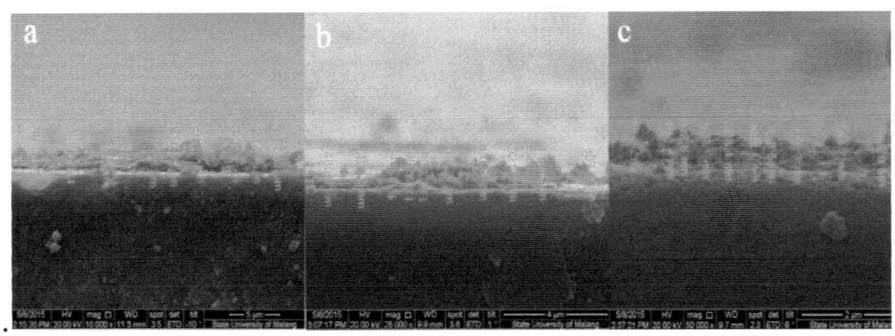

Figure 2. Morphology of (a) 1 coating, (b) 2 coatings and, (c) 3 coating of TiO_2 Thin Films

IOP Conf. Series: Journal of Physics: Conf. Series **853** (2017) 012035 doi:10.1088/1742-6596/853/1/012035

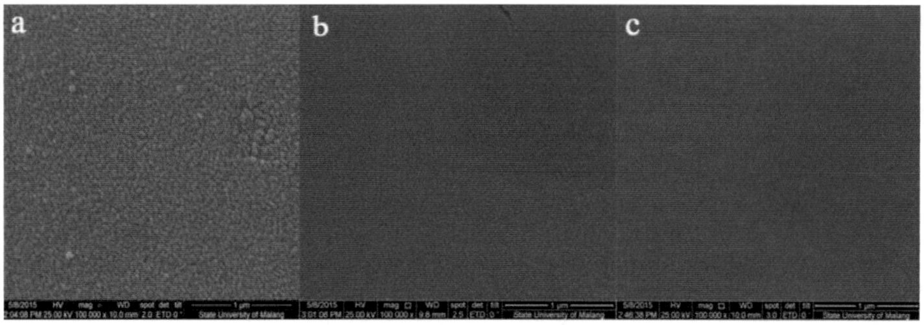

Figure 3. Morphology SEM Thickness (a) 1 coating (b) 2 coating and (c) 3 coating of TiO_2 Thin Films

The result of SEM can be seen in Figure 2. Based on the existing scale, the average thickness of one coating surface was 0.2 μm. There were white particles indicating that the impurities did exist during the coating process. The thickness average of the two coating surface was 0.3 μm,while the thickness average of the three coating surface was 0.5 μm

Table 1. The crystal grain sizes

Thickness TiO_2 thin films	1 coating	2 coating	3 coating
Crystal grain sizes	235 nm	179 nm	137 nm

Table 1 and Figure 3 show that the sample with a crystal grain size was small since it was basically one big crystal with an orientation at producing diffraction peaks approaching the vertical axis. Small-sized crystals that generated peak width as small crystals have a limited X-ray reflectance field. The diffraction peaks generated by constructive interference light was reflected by the crystal fields. The thicker the thin film, the smaller the grain size of crystals formed. The smaller the crystal grain size, the more the cavities formed, so that there would be more surface area to absorb free radicals.

Figure 4 shows the relationship between the $\alpha h\nu^n$ as a function of $h\nu$of TiO_2 thin film at different coating variations. The value of $\alpha h\nu^n$ obtained from absorbance measurement by using UV-Vis spectrophotometer in the wavelength ranged between 320 nm and 800 nm. The optical band gap calculation was started from the determination of the maximum and minimum transmittance, thick and thin film absorption coefficient and ended by applying the method of Tauc plot calculation. Transmittance measurements performed at the wavelength of 320 nm - 800 nm started with the ultraviolet light to visible light.

$$\alpha hv = b(hv - E_g)^n \tag{2}$$

Where α = absorption coefficient, hv = photon energy (eV), b = constants. Value n (n = 2) or (n = 1/2)

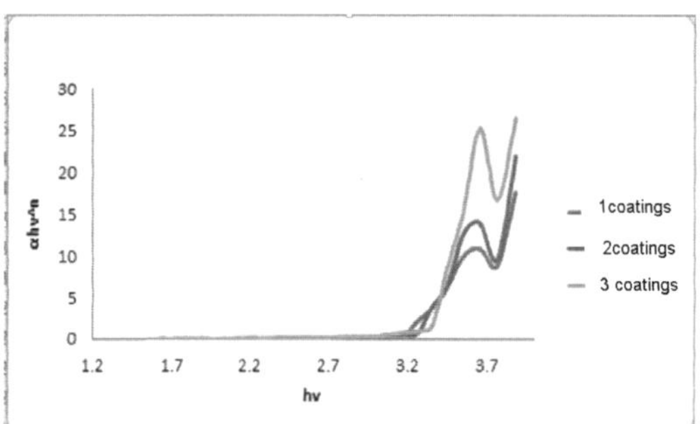

Figure 4. The $\alpha h\nu^n$ as function of $h\nu$at different coating variations

On this calculation, the largest indirect band gap is about 3.30 eV, 3.33 eV and 3.33 eV as shown in Figure 4.

IOP Conf. Series: Journal of Physics: Conf. Series **853** (2017) 012035 doi:10.1088/1742-6596/853/1/012035

Figure 5. Contact angle before sun-drenched of (a) 1 coating, (b) 2 coatings, (c) 3 coatings, and after sun-drenched (c') 1 coating, (b') 2 coatings, and (c') 3 coatings

The contact angle measurement was performed by using water droplets. The contact angle between the water and the material was analyzed by using *ImageJ* software. Small contact angle could be achieved when a stable form hydroxyl grouped on the surface. Hence, when the water appeared, it could be directly bound to these hydroxyl groups. The results obtained from this process are shown in Figure 5. The water contact angle and thin films surface between 20°-40° were still within the range of hydrophilic properties.

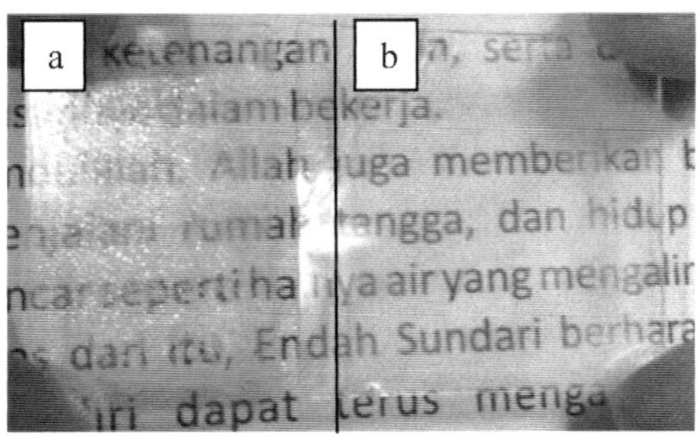

Figure 6. Test as *anti fogging* Thin Film TiO_2 at a) glass b) TiO_2 film/glass

Figure 6 is the result of anti-fogging test done by spraying water onto the surface of the glass substrate which has successfully grown TiO_2 thin film and the glass substrate which was not coated at all. Hydrophilic nature which was seen on the substrate could grow TiO_2 thin films, direct water spread to form a layer (not formed / droplet of water), whereas the uncoated glass piece was seen on granules / water droplet. Thus, it is evident that the glass thin film which grows TiO_2 has hydrophilic properties, so as to form a hydrophilic layer of water.

4. Conclusion

The TiO_2 thin films are successfully grown on glass substrate via sol-gel method. The crystal structure formed through this process is anatase on a 101 plane. Along with the addition of the thickness of a film, the absorption increases at the same time, and the band gap obtained is approximately 3.33 eV. Conforming to the big band gap, the thin film surface possess a hydrophilic nature so that it could be used as anti-fogging.

Acknowledgments
This research was financially supported by the Directorate General for Higher Education through Penelitian Unggulan Perguruan Tinggi 2017.

References

[1] Brauer D S, Gentleman E, Farrar D F, Stevens M M, Hill R G 2011 Benefits and drawbacks of zinc in glass ionomer bone cements *Biomed. Mater.* **6** 045007. doi:10.1088/1748-6041/6/4/045007

[2] Cedillo-González E I, Riccò R, Montorsi M, Montorsi M, Falcaro P, Siligardi C 2014 Self-cleaning glass prepared from a commercial TiO_2 nano-dispersion and its photocatalytic performance under common anthropogenic and atmospheric factors *Build. Environ.* **71** 7–14. doi:10.1016/j.buildenv.2013.09.007

[3] Sanderson K and Goer B 2016 Glass, Self-Cleaning *in: Ref. Module Mater. Sci. Mater. Eng., Elsevier* http://www.sciencedirect.com/science/article/pii/B978012803581801852X.

[4] Gherardi F, Colombo A, D'Arienzo M, Di Credico B, Goidanich S, Morazzoni F, Simonutti R, Toniolo L 2015 Efficient self-cleaning treatments for built heritage based on highly photo-active and well-dispersible TiO_2 nanocrystals *Microchem. J.* **126** 54–62. doi:10.1016/j.microc

[5] Wang H, Zhang R, Yuan Z, Shu X, Liu E, Han Z 2015 A Comparative Study of The Corrosion Performance of Titanium (Ti), Titanium Nitride (Tin), Titanium Dioxide (TiO_2) And Nitrogen-Doped Titanium Oxides (N– TiO_2), As Coatings For Biomedical Applications *Ceram. Int.* **41** 11844–11851. doi:10.1016/j.ceramint.

[6] Kazemi M, Mohammadizadeh M R 2011 Superhydrophilicity and Photocatalytic Activity Of Sol–Gel Deposited Nanosized Titania Thin Films *Thin Solid Films.* **519** 6432–6437. doi:10.1016/j.tsf.2011.04.216.

[7] Zahedi F, Behpour M, Ghoreishi S M, Khalilian H 2015 Photocatalytic Degradation Of Paraquat Herbicide In The Presence TiO_2 Nanostructure Thin Films Under Visible And Sun Light Irradiation Using Continuous Flow Photoreactor *Sol. Energy.* **120** 287–295. doi:10.1016/j.solener.2015.07.010.

[8] Kazemi M and Mohammadizadeh M R 2011 Superhydrophilicity and Photocatalytic Enhancement Of Titania Nano Thin Films *Appl. Surf. Sci.* **257** 3780–3785. doi:10.1016/j.apsusc.2010.11.146.

[9] Chu L, Qin Z, Zhang Q, Chen W, Yang J, Yang J, Li X 2016 Mesoporous Anatase TiO_2 Microspheres With Interconnected Nanoparticles Delivering Enhanced Dye-Loading And Charge Transport For Efficient Dye-Sensitized Solar Cells *Appl. Surf. Sci.* **360** Part B 634–640. doi:10.1016/j.apsusc.2015.11.034.

[10] Di Paola A, Cufalo G, Addamo M, Bellardita M, Campostrini R, Ischia M, Ceccato R, Palmisano L 2008 Photocatalytic Activity of Nanocrystalline Tio2 (Brookite, Rutile And Brookite-Based) Powders Prepared by Thermohydrolysis Of Ticl4 In Aqueous Chloride Solutions *Colloids Surf. Physicochem. Eng. Asp.* **317** 366–376. doi:10.1016/j.colsurfa.2007.11.005.

[11] Danish M, Ambreen S, Chauhan A, Pandey A 2015 Optimization and Comparative Evaluation of Optical and Photocatalytic Properties of TiO_2 Thin Films Prepared Via Sol–Gel Method, *Spec. Issue Nanomater. Photocatal.* **19** 557–562. doi:10.1016/j.jscs.2015.05.010.

[12] Photocatalytic Antibacterial Performance of Glass Fibers Thin Film Coated with N-Doped SnO 2/TiO_2, Hindawi Publ. Corp. vol.2014 (2014) 9 pages. doi:10.1155/2014/869706.

[13] Sobczyk-Guzenda A, Pietrzyk B, Szymanowski H, Gazicki-Lipman M, Jakubowski W 2013 Photocatalytic Activity of Thin Tio_2 Films Deposited Using Sol–Gel And Plasma Enhanced Chemical Vapor Deposition Methods *Ceram. Int.* **39** 2787–2794. doi:10.1016/j.ceramint.2012.09.046.

IOP Conf. Series: Journal of Physics: Conf. Series **853** (2017) 012036 doi:10.1088/1742-6596/853/1/012036

Hyaluronic Acid (HA)-Polyethylene glycol (PEG) as inject-able hydrogel for intervertebral disc degeneration patients therapy

Cityta Putri Kwarta[1], Prihartini Widiyanti[2*] and Siswanto[3]

[1] Biomedical Engineering Program, Faculty of Science and Technology, Universitas Airlangga, Indonesia
[2] Institute of Tropical Diseases, Universitas Airlangga, Indonesia
[3] Physics Department, Faculty of Science and Technology, Universitas Airlangga, Indonesia

*Corresponding email address: pwidiyanti@fst.unair.ac.id

Abstract. Chronic Low Back Pain (CLBP) is one health problem that is often encountered in a community. Inject-able hydrogels are the newest way to restore the disc thickness and hydration caused by disc degeneration by means of minimally invasive surgery. Thus, polymers can be combined to improve the characteristic properties of inject-able hydrogels, leading to use of Hyaluronic Acid (a natural polymer) and Polyethylene glycol (PEG) with Horse Radish Peroxide (HRP) cross linker enzymes. The swelling test results, which approaches were the ideal disc values, were sampled with variation of enzyme concentrations of 0.25 μmol/min/mL. The enzyme concentrations were 33.95%. The degradation test proved that the sample degradation increased along with the decrease of the HRP enzyme concentration. The results of the cytotoxicity assay with MTT assay method showed that all samples resulted in the 90% of living cells are not toxic. In vitro injection, models demonstrated that higher concentration of the enzymes was less state of gel which would rupture when released from the agarose gel. The functional group characterization shows the cross linking bonding in sample with enzyme adding. The conclusion of this study is PEG-HA-HRP enzyme are safe polymer composites which have a potential to be applied as an inject-able hydrogel for intervertebral disc degeneration.

1. Introduction

Chronic Low Back Pain (CLBP) is one of some health problems, which often encountered in a community. Almost everyone experiences low back pain regardless of age and gender differences. Approximately 60-80% of the world's population has ever experienced low back pain at least once during their lifetime (lifetime prevalence). Pain Study Group of PERDOSSI (Nerve Specialist Doctors Association of Indonesia) conducted a study in 14 teaching hospitals in May 2002. The results showed that the incidence of LBP included 18.37% of all pain cases that had ever been handled. The primary causes of low back pain vary, one of which is intervertebral disc degeneration [1].

There are different ways to cure lower back pain, such as conservative measures (bed rest, anti-pain medication, anaesthesia, medical rehabilitation) and surgical (operative) measures. Total disc replacement surgery can only be performed when conservative therapies fail to overcome the severe pain which resulting limited movements and pain attacks that would occur repeatedly to the patients.

Complications, caused by the operative measure, include allergies, infection, the increase of motoric deficit, and the failed back syndrome [2]. Inject-able hydrogel is the newest way to restore the disc thickness and hydration caused by disc degeneration by means of minimally invasive surgery [3]. Inject-able hydrogel is easily formed, fills the defect perfectly, and has good permeability [4].

Frith *et al* in 2013 used the enzyme horse radish peroxide (HRP) catalysed in polysaccharides and it was proven to increase the degradation rate linear with the increasing of the concentrations of the enzyme. Tyramine, as a conjugate polymer, was added in the reaction between HRP and polysaccharides, and the HRP catalysed phenol and aniline by tying H_2O_2 chain [5]. This study synthesized Hyaluronic Acid (HA) and Polyethylene Glycol (PEG) by using enzymatic cross linking method by adding enzymes to speed up the formation of the hydrogels and to increase the degradation rate of the hydrogels. The characterization used was the swelling test, degradation test, cytotoxicity assay, an in vitro study of injection models, and functional group test.

2. Materials and methods

2.1. Materials

The materials used were Hyaluronic Acid (HA) 215 KDa, Polyethylene Glycol (PEG) 10000 KDa, enzyme Horse Radish Peroxide (HRP), and tyramine hydrochloride, which derived from Sigma Aldrich. This research also used 3-4-hydroxyphenyl Propionic Acid (HPA) from Sigma Aldrich, Dimethyl Aminopropyl-ethyl carbodiimide hydrochloride (EDC), Sulpho-Hydroxysuccinimide (NHS), MES buffer, NaOH, distilled water, a solution of H_2O_2, Phosphate Buffered Saline (PBS), and agarose.

2.2. Variables

The control variable of this research is concentration of HA (15 gram) and PEG (16,5 gram). The released variable is concentration of HRP enzyme (0 ; 0,15 ; 0,25 ; 0,35 ; 0,45 μmol/min/mL). While the dependent variable is characterization of *inject-able hydrogel.*

2.3. Methods

The first solution which was prepared as a stock solution was Hyaluronic Acid (HA) and Tyramine Hydrochloride (Tyr) mixed with a 1:10 ratio in 0.1 MES of Buffer. After that, it was added with Aminopropyl Dimethyl-Ethyl Carbodiimide Hydrochloride (EDC) and Sulpho-Hydroxysuccinimide (NHS) with a composition of 1 M of NHS and 10 M of EDC. The HA-Tyr solution of 1 M of NaOH solution was added to increase the pH from 4.0 to 6.0 and reacted at room temperature with gentle agitation afterwards. Finally, a solution of HA-Tyr was freeze-dried.

The second solution which was prepared as a stock solution was 3-4-hydroxyphenyl Propionic Acid (HPA) dissolved in 0.1 M MES Buffer, added with 50 μM of NHS and 500 μM of EDC to be reacted. Furthermore, Polyethylene Glycol (PEG) was dissolved in 0.1 M of MES Buffer. After all of the solutions were prepared, HPA was reacted with PEG at room temperature with gentle agitation. In the last of it, the solution of PEG-HPA was freeze-dried.

The Freeze-drying result of HA-Tyr was reacted with PEG-HPA in a micro tube. The Horse Radish Peroxide enzyme (HRP) was added to a solution of Phosphate Buffered Saline (PBS) and the concentration ratio which was varied in accordance with the ratio of HA: PEG: HRP (gram: gram: μmol/min/mL), namely 15: 16.5: 0 ; 15: 16.5: 0.15; 15: 16.5: 0.25; 15: 16.5: 0.35; and 15: 16.5: 0.45. The solution was mixed with a vortex. Then, 2.5 mM of H_2O_2 was added for gelation, and the vortex was performed.

3. Results and discussion

On this research, there has been successfully synthesized hydrogel made from Hyaluronic Acid (HA) and Polyethylene Glycol (PEG) by the method of enzyme using enzymatic cross linking Horse Radish

Peroxide Type I. Samples carried out five variations of composition of enzymes(Figure1), (a) HA-PEG-0 HRP, (b) HA-PEG-0.15 HRP, (c) HA-PEG-0.25 HRP, (d) HA-PEG-0.35 HRP, and (e) HA-PEG-0.45 HRP. And thus the hydrogel is white.

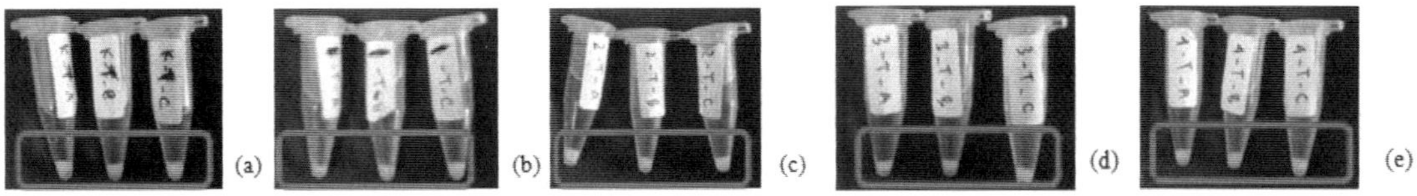

Figure 1. Hydrogel Research Results

3.1. *Swelling test*

In the swelling test, hydrogels were freeze-dried, and then the initial weight of all samples were weighed and soaked in a solution of Phosphate Buffer Saline (PBS) for 24 hours to determine the ability of the material to absorb the PBS solution. Inject-able hydrogels that were qualified to be applied to the intervertebral disc ranged from 21-27% to meet the requirements of occupying disc defect and that it would not spread to the avascular tissues around the disc nucleus pulposus (Firth *et al*, 2013). The result in Figure 2, obtained after weighing the samples, was that those three samples demonstrated decreasing swelling graphs when the concentration of HRP enzyme increased. The result that was nearly ideal was the concentration of the enzyme at 0.25 μmol/min/mL was 33.95%.

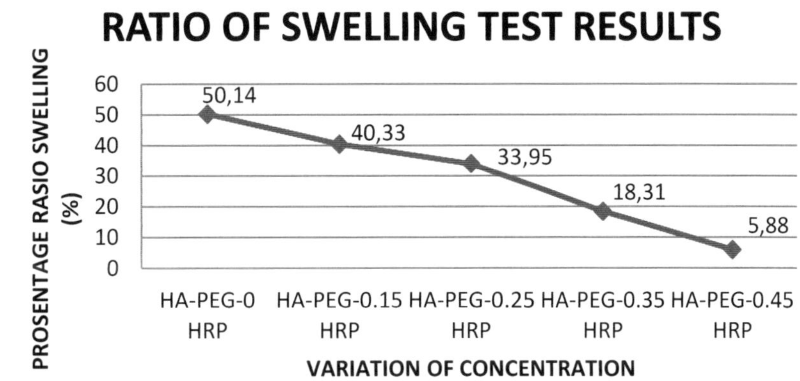

Figure 2. Line Graph of Swelling Test Results

3.2. *Degradation test*

Hydrogel in degradation testing PBS during the time period 1, 7 and 14 days in vitro simulation was done as for the observed degradation hydrogel in body fluids on the physiological PH with normal body temperature; 37◦C. Soaking solution of PBS is the simulation of bioactive material. Based on data analysis (Figure 3), the best results are in a composite sample of HA-PEG-0.45 HRP. The rate of degradation is decreased if there is addition of enzyme concentration cross linking bonds due to the presence of hydrogel. This sample has a steady degradation rate and total degradation during long 8.5 weeks (2.5 months). Long degradation of total on this sample varies with the time of maximum 3-month hydrogel according to Firth *et al* (2013) due to the difference in the use of the polymer on the research. In the previous research used, the polymer molecular weight was PEG 40000 KDa, where as on this research, used molecular weight was PEG 10000 KDa. This caused a difference in the strength of the bond that formed between polymers and enzymes. The larger molecular weights polymer, then ties the chain which is formed and will be getting longer, so the bond with the other ingredients will be more powerful and hard-degraded [6].

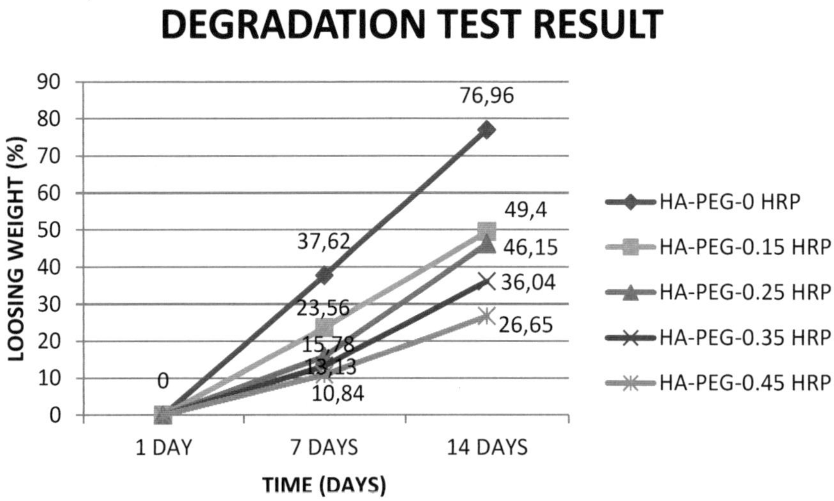

Figure 3. Line Graph of Degradation Test Results

3.3. Cytotoxicity assay

Cytotoxicity assay using [3-(4,5-Dimethylthiazol-2-Yl) -2,5-Diphenyltetrazolium Bromide] method (MTT Assay) with liver cells Hepatosite (Dhuha-7). Samples were incubated at Dhuha-7 cells in Eagle medium for 24 hours, and then, the absorbance wavelength used ELISA Reader.

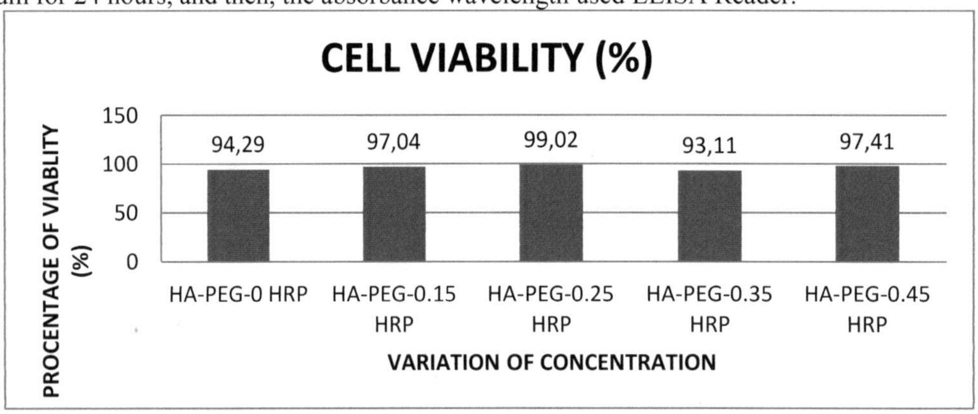

Figure 4. Bar Graph of Cytotoxicity Assay Results

From the results of cytotoxicity testing (Figure 4), they against the samples of all the variations, the percentage of live cells obtained above 90%, the highest living cell was found in HA-PEG-0.25 HRP which was 99.02% and the lowest living cell about 93.11% when concentration of HRP was 0.35µmol/min/mL. The results showed that all types of sample hydrogel are not toxic because it was far above the limit of the number of toxic substances, i.e. 50% [7]. In the sample, the covalent bond intermolecular with phenol polymer cluster which does not cause damage to living cells mitochondria so that resulting test results proving the old purple creature compounds are not toxic.

3.4. In vitro injection model

The testing of In Vitro Injection Model was qualitative to see the changes of hydrogel shapes before and after being put into agarose as an intervertebral disc simulation [8]. The result shows the following in Figure 5 which are hydrogel when injected into the agarose will (a) occupy the space of agarose or (b) a portion of the whole, while the form of hydrogel when released from the agarose hydrogel (c) intact or (d) rupture.

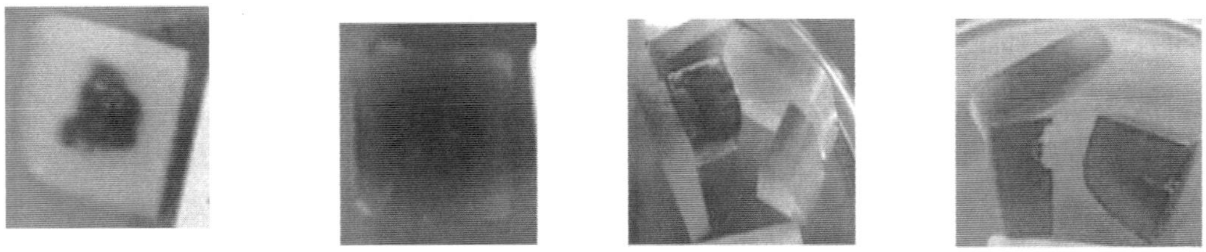

a b c d

Figure 5. In Vitro Injection Model Results

The higher the concentration of HRP enzyme, the better the hydrogels occupied rooms in agarose and the shape of the hydrogels would be whole when released from agarose (Table 1). This shows that there was an increase in the cross linking bonding by Horse Radish Peroxide enzyme towards Hyaluronic Acid and Polyethylene Glycol.

Table 1. In Vitro Injection Model Testing Results

Variation	Samples	Hydrogel			
		Occupy the Space of Agarose		Form Of Hydrogel When Released From The Agarose	
		Portion	Whole	Intact	*Rupture*
HA-PEG-0 HRP	K-M-1		√		√
	K-M-2		√		√
	K-M-3		√	√	
HA-PEG-0.15 HRP	1-M-1		√		√
	1-M-2		√		√
	1-M-3		√		√
HA-PEG-0.25 HRP	2-M-1	√			√
	2-M-2	√		√	
	2-M-3		√	√	
HA-PEG-0.35 HRP	3-M-1	√		√	
	3-M-2		√	√	
	3-M-3	√		√	
HA-PEG-0.45 HRP	4-M-1	√		√	
	4-M-2	√		√	
	4-M-3	√		√	

3.5. Functional groups testing

Confirmation tests on functional groups that were conducted using Fourier Transform Infra-Red (FTIR) in both samples (samples with a concentration of 0.45 μmol/min/ml could not be tested due to constrained availability of materials) demonstrated that an indication of cross linking was successful because a functional group was found on both samples, which was a functional group of enzyme Horse Radish Peroxide. The characterization of functional groups aim to find out the number and percentage of transmittance wave at each hydrogel. On the spectrum of Figure 6, it retrieved on absorption peak wave number 1348 cm-1. The absorption intensities of the functional groups is carboxyl, the result of a cross link between HA and PEG. Spectrum formed indicative of cross-linking of polymer bonds as covalent bonds and ties are reversible.

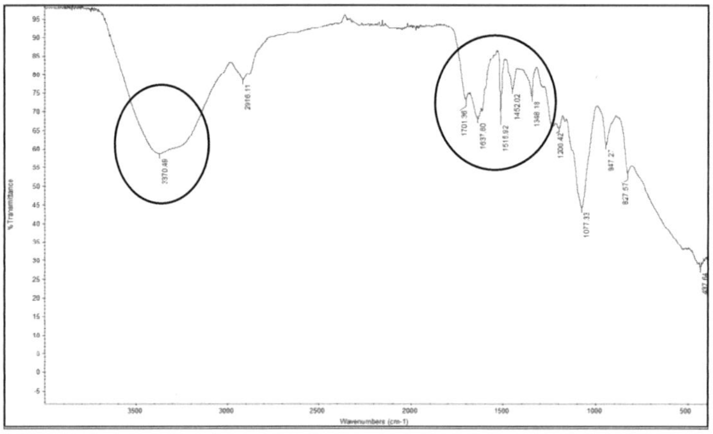

Figure 6. IR Spectrum of Hydrogel With Variations of HA-PEG-0 HRP

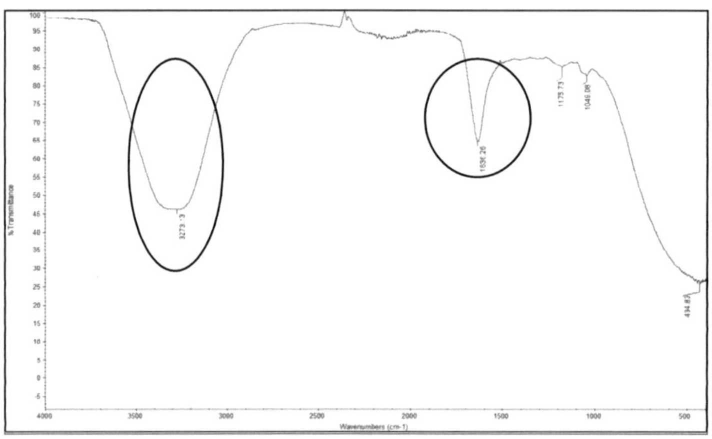

Figure 7. IR Spectrum of Hydrogel With Variations of HA-PEG-0.25 HRP

Based on the analysis of functional groups in the IR spectrum of the enzyme composition variations with hydrogel 0μmol/min/mL (Figure 6) and 0.25 μmol/min/mL (Figure 7), indicate that the existence of the new cluster experienced a chemical cross linking (Table 2), i.e., hydrocarbons, amine, amide II and carboxyl. HRP in the presence of a catalyst of hydrogen peroxide of hydrogen atom undergoes on a cluster of phenol hydroxyl resulting in oxygen atoms on the phenol hydroxyl HA cluster frees electrons and have the nature of reactive free radicals, evidenced by the widening of the area transmittance absorption on the 3300 cm^{-1}, resulting in a molecular results between cross linking HA with a PEG. The formation of free radicals results in polymer molecular distances are relatively short HRP and polymer, so that it can bind with the addition of the HRP enzyme, stronger cross linking. Cross linking which occurs on intermolecular protein crosslinking hydrogel is so functional group generated between before and after the addition of the enzyme will be different [8].

Table 2. Results of Functional Group Test

Samples	Wave Number (cm^{-1})						
	Hydroxyl	Hydrocarbon	Ether	Amide I	Amide II	Amino	Carboxyl
	O-H	C-H	C-O-C	C=O	C=O	NH2	COOH
HA-PEG-0 HRP	3370.49	2916.11	1200.42	1637.6	1452.02	1515.92	1348.18
HA-PEG-0.25 HRP	3273.13		1175.73	1636.26			

4. Conclusion

The combination of Hyaluronic Acid (HA) -Polyethylene Glycol (PEG) has a potential to be the candidate for inject-able hydrogels for intervertebral disc degeneration. This is proven through swelling test, degradation test, cytotoxicity assay, an in vitro study of injection models, and functional group test.

Acknowledgments

The author would like to thank the Ministry of Research and Technology of Indonesia on the funding of this research.

References

[1] Purwanto E T 2003 Hernia Nukleus Pulposus Lumbalis Dalam Nyeri Punggung Bawah *Kelompok Studi Nyeri Perhimpunan Dokter Spesialis Saraf Indonesia (PERDOSSI)* ;**133**- 48

[2] Ethical Digestion 2008 *Jurnal Kedokteran* No **52** Tahun VI Juni

[3] Chan, Samantha, Alexander Burki, Loein Benneker 2013 Papain-Induced In Vitro Disc Degeneration Model For The Study Of Inject-able Nucleus Pulposus Therapy *The Spine Journal* **13** (2013) 273–283

[4] Sivashanmugam, R Arunkumar, M Vishnu Priya, Shantikumar V Nair, R Jayakumar 2015 An Overview Of Inject-able Polymeric Hydrogels For Tissue Engineering *European Polymer Journal*.

[5] Frith, Jessica E, Andrew R Cameron, Donna J Menzies 2013 An Inject-able Hydrogel Incorporating Mesenchymal Precursor Cells *Biomaterials* **34** (2013) 9430e9440.

[6] Borzacchiello, A Gloria 2011 *Spinal Disc Implants Using Hydrogels* National Research Council, Italy Woodhead Publishing Limited

[7] Spielmann, H Hoffmann, S Botham 2007 *The ECVAM International Validation Study on In VitroTests for Acute Skin Irritation* Germany: ATLA.

[8] Heck, Tobias, Greta Faccio, Michael Richter, Linda Thöny-Meyer 2013 Enzyme-Catalyzed Protein Crosslinking *Appl Microbiol Biotechnol* (2013) **97**:461–475

IOP Conf. Series: Journal of Physics: Conf. Series **853** (2017) 012037 doi:10.1088/1742-6596/853/1/012037

Preparation of molecular sieve from natural pyrophyllite and characterization of its Al/Si ratio, crystal structure, and Porosity

Riris Idiawati[1], Abdulloh Fuad[1,2*], Nandang Mufti[1,2], Hartatiek[1], Syamsul Bahtiar[1], Subakti[1,2], and Ahmad Taufiq[1,2]

[1]Department of Physics, Faculty of Mathematics and Natural Sciences, Universitas Negeri Malang (State University of Malang), Jl. Semarang No. 5, Malang 65145, Indonesia
[2]Central Laboratory for Mineral and Advanced Materials, Faculty of Mathematics and Natural Sciences, Universitas Negeri Malang (State University of Malang), Jl. Semarang No. 5, Malang 65145, Indonesia

*Corresponding email address: *abdulloh.fuad.fmipa@um.ac.id*

Abstract. Pyrophyllite is one abundant mineral in Indonesia which has not been optimally processed. Hence, this study further processed natural pyrophyllite to be an advanced material usable for industrial sector as a molecular sieve (MS). Natural pyrophyllitewas chosen due to its ability to filter gas or liquid selectively. The MS made from natural pyrophyllite was prepared by using a simple method, in short time and with less cost via leaching process. NaOH was varied to 10 M (MS1), 15 M (MS2), and 20 M (MS3) of molarity. Such solution was subsequently dissolved in distilled water and followed by decantation and filtration processes to obtain the deposit. Eventually, the deposit was drained to form MS powders. The BET characterization showed that the respective surface areas of MS1, MS2, and MS3 are 0.350 m^2/g, 2.869 m^2/g, and 1.176 m^2/g with the pore sizes of 30Å, 542 Å, and 550 Å, respectively. The XRF characterization results showed that the Al/Si ratio of MS10, MS15, and MS20 are 2.4, 2.2, and 2.3, respectively. Meanwhile, the XRD investigation pointed out that the primary phase of MS10 and MS15 samples wassodalite with cubic crystal system, quartz with hexagonal crystal system, and diaspore with orthorhombic crystal system, while the MS20 phase was pure in the form of sodalite phase. Moreover, the results of FITR characterization showed that the synthesized MS has a functional group of Si-O-Si bending, Si-O-Al, Si-O, Si-O normal to the plane stretching, inner surface Al-OH deformation, Si-O-Si siloxine, H-O-H, -OH, C-H stretching, and H-O-H bending water.

1. Introduction

Indonesia is a mineral-rich country since mineral stones are spread in almost all areas across the country. However, unfortunately, such plentiful minerals has not been optimally benefited. Thus far, the minerals are not processed to be advanced materials yet. The advanced materials provide larger benefits, either for economic, scientific, or technological development. Therefore, a mineral processing to produce materials that are more advanced is highly necessary. One sample of richmineral available in Indonesia which has big potential to be processed as theadvanced material is pyrophyllite.

Chemically, pyrophyllite is aluminum silicate hydroxide synthesis with the formula of $Al_2Si_4O_{10}(OH)_2$ and has similar characteristics with zeolite [1] or kaolinite [2] which has relatively high silica and alumina contents [3]. In general, pyrophyllite in Indonesia has been quarried conventionally and sold as raw material without any further processes. The findings of recent research performed by Fuad et al. states that Indonesian pyrophyllite contains relatively high silica and alumina contents of approximately 78% and 15.4% [4].

Technically, pyrophylliteis formed in advanced argillic alteration zone, such as kaolinite. Nonetheless, it can be formed at high temperature and acidic pH. Molecular sieve synthesis requires raw materials with high silica and alumina contents. Ugal, J.r. successfully synthesized type 4A molecular sieve ($Na_{12}(AlO_2)_{12}(SiO_2)12.27H_2O$) from kaolinite by refluxing at the temperature of 90 °C and with 1:5 comparison between kaolinite and sodium hydroxide [5]. A similar study was also conducted by Ivan Petrov who fabricated type A MS from zeolite via hydrothermal method at the temperature of 100 °C. Zeolite and kaolinite have comparable Al/Si content [6]. Unfortunately, however, there is not a report of molecular sieve synthesis from pyrophyllite. Thus, we proposed pyrophyllite utilization to form molecular sieve.

Molecular sieve synthesis has been executed in several methods, such as reflux [7] and hydrothermal [8] methods. Even though reflux method offers an advantage as being able of extracting rough-textured samples, this method has a weak point since it needs a great volume of solvent and some manipulations from the operator. In line with the hydrothermal method, although it requires a simple preparation, it requires highly costly equipment, and it is hard to control its stoichiometry [9, 10]. Thus, the development of molecular sieve method becomes essential to overcome the deficiencies of the two approaches. In this research, we propose pyrophyllite-based molecular sieve synthesis advancement through the means of simple leaching method and NaOH. Furthermore, the crystal structure, porosity, particle size and chemical bond of the synthesized molecular sieve would be observed to identify the success of the advancement.

2. Method

The sample of molecular sieve in this research was prepared by using NaOH leaching method. The first stage was to prepare the raw material by turning the pyrophyllite taken from Malang, Indonesia into powder. The powdered pyrophyllite was then refined by using filter paper with 200 mesh size. Then, the researchers performed a reaction obetween40-gramspyrophyllite powder and 10M NaOH through a stirring process with a stirrer accompanied by heating at the temperature of 120 °C for 24 hours. After that, the material was dissolved in distilled water and decanted for 24 hours followed by the filtration process to obtain the deposit. The deposit was dried at the temperature of 250 °C for 21 hours. The same procedure was done to the other variations of NaOH molarity; i.e.15 M and 20 M. The powdered sample obtained was then smoothened for the characterizations of X-Ray Fluorescence (XRF), X-Ray Diffractometry (XRD), Brunauer Emmett Teller (BET), and Fourier Transform Infra-Red (FTIR) Spectroscopy.

3. Results and discussion

Figure 1 presents the results of XRD characterization on the pyrophyllite. According to the findings of qualitative data analysis, pyrophyllite has corresponding peaks with the pyrophyllite phase, while the results of quantitative data analysis point out that pyrophyllite contains three elements, namely the pyrophyllite itself of as much as 48%, diaspore element of 12.9%, and 39.1% SiO_2-Quartz element. XRF characterization was conducted subsequently to ascertain the pyrophyllite contents. The results of XRF characterization of pyrophyllite and the synthesized MS sample are subsequently presented in Table 1 and Table 2.

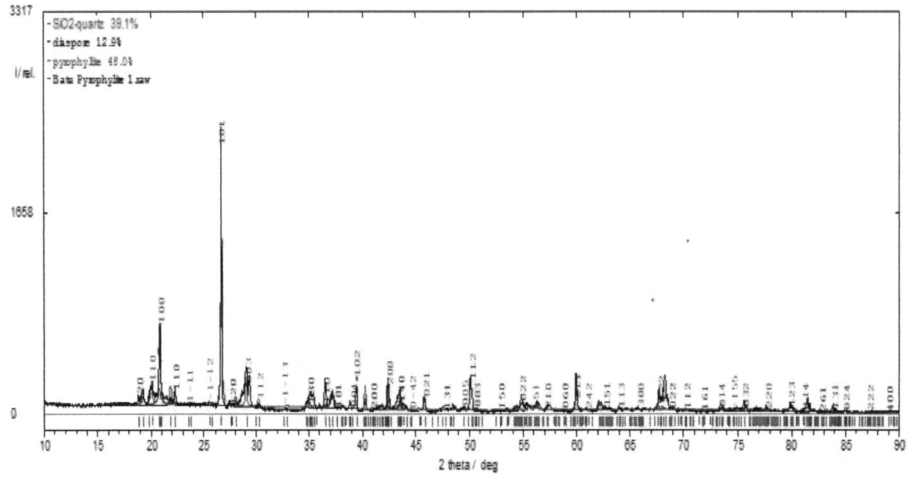

Figure 1.X-ray Diffraction Pattern of Natural Pyrophyllite

Table 1. Results of XRF Characterization of Natural Pyrophyllite

Elemental content	Al	Si	Ca	Ti	Other elements
Percentage (%)	17.6	77.6	0.96	2.08	1.7

From the XRF characterization results, it is identified that the biggest pyrophyllite constituent is Si with the percentage of 77% and A1 with the rate of 17.6%. The two elements are the main components with the highest concentration, while there are other minor elements, namely Ti, Ca, V, Cr, Fe, Ni, and Cu. Thus, pyrophyllite is a kind of mineral which contains high concentration of silica or quartz.

Table 2. Comparison of Pyrophyllite Concentration in MS10, MS15, and MS20

Sample	Elemental content (%)				
	Al	Si	Ca	Ti	Other elements
Pyrophyllite	17.6	77.6	0.96	2.08	1.71
MS10	23.9	56.5	1.8	12.9	4.95
MS15	23.0	51.2	2.2	17.0	6.42
MS20	21.0	48.0	2.8	18.4	9.78

Based on the results of XRF characterization presented in Table 2, it can be described that there was a declining concentration of silica which is represented by relatively significant Si element so that there is an increase of alumina (A1) level. The levels are suitable since it produced an almost similar ratio comparison between silica and alumina as shown in Table 3.

Table 3. Comparisons of Si/Al Ratio in MS10, MS15, and MS20

Sample	Al	Si	Al : Si
MS10	23.9	56.5	1 : 2.4
MS15	23.0	51.2	1 : 2.2
MS20	21.0	48.0	1 : 2.3

According to Table 3, it can be concluded that Al/Si has an almost similar ratio comparisons between 1:2.2 and 1:2.4. In other words, the addition of greater NaOH concentration did not significantly influence the A1/Si ratio in the sample. The highest ratio was found in MS1, i.e. 1:2.4, and the lowest ratiois found in MS2, namely 1: 2.2. Therefore, the A1/Si in this study is included in the S materials (Sodalite/SOD), namely the minerals contained in the zeolite structural classification since it is categorized as aluminosilicate mineral [11]. These research findings are also supported by a theory proposed by A. Derkowski which states that Si/A1 ratio of sodalite is 1: 1.21 [12].

Figure 2 reflects the results of X-ray diffraction patterns of the MS samples via NaOH leaching method with the variations of NaOH morality of 10 M, 15 M, and 20 M.

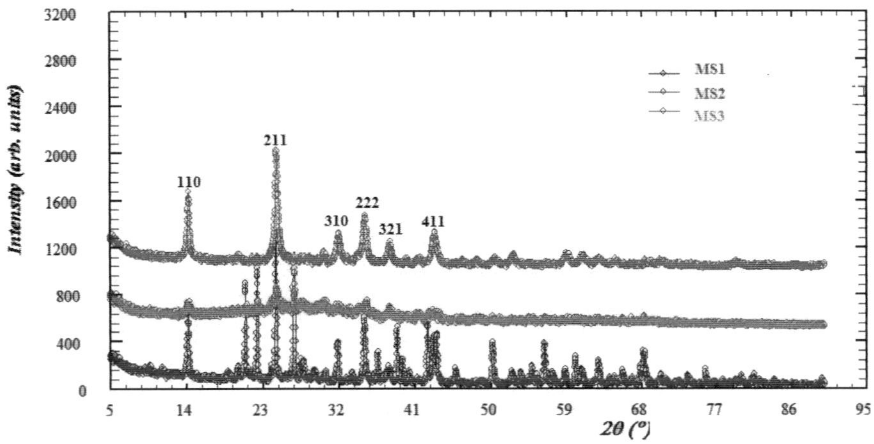

Figure 2.X-Ray Diffraction Patterns of MS10, MS15, and MS20

From the results of refinement analysis, the MS10 and MS15 samples have three phases, namely quartz phase with hexagonal crystal system, diaspora phase with orthorhombic crystal system, and sodalite phase with cubic crystal system, while there was one pure phase of sodalite found in MS20 sample. Detailed results of X-ray diffraction data analysis for all samples are demonstrated in Table 3.

Table 4. Results of X-Ray Diffraction Data Analysis on MS10, MS15, and MS20

Sample	Compositions of Phase	Parameter							
		FoM			Lattice Parameter			%wt	V (Å^3)
		Rp (%)	Rwp (%)	Rexp (%)	a(Å)	b(Å)	c(Å)		
MS10	Sodalite	18.96	27.21	1.75	8.897	8.897	8.897	72.4	704.1 85
	Quartz	18.96	27.21	1.75	4.919	4.919	5.405	26.8	130.749
	Diaspore	18.96	27.21	1.75	4.402	9.432	2.847	0.80	118.195
MS15	Sodalite	10.11	13.14	0.28	8.870	8.870	8.870	94.7	697.770
	Quartz	10.11	13.14	0.28	4.906	4.906	5.423	5.30	130.536
	Diaspore	10.11	13.14	0.28	4.301	9.525	2.890	0.00	118.402
MS20	Sodalite	11.6	15.21	2.64	8.865	8.865	8.865	100	696.755

The data analysis findings of crystal structure in Table 3 display the formation of three phases which are quite similar with the cubic crystal model of sodalite phase from ICSD with the database code of 072059, p-43n space group number 218, quartz phase with hexagonal crystal model from ICSD with the database code of 027832, P3221 space group number 154, and diaspore phase with orthorhombic crystal model from ICSD with the database code of 066800, Pbnm space group number 62. Furthermore, the results of data analysis also demonstrate that the NaOH composition in the related parameter of crystal lattice in each sample has the tendency to be stable. Such stability has a consequence of having stable crystal volume. Hence, the variations of NaOH molarity do not affect the crystal volume in each sample.

All of the samples were characterized by using the BET technique to determine the physical properties of their pore size and surface area. Technically, the surface area was obtained through an N_2 adsorption test. This method utilized Quadrasorb SI analyzer complemented with QuadraWin software. The test results in the form of average pore size and surface area of MS10, MS15, and MS20 samples are presented in Table 5.

Table 5. Pore Diameter and Surface Area of the Samples

Sample	Pore Diameter (Å)	Surface Area (m^2/g)
MS10	30	0.350
MS15	542	2.869
MS20	550	1.176

As reported in Table 5, MS10 sample has 30 Å pore diameter. Thus it is classified as mesoporous material. Additionally, the respective pore diameters of MS15 and MS20 samples are 542 Å and 550 Å, so that they are categorized as macroporous materials. This study shows that the higher the molarity concentration of NaOH, the larger or the more porous the pore diameter. Based on a theory proposed by Rosemane Szostak, BET method is used not only to identify the pore diameter but also to determine the surface are of the samples [13]. The surface area of MS10 was identified to be 0.350 m^2/g, while MS15 has an area of2.869 m^2/g, and MS20 has a surface area of1.176 m^2/g. The surface area of an adsorbent determines its adsorption ability. A large surface area leads to a high performance of adsorption. Thus, the variations of NaOH compositions have are highly potential to turn the samples into a molecular sieve.

The functional group formed in all samples with variations of NaOH molarity is characterized through the means of FTIR. This characterization identified a graphic of correlation between the transmittance and wave number as can be seen in Figure 3. The black curve shows the characterization result of MS10, the red curve displays the MS15 characterization result and the blue curve presents MS20 characterization result.

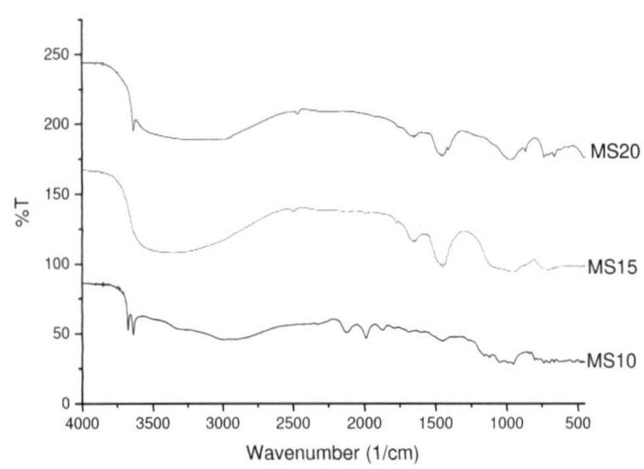

Figure 3.FTIR Spectra of MS10, MS15, and MS20

The IR Spectrum of MS10 on the wave number of 482.2 cm^{-1} shows the formation of Si-O-Si bond [14]. The Si-O-Al stretching bond [14] was formed at the wave number ranging from 540.07-665.44 cm^{-1}. Si-O quartz bond [14] was found at the wave number ranging from 696.3-798.53 cm^{-1}. There was inner surface A1-OH deformation bond [15] found at the wave number of 950.91 cm^{-1}. The wave number of 985.62 cm^{-1} showed Si-O planar stretching bond [15]. There was Si-O-Si siloxane bond [16] at the wave number ranging from 1047.35-1120.64 cm^{-1}. There was Si-O normal to the plane stretching bond [15] in the wave number of 1159.22 cm^{-1}. H-O-H bending bond can be foundat the wave number of 1454.33 cm^{-1}. Aromatic nitrate bond [14] was foundat the wave number of 1598.99 cm^{-1}. The wave number of 1687.71 cm^{-1} showed H-O-H bending water bond [14]. C-H stretching bond[14] could be found at the wave number ranging from 2846.93-2945.3 cm^{-1} and–OH silanol bond [16] also was found at the wave number ranging from 3527.8-3672.47 cm^{-1}.

On MS15, the wave number ranging from 489.92-665.44 cm^{-1} exhibited the existence of Si-O-Si bond [14]. Si-O quartz bond [14]was identified at the wave number ranging from 702.09-713.66 cm^{-1}. Si-OH silanol bond [16] was determined at the wave number ranging from 869.9-881.47 cm^{-1}. Inner surface Al-OH deformation bond [15] was found at the wave number of 948.98 cm^{-1}. Al-O as Si cage (TO4) bond[14]was found at the wave number of 1396.46 cm^{-1}. The wave number ranging from 1415.75-1435.04 cm^{-1} showed H-O-H bending bond[15]. The wave number of 1633.71 cm^{-1} showed O-H (water molecule) bond [16]. –OH silanol bond [16]was found at the wave number ranging from 3269.93-3379.29 cm^{-1}. Meanwhile, on the MS20, the wave number ranging from 430.13-665.44 cm^{-1} marked the formation of Si-O-Si bending bond [14]. The wave number of 709.8 cm^{-1} showed Si-O quartz bond [14]. Si-OH silanol bond [16]was identified at the wave number of 866,04 cm^{-1}. The wave number ranging from 958.62-979.84 cm^{-1} showed the inner surface Al-OH deformation bond [15]. The wave number ranging from 1413.82-1454.33 cm^{-1} showed H-O-H bending bond[15]. The wave number of 2978.09 cm^{-1}identified C-H stretching bond[14]. There was OH silanol bond [16] at the wave number ranging from 3010.88-3635.82 cm^{-1}.

The results of the FITR characterization show that the MS10 produced a greater number of bonds than MS15 and MS20. However, as a consequence, there can be a higher level of impurity contained in the samples. Such results are in accord with the XRD data analysis results which state that MS10 has three phase contents. Hence, it is highly possible that there was a functional group of the impurity which was accidentally adsorbed by the ultraviolet ray during the FITR analysis.

4. Conclusion

The molecular sieve has been prepared from natural pyrophyllite via leaching route. The samples in the forms of MS1, MS2, and MS3 have respective Si/Al ratio of 2.4, 2.2, and 2.3. MS1 and MS2 were formed in sodalite phase with the cubic crystal structure, quartz phase with the hexagonal crystal structure, and diaspore phase with the orthorhombic crystal structure, while the MS3 was formed in one pure sodalite phase with cubic crystal phase. Moreover, the respective pore sizes of MS1, MS2. And MS3 are 30 Å, 542 Å, and 550 Å. Meanwhile, the results of FITR characterization showed that the MS10 had more bonds compared with MS15 and MS20 due to the existence of impurity.

References

[1] Hu Y, Wei J, Liang Y, Zhang H, Zhang X, Shen W, Wang H 2016 Zeolitic Imidazolate Framework/Graphene Oxide Hybrid Nanosheets as Seeds for the Growth of Ultrathin Molecular Sieving Membranes *Angew. Chem. Int. Ed.***55** pp2048–2052

[2] Gougazeh M, and Buhl J 2014 Synthesis and Characterization of Zeolite A By Hydrothermal Transformation of Natural Jordanian Kaolin *J. Assoc. Arab Univ. Basic Appl. Sci.***15**pp35–42

[3] Bhavan I 2015 Indian mineral yearbook 2013.

[4] Fuad A, Mufti N, Diantoro M, Subakti, Rupiasih N N, Suharta W G, Suyanto H 2016 Synthesis and Characterization of Highly Purified Nanosilica From Pyrophyllite Ores *AIP Conf. Proc., AIP Publishing* p 030020.

[5] Ugal J R, Hassan K H, Ali I H 2010 Preparation of Type 4A Zeolite From Iraqi Kaolin: Characterization and Properties Measurements *J. Assoc. Arab Univ. Basic Appl. Sci.***9** pp2–5

[6] Petrov I andMichalev T 2012 Synthesis of Zeolite *A: A Review.* 51

[7] Liu X andWang R 2017 Effective Removal of Hydrogen Sulfide Using 4A Molecular Sieve Zeolite Synthesized From Attapulgite *J. Hazard. Mater* **326** pp157–164

[8] Kazemimoghadam M 2016 Comparison of Kaolin and Chemical Source for Preparation of Nano Pore Naa Zeolite Membranes *J. Water Environ. Nanotechnol* **1** pp45–53

[9] Gupta S M andTripathi M 2012 A review on the Synthesis Of Tio_2 Nanoparticles by Solution *RouteCent. Eur. J. Chem.* **10** pp279–294.

[10] Feng L, Liu A, Ma Y, Liu M, Man B 2010 Fabrication, Structural Characterization and Optical

IOP Conf. Series: Journal of Physics: Conf. Series **853** (2017) 012037 doi:10.1088/1742-6596/853/1/012037

Properties of The Flower-Like Zno Nanowires *Acta Physiol Pol.* 117
[11] Smith J V 1963 Structural Classification of Zeolites*Miner. Soc Am Spec Pap.* 1 281–290.
[12] Derkowski A, Franus W, Beran E, Czímerová A 2006 Properties and Potential Applications of Zeolitic Materials Produced From Fly Ash Using Simple Method of Synthesis *Powder Technol.* **166** 47–54
[13] Szostak R 1989 Molecular sieves principles of synthesis and identification *Springer*
[14] Saikia B J and Parthasarathy G 2010 Fourier Transform Infrared Spectroscopic Characterization of Kaolinite from Assam and Meghalaya, Northeastern India *J. Mod. Phys.* **1** 206.
[15] Aroke U and El-Nafaty U 2014 XRF, XRD and FTIR properties and characterization of HDTMA-Br surface modified organokaolinite clay, Int. J. Emerg. Technol. Adv. Eng. **4** 817–825.
[16] Munasir, Triwikantoro, Zainuri M, Darminto 2015 Synthesis Of Sio2 Nano Powders Containing Quartz And Cristobalite Phases From Silica Sands Mater *Sci.-Pol.* **33** 1 pp47–55

An in vivo photodynamic therapy with diode laser to cell activation of kidney dysfunction

Suryani Dyah Astuti[1,2,*], Brahma Indra Prasaja[1] and Tri Anggono Prijo[1]

[1]Department of Physics, Faculty of Science and Technology, Airlangga University
[2]Department of Biomedical Engineering Post Graduate School, Airlangga University

*Corresponding email address: suryanidyah@gmail.com

Abstract. This study aims to analyze the effect of photodynamic therapy (PDT) low level laser therapy (LLLT) 650 nm in the experimental animals mice (*Musmuculus*) suffering from kidney organ damage in mice (*Musmuculus*) in vivo. Exposure laser acupuncture was performed on the kidney BL-23. The conditioning of kidney damage in mice used carbofuraan 35 at a dose of 0.041697 mg/mice. LLLT 650 nm exposure was done on a wide variety of energy (0.5; 1.0; 1.5; 2.0; 4.0; 5.0; 6.0; 7.0) J. The histopathological kidney cells in mice renal impairment showed that exposure to 650 nm laser energy 1 Joule resulted in the reduction of damaged cells (necrosis) and normal cells were increased with the improvement of renal tubular cells (64.14 ± 8:02)%. Therefore, exposure to 650 nm LLLT on acupuncture points Shenshu (BL-23) has the ability to proliferation of renal tubular cells of mice.

1. Introduction

Since 2008, biostimulation laser, which is now referred to as Low Level Laser Therapy (LLLT), has been applied to medical therapy [1]. He-Ne laser is the first laser used for wound healing. Biostimulation laser is a photobiology phenomenon that involves chemical reactions and enzymes [2]. Photodynamic activation for biostimulation cells using LLLT is laser stimulation of the skin with low intensity and a non-thermal laser irradiation. Deep transmission of laser energy, wavelength, and laser energy are the important factors on therapy. In additional, scattering and absorption by various tissues of different skins need to be considered. The skin is a network of multi-layer in a homogeneous, anisotropic structure; so that the optical properties are very complex [3]. The intensity of light through the skin decreases exponentially due to the scattering and the absorption. Ultraviolet and visible spectrum (~400-600) nm will be absorbed by hemoglobin and melanin, so, a little amount of the light is transmitted back. For blue light, the irradiance depth would be decreased 99% to ~700 μm. Wavelengths above 1400 nm is the area of water absorption that results on less transmission. Red and infrared spectrum are the most widely used for photodynamic biomodulation because the absorption is very small and the transmission is large [4]. However, scattering by collagen is still considered because it could reduce transmission.

Kidney Disease Outcomes Quality Initiative of The National Kidney Foundation defines that chronic renal failure is a state of the occurrence of kidney damage or glomerular filtration rate (GFR) <60 mL/min/1.73 m^2 within 3 months or more [5]. The disorder is caused by dysfunction of the glomerular filter or blood and protein filter in the primary urine. Generally, dialysis or hemodialysis could be upgraded to kidney transplant in order to treat chronic kidney failure. Alternative therapy to restore organ function is photodynamic activation through photobiomodulation of kidney cells with a laser exposure. This laser exposure can be performed on acupuncture points, which also called Laserpunktur therapy [6]. Laser therapy acupuncture has many advantages. It is non-invasive, free cause of infection, and environment-friendly. It does not cause pain because of its needle pricks, reduces the risk of disease transferred through a needle, and radiates warm feelings during the therapy. Furthermore, it also causes high immune system response, high growth factor, and is safe for all ages [7]. Based on research, specific acupuncture points used for kidney treatment is shensu (BL-23) [8]. This paper discusses the influence of laser phototherapy at 650 nm in testing animals, which are mice (*Musmuculus*) suffered from kidney disorders, on the kidney acupuncture points BL-23.

2. Materials and methods

Animals used in this study were mice (*Musmusculus*), male, 3-4 week old, and weighing 25-35 grams. *Musmuculus* were conditioned with damaged kidney by giving carbofuran 3% on the dosage of 0.5 mg/kg weight *Musmuculus* [9]. LD_{50} fraction did not cause the death of the tested animals, yet caused kidney damage is 1/24 LD_{50} (0.0208 mg / kg), 1/12 LD_{50} (0.0417 mg/kg) and 1/6 LD_{50} (0.0833 mg/kg). Therefore, kidney of mice (*Musmusculus*) were set to damaged condition by using Carbofuran 3% with a dose of 0.041697 mg/mice.

2.1. Light sources

Laser irradiation was carried out using diode lasers with output wavelengths of (650.00±0.01) nm. The power outputs were (16.53 ± 0,06) mW. Those data were the result of characterization using Jasco monochromator CT-10 and OPM (Optical Power Meter) Thorlabs PM100D which were capable of detecting power between 100 pW – 200 W and a wavelength of 185 nm – 25 μm. The laser energy was calculated from the output power multiplied by the length of time the laser exposure [11].

2.2. Laser treatment

To determine the laser effect on *Musmuculus*, the samples were distributed into 3 groups, as follows: (1) Groups A was treated using carbofuran and laser 405 nm (C+L+), (2) Groups B was treated only with carbofuran (C+L-), and (3) Group C had no treatment of everything (C-L-). For each group the experiment was repeated at least 3 times. Group A was treated with various energy irradiating of diode laser (1.0; 1.5; 2.0; 4.0; 5.0; 6.0; 7.0) J to acupuncture point kidney (BL-23). Figure 1 shows the acupunctured point BL-23.

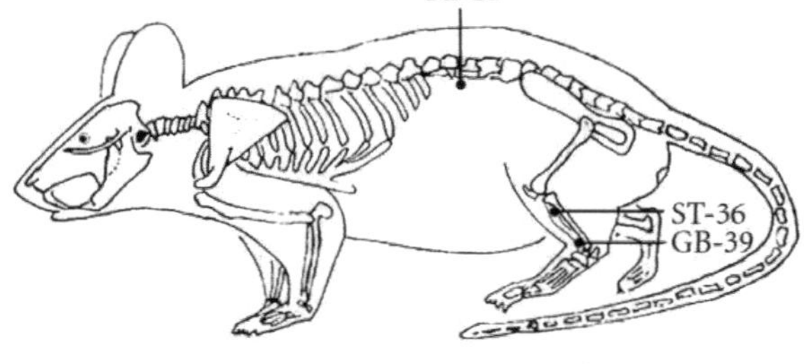

Figure 1. Accupuncture point of kidney BL-23

2.3. Histopathological anatomy test

Histopathology test was used to determine the structure of the cells of the renal tubules of mice that had been given laser irradiation treatments as well as healthy and pain control. Analytical calculation of renal tubular cell was conducted using a visual field [10]. The formula to calculate the percentage of regenerated normal cell is by using equation:

$$\%\text{ of normal tubular cells} = \frac{\bar{x}\text{ control diseased cells} - \bar{x}\text{ number of normal cells}}{\bar{x}\text{ control diseased cells}} \times 100\%$$

3. Results and discussion

Laser characterization results in this study are showed in Figure 2 and 3.

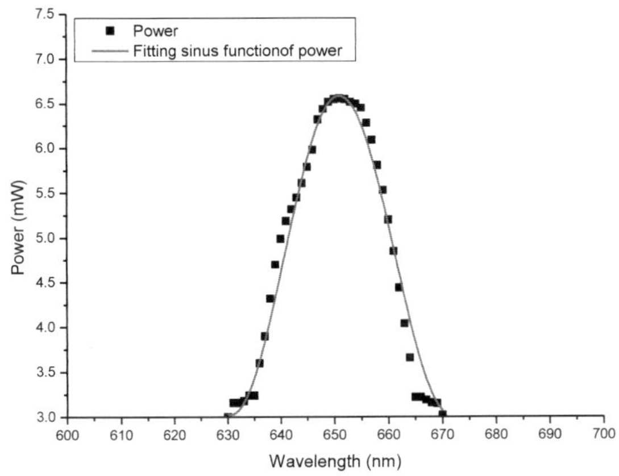

Figure 2. Spectrum of 650 nm diode laser

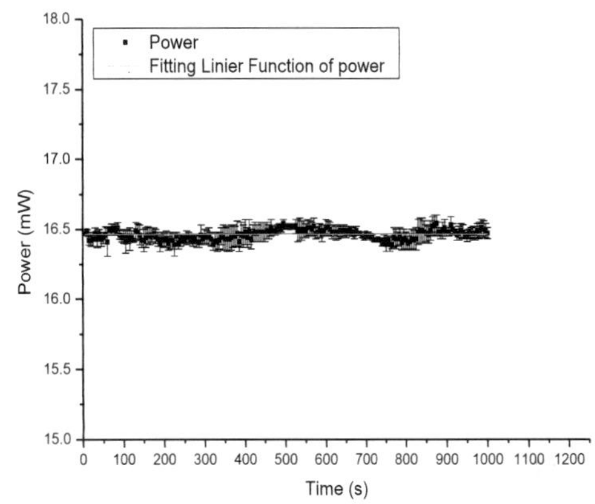

Figure 3. Graph of 650 nm laser power stability versus time

The characterization results in Figure 2 show that the diode laser light source has a spectrum (650.00 ± 0.01) in nm. Figure 3 shows the data output on power stability (16.53 ± 0.06) mW

3.1 Histopathological analysis results

Giving carbofuran orally in mice resulted in infiltration of inflammatory cells and necrosis in renal cells which are indicated by damage within kidney cells [10]. Necrosis is the death of cells in living tissue due to the toxins which enter the body. Inflammation is a complex biological response of vascular tissues to harmful stimuli, such as pathogenic agents, cell death/damaged, or irritation. Inflammation is the body armor attempting to deal with adverse stimuli and to initiate the healing process [12].

In this study, to stimulate the healing process of renal cell, the therapy applied 650 nm laser diode. This therapy was examined through seven variation of energy dose with considerably similar laser beam output because the laser mounted on Shensu acupuncture point and BL-23 had the same cross-sectional area. Figure 4a to 4f illustrate histopathological picture of *Musmusculus*' kidney glomerulus for normal cells, diseased cells, and cells that were exposed to a laser with a magnification of 400X.

Exp: A=glomerular, B=tubulus, C=normal cell, D=Necrosis
Figure 4a. Histophatology normal kidney cells

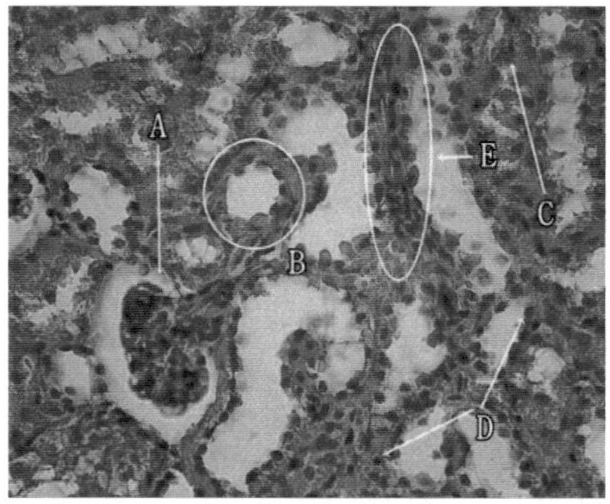

Exp: A=glomerular, B=tubulus,C=normal cell, D=Necrosis, E=Inflamed cell
Figure 4b. Histophatology disease kidney cells

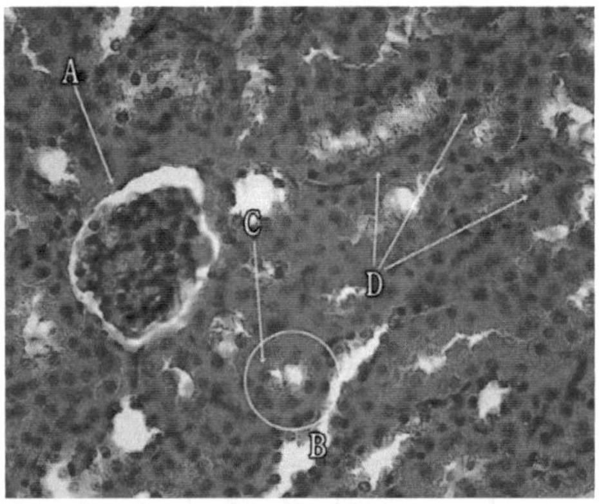

Exp: A=glomerular, B=tubulus, C=normal cell, D=Necrosis
Figure 4c.Histophatology kidney cells with treatment laser 650 nm 0.5 J

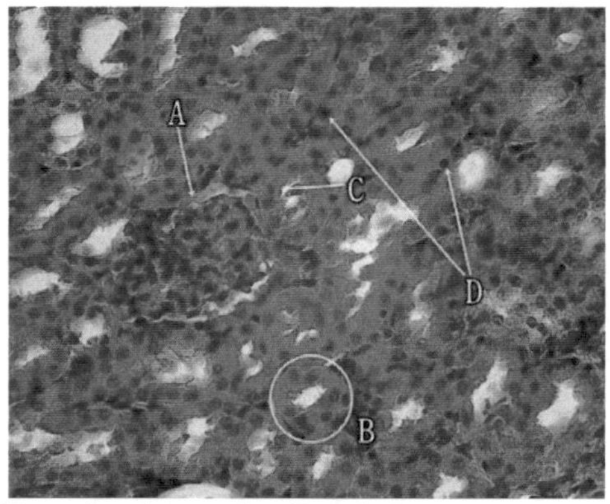

Exp: A=glomerular, B=tubulus, C=normal cell, D=Necrosis
Figure 4d. Histophatology kidney cells with treatment laser 650 nm 1.0 J

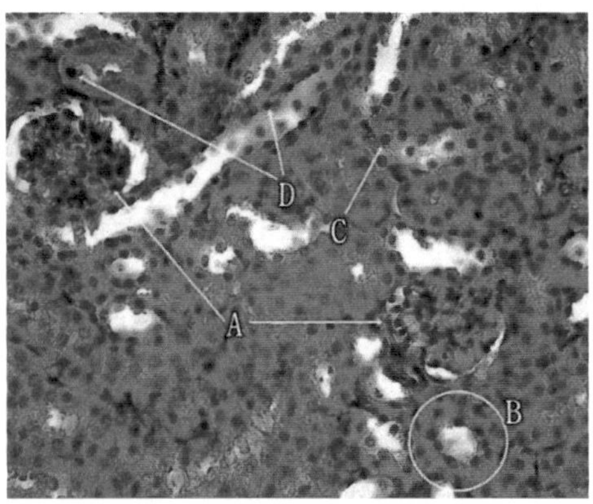

Exp: A= glomerulus, B = Tubulus, C = normal cell, D = Necrosis
Figure 4e. Histophatology kidney cells with treatment laser 650 nm 1.5 J

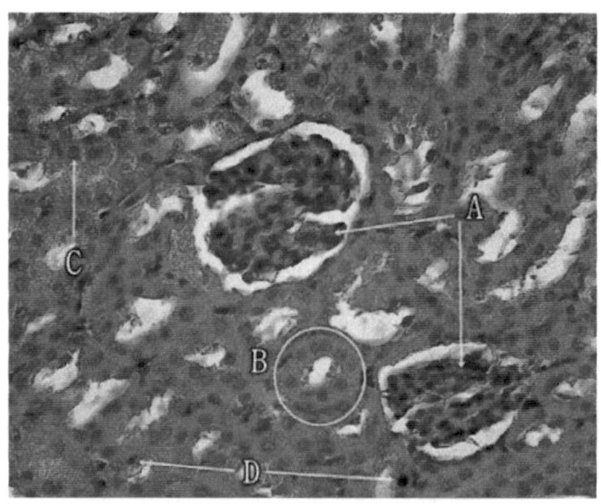

Exp: A= glomerulus, B = Tubulus, C = Normal cell, D = Necrosis
Figure 4f. Histophatology kidney cells with treatment laser 650 nm 2.0 J

Figures 5a and 5b demonstrate the graph of the number of normal cells and damaged at various laser exposure.

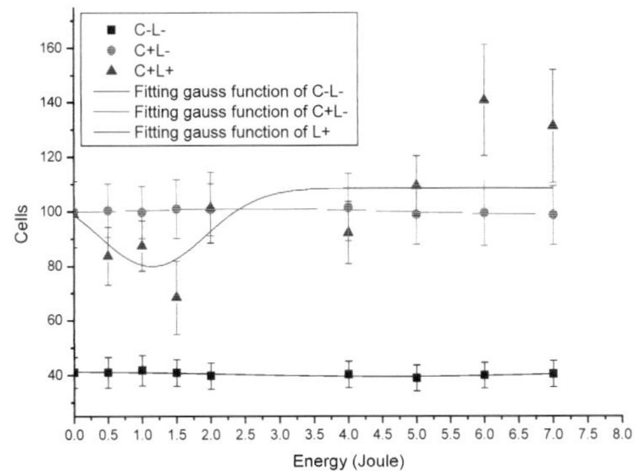

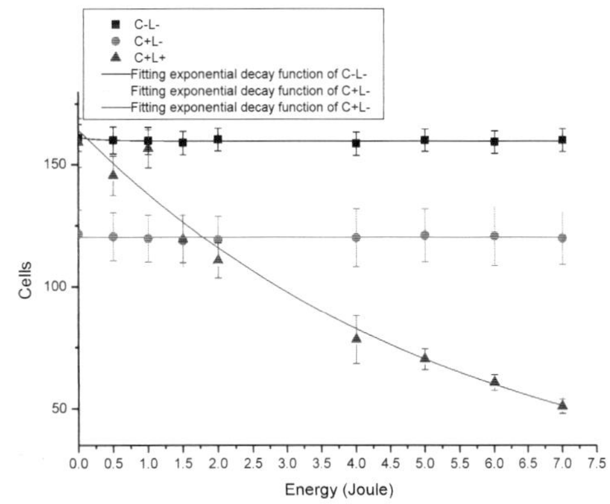

Figure 5a. The effect of various energy laser exposure on the number of damaged renal cells in mice

Figure 5b. The effect of various energy laser exposure on the number of normal cells in mice

Note:
C+L+ treated with carbofuran and laser 650 nm
C+L- treated with carbofuran and no laser treatment
C-L- no treatment laser and carbofuran

Figure 5a shows the increase in the number of normal renal cells of mice as a Gaussian function caused by exposure to 650 nm laser for various energy. The laser exposure of 650 nm wavelength and 1.5 J energy dose at acupuncture point BL-23 determines a decrease in the damaged cells the most. As indicated by the Gaussian function, $y = y_0 + \frac{A}{w\sqrt{\pi/2}} e^{-2\frac{(x-x_c)^2}{w^2}}$.

Meanwhile, the number of normal cells in the kidneys of diabetic mice as a result of exposure to 650 nm laser decreases exponentially compared to the number of normal cells if exposed toward the laser with precisely1.0 J energy. The exponential function is $y = y_0 + A_1 e^{-x/t_1} + A_2 e^{-x/t_2}$ with R^2 value of 0.96.

Figure 6 describes the percentage of renal cell improvement based on histopathological picture.

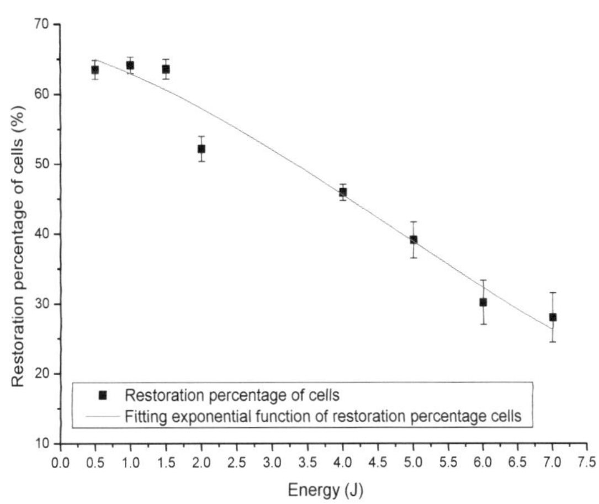

Figure 6. Growth percentage of renal tubular cells

Figure 6 figures out that the use of a 650 nm laser with 1 Joule energy has improved its percentage (64.14 ± 8.02)%, which is the most optimum result in the improvement of renal tubular cells. The exponential function is $y = y_0 + A_1 e^{-x/t_1} + A_2 e^{-x/t_2}$with R^2 value of 0.93.

The results of this study showed that a low power laser (LLLT) at a wavelength of 650 nm with 1 J energy at acupuncture points shenshu (BL-23) had the ability to improve renal tubular cells. This is consistent with Peplow's research that shows that LLLT is able to modulate cell proliferation [13, 14]. The results also showed biomodulation effects of LLLT in myofibroblast T and B cells during wound healing. LLLT facilitates differentiation myofibroblastic during the early stages of the cicatricialrepair process. Additionally, LLLT also appears to modulate the inflammatory response by down regulating lymphocyte proliferation during the wound healing process [15].

Wavelength laser parameters also affect the ability of cell proliferation. Laser wavelength associated with the depth of penetration into skin. Some research show that wavelength affects on the biological response of HeLa cells, which are irradiated with light of 580-860 nm [16]. Results of other studies also demonstrate the use of laser 780 nm is not as effective as 660 nm laser, yet it has a positive effect on the early stages of the emergence and development of inflammation. At the end of the experimental period, major effect is seen on the amount and quality of granulation tissue. If laser 660 nm is applied daily, it will be more effective than the 780 nm laser to improve third-degree burnt healing in diabetic rats started in the early stages of post-combustion [17-18].

Senshu acupuncture points (BL-23) is a bladder meridian acupuncture points related to the kidney [19]. The bladder is associated with the internal and external kidneys. Stimulation of the points BL-23 provides strengthening effect on the kidneys and kidney bioenergy balance, which in turn will restore kidney function.

4. Conclusion

Based on the results of histopathological kidney cells in mice with renal impairment, it can be concluded that exposure to 650 nm laser with 1 J energy results in a reduction of damaged cells (necrosis) and increase of normal cells with the improvement of renal tubular cells (64.14 ± 8.02)%. Therefore, the exposure to 650 nm LLLT on acupuncture points Shenshu (BL-23) has the ability to proliferation of renal tubular cells of mice.

References

[1] Karu T I, 1999 Primary and secondary mechanisms of action of visible-to-near IR radiation on cells J. Photochem. Photobiol B Biol. **49**:1–17.

[2] Anderson R R 1993 Optics of the skin. In: Lim HW, Soter NA (eds) Clinical photomedicine Marcel Dekker New York pp 19–35

[3] Kola ́ rova ́ H, Ditrichova ́ D and Wagner J 1999 Penetration of laser light into the skin in vitro *Lasers Surg Med* 24:231–235 DOI 10.1002/(SICI)1096-9101

[4] Kidney Disease Outcomes Quality Iniatiative 2002 Clinical Practice Guidelines For Chronic Kidney Disease: Evaluation, Classification and Stratification New York National Kidney Foundation Inc

[5] Gerhard L and Gerhard O 2012 Technical Parameters for Laser Acupuncture to Elicit Peripheral and Central Effects: State of the Art and Short Guidelines Based on Result from the Medical University of Graz*, the German Academy of Acupuncture, and the Scientific Literature, Research Article* **vol. 2012** Hindawi Publishing Corporation Germany.

[6] Baxter, G David, C Bleakley and S McDonough 2008*Clinical Effectiveness of Laser* Acupuncture A Systematic Review *Journal of Acupuncture and Meridian Studies* **1**(2): 65-82

[7] He, Tian-Feng, Wen-Jia Yang, Shu-Hui Zhang, Chun-Yan Zhang, Lian-Bo Li and Yun-Fei Chen 2011 Electroacupuncture Inhibits Inflammation Reaction by Upregulating Vasoactive Intestinal Peptide in Rats with Adjuvant-Induced Arthritis Shanghai: Shanghai University of

Traditional Chinese Medicine.

[8] Kamboj S S, Kumar, A Kamboj and R Sandhir 2008 Mitochondrial Oxidative Stressand Dysfunction in Rat Brain Induced by Carbofuran Exposure Cell Mol Neurobiol **28**: 961-969

[9] Baker J H, Lindsey JK, Steven HW 1979 *The Laboratory Rat: Biology and Disease.* **1.** Unisted States of America: Academic Press.

[10] Neimz, Markolf H 2007 Laser-Tissue Interactions Fundametals and Applications Third, Enlarged Edition. Germany: Springer

[11] Kaur, Bhupindervis, Alka Khera and Rajat Sandhir 2012 *Attenuation of Cellular Antioxidant Defense Mechanisms in Kidney of Rats Intoxicated with Carbofuran* India Journal of Biochemical and Molecular Toxicoogy. **26**

[12] Philip V Peplow, Tzu-Yun Chung, and G David Baxter *Laser Photobiomodulation of Proliferation of Cells in Culture: A Review of Human and Animal Studies* Photomedicine and Laser Surgery August 2010, **28**(S1): S-3-S-40. doi:10.1089/pho.2010.2771

[13] Philip V. Peplow, Tzu-Yun Chung, and G David Baxter, *Laser Photobiomodulation of Wound Healing: A Review of Experimental Studies in Mouse and Rat Animal Models* Photomedicine and Laser Surgery. June 2010, **28**(3): 291-325. doi:10.1089/pho.2008.2446.

[14] Maria Amalia Gonzaga Ribeiro, Ricardo Luiz Cavalcanti Albuquerque, Jr Luciana Maria Pedreira Ramalho, Antonio Luiz Barbosa Pinheiro, Leonardo RigoldiBonjardim, and Samantha Seara Da Cunha, *Immunohistochemical Assessment of Myofibroblasts and Lymphoid Cells During Wound Healing in Rats Subjected to Laser Photobiomodulation at 660 nm, Photomedicine and Laser Surgery* March 2009, **27**(1): 49-55. doi:10.1089/pho.2007.2215.

[15] T I Karu, S F Kolyakov, *Exact Action Spectra for Cellular Responses Relevant to Phototherapy*, Photomedicine and Laser Surgery 2005, **23**(4): 355-361

[16] G C S Meireles, J N Santos, P O Chagas, A P Moura, A L B Pinheiro, *Effectiveness of Laser Photobiomodulation at 660 or 780 Nanometers on the Repair of Third-Degree Burns in Diabetic Rats* Photomedicine and Laser Surgery 2008 **26**(1):47–54 DOI: 10.1089/pho.2007.2051

[17] Brodie, L M 2003 *Welding of Skin using Nd:YAG Laser with Bipolar Contact Applicators. Queensland* tesis of University of Southern Queensland Faculty of Engineering and Surveying

[18] Z Yanfu, Chinese Accupuncture and Moxibustion, Shanghai University of Traditional Chinese Medicine2002

IOP Conf. Series: Journal of Physics: Conf. Series **853** (2017) 012039 doi:10.1088/1742-6596/853/1/012039

Antimicrobial effect of *pleomeleangustifolia* pheophytin A activation with diode laser to *streptococcus mutans*

Sinari Alfat Sunarko[1], Wiwied Ekasari[2], and Suryani Dyah Astuti[1,3,*]

[1]Department of Biomedical Engineering Post Graduate School, Airlangga University
[2]Faculty of Pharmacy, Airlangga University
[3]Department of Physics Faculty of Science and Technology, Airlangga University

*Corresponding email address: suryanidyah@gmail.com

Abstract. The main purpose of this research is to identify potential of Pheophytin A. as photosensitizer a agent to inactivate *Streptococcus muttans* using laser diode of 405nm. Pheophytina is known as chlorophyll derivate that losses magnesium ion at the center of porphyrin ring structure. In this research, phrophytin was extracted from Suji leaf (Pleomeleangustifolia). To determine the antimicrobial effect of treatments on *S.mutans*, samples were divided into three groups as follows: (1) Groups A(treated with Pheophytin A. and laser 405 nm at varying energy density of 2.5; 5, 7.5; 10.0; 12.5; 15.0; 17.5 and 20.0 J/cm^2), (2) Group C^-(negative control, no treated), (3) Group C^+ (treated only with pheophytin). The experiments were repeated at least three times for each group. The results were analyzed using analysis of variance and the Tukey test. A P value $\leq$0.05 was considered to indicate a statistically significant difference. The decrement of percentage of number of bacterial colonyes growth was defined as: $|$ (Σ sample colony - Σ control colony)/ Σ control colony $|$ x 100%. The result showed that the incubation of Pheophytin A. using irradiation from laser diode of 405nm have a significant effect towards the decrement in bacterial growth. The most decreased percentage colony of *S. mutans* occurred on the incubation of pheophytin a treatment and laser irradiation 405nm with density 20 J/cm^2 is 61.9%. This showed that pheophytin a functions as a photosesitizer activator to inactivate *S. mutans* bacteria.

1. Introduction

Dental caries is a progressive pathological process of teeth's destruction, which is caused by combinations of diet, host, micro flora and exposure time [1]. Micro flora such as *Streptococcus mutans* is an acidogenic bacteria that colonized at teeth surface. Exposure time is the duration of teeth exposition towards acid produced by bacterial which causes teeth plaque. The main bacterial causing dental caries is a group of streptococci mainly *S. mutans*. *S. mutans,* which is a normal flora in oral. However, *S. mutans* might increase significantly at favorable environment and change to be pathologic [2].

aPDT is a non-antibiotic approach that was developed to inactivate microorganism, and it is a potential alternative compared conventional antibiotic [3]. aPDT combines a non-toxic photosensitizer and visible light resulting singlet oxygen and free radical that caused microbial cell destruction [4]. The main target of aPDT is an external microbial structure. Suitable adhesivity at bacterial structure causes destruction activated by light. Photosensitizer doesn't need to get into inside of the microorganism, therefore no resistance of microorganism [5].

Photosensitizer used in aPDT must have photophysical, photochemical, and photobiology characteristics to work optimally [6]. The basic structure of ensitizer consists of tricyclic dying with different mesoatom (acridine, proflavine, riboflavin, metilen blue, flurescein and eritrosin), tetrapyrolle (porphyrin and its derivate, chlorophyll, phylloerythrin and phthalocyanines and Furocoumarins (Psoralen and its methoxy derivate, xanthotoxin and bergaptene) [7].

Porphyrin is a class of aromatic heterocyclic that can be found easily in nature. Porphyrin has important biochemical process that functions including transportation (haem) and photosynthesis (chlorophyll). Because of its physical and chemical properties, porphyrin is used in various application such as artificial photosynthesis, oxidation catalysis, sensor, non-linear optic, microorganism photo in activation, and nanomaterial for cancer PDT [8]. Each of porphyrin molecule has the ability to absorb light at the specific wavelength. Light exposure at proper dose with wavelength spectrum corresponds to photosensitizer absorption spectrum of porphyrin, resulting photo inactivation of bacterial cell [9]. Photosensitization depends on the type and quantity of porphyrin and suitability between light spectrum and photosensitizer absorption spectrum [10].

Chlorophyll molecule consists of a porphyrin (head structure) having polar properties and a fitol in the tail. Porphyrin is formed from tetrapyrolle ring with a magnesium ion at the centre of tetrapyrolle ring. Phorphyrin ring in chlorophyll rules as light absorber (photosensitizer), therefore chlorophyll and its derivates are usefull as photosensitizer agents [11]. Pheophytina is a derivate of chlorophyll that lost magnesium ion at the centre of porphyrin ring structure. Pheophytin A. have two maximum wavelength absorptions in soret band area (408,9 nm) and Q band area (665,4 nm) [12].

According to previously research, diode laser is used as light source in photodynamic therapy because of the small beam, focused beam and high coherence. Finally, the purpose of this research is to explore the potential Pheophytin A. of Suji leaves (Pleomele angustifolia) as derivative chlorophyll and agent of photosensitizer to the process of inactivation of *S. Mutans* with laser diodes 405 nm.

2. Methods

2.1. Bacterial strain and culture conditions

The sample strains used on this research was pure culture bacteria from *Streptococcus mutans*. The bacteria were collected from tooth of patients diagnosed with dental caries in Dental Hospital Airlangga University Surabaya Indonesia.

2.2. Materials

Pheophytin A. was extracted from Suji leaves (*Pleomeleangustifolia*). Figure.1 showed the character of Pheophytin A.. A standard solution 1mg/ml was made by using Dimethylsulfoxide (DMSO). Pheophytin A. solution with concentration 20% was made inside sodium hydrochloride (NaCl).

2.3. Light source

Laser irradiations were carried out using diode lasers with wavelength output of 405.52±0,23nm. The power outputs were 49,50±0.20mW with wide size: 0.39±0.01 cm^2.

2.4. Antimicrobial effect of treatments against a. actinomycetemcomitans with total plate count (TPC)

To determine the antimicrobial effect of treatments on *A. actinomycetemcomitans*, samples were distributed to 3 groups as follow: (1) Groups A(treated with Pheophytin A. and laser 405 nm at varying energy density of 2.5; 5, 7.5; 10.0; 12.5; 15.0; 17.5 and 20.0 J/cm^2), (2) Group C^-(negative control, not treated), (3) Group C^+ (treated with pheophytin only). The experiments were repeated at least three times for each group. The suspension was planted on Tryptocase Soy Agar (TSA) (Merck, Darmstadt, Germany) sterile media and incubated at a temperature of 37°C for 24 hours. After incubation, the number of colony-forming units per milliliter (CFU/ml) was determined. The results were analyzed by analysis of variance and the Tukey test. A P value ≤0.05 was considered to indicate a statistically significant difference. The percentage decrease in the number of bacterial colonies growth defined as: | (Σ sample colony - Σ control colony)/ Σ control colony | x 100%.

3. Results and discussion

Pheophytin A. that used in this research was extracted from Suji leafs (*Pleomeleangustifolia*). The result of characterization Pheophytin A. using UV-Visible Spectroscopy is presented on figure.1.

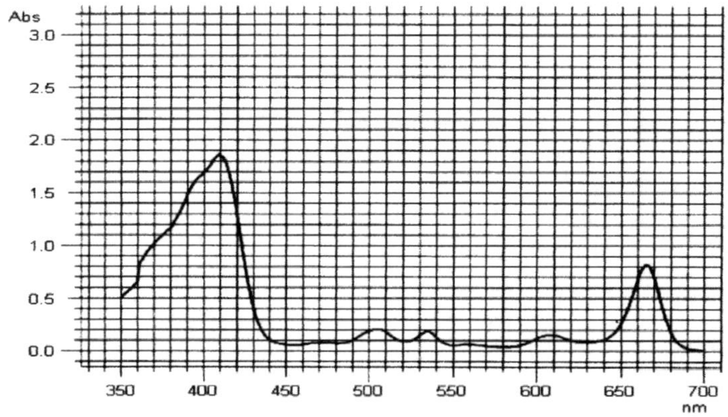

Figure 1. Spectrum Profile of Pheophytin A. from *Pleomeleangustifolia* Leaves

This research used a Pheophytin A. as a photosensitizer to inactivate *S. mutans* bacteri using laser diode 405nm. Pheophytin A. is a derivative from chlorophyll that has an ion magnesium loss in the center of porphyrin. Chlorophyll is a green pigment that can be found in every greenery that has a role in photosynthesis. *Pleomeleangustifolia* leaf, which is known as Suji leaf or *Pandan Betawi*, contains much chlorophylls that is commonly used in Indonesian society as a natural food coloring. The photosensitizer was Pheophytin A. solution with concentration 20% in NaCl 0.9% from standard pheophytin solution 1mg/ml DMSO. Before the treatment, absorption test of Pheophytin A. was done by using UV-Vis spectrophotometer to obtain the percentage of absorption on Pheophytin A. 20% solution. Figure.1 shows that maximum absorption of soret band and Q band located at 405 nm and 665 nm corresponds to literature from Masami kobayaki *et.al.*, 2013 located at 408.9 nm and 665.4 nm. It showed that profile spectra Pheophytin A. of *Pleomeleangustifolia* leaf has a similar characteristic of maximum absorption of soret band and Q band similarly to the previous research.

The absorbance of Pheophytin A.. at laser diode 405nm is equal to log (1/transmittance). Percentage of Pheophytin A. absorption is (1-transmittance) x 100%. Result measurement of absorbance of Pheophytin A. used UV-VIS Spectroscopy at spectrum 405nm is 2.3082. Percentage of Pheophytin A. absorption is 99.5%. The dose that was capture by Pheophytin A. laser diode can be calculated by Dose = (output power/wide beam) x irradiation time x percentage of Pheophytin A. absorption at 405nm. The dose that were (2.5; 5, 7.5; 10.0; 12.5; 15.0; 17.5 and 20.0) J/cm^2.

3.1. Photoinactivation study of s. Mutans incubated Pheophytin A. using 405 nm diode laser

Instruments diode laser light source was used in this study. Performance spectrum laser test showed peak spectrum at 405.52±0,23nm. The output power at 49,50±0.20 mW with broad beam 0.39±0.01 cm^2. The measurement of output power at 2 cm distance showed that the stability power up to measurement of 1000 seconds. The measurement of temperature during exposure to the sample showed the elevated temperature within the range of bacteria *S.mutans* growth. Thus, the death of bacteria were actually caused by laser irradiation treatment not the influence of the environment.

Result of the effect of laser irradiation of various energy laser 405nm with photosensitizer Pheophytin A. was examined by using ANOVA Factorial Test. Table 1 shows summary of the result of ANOVA Factorial Test.

Table 1. Analysis result of treatment of laser diose 405nm towards *S. mutans* bacteria that were incubated with Pheophytin A.

Group	Dose (J/cm^2)	N	% *S. mutans* colonyes reduction		Factorial anova
			Average	SD	significantly
Laser	2.5[a]	4	0.3	0.0	P=0
	5.0[b]	4	4.1	0.4	
	7.5[bc]	4	5.7	0.6	
	10.0[d]	4	9.1	0.6	
	12.5[de]	4	10.9	1.2	
	15.0[ef]	4	12.3	1.2	
	17.5[fg]	4	13.8	1.1	
	20.0[fg]	4	14.0	0.6	
Laser with Pheophytin A.	2.5[c]	4	6.5	0.5	P=0
	5.0[g]	4	15.3	1.1	
	7.5[g]	4	15.2	0.3	
	10.0[h]	4	25.6	0.4	
	12.5[i]	4	45.0	0.9	
	15.0[j]	4	42.9	0.6	
	17.5[k]	4	57.9	0.7	
	20.0[l]	4	61.9	0.6	
	Total	64	21.3	19.1	

Description: N = sample size. SD = standard deviation. The same superscript showed no significant difference from the results of the Tukey test.

Result of this test showed that laser irradiation, with incubation of Phoephytin A. or not, effected the reduction of amount of colonies of bacteria *S.mutans*. Multi comparison test used was *Post Hoc Multiple Comparison Tukey* to show and analyse the difference group of treatments.

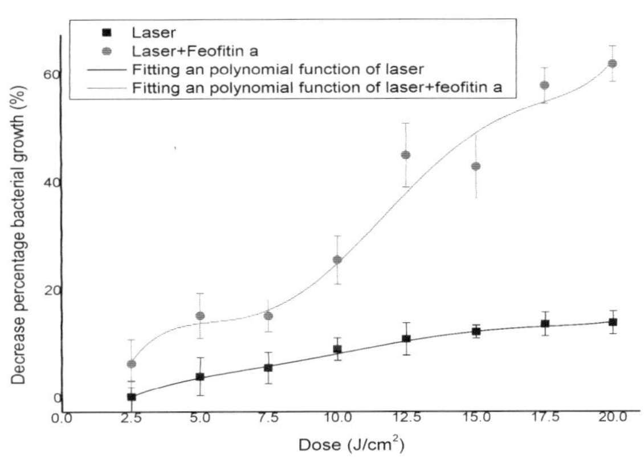

Figure 2. The result of laser diode 405 nm treatment of *S. mutans* with and without Pheophytin A.

Figure 2 shows polynomial graph of the result of bacteria *S.mutans* treatment of laser diode 405nm with Pheophytin A. (R^2=0.92) and without Pheophytin A. (R^2=0.99). Percentage of the decrement of *S.*

mutans colonies that were incubated with Pheophytin A. with laser irradiation 405nm and energy 5.0 and 7.5 J/cm^2 proportional trough the decrement by using energy 17.5 J/cm^2 and 20.0 J/cm^2.

The result of ANOVA test of the treatment of *S. mutans* group that was incubated with Pheophytin A. and laser irradiation showed significant differences toward decrement percentage of colony bacteria *S. mutans* with treated only using laser. The dose increase from 10 J/cm^2 until 20 J/cm^2.

The most significant decrement of percentage colony *S. mutans* toward the incubation of Pheophytin A. treatment and laser irradiation 405nm with energy 20 J/cm^2 was 61.9%. It showed that Pheophytin A. Functions as a photosensitizer activator to inactivate *S.mutan* bacteria.

4. Conclusion

The result of this research showed that incubation of Pheophytin A. with irradiation of laser diode 405nm has a significant effect toward bacterial growth. The most significant decrement of percentage colony of *S. mutans* toward the incubation of Pheophytin A. treatment and laser irradiation 405nm with energy 20 J/cm^2 was 61.9%. It showed that Pheophytin A. functions as a photosensitizer activator to inactivate *S.mutan* bacteria.

Reference

[1] Featherstone J D B 2000 The Science and Practice of Caries Prevention *Journal of American Dental Association* vol **131** 887-899

[2] Aas J A, Griffen A L, Dardis S R, Lee A M, Olsen I, Dewhirst F E, Leys E J, and Paster B J 2008 Bacteria of Dental Caries in Primary and Permanent Teeth in Children and Young Adults *Journal of Clinical Microbiology* **46** (4) 1407–1417 doi:10.1128/JCM.01410-07

[3] Huang L, Dai T, and Hamblin M R 2010 Antimicrobial Photodynamic Inactivation and Photodynamic Therapy for Infections *Methods Mol Biol* 635: 155–173. doi:10.1007/978-1-60761-697-9_12.

[4] Minnock A, Vernon D I, Schofield J, Griffiths J, Parish J H and Brown S B 1996 Photoinactivation of bacteria: Use of a cationic water-soluble zinc phthalocyanine to photoinactivate both gram-negative and gram-positive bacteria *J Photochem Photobiol B* **32**:159.

[5] Tavares A, Carvalho C M B, Faustino M A, Neves M G P M S, Tomé J P C, Tomé A C, Cavaleiro J A S, Cunha Â, Gomes N C M, Alves E and Almeida A 2010 Antimicrobial Photodynamic Therapy: Study of Bacterial Recovery Viability and Potential Development of Resistance after Treatment, *Mar. Drugs*, *8*, 91-105; doi:10.3390/md8010091

[6] Plaetzer K, Krammer B, Berlanda J, and Berr F 2009 Photophysics and Photochemistry of Photodynamic Therapy: Fundamental Aspects *Journal of Laser Medical Sciences* **24**: 259-268

[7] Scheer H 2006 In Chlorophylls and Bacteriochlorophylls : Biochemistry, Biophysics, Functions and Applications. Eds *Advances in Photosynthesis and Respiration* **25**:1-26.

[8] Derosa M C and R J Crutchley 2002 Photo-sensitized Singlet Oxygen and Its Application *Coord. Chem. Rev.*2002 **233**:351-371

[9] Astuti S D, Zaidan A H, Setiawati E M, and Suhariningsih 2016 Chlorophyll mediated photodynamic inactivation of the blue laser on Streptococcus mutans *AIP Conf. Proc. (2016)* **1718**, 120001-120008. doi: 10.1063/1.4943353

[10] Nitzan, M Salmon Divon, E Shporen, and Z Malik ALA induced photodynamic effects on gram positive and negative bacteria *Photochem. Photobiol. Sci.* vol. **3** no. 5 pp. 430–435

[11] Song B H, Lee D H, Kim B C, Ku S H, Park E J, Kwon I H, Kim K H, and Kim K J 2014 Photodynamic Therapy using Chlorophyll-a in the Treatment of Acne Vulgaris: A Randomized, Single-Blind, Split-Face Study *J Am Acad Dermatol* **71** 4:764-71.

[12] Sanja M Milenković, Jelena B Zvezdanovic, Tatjana D Anđelkovic, and Dejan Z Marković The Identification of Chlorophyll and its Identification Derivatives in The pigment mixtures: HPLC-Chromatography Visibles and Mass Spectroscopy Studies, Advanced technologies,

2012; 1, 1:16- 24
[13] Kobayashi M, Akutsu S, Fujinuma D, Furukawa H, Komatsu H, Hotota Y, Kato Y, Kuroiwa Y, Watanabe T, Ohnishi-Kameyama M, Ono H, Ohkubo S and Miyashita H Physicochemical Properties of Chlorophylls in Oxygenic Photosynthesis — Succession of Co-Factors from Anoxygenic to Oxygenic Photosynthesis, agricultural and biological science, chapter 3, 2013, 47-90, http://dx.doi.org/10.5772/55460

Influences of the calcination temperature and polymethyl methacrylate templates to characteristic pore membrane of bioceramic titanium dioxide

Jan Ady[1*] ,Eka Viandari[2] , and Dewi W Hasanah[2]

[1]Department of Physics, Faculty of Science and Technology, Airlangga University, Surabaya, Indonesia
[2] Department of Biomedical Engineering, Faculty of Science and Technology, Airlangga University, Surabaya, Indonesia

*Corresponding email address: jan-a@fst.unair.ac.id

Abstract. The synthesis of the bioceramics titanium dioxide on its template by polymethyl methacrylate to be a sample with forming TiO_2/PMMA was made with sol-gel process, and its pore membrane characteristics has also been studied. Different calcination temperatures 500^0C, 550^0C, and 600^0C were given to sample for 17 hours. This purpose themselves was to fill TiO_2 fissures with PMMA on different concentrations of 0.0, 2.0 wt%, 3.0 wt%, 4.0 wt%. Template leaching technique was used to remove PMMA from samples, and it was then sequentially found of the pore size of the membrane in approximate ranges (1900 nm - 2000 nm), (860.5 nm – 1669 nm), (312.8 nm-382.5 nm), and (136.1 nm - 269.7 nm). SEM test using and fourth it has average thickness in about 6.7 nm with Ellipsometer measurement. The percentage values of titanium and oxygen atoms are found by SEM-EDX from 3.03 at.% up to 66.81 at.% and there is amount 99.99% of the sample in anatase phase forming at 550 0C with angle of diffraction is 25.410 and it was prepared by XRD measurement.

1. Introduction

The bioceramics titanium oxide (titania) as a matrix-forming compound to be a bioceramic-biocompositetitania hydroxyapatite (TiO_2/HA) or its bioglass has been synthesized by means of sol-gel process to many needs in clinical applications especially for humans. Its characteristics were also studied, for instance, its role in the human body in playing the most important role in regenerating osteoblast bone [1]. This is similar to its titanium dioxide as a matrix-fiber of bioceramic-biocomposite which is qualified properties such as biomaterials bioinert, bioresorbable, and bioactive [2]. Titanium dioxide is also non-toxic, and it has photo catalytic effect reactions with UV light which makes it safe to use for in membrane bio composite [3].

Synthesis titanium dioxide can be prepared by means of the precursor matrix solution into the tangible gelatine known as sol-gel process [4]. In the synthesis, it does not require high temperature conditions, or in other words, it could be performed at room temperature. The crystalline grains produced by sol-gel always produce units to nanometer.

The membrane is portrayedas both selective and semipermeable thin layer types lying between of the concentrate or permeate phase [5]. Substance of the concentrate would be retained by its membrane because the grains size is too large than the size of the membrane pore. On the contrary, substance of the permeate phase is too small which makes it able to go through the membrane.

Characteristics of the membrane poresize mostly depend on the sample arrangement on its substrate such as titanium dioxide on the polymethyl methacrylate (PMMA) substrate which becomes TiO_2/PMMA. It aims to discover the characteristics of pore membrane such as macroporous (pore size > 50 nm), mesoporous (2 nm – 50 nm), and microporous (pore size < 2 nm) [6].

Influences of the calcination temperature given to the sample have been studied in order to find the characteristics of crystalline phases of titanium dioxide for 500 ^{0}C, 550 ^{0}C, and 600 ^{0}C including its phases such titania-rutile, titania-anatase, and titania-brookite [7]. From the third phase, only titania-rutile and titania-anatase were stable with their tetragonal structure, and their lattice constant value is different from one another. Contrary tobrookite phase, it is an unstable structure, and the characteristics are usual. The differences of both of titania-rutile and titania-anatase are in their octahedron and octahedral atomics arrangement for each oxygen or titanium cations in its bonding form Ti – O – Ti and with triangle form in about 900, 1200, and 1800 [8].

The purpose of this study is to demonstrate the synthesis of titania bioceramic by means of sol-gel process and characterize the pore membrane under the influences of the calcination temperature from its crystalline structure phasesin titania-rutile (500 ^{0}C), titania-anatase (550 ^{0}C), titania-brookite (600 ^{0}C).Then it was tobe added to different PMMA in order to make a template membrane such as in the form of TiO_2 / PMMA.

2. Research method

2.1. Equipment and materials
The apparatus used in measurement in this study included scanning electron microscope (SEM) and SEM-EDX such as Ellipsometer and X-ray diffraction (XRD). The tools used included Furnance, Beaker glass, pipette, and magnetic stirrer. The materials used included titanium tetraisopropoxide with98% (Sigma-Aldrich;St.Louis,MO,USA), Poly Methyl methacrylate (PMMA), glacial acetic acid, acetone, ethanol 96%, glass substrate preparations, and chloride acid (HCL) 37%.

2.2. Preparation Sol – Gel TiO_2 (Titania) and PMMA Template
The prior glass should be cleaned using ethanol, and it was cut into small pieces with 1cm x 1cm in dimension. Next, it was dipped into PMMA solution and left to dry for PMMA coating on a glass to be a template for the sample.

The synthesis of titania with sol-gel process has been described in [4]. Prior to that is synthesis of sol-titania by mixing ethanol 225 ml and 15 ml of acetic acid for four minutes using a stirrer, then it was added with 6.0 ml of $Ti(OH)_4$ or tetraisopropoxyde (TTIP). Next, 0.4 ml of HCL was added, and it was stirred for two hours. After that, the mixture was shaken for 24 hours and held up to 60 hours. From those processes was found precipitation of crystalline in the form of TiO_2. Steps of the chemical reactions are portrayed as below:

$$Ti(OH)_4 \rightarrow -Ti-O-Ti- + H_2O \quad (1)$$

$$-Ti-O-Ti- \rightarrow TiO_2 + 2H_2O \quad (2)$$

Thermal treatment was given to remove titania from the gelatin, and it was followed by equation (2).

2.3. Preparation of the synthesis of TiO_2 (Titania) membrane
PMMA template was previously prepared with immersion process for one hour in order to create sol-titania. Next, it was dried at room temperature 25^{0}C for three hours. After that, titania was interstitial to the PMMA template and calcinated at different temperatures of 500 ^{0}C, 550 ^{0}C, and 600 ^{0}C for four hours to remove PMMA template from previous sample in the form of TiO_2/PMMA. Subsequently, TiO_2membrane was found with desired pore size.

3. Results and discussion

The pore membrane of titania was molded using template leaching technique with PMMA as a polymer, a primary component to form pore membrane. Titania is created with sol-gel process with titanium-tetraisopropoxyde (TTiP) precursor, a solution which begsn to be colloidal suspension particles through its hydrolysis and its condensation reactions [9], when cast by alcohol. The influences of calcination temperatures could be seen via XRD test for each temperature of 500 ^{0}C, 550 ^{0}C, 600^{0}C as shown in figure 1. 88.9% of anatase-titania was resulted at 500 ^{0}C,and 99.9% of it was resulted at 550 ^{0}C as shown in Figure 1(a) and (b). Meanwhile, rutile-titania was resulted at 600 ^{0}C with 44.15% of its anatase phase and then 55.85% in its rutile phase as shown in Figure 1.(c) [10]. The phases of the crystalline forming mostly depended on the photochatalysis which were also part of calcination process.

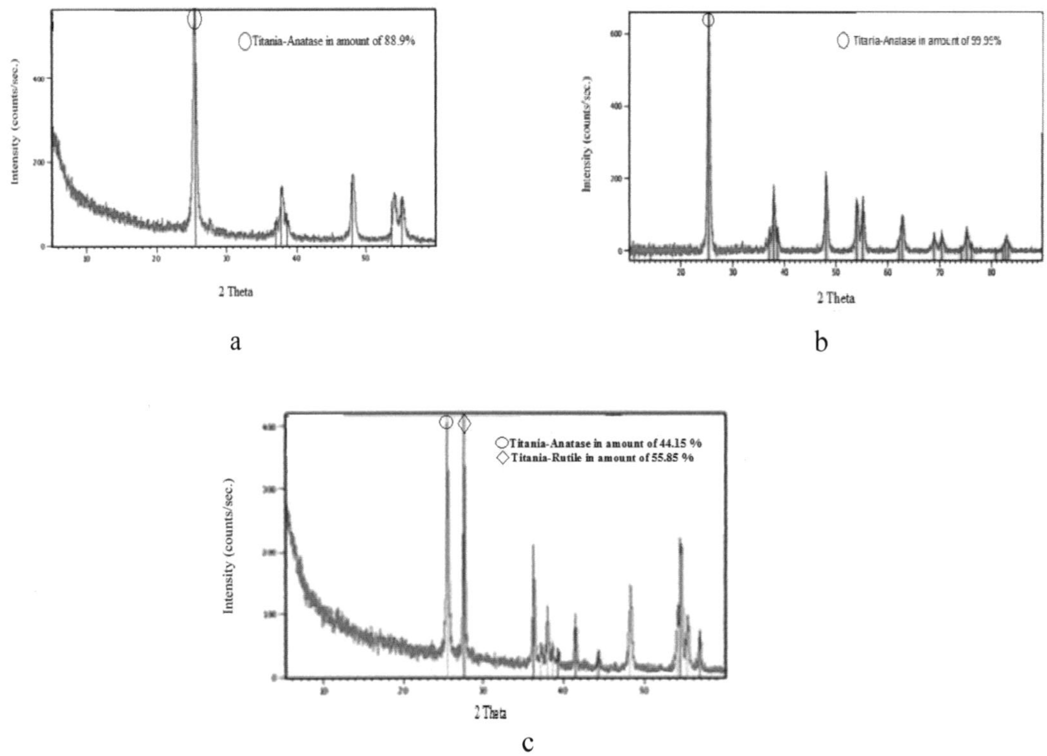

Figure 1. XRD test results of the samples for each temperature calcination are shown **(a).** Amount of anatase TiO_2-88.9% at 500^{0}C, **(b)**. amount of anatase TiO_2-99.9% at 550^{0}C, and **(c)**. amount of anatase TiO_2-44.15% and amount of Rutile TiO_2-55.85% at 600^{0}C.

The presence of Titanium (Ti) and Oxygen (O) in its membrane is evidenced by Energy Dispersive X-Ray Spectroscopy (EDX), and it was molded on a glass in order to create a thin layer. Results of this test are shown in Table 1 and the Figure 2.

Table 1. The elements (At.%) and weight (Wt.%) of the sample indicated by Energy Dispersive X-ray (EDX) test

Types of Component	Wt. (%)	At. (%)
O-K	53.12	66.81
Na-K	21.39	18.72
Mg-K	06.98	05.78
Ca-K	11.29	05.67
Ti-K	07.21	03.03

The forming process of membrane pores using PMMA template is based on the effect of temperature on the PMMA structure. When subjected to heat with some temperatures like in this study, PMMA shrank which eventually made TiO_2 particles agglomerate each other. In Figure 3(a), (b), (c), and (d) is shown SEM image showing a lot of pores whose emergence was influenced by PMMA.

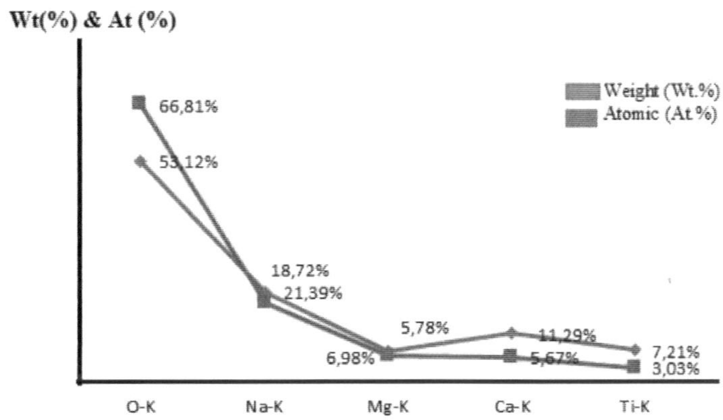

Figure 2. The weight (Wt.%) and Atomic (Wt.%) of elements or compounds in its membrane form are evidenced by Energy Dispersive X-Ray Spectroscopy (EDX) test

Membrane such as in Sample1 (S1) as shown in Figure 3(a) is without PMMA, and it cracked and fractured with broad spaces. The cracks occurred because the sample's solution contained high solvent [11]. Moreover, it did not interact with other elements or compounds which evaporated, hence huge cracks[12]. For sample 2(S2), sample 3(S3), and sample 4 (S4) as shown in Figure 3(b), (c), and (d) are membranes of titania into which PPMA was added using sol-gel process. However, they differed each other.

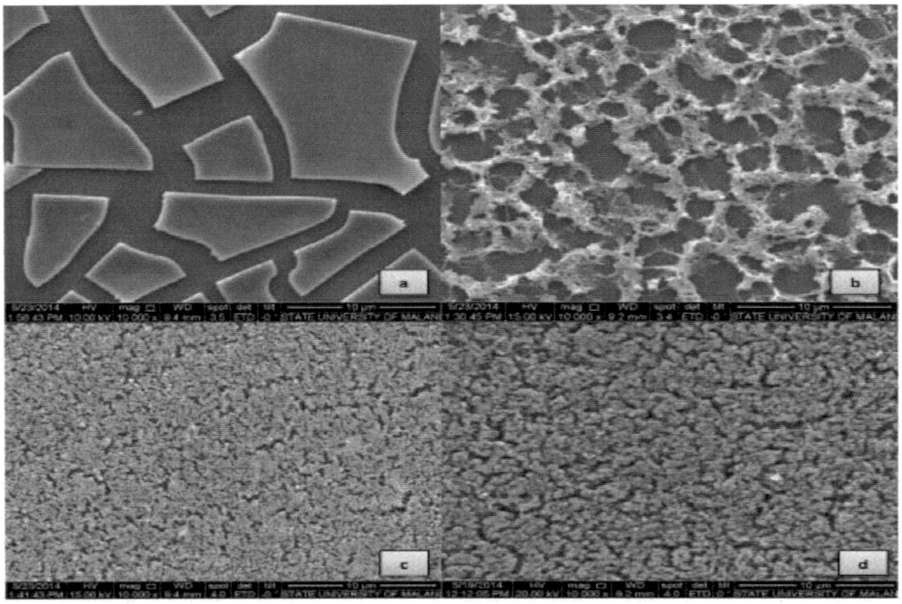

Figure 3. SEM image of TiO_2 Membrane Morphology with various PMMA for enlargement 10.000x (a). S1 Without PMMA, (b). S2 PMMA 2.0wt%, (c). S3 PMMA 3.0wt%, (d). S4 PMMA 4.0wt%

The pores formed were smaller, especially in Figure 3 (c) and (d) because the PMMA shrank after calcination with different temperatures. It was in membrane pore size, about (1900nm - 2000 nm), (860.5nm - 1669nm), (312.8nm-382.5nm), (136.1nm - 269.7nm). Additionally, the average thickness was 6.7nm with cracks or pinholes as measured using SEM test and ELLIPSOMETER. That is shown in Figure 4.

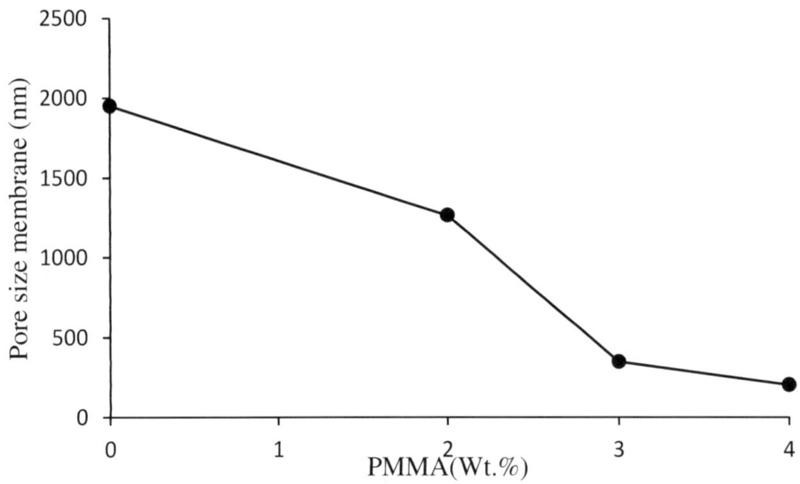

Figure 4. Pore size of TiO_2 Membrane for samples: (S1) without PMMA, (S2) PMMA 2.0wt%, (S3) PMMA 3.0wt%, and (S4) PMMA 4.0wt%. Its thickness was 6.7 nm as measured using ELLIPSOMETER

4. Conclusion

Bioceramic titania in TiO_2/PMMA formwith the sol-gel process was studied through its pore membrane characteristics, where the influence of calcination temperature enabled titania to form became perfectly intonanometer size. The pore size of titania membrane decreased along with the increase of PMMA concentrations in the samples. Furthermore, its optimum form with 4.0 Wt% of PMMA with anatase-titania phase at 550^0C.

References

[1] Larry L 1991 Bioeramics: From concept to clinic *J. Am. Ceram. Sos.***74** (7) 487-510

[2] RamiresPA, Romito A, Cosentino F, Milella E 2001The influence of titania/hydroxyapatite composite coatings on in vitro osteoblasts behaviour *Biomaterials* **22** 146-1474

[3] Gjevori A 2010 Phase Formation of Photoactive TiO_2 Thin Films by Metal plasma Immersion Ion Implantation and Deposition (Faculty of Natural Sciences University of Tyrana)

[4] Septina W 2007 Sintesa Nanokristal Mesopori TiO_2 dengan Metoda Sol-gel (Unpublished Thesis Bachelor of Physics ITB)

[5] Pinem J A and Angela R 2011 Sintesis Dan Karakterisasi Membran Hibrid PMMA / TEOT : Pengaruh Konsentrasi Polimer:*Proc. Seminar Nasional Teknik Kimia Kejuangan* ISSN 1693 – 4393.

[6] Gupta KK, JassalM, Agrawal AK 2008 Sol-gel Derived Titanium Finishing of Cotton Fabric For Self Cleaning *Indian J. of Fibre & Textile Research* **33**pp 4430450

[7] Afrozi AS 2010 Sintesis dan Karakterisasi Katalis Nanokomposit Berbasis Titania Untuk Produksi Hidrogen Dari Gliserol Dan Air(Unpublished Thesis Bachelor of Chemistry Engineering University of Indonesia)

[8] Setiawati T, Amalia I S, Wisnu A A 2006 Sintesis Lapisan Tipis TiO_2 Dan Analisis Sifat Fotokatalisnya *J. Sains Materi Indonesia* pp 141 – 146 ISSN : 1411-1098

[9] Septina W 2007 Sintesa Nanokristal Mesopori TiO_2 dengan Metoda Sol-gel(Unpublished ThesisBachelor of Physics Engineering ITB)

[10] Byranvand M M, Kharat N A, Fatholahi L, Beiranvand L M 2013 Review on Synthesis of Nano-TiO_2 Via Different Methods *J. of Nanostructure* JNS 3 1-9

[11] Hyodo T, Sasahara K, Shimizu, Y, Egashira, M 2005 Preparation of Macroporous SnO_2 films using PMMA Microspheres and Their Sensing Properties to NOx and H_2 *J. of Sensor and Actuators B* **106** pp 580-590

[12] Zhang H, DongF, Fang S, Ye C. Wang M, Cheng H, Han Z, Zhai S 2012 Fabrication of Macroporous Titanium Dioxide Film Using PMMA Microspheres as Template *J. of Colloid and Interface Science* **386**

Fabrication and characterization of Zinc Oxide (ZnO) nanoparticle by sol-gel method

Siswanto[1,*], Nurul T. Rochman[2], Putri Riski Akwalia[1]

[1]Physics Department, Faculty of Science and Technology, Airlangga of University
[2]Indonesian Nano Society , Serpong, Jakarta.

Corresponding email address: siswanto_fst@yahoo.co.id

Abstract. Currently, nanomaterial is an interestingfield of study. This is due to its chemical and physical properties that are superior to that of large-sized materials.One nanomaterial widely studied is zinc oxide (ZnO). In this study, a synthesis of ZnO nanoparticles made by Sol-Gel method was conducted.The process parameters used are variations in pH, in increasing order, of 7; 8; 9; 10; 11; and 12. There are two principal reactions to produce a compound oxide, namely hydrolysis and condensation. NaOH is an agent for the hydrolysis of $(CH_3COO)_2$ Zn resultingin Zn $(OH)_2$. Subsequently, condensation produces ZnO. Calcination was carried out at a temperature of 80^0C for 1 hour. The ccharacterization of the samples showed that the condition of pH 12 produced the best sample with a size of 73.8 nm and ZnO percentage of 100%. Although pH 7 produced a particle size of 1.3 nm, the percentage of ZnO formed was only 42.9%. The calcination process was performed to remove CH_3COONa. However, the process can lead to aggregation of ZnO particles to each other, which increases the particle size.

1. Introduction

Nanoscale material is a material which is very attractive because it has properties that are very different than those shown on the macroscopic scale. Various attractive quantum phenomena arise from the downsizing of material into nanoscale dimensions. Platinum bulky metal, known as inert material, can be turned into a catalytic material if its size is reduced to the nanoscale. Stable materials such as aluminum can be flammable and insulating materials turn into a conductor[1]. With nanotechnology, any substance or material will allow for a weight reduction accompanied by increased stability and improved functionality.

There are two methods of synthesis of nanoparticles, namely top down and bottom up. Top down synthesis process does not involve a chemical reaction. The process that occurs is only major material splitting into nanometer-sized material; the obtained new material that has better performance and is different from the original bulk material. Meanwhile, the bottom-up synthesis process involves a chemical reaction of a starting material to produce another material which is nanometer-sized.. Top down methods include mechanical milling, repeated quenching, and lithography, whereas bottom up methods include the sol gel process, aerosol-based process, chemical vapor deposition (CVD), atomic condensation, gas phase condensation, and fluid supercritical synthesis[3].

One of the many synthesized materials into nano-sized particles is Zinc Oxide. Zinc Oxide is a material that is currently widely studied. This is because ZnO exhibit optical, acoustic, and electrical properties allowing it a number of potential applications in the fields of electronics, optoelectronics, and sensors[2].Zinc Oxide is widely applied as an anti-reflection coating material, transparent electrodes in solar cells, gas sensors, varistors, electroluminescence material, and material Photoluminescence. Semiconductor ceramic materials are attractive because of commercial demand for optoelectronics such as transparent conductivity electrode, ultraviolet (UV) light emitters, and electron spin that are semiconductors with a wide energy gaps and binding energies on optical processes. Additionally, Zinc Oxide has several advantages including its stable chemical structure, non-toxicity, and its utility as an additive in a variety of materials, as well as its abundant availability in nature that makes its price cheaper.

From a number of Zinc Oxide nanoparticles synthesis method, the sol-gel method is a method that is relatively simple and works at lower temperature when compared to other methods such as chemical vapor deposition (CVD), hydrothermal, and plasma which require high temperatures in the process of synthesis. The sol-gel method has a principle similar to the solid state reaction. The process involves heating the mixture to stoichiometric reagents to make the ions move so that the desired solid product is obtained.

Materials that are commonly used in the sol-gel process is metal oxide. Furthermore, sonicator is used for mixing, and Zinc Oxide particles break. Zinc Oxide precipitates further heat in order to obtain a powder of nanoparticles of Zinc Oxide. The sol-gel method uses a lot of chemicals so it works on varying levels of acidity (pH) in the solution. The effect of pH contribute to the effects of hydrolysis and condensation during the process of forming a gel and the morphology of the resulting Zinc Oxide [5]. In addition, the pH can change the value of atomic nuclei and the development unit. The result of the synthesis of zinc oxide nanoparticles will be characterized using Particle Size Analyzer for observation of particle size and X-Ray diffraction for composition observation.

2. Materials and methods

The materials used in this study include: Zn $(CH_3COOH)_2$ (zinc acetate dihydrate), NaOH (sodium hydroxide), CH_3OH (methanol), and aquabidest. Some $(CH_3COOH)_2Zn.2H_20$ powder as much as 4.39 grams was dissolved in 100 ml of methanol and stirred with 750 watt sonicator for 30 minutes to obtain a homogeneous solution with a concentration of 0.2 M. Separately, 1.0 Mof NaOH were dissolved in 500 ml aquabidest. Furthermore, the optimization of NaOH titration was performed. The results of the titration optimization was used for titration of NaOH into a solutionof $(CH_3COOH)_2Zn.2H_20$, thereby changing the pH value. In order to determine the effect of pH on the synthesis results, the researchers then made a colloid with five variations in pH of 7, 8, 9, 10, 11, and 12.When the solution has formed a milky white, they were then stirred again by using a sonicator for 30 minutes. Then, they were allowed to stand for several days for observation. pH 7 and pH 12 has the fastest time and the longest colloids to settle, and then the results of both were analysed by the PSA. They were then centrifuged to separate the sediment at 3000 rpm for 30 minutes, and thesediment was taken. The precipitate was heated so that H_2O precursors and other compounds can be lost using the oven at 800C for 1 hour. Furthermore, ZnO solids were crushed using a mortar and the powder obtained consist ZnO nanoparticles. Then, ZnO sample was analyzed by the XRD to determine the composition of the sample.

3. Results and discussion

The mix $(CH_3COO)_2Zn.2H_20$ and $NaOH.H_20$ used titration. Optimization of titration was used in terms oftime of hatching, temperature, and stirrer speed. The optimization of the time variation of hatching was performed at 0, 1, 20, and 60 seconds. The optimization results show that the longer the time of hatching,the longer the time a solution to settle will be. The optimization of temperature was performed at 0^0C, 25^0C, and 50^0C.The results showed that the longest deposition occurs at room temperature (25^0C). The third optimization focused on variations of the magnetic stirrer speeds of 100 rpm, 300 rpm, and 600 rpm. The results showed that the solution takes most time to settle when magnetic thestirrer speed of 600 rpm. Thus titration parameters are optimized to hatching time of 60 seconds, the room temperature 25^0C, and 600 rpm speedof magnetic stirrer. Then, the titration of

IOP Conf. Series: Journal of Physics: Conf. Series **853** (2017) 012041 doi:10.1088/1742-6596/853/1/012041

NaOH into 100 ml of $(CH_3COO)_2Zn.2H_20$ was performed manually using a pipette until the pH of the solution turned into 7, 8, 9, 10, 11, and 12. The colloidal ZnO formed was of milky whitecolor. The colloidal ZnO is fastest and most long to settle that at pH 7 and pH 12 (Table 1)

Tabel 1 The relationship between pH with deposition time

pH	Deposition time (minutes)
7	4320
8	1440
9	600
10	480
11	240
12	120

Furthermore, the solution has had various pH 7, 8, 9, 10, 11, and 12 centrifuged at 3000 rpm for 30 minutes. Centrifuges were used to separate the precipitate from the solution. The position of each solution was taken and placed in crucible, then calcined in an oven heated to 800C for 1 hour. The calcination of the Zinc Oxide will produce a white blob and then the white blob was crushed using a mortar to produce Zinc Oxide powder. Calcination in this case serves to remove the element contained in the compound, and the compound resulting from shrinking volume.The depreciation of compounds that happen cause re-unification of these elements so that the particles get bigger. Furthermore, ZnO powder were tested using PSA to determine its size and XRD to determine the composition of the compounds contained in the ZnO.

The results showed that among the variations of the deposit of zinc oxide, the one from the solution of pH 12 obtained the fastest settling time, whereas the solution of zinc oxide obtained at pH 7 solution_ was the slowest to settle. Therefore, we tested the colloidal ZnO of pH 12 and pH 7 using PSA (Figure 1). The average particle size of ZnO of the pH 7solution is 1.3 nm while for the pH 12 solutionit is73.8 nm. The greater the pH of the solution, the less time it takes to settle so that the solution of the resulting particle size increases. Conversely, the smaller the pH of the solution, the longer time it takes to settle and the smaller is its resulting particle size (Figure 2)

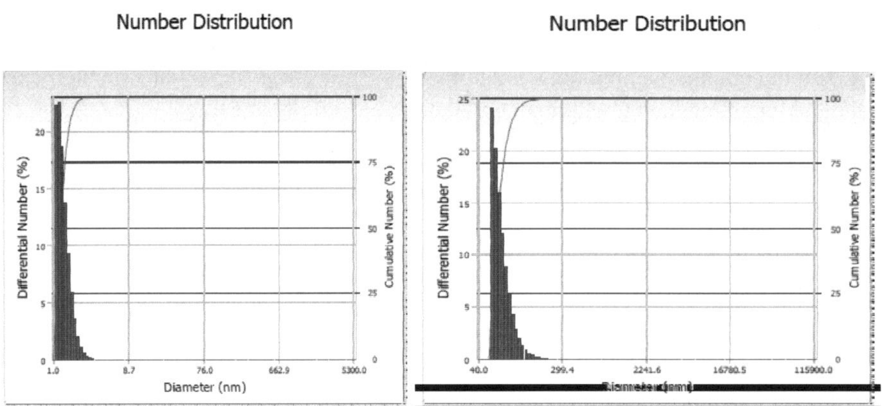

(a) (b)

Figure 1. Number distribution of sample at (a) pH 7 and (b) pH 12

The XRD observation results for the sample pH 7, 8, 10, and 12 were shown in Figure 2. XRD observation data were analyzed using two approaches, namely by fitting the data using Microsoft Excel and by the X-powder program. Let us turn first to fitting the data using Microsoft Excel. In this case, the test results of XRD to pH 7, 8, 10, 12 are matched to the database manually with Microsoft Excel, i.e.by matching peaks of database ICDD Pcpdfwin 1998 ZnO with peaks that are indicated by ZnO synthesized at various pH. From the results of Figure 2 it appears that at pH 7 ZnO nanoparticles are still not fully formed, same with pH 8 which also has not formed ZnO and other compounds. At a pH of 10 formation ZnO nanoparticles began, whereas at pH 12 pure ZnO nanoparticles had formed.

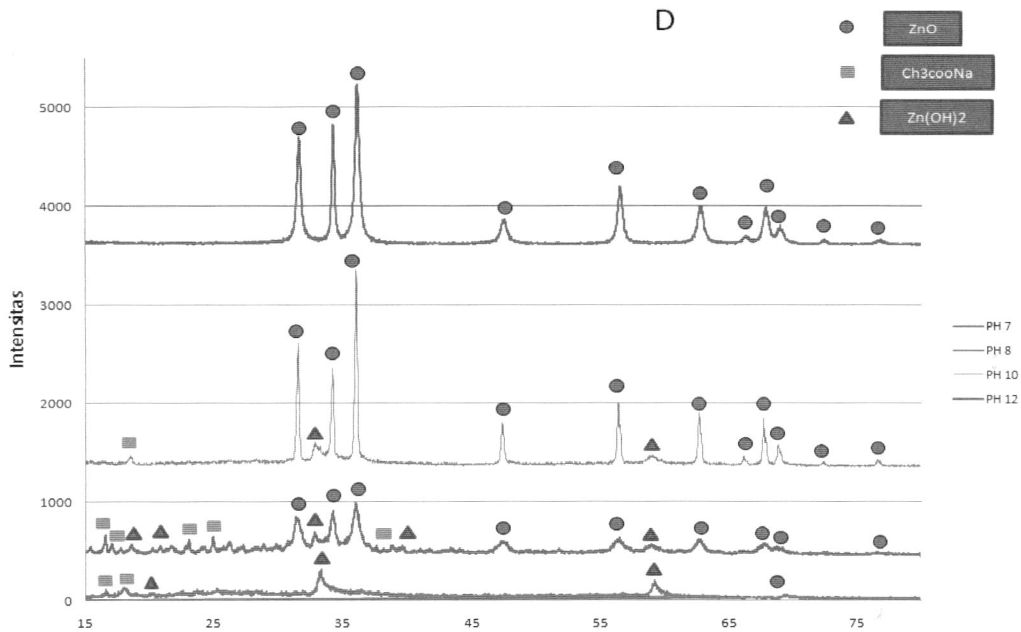

Figure 2. Quantitative analysis of the sample at pH 7, 8, 10, dan 12.

The second approach was performed using the X-powder. The diffraction patterns in the form of test results of the XRD spectra provide information about the diffraction angles (2θ) on the horizontal axis and the intensity generated on the vertical axis. The XRD testing aims to determine the composition of the compounds formed in ZnO. Determining the composition of these compounds is done using a search match with X-powder program. The search results matched the X-powder in a way fitting the data obtained by the percentage weight of the compound of synthesized ZnO at pH 12, 10, 8, and 7 which are shown in Table 2.

Table 2. Influence of pH to the procentage of ZnO

pH	% ZnO
7	42.9
8	62.2
10	64.7
12	100

From the XRD observation data, using the program X-powder and Williamson-Hall plot analysis he crystal size can also be determined (Table 3). From table 3, it appears that the greater the pH value, the greater the size of the crystal. Conversely, the smaller the pH value, the smaller the size of the crystalis. This is consistent with the results of the particle sizes of ZnO that were formed. The size of the crystals can cause dilated calcination so that the particle agglomeration will grow back.

Tabel 3. Relation between pH with size of crystal

pH	Ukuran kristal (nm)
7	10.94±0.99 nm
8	17.44±5.36 nm
10	38.27±2.14 nm
12	74.04±41.77 nm

4. Conclusion

One of the factors that influences the size of the diameter of the nanoparticles using thesol-gel method is pH. The greater the pH of the sol-gel process, the greater the agglomeration so that the particle size gets bigger and vice-versa. The ZnO particle size characterization of the results obtained by the PSA at pH 7 and pH 12 was 1.3 nm and 73.8 nm. The results of XRD analysis with X-Powder program obtained the degree of purity of ZnO produced at pH 7, pH 8, pH 10 and pH 12 at42.9%, 62.2%, 64.7%, and 100%, respectively. The study posits that using sol-gel method is the best conducted ata pH of 12 which resulted in a particle size of 73.8 nm with a composition of 100% ZnO.

5. References

[1] Mikrajuddin Abdullah, Yudistira Virgus, Nirmin, Khairurrijal, 2008, Review : Sintesis Nanomaterial *Jurnal nanosains dan nanoteknologi* Fakultas Matematika dan Ilmu Pengetahuan Alam Institut Teknologi Bandung Bandung.

[2] Pusfitasari and Eka Dian 2009 Sintesis dan Karakterisasi Nanopartikel Zinc Oxide (ZnO) dengan menggunakan Metode Sol Gel Berdasarkan Variasi Suhu Jurusan Kimia Skripsi S1 UNAIR Surabaya

[3] Rochman, NT 2009 Handbook Nanopartikel Pusat Penelitian Ilmu Pengetahuan dan Tekhnologi-LIPI, Serpong

[4] Neppolian, Q Wang, H Jung and H Choi 2009, Ultrasonic-assisted sol-gel method of preparation of TiO_2 nano-particles: Characterization, properties and 4-chlorophenol removal application, Journal of Alloys and Compounds.

[5] Alias SS, AB Ismail, AA Mohamad, 2010, Effect of pH on ZnO nanoparticle properties synthesized by sol–gel centrifugation, Journal of Alloys and Compounds, School of Materials and Mineral Resources Enginering University Sains Malaysia, 14300 Nibong Tebal, Penang, Malaysia.

The synthesis and characterization of hydrogel chitosan-alginate with the addition of plasticizer lauric acid for wound dressing application

Djony Izak Rudyardjo* and Setiawan Wijayanto

Department of Physics, Faculty of Science and Technology, Airlangga University Mulyorejo Street Surabaya 60115 Indonesia

*Corresponding email address: djoni_unair@yahoo.com

Abstract. The writers conducted a study about the synthesis and characterization of hydrogel chitosan-alginate by addition plasticizer lauric acid for wound dressing application. The purpose was to find out the impact of lauric acid concentration variation on hydrogel chitosan-alginate to get the best mechanical and physical properties to be applied as wound dressing in accordance with existing standards. This study used commercially chitosan from extract of shells crab, commercially-available alginate from the extract of sargassum sp, and commercial lauric acid from palm starch. The addition of lauric acid was aimed to repair mechanical properties of hydrogel. The composition of chitosan-alginate is 4:1 (v/v), while the lauric acid concentration variations are 0%, 1%, 2%, 3%, 4%, and 5% w/v. The characterization of mechanical properties test (Tensile strength and Elongation at break) at hydrogel showed the hydrogel chitosan-alginate-lauric acid have the characteristic which meets the standard of mechanical properties for human skin. The best performance of hydrogel chitosan-alginate-lauric acid was obtained by increasing luric acid concentration by 4%, which has a thickness value of 125.46±0.63 μm, elongation 28.89±1.01 %, tensile strength (9.01±0.65) MPa, and ability to absorb liquids (601.45 ±1.24) %.

1. Introduction

Wound healing and tissue restoration is defined as a reparative process which usually consists of a series of continuous inflammation and repair. During this process, epithelial cells, endothelial cells, inflammatory cells, platelets, and fibroblasts interact to restore normal functions. Recently, a lot of research is conducted to find a method of regeneration and wound healing through the use of various cover materials (dressing) to facilitate proper wound management [1]. The research on wound closure has focused on accelerating wound repair by improving the design of the cover material. One type of wound closure material that is being developed is hydrogel. Hydrogels have the advantage to work well in a moist environment. This material is a natural absorbent of biodegradables through decomposition by microbes [2]. The use of a blend of two or more substances will provide hydrogel with good mechanical properties. The use of biological materials such as chitin and chitosan derivatives, of which itself has a fairly medical extensive uses [3]. The use of chitosan in wound management has many advantages due to its biocompatibility and molecular biodegradability that do not harm the environment. When chitosan is used as wound covering, besides biocompatible chitosan will also be biodegradable by lysozyme, chitinase, and kitosanase into oligomers and monomers (sugar amines). making it harmless and completely absorbable by the body [1]. Chitosan is a hemostat that assist unnatural blood clotting, but this material has brittle and can be easily torn. The modification of the materials for better characteristics can be done by adding alginate. Alginate and chitosan arepolyanionic and polycationic when dissolved under tthe right conditions and can interact with each other via the carbonyl group of the amino group of the alginate and chitosan. Alginate also has the advantage of being non toxics, biodegradable, biocompatible, and does not cause allergies.

One previous study identified chitosan-alginate membrane potential as a candidate for hemodialysis membrane with the composition of 4:1 [4]. In the study, the elongation values obtained were not included in the standard range of mechanical properties of human skin. Increasing the elongation of the composite can be done by the addition of plasticizers. Plasticizer has a function to increase the flexibility and extensibility of the gel, avoid the gel from cracking, and improve the elasticity of the gel to increase the elongation of the hydrogel. One material that could potentially improve the mechanical properties of the hydrogel dressings is lauric acid [5]. Lauric acid is a medium chain saturated fatty acids found to be antimicrobial (antiviral, antibacterial, and antifungal). Lauric acid in the body and on the skin will turn into monolaurin, which can kill bacteria and thus speed up the metabolism of cells in the skin. This created a base for the present study on the synthesis and characterization of chitosan-alginate hydrogel with the addition of lauric acid plasticizer for the application of wound dressings.

2. Experiment design

In this study, the materials used are as follows: chitosan, Na-alginate, lauric acid, glacial acetic acid, distilled water, PBS, KBr, and 96% ethanol. The tools used are glass beaker, spatulas, paper weights, pipettes, stirrers, aluminum foil, measuring cups, glass mold, digital balance, and magnetic stirrer. Equipment for analysis of quantitative and qualitative Coating Thickness Gauge is a type TT 210, Bruker Tensor 27 FTIR and IMADA Tensile machine type HV-1000N.

The materials are chitosan-alginate prepared in a ratio of 4: 1 with the addition of lauric acid (plasticizer) solution of 0% w / v, 1% w / v, 2% w / v, 3% w / v, and 5% w / v. The preparation of alginate required 3 grams of alginate powder that was mixed with 100 ml of distilled water. Furthermore, the stirring was conducted at room temperature until the solution became homogeneous. Alginate was added slowly while stirring occurs. The chitosan solution required 2 ml of glacial acetic acid. They were mixed with 3 grams of chitosan powder ready to use along with 98 ml of distilled water, which were then stirred for 45 minutes. A 10 ml alginate solution was then added gradually into a solution of 40 ml chitosan which was still in a state of stirring. The solution was then added with lauric acid (plasticizer) with a composition variation of 0% w / v, 1% w / v, 2% w / v, 3% w / v, and5% w / v which had been melted at a temperature of 44 ° C, which is the melting point of lauric acid Lauric acid is a saturated fatty acid that is required to dissolve 96% ethanol. After all the ingredients were mixed, stirring was continued for 30 minutes in order to obtain a homogeneous solution.

The hydrogel solution that has formed was poured and leveled on a glass plate with a thickness of 2-4 mm. Then, the solution that had been leveled at the glass plate was dried at room temperature for 7 days until the hydrogel was completely formed in sterile condition.

2.1. Thickness test

The thickness test of the test sample was performed using the Coating Thickness Gauge TT type 210. The Coating Thickness Gauge type TT 210 was put above the hydrogel samples in order to obtain the data. These measurements aimed to determine the effect of variations in the composition of the chitosan-alginate with lauric acid addition to the thickness of the particular test sample produced.

2.2. Hydrogel tensile strength test

The mechanical characteristics observed from a chitosan-alginate hydrogel material with the addition of lauric acid plasticizer were measured with a tensile test using an IMADA type HV-1000N tensile machine. The tensile test was conducted to determine the mechanical properties such as strength, elasticity, rigidity, and plasticity. The determination of mechanical properties was done by cutting the hydrogels of the appropriate size. The ends of hydrogels were linked to test equipment and towing load mounted on the unit load N. The hydrogel was drawn at a certain speed to cause drop. The maximum towing load (fmax) and the change in length hydrogels at break was recorded. The results obtained by the value of the stress and elongation at break were calculated using the equation [7]:

$$\sigma = \frac{F}{A} \quad (1)$$

with σ = Stress (N / m^2), F = Load (N), A = surface area (m^2).

$$\varepsilon = \frac{L - L_0}{L_0} x\ 100\% \quad (2)$$

with ε = elongation (%), L = length of the end of the test specimen (cm), Lo = initial length of the test specimen (cm).

Hydrogel of chitosan, alginate and lauric acid as plasticizer can be used as a wound cover (Wound Dressing) if it meets the standard of certain mechanical properties.

2.3. Absorption ability test

The absorption capacity of the hydrogel was determined by incubating the hydrogel at pH 7.4 in Phosphate Buffer Saline (PBS) at room temperature. Hydrogel wet weight was calculated for several times to let sponge filter paper remove the water adsorbed on the surface and then immediately weighed with digital scales. The amount of water absorbed by the hydrogels can be calculated using the equation:

$$E = \frac{m_e - m_o}{m_o} x 100\% \quad (3)$$

where E is the percentage of absorption of water in the hydrogel. m_e stands for the weight of hydrogel that had absorbed PBS, while m_o is the initial weight.

2.4. Fourier Transform Infra Red (FT-IR) Test

The FT-IR spectroscopy test was conducted by eroding the chitosan-alginate hydrogel-lauric acid together KBr with a ratio of 1:10 (w / w). The mixture was then pressed by using presses at a pressure of 10 tons, so that the pellet became solid. These pellets are then analyzed using Bruker Tensor spectroscopy type 27. The results of FT-IR spectra was obtained in the form of illustrations of the %value of transmittance and wave number, so it can be any functional group contained in chitosan-alginate hydrogel-lauric acid.

3. Results and discussion

3.1. Fourier transform infra red (FT-IR) test

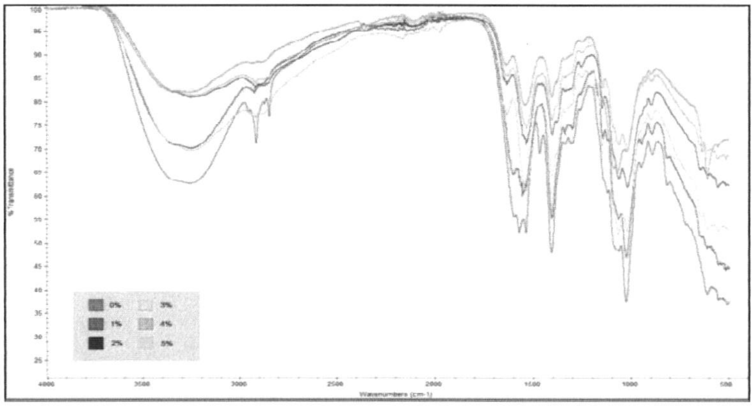

Figure 1. The IR spectrum-Chitosan Hydrogel with Alginate Lauric Acid Addition

Table 1. Data of absorption spectrum

Functional Group	Wave Number (cm^{-1})								
	Chitosan	Alginate	Lauric Acid	Hydrogel 0%	Hydrogel 1%	Hydrogel 2%	Hydrogel 3%	Hydrogel 4%	Hydrogel 5%
O-H (*Stretching*)	3291.10	3221.74	3319.98	3249.78	3259.75	3252.99	3245,05	3272.40	3258.09
O-H	2870.83	-	-	-	-	-	-	-	-
C=C	2356.73	-	-	-	-	-	-	-	-
N-H (*Bending*)	1647.73	-	-	1557.12	1539.13	1538.36	1538.19	1541.34	1538.97
C=O (*Stretching*)	1419.94	-	1705.80	1603.17	1596.83	1633.84	1632.71	1407.39	1635.19
C=O	-	1596.00	-	-	-	-	-	-	-
C-O (*Stretching*)	1088.59	1089.69	1088.59	1064.91	1064.91	1064.80	1065.37	1072.34	1066.58
C-H (*Stretching*)	-	-	2927.13	2923.48	2916.57	2917.99	2918.26	-	2913.98
C-H (Bending)	-	1405.70	1448.64	-	-	-	-	-	-
N=C=O	-	-	-	-	-	-	-	2356.73; 2334.56	-

The shift in the wave number of functional groups was due to an increase or decrease in the radiation energy absorbed by the molecules. In chitosan-alginate hydrogel with the addition of 4% lauric acid, there was no C-H groups, but there was a new functional group, namely N = C = O at wavenumbers of 2356.73 cm^{-1} and 2334.56 cm^{-1}. The new functional group was formed by the reaction between the primary amine group (NH_2) on chitosan and carboxyl group (COO) on alginate and lauric acid. In addition, the formation of two peaks was due to their symmetrical and asymmetrical vibration. The interaction between the primary amine group with a carboxyl group formed an amide bond (-NH-C = O). The amide bond underwent rearrangement by shifting electrons forming a new functional group (-N = C = O) that they can be turned into electron structures of (N CO-), (-N = C = O + R +), (-N + CO-), and (-N = C = O). Thus, the crosslink forming reaction between chitosan-alginate by the addition of lauric acid took place more quickly so that the resulting structure became more dense.

The chitosan-alginate hydrogel with the addition of 4% lauric acid is the chemical compound with the advent of a new functional group so that the hydrogel has a chemical structure that is much different from their constituent components. A greater concentration of the resulting bond of the new group made the molecules of the polymer blend to be denser. This meeting of molecules may cause effects on the macroscopic properties of the hydrogel.

Table 2. Characterization of chitosan-alginate hydrogel-lauric acid

Sample (%)	*Thickness* (μm)	*Elongation (%)*	*Tensile Strength (MPa)*	*Absorption Percentage (%)*
0	*80.96 ± 2.54*	*6.90 ± 0.89*	*27.91 ± 1.55*	*400.98 ± 1.50*
1	*92.66 ± 1.22*	*11.11 ± 0.93*	*10.81 ± 1.18*	*482.85 ± 1.45*
2	*103.76 ± 0.95*	*13.66 ± 0.94*	*9.87 ± 0.94*	*485.59 ± 1.34*
3	*115.46 ± 0.64*	*17.78 ± 0.96*	*9.51 ± 0.76*	*529.71 ± 0.93*
4	*125.46 ± 0.63*	*28.89 ± 1.01*	*9.01 ± 0.65*	*601.45 ± 1.24*
5	*128.46 ± 0.29*	*16.29 ± 0.95*	*10.18 ± 0.62*	*340.12 ± 0.91*

3.2. Thickness test
Based on Table 2 description of chitosan-alginate hydrogel-lauric acid values of different thickness, it can be explained that along with the increase in the addition of plasticizer concentration of lauric acid, the thickness of the hydrogel has a tendency to increase. The more of lauric acid is added in the hydrogel,the more water is trapped in the starch granules because lauric acid inhibits the evaporation of water during the drying so that the resulting hydrogel gets thicker. Hydrogel can be used as wound coveringsbased on this finding.

3.3. Hydrogel tensile strength test
The elongation values of hydrogel showed that with increasing concentration of lauric acid, hydrogels' elongation values also increase. At a concentration of 5%, elongation values declined, but the value is still above the value of the elongation hydrogels without the addition of lauric acid. In Riki Siswanto (2013), chitosan-alginate membrane consisting of a mixture of chitosan and alginate and without the addition of lauric acid plasticizer produced small elongation values in the range 3.1% - 13.1%. This proves the effect of the addition of lauric acid which serves to increase the value of elongation hydrogel. The amount of elongation determines ductility of a material, and if its value is close to zero, the material is a brittle material [6]. Increasing elongation value resulted in more elastic hydrogel when applied to the skin surface. Increased elongation of chitosan-alginate hydrogel-lauric acid occurs because the plasticizer molecules of lauric acid has a strong enough force interaction with polymer hydrogels that plasticizer molecules diffuse into the polymer chain (between the polymer chitosan and alginate). The addition of lauric acid plasticizer of more than 4% shows the results of the elongation decreased. This occurred because the addition of lauric acid was exceedingly saturated so excessive plasticizer molecules are in a separate phase outside the phase of the hydrogel. The situation causes a decrease in the intermolecular force between the chains.

Values for tensile strength is inversely proportional to the value of elongation, which is more and more increasing concentration of lauric acid, then the value of tensile strength of the hydrogel decreases. However, at concentrations of 5% tensile strength values increase again. This is in line with the value of elongation at 5% concentration. The tensile strength of a material is influenced by its constituent chemical bonds. According to Park et al. [7], the shape, the number of carbon atoms in the chain, and the number of hydroxyl groups contained in the plasticizer molecules will affect the mechanical properties (tensile strength and elongation) of a hydrogel. Plasticizer will decrease the hydrogen bonds within the hydrogel to increase the flexibility of the hydrogel where the increase flexibility of the tensile strength of the hydrogel will be smaller [8].

Chitosan-alginate hydrogel-lauric acid can be used as a medical material if it meets a certain standard of mechanical properties as shown in Table 1. Based on the research Jansen and Rottier

(1958), that threshold for medical materialists an elongation value between 17% - 207% and the value of tensile strength of between 1 MPa - 24 MPa. From the data analysis, then the chitosan-alginate hydrogel-lauric acid with a composition ratio of 4: 1: 3 and 4: 1: 4 (v / v) was selected as the best sample because it has a value of elongation of (17.78 ± 0.96) % and (28.89 ± 1.01) % with a value of tensile strength of (9.51 ± 0.76) MPa and (9.01 ± 0.65) MPa, fitting in the range as a medical material wound closure proposed by Jansen and Rottier (1958), wherein the wound closure can be applied to human abdomen.

3.4. Absorption ability test

Hydrogel dressings formed from the ring insoluble polymers. It has a high water content which allows this material to be ideal as cover for wounds [9]. Hydrogel does not dissolve in water because it only not only absorbs and releases water, but also provides nutrients proportionately when needed. Hydrogels can absorb water as much as 1000 times the weight of the hydrogel itself.

The addition of lauric acid in chitosan-alginate hydrogel made solution of PBS (Phosphate Buffer Saline) absorbed to increase, but the addition of lauric acid more than 4% of the hydrogel decrease its ability to absorb. This happens because of the addition of lauric acid that passes through the saturation limit causes excessive plasticizer molecules that are in a separate phase from hydrogel. Hydrophobic properties contained lauric acid result in a less efficiency, making it not compatible. This is supported by the results of tensile strength tests (tensile strength) which indicates the effective value of elongation of hydrogel with the addition of lauric acid is not more than 4% (w / v).

4. Conclusion

Based on the results obtained in this study, it can be concluded that:

1. Variations in the composition of the lauric acid plasticizer in chitosan-alginate hydrogel yields effect to the mechanical and physical properties of the material. The higher the addition of lauric acid plasticizer, the higher the thickness, elongation, and the ability to absorb the chitosan-alginate hydrogel-lauric acid increased, while the value of tensile strength decreased.
2. The most suitable characteristics of chitosan-alginate hydrogel-lauric acid for standards to be applied as a wound dressing is shown by chitosan-alginate hydrogel with the addition of 4% lauric acid, wherein the hydrogel has a thickness of (125.46±0.63) μm, elongation of (28.89 ±1.01)%, tensile strength of (9.01± 0.65) MPa, and capability of absorbing fluid of (601.45±1.24) %.

5. References

[1] Lim Chin K and Halim Ahmad 2010 Biomedical-Grade Chitosan in Wound Management and Its Biocompatibility In Vitro Chapter 2 p:19 Malaysia.

[2] Istiqomah Nurul 2012 *Pembuatan Hidrogel Kitosan-Alginat untuk Aplikasi Penutup Luka secara In Vivo* Skripsi Progam Studi Teknobiomedik Fakultas Sains dan Teknologi UNAIR Surabaya.

[3] Goosen MFA 1997 Applications of Chittin and Chitosan *Techonimic Publishing Co Inc Lancaster*.

[4] Siswanto Riki 2013 *Sintesis dan Karakterisasi Biokomposit Kitosan-Alginat sebagai Kandidat Membran pada Aplikasi Hemodialisa* Skripsi Progam Studi Teknobiomedik Fakultas Sains dan Teknologi UNAIR Surabaya.

[5] Kurnia W 2010 *Sintesis dan Karakterisasi Edible Film Komposit dari Bahan Dasar Kitosan, Pati dan Asam Laurat* Skripsi Progam Studi Fisika Fakultas Sains dan Teknologi UNAIR Surabaya.

[6] Van Vlack LH 2004 *Elemen-Elemen Ilmu dan Rekayasa Material* edisi ke-6 terjemah Sriati Djaprie Erlangga Jakarta.

[7] Park HJ et al 1993 Permeability And Mechanical Properties of Cellulose Based Edible Films *Journal Food Science* **53** (6): p1361-1370.

[8] Krochta JM 1994 Edible Coatings and Films to Improve Food Quality Pennsylvania: *Technomic*

Publising Co Inc.

[9] Dai Tianhong, Tanaka Masamitsu, Huang Ying-Ying and Hamblin Michael R 2012 *Chitosan Preparations for Wounds and Burns* : Antimicrobial and Wound Healing Effect National Institute of Health Public Access.

[10] Annaidh, Aisling Ni, Bruyere, Karine, Destrade, Michel Gilchrist, Mishael D, Ottenio and Meilanie 2011 Characterization of the Anistropic Mechanical Properties of Excised Human Skin *Elseiver* Ltd. All Rights Reserved.

Implementation of density functional theory method on object-oriented programming (C++) to calculate energy band structure using the projector augmented wave (PAW)

E Alfianto[1], F Rusydi[2*], N D Aisyah[2], R N Fadilla, H K Dipojono[3], M A Martoprawiro[4]

[1]Dep. of Computer Systems, Institut Teknologi Adhi Tama Surabaya, Jl Arif Rachman Hakim 100, Surabaya, Indonesia.
[2]Theoretical Physics Research Group, Dep. of Physics, Fac. of Science and Technology, Universitas Airlangga, Jl. Mulyorejo, Surabaya, Indonesia 60115
[3]Laboratory of Computational Material Design, Research Group of Engineering Physics, Institut Teknologi Bandung, Jl Ganesha 10, Bandung, Indonesia, 40132,
[4]Inorganic and Physical Chemistry Division, Dep. of Chemistry, Institut Teknologi Bandung, Jl Ganesha 10, Bandung, Indonesia, 40132

[*]Corresponding email address: rusydi@fst.unair.ac.id

Abstract. This study implemented DFT method into the C++ programming language with object-oriented programming rules (expressive software). The use of expressive software results in getting a simple programming structure, which is similar to mathematical formula. This will facilitate the scientific community to develop the software. We validate our software by calculating the energy band structure of Silica, Carbon, and Germanium with FCC structure using the Projector Augmented Wave (PAW) method then compare the results to Quantum Espresso calculation's results. This study shows that the accuracy of the software is 85% compared to Quantum Espresso.

1. Introduction

The research on material science increased rapidly for the last twenty years. It is caused by the increasing numbers of calculation and simulation method.[4]. They are Molecular Dynamic, Density Functional Theory, Hatree Fock, etc. All of these methods use the numerical method. Thus, we need the support of advanced computers with good software [6].

Generally, software is created using Fortran language [9]. Fortran language has advantages in numerical calculation. However, currently the Fortran language is less popular because of its complex usage [6]. This research presents a software which uses C++ programming language with Object-oriented programming rules. The created code is similar to mathematical formula (expressive software), hence the user will be able to understand each calculation process easily [8].

2. Computational method and implementation

2.1. Algorithm

Figure 1 shows a complete flowchart of eDFT, the name of this software. The process start from reading the input file and continue to define the basic variable. The basic variable contains basic information of calculation: the atomic number, the atomic mass, the number of electrons etc. After getting the basic variable information, we are able to calculate the sampling points and reciprocal lattice vector. The next process is to calculate potential and ewald energy.

In order to solve the Kohn Sham (KS) equation [3], we need to give an initial density. Initial density is needed to construct an effective potential. After solving KS equation, we will be able to get a new density. Then, the program will compare the new density to the initial density [2]. The comparison process is called evaluation. If the deviation is less than 0.000001, the program continue to calculate the total energy and eigenvalue.

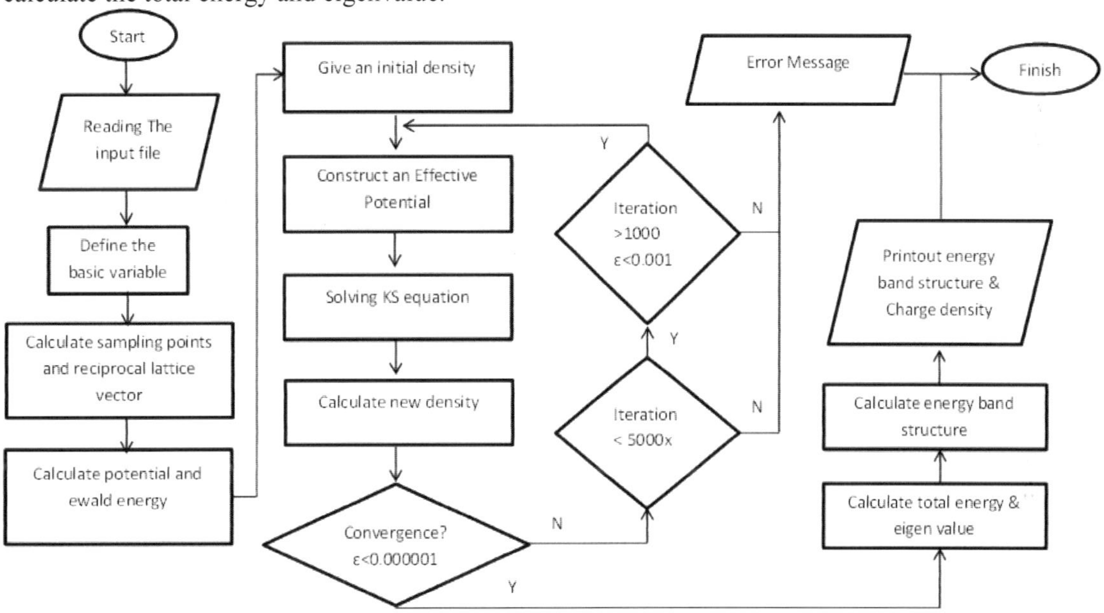

Figure 1. Algorithm of the eDFT

In order to find an energy band structure, we applied the Projector Augmented Wave (PAW) method for the calculation process. After the calculation is finished, the calculation results were printed in an output file. The output file contains the energy band structure and charge density data.

If the first evaluation process showed that the deviation is more than 0.000001, the second evaluation would be applied. The second evaluation wasto evaluate the iteration process. In this process, the iteration must be less than 5000 times. If the iteration wasmore than 5000 times, the calculation would stop and give an error message. Besides the evaluation, there was deviation checking. The deviation checking gives information about convergence: If the iteration was more than 1000 times and the deviation was more than 0.001, the calculation process would stop and give an error message. The aim of giving restriction on the iteration and the deviation is to avoid producing divergent values as the continuous calculation result.

2.2. Code Structure

eDFT is designed using object-oriented programming rule. The program is divided into many parts. The first part is master program and the second part is mathematical modules and installation instruction using Makefile. The mathematical module is divided into 6 files. Each file contains mathematical instructions and method to solve specific mathematical problem.

2.2.1. Master Program

Master Program is a code that handles all processes in the calculation. The master program is called eDFT. eDFT is divided into 3 part, eDFT.cpp, eDFT.h and eDFT.x. eDFT.cpp is a code based on C++ programming language which contains all calculation processes.

The expressive software is used to write the code. For example, we use the code of "std::vector<std::vector<int>> R=pac_Diag(Latice)" to define lattice vector. It means that we create vector into vector which contains integer variable "R". The letter R means diagonal value of Lattice matrix. Another example is we use "poisson(r,S,R,G2)" to calculate Poisson solver.

The control setting for output is also available in the master program. We are able to show or hide the calculation process when the job is run. This part is designed to be a simple tree. A simple tree means simple code structure. The aim of designing the simple tree is to facilitate the user to modify the program easily.

eDFT.x is a compiled file or binary file which is the result of compiling all codes. eDFT.h contains lists of header used in all calculation process. The header lists are from the standard C++ and library from the third party. The header lists are:

```
#include <vector>
#include <iostream>
#include <stdlib.h>
#include <math.h>
#include <complex.h>
#include <fftw3.h>
#include <string>
#include <fstream>
#include <sstream>
#include "OpenBlas/include/cblas.h"
```

2.2.2. Mathematical Module

Consisted of 7 files which were specifically designed for the calculation using DFT method. The first was"opmat.cpp". opmat stands for "Operasi Matriks (Matrix operation)". This file contained module for matrix operation, such as opmat_invers, opmat_kali, opmat_transpose, opmat_vmin etc. All the functions had opmat_function-name as the format.

The package for defining basic variable, sampling points and reciprocal lattice vector was poisson.cpp. The calculation process in the poisson.cpp used Matrix and Vector. We used opmat.h package to support calculation in poisson.cpp part. The use of opmat package would be more efficient if we compare it to the writing of every calculation code.

The DFT calculation process is contained in the ewald.cpp. The calculation process began with finding r at every point. In this calculation, we used matrix operation because all of the data were matrix and vector. Fast Fourier Transform was also used in this calculation to find "n" from many vector positions. Unfortunately, this calculation used complex numbers, so FFTW package from BLAS was used to calculate complex numbers[7].

The other support file was package.cpp.This part contains a package to calculate the value in the matrix "Prod", finding diagonal matrix "Diag" etc. This package is only aimed for support the DDFT calculation, but it is rarely used.

To get the print out of calculation result, out.cpp is created. This package uses standard output file in C++ (iostream). We modified the standard output (integer and double) to be a matrix and vector output. The illustration of file structure is shown in Figure 2

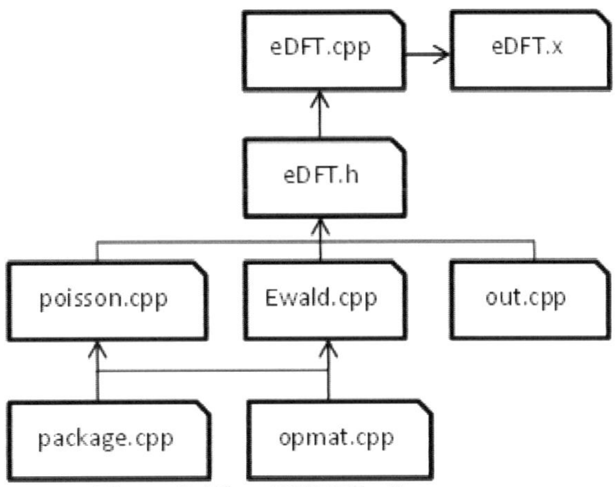

Figure 2. The eDFT File Structure

2.3. Finding sampling points and Reciprocal Lattice Vector

We have to find sampling points to calculate Reciprocal Lattice Vector [5]. The algorithm to define the sampling points and to calculate Reciprocal Lattice Vector is shown in Table 1.

Table 1. Algorithm to define Sampling points and to calculate Reciprocal lattice vector

Real-space Sampling points and Reciprocal lattice vector (RLV)
• Read vector S as an input file • Calculate capacity of vector in package.cpp (pac_prod(S)) • Find matrix M & N as a 3 dimensional representation of system • Calculate value of Sampling points (SP) as a matrix "r" and RLV as "G" matrix • Build matrix dimension from vector S as representation of input system r=M*pac_Diag(S) • Calculate RLV using G=2*PI*N/R

We need to calculate the charge density "n" to get information about charge distribution [5]. The solution of the charge density is

IOP Conf. Series: Journal of Physics: Conf. Series **853** (2017) 012043 doi:10.1088/1742-6596/853/1/012043

$$n = g_1(r) - g_2(r) = \frac{e^{-\frac{r^2}{2\sigma_2^2}}}{(2\pi\sigma_2^2)^{\frac{3}{2}}} - \frac{e^{\frac{r^2}{2\sigma_1^2}}}{(2\pi\sigma_1^2)^{\frac{3}{2}}}\sigma, \tag{1}$$

$$n = g_1(r) - g_2(r) = \frac{e^{-\frac{r^2}{2\sigma_2^2}}}{(2\pi\sigma_2^2)^{\frac{3}{2}}} - \frac{e^{\frac{r^2}{2\sigma_1^2}}}{(2\pi\sigma_1^2)^{\frac{3}{2}}}\sigma,$$

where r is space between two atoms from the proton, $g_1(r)$ and $g_2(r)$ are normalization from 3 dimensional Gaussian, and σ is radius of the system. The algorithm to calculate poisson solver is shown in Table 2.

Table 2. Algorithm to calculate Poisson solver

Poisson solver
• Matrix allocation "pac_prod(S) • Create matrix "one". The value of all element is 1. • Find value of "Sr" by times the matrix with 0.5. • Times matrix "one" with the "Sr" • Find Gaussian normalization "G" • Calculate Phi in reciprocal using FFTW • Find real number of Phi from reciprocal

2.4. Adding PAW

We added "eband.cpp" to find energy band structure. eband.cpp is a code based on Plane Augmented Wave (PAW). The calculation begins with solving the Schrodinger-like equation that have been finished in "ewald.cpp" and "poisson.cpp". The wave function with plane wave is given by

$$\psi_{nk} = \sum_G c_{nk}(G)e^{i(G+k)r}, \tag{2}$$

$$\psi_{nk} = \sum_G c_{nk}(G)e^{i(G+k)r},$$

where ψ_{nk} is wave function, n is band number, k is point in Brillion Zone, G is reciprocal Lattice Vector.

Wave function in real space is given by

$$\rho(r) = \sum_n \sum_k \psi_{nk}(r)^* \psi_{nk}(r), \tag{3}$$

$$\rho(r) = \sum_n \sum_k \psi_{nk}(r)^* \psi_{nk}(r),$$

where $\rho(r)$ is wave function in real space. Then, we calculated Hamiltonian from the plane wave matrix element, which is notated in dirac as follows:

$$\langle k+G'|H|k+G\rangle = |k+G|^2 + \langle k+G'|\hat{V}_H + \hat{V}_{xc} + \hat{V}_{pp}|k+G\rangle. \tag{4}$$

$$\langle k+G'|H|k+G\rangle = |k+G|^2 + \langle k+G'|\hat{V}_H + \hat{V}_{xc} + \hat{V}_{pp}|k+G\rangle. \tag{5}$$

$$\langle k+G'|H|k+G\rangle = |k+G|^2 + \langle k+G'|\hat{V}_H + \hat{V}_{xc} + \hat{V}_{pp}|k+G\rangle.$$

IOP Conf. Series: Journal of Physics: Conf. Series **853** (2017) 012043 doi:10.1088/1742-6596/853/1/012043

The symbol $\hat{V}_H$ is Hartree Potential, $\hat{V}_{xc}$ is potential from exchange and correlation, and $\hat{V}_{pp}$ is empirical pseudopotential. Then the Hartree potential in reciprocal space is

$$\langle k + G'^{|\hat{V}_H|} k + G\rangle = \frac{8\pi\rho(G' - G)}{|G' - G|^2} \tag{6}$$

$$\langle k + G'^{|\hat{V}_H|} k + G\rangle = \frac{8\pi\rho(G' - G)}{|G' - G|^2}$$

The algorithm to get the energy band structure is shown in Table 3.

Table 3. Algorithm to calculate Energy band structure

Energy Band Structure
• Calculate G vector
• Determine k-points
• Insert the initial charge density (Using Gaussian function)
• Calculate Reciprocal space using Fourier transform
• Calculate using self-consistency method (Calculate Hamiltonian)
o Calculate Hartree potential
o Calculate exchange and correlation potential
o Calculate pseudopotential
• Calculate band energy in every k-points.

3. Result and Discussion

The first calculation process to create G Vector which was shown in Table 1. The aim of this calculation is to plot the value in three dimension coordinate. The complete test of G vector is shown in Figure 3. The value test in Figure 3 used the length of cell of 6 Bohr and matrix input S = [20;25;30]. The calculation result shows that G vector has three-dimensional value.

The next test is the test for Poisson solver calculation package. The algorithm of this calculation is shown in Table 2. The result of Poisson test is shown in Table 4.

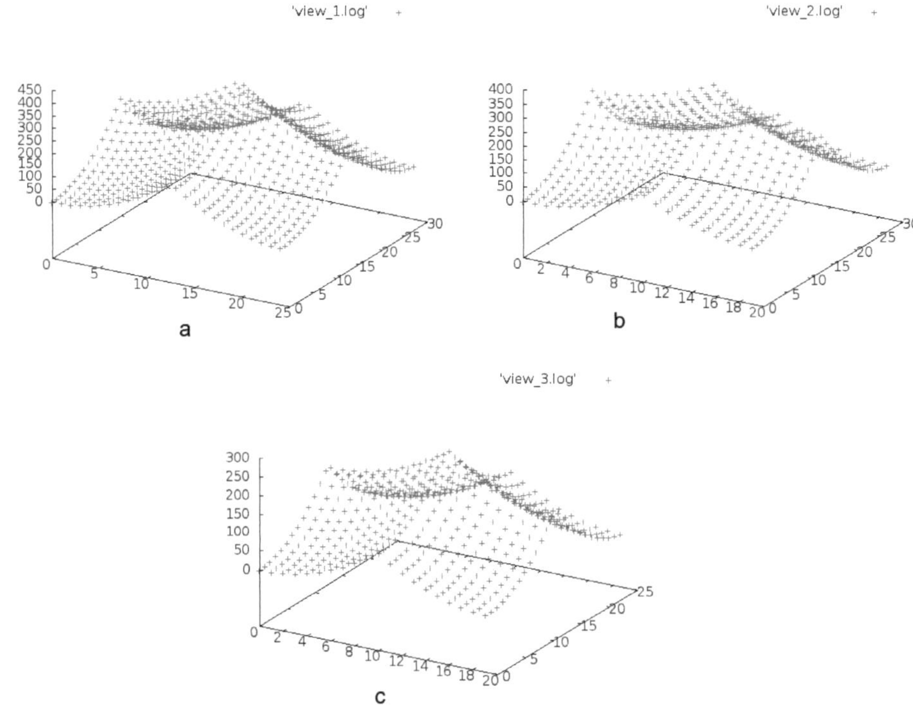

Figure 3. (a). G vector in slice x. (b) G vector in slice y. (c) G vector in slice Z

Table 4. Poisson calculation test results

	Analytic	eDFT
Gaussian Normalization	1	0.999781
Coulomb Potential energy	0.0551425	0.0551405

The final calculation was to find the energy band structure of Silicon as a system tests. The calculation result is shown as energy band structure graph at every k-points. The Silicon energy band structure graph as the result of eDFT calculation is similar to the energy band structure graph as the result of Quantum Espresso calculation. The similarities between them reaches 85%. Figure 4 shows us the energy band structure as the result Quantum Espresso calculation and eDFT. The trend of the graph is similar.

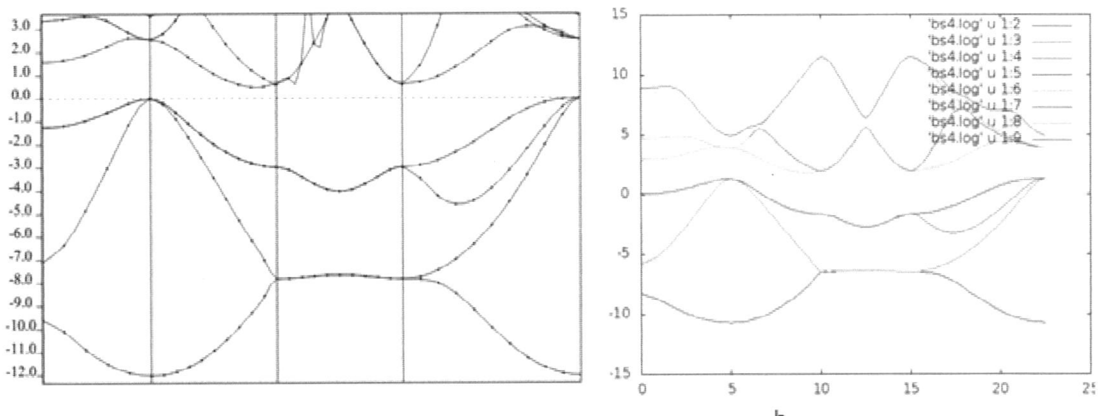

Figure 4. (a) Energy band structure of silicon calculate by Quantum Espresso, (b) Energy band structure of Silicon calculate by eDFT

4. Conclusion

We have implemented DFT which uses PAW method to calculate energy band structures. Silicon is chosen as the sample for the calculation. We have written the software using expressive software paradigm which can be easily understood by users. The result shows that our eDFT calculation result is 85% similar to Quantum Espresso calculation result.

Acknowledgement

This calculation of this work is supported by computer research facility of Computational Material Design laboratory, Institut Teknologi Bandung, Indonesia.

References

[1] Born M and Huang K 1988 *Dynamical Theory of Crystal Lattices* Oxford Classic Texts in the Physical Sciences Clarendon Press Oxford.
[2] Hohenberg and W Kohn 1964 *Inhomogenous electron gas* Phys Rev 136(3B):B864â871
[3] Kohn and L J Sham 1965 Self-consistent equations including exchange and correlation effects Phys Rev. 140 4A:A1333â1138
[4] SS David 2009 *Density Functional Theory A practical introduction* John Willey and sons Inc.
[5] S Ismail-Beigi and T A Arias Comp. Phys. Comm. 128 1.
[6] R Carry, Stevlana and G Shasharina 196 Comparison of C++ and Fortran 90 for Object-Oriented scientific programming *Computer Physics Communications* 105 (1997) 20-36
[7] Fathurahman F, Alfianto E, Dipojono, H K, and Martoprawiro, M A (2014, September). Acceleration of norm-conserving Pseudopotential Plane-Wave-Based DFT Calculation on GPU using CUDA. In *Proceeding of 3rd International Conference on Computation for Science and Technology* Vol. 2
[8] Alfianto, E. 2015 Implementasi Metode Teori Fungsional Kerapatan pada bahasa C untuk menemukan energi keadaan dasar berbagai atom. *Jurnal Arus Elektro Indonesia*, *1*(3).
[9] Ratcliff, L E, Grisanti, L, Genovese, L, Deutsch, T, Neumann, T, Danilov, D and Cornil, J 2015 Toward Fast and Accurate Evaluation of Charge On-Site Energies and Transfer Integrals in Supramolecular Architectures Using Linear Constrained Density Functional Theory (CDFT)-Based Methods. *Journal of chemical theory and computation* 11(5)

IOP Conf. Series: Journal of Physics: Conf. Series **853** (2017) 012044 doi:10.1088/1742-6596/853/1/012044

Study on conventional carbon characteristics as counter electrode for dye sensitized solar cells

Muhammad Noer Fajar and Endarko*

Physics Department, Institut Teknologi Sepuluh Nopember, Kampus ITS, Sukolilo, Surabaya, 60111, East Java, Indonesia

*Corresponding email address : endarko@physics.its.ac.id

Abstract. Activated carbon (AC), black carbon (BC), and graphite were deposited onto ITO (Indium Tin Oxide) glass for counter electrode application in Dye-Sensitized Solar Cells. SEM-EDX was used to observe and analyse the morphology and composition of electrodes. The results showed that the particle distribution of the graphite electrode observed was approximately 34% with a size of 1 to 2 μm and BC electrode about 20% have a size of 0.5 to 1 μm, while AC electrode has a size of 0 – 0.5 μm observed around 20%. AC electrode has a more porous and uniform particle aggregates compared to BC and graphite electrodes. The efficiency of the counter electrode was measured using the solar simulator. The highest efficiency was at 0.011516% for the counter electrode that was fabricated by AC. Meanwhile, black carbon and graphite electrodes were achieved at 0.008744% and 0.010561%, respectively. The results proved that the porosity and the uniform aggregate of the particles were the most significant factors to improve the performance of DSSC.

1. Introduction

Recently, DSSC has gained extensive attention due to their environment-friendliness, simple fabricating procedure, and low-cost compared to conventional Si solar cells [1-2]. A typical DSSC is composed of a nanoporous TiO_2 photoanode and a Pt counter electrode separated by an iodide/triiodide electrolyte [3]. The photoanode is usually prepared from TiO_2 nanoparticles on a transparent conducting oxide (TCO), while the counter electrode is a thin layer of Pt deposited on another TCO substrate [2].

As one of the most important components, counter electrode should possess mesoporous [4], high catalytic activity [5] and chemical stability against the corrosive electrolyte in the DSSCs [6]. Pt counter electrode has been extensively used as an efficient catalyst for reduction of I^{3-} ions in DSSC. Pt exhibits high electrochemical catalytic activity for triiodide reduction. However, Pt is scarce, expensive metal and has poor stability in corrosive I^-/I^{3-} redox system [2][7].

Recently, a various carbon-based material including graphene [5], graphite [8], and carbon nanotubes [9] have been studied as a low-cost replacement for Pt counter electrode. Carbon materials attracted much attention due to their high porosity, high electrochemical catalytic activity, low cost, and excellent corrosion resistance against the redox couple [10-12]. In this present study, we reviewed some carbonaceous material including graphite, activated carbon, and black carbon used as counter electrode and compared their performance in DSSC system.

2. Experimental design

2.1. Preparation of carbon counter electrodes

The carbon powders that were used in this research are graphite, black carbon, and activated carbon. Each carbon paste was prepared by mixing carbon powder with ethanol and was then stirred for 1 hour. The carbon pastes were deposited onto a conductive glass substrates-ITO glass using doctor blade method with a *thickness* around ~*10 μm*. The electrode was heated at 250°C for 30 minutes to remove the organic residue and stabilize the material.

2.2. DSSC's fabrication

TiO_2 photoanode was prepared according to the procedure previously reported by Lee *et.al*[12]. 0.7 g of TiO_2 powder is finely ground in a mortar, then 1.4 mL of distilled water were added. After10 minutes, 0.3 g of PEG, 0.7 mL of acetic acid, along with 1 mL and 0.7 mL Acetylacetone triton X-100were added. The TiO_2 paste was coated on 2.5 cm^2 area of ITO by a doctor blade technique with a thickness around10 μm, and was then sintered at 450°C for 30 minutes. The electrode was immersed overnight in a 0.05 mM ethanol solution of the N749 dye at room temperature. A sandwich-type DSC was fabricated with the dye-sensitized TiO_2 photoanode, electrolyte gel, and carbon as the counter electrode. **Figure 1** shows the structure of the DSSC's cell. The electrolyte consisted of 7 g PEG 1000, 25 mL chloroform and a liquid electrolyte homogeneously mixed and stirred with a magnetic stirrer while heated at80°C for 1 hour.

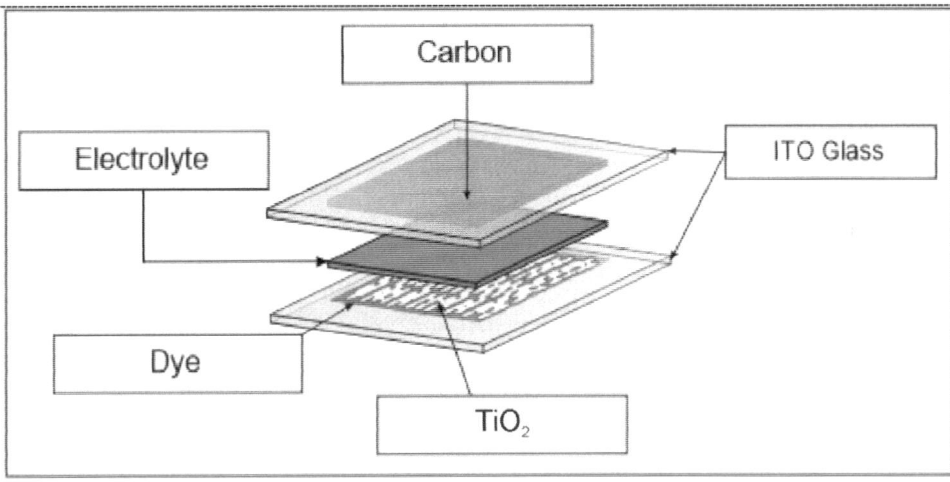

Figure 1. DSSC's formation

The surface morphologies and compositions of the counter electrodes were observed and analyzed using SEM-EDX whilst the current density-voltage (*J-V*) characteristics of assembly DSSC were measured using an I-V Keithley meter, under illumination at 100 mW/cm^2

3. Results and discussion

Figure 2 shows the top view SEM images of activated carbon (AC), black carbon (BC), and graphite electrode. These images confirmed that particle shape of all samples is spherical irregular particles. It also can be seen that porous structure was observed in activated carbon. The morphology of BC and graphite electrodes showed somewhat large intensity inhomogeneous agglomerations. These agglomerations were decreased by electrical contact, making the resistance of charge transfer increase[13].

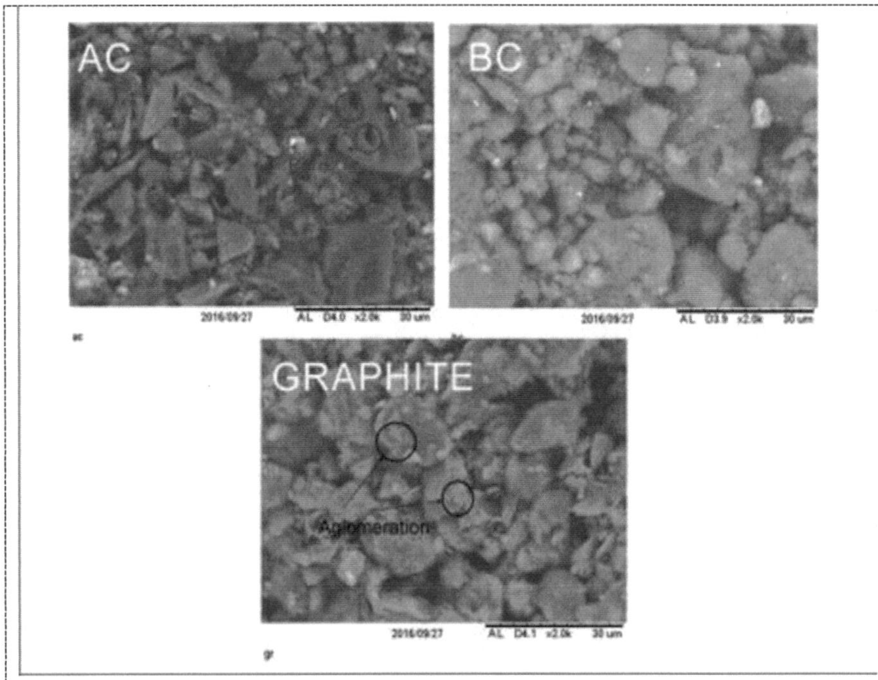

Figure 2. SEM images of AC, BC, and Graphite counter electrode

Figure 3 shows the particle size distribution observed in any carbon materials used in the present study. The distribution of particles with a size between 1 to 2 μm around 34% observed in graphite electrode and the size of 0.5 to 1 μm for 20% occurred in BC electrode, whilst a size of 0 to 0.5 μm for 20% was observed in AC. The distribution of particle gave large surface area so that the surface contact between carbon materials and electrolyte will increase the transport of electrons to the dye contained in photoanode [10]. The large particles resulted in limited transport of electron process from the outside toward the inner circuit and affect the performance of DSSC.

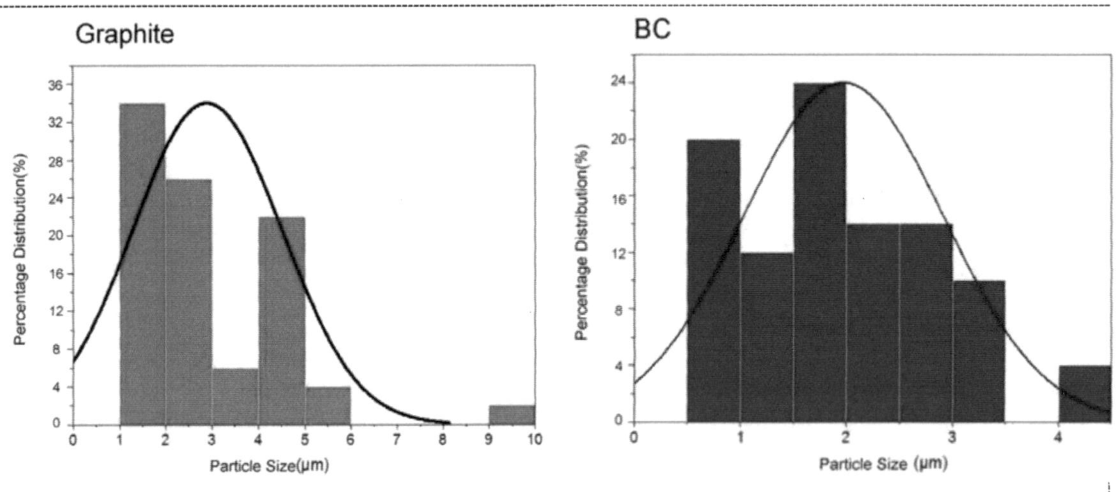

IOP Conf. Series: Journal of Physics: Conf. Series **853** (2017) 012044 doi:10.1088/1742-6596/853/1/012044

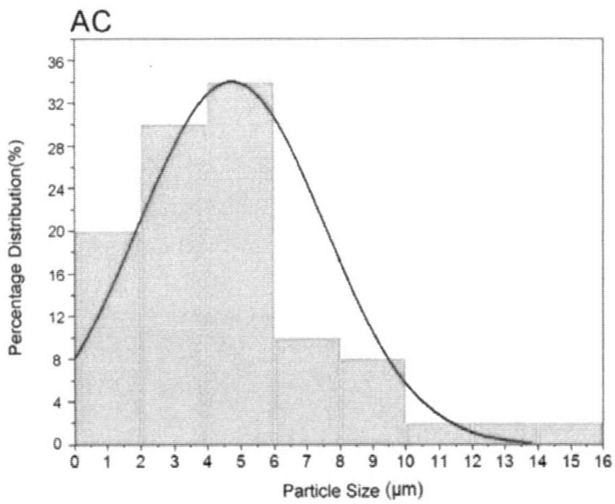

Figure 3. Distribution particle size of Graphite, BC, and AC counter electrodes

From Table 1, it can be deduced that AC electrode gave the highest efficiency compared to BC and graphite electrodes. This result agrees with the result that obtained by [14-15]. The best values for fill factor was shown by AC electrode. Fill factor is the performance values from DSSC obtained from short circuit current density and open circuit voltage. The fact proved that the performance of AC electrode is better than other carbons used in this study. This is because large particles would slow the rate of I3 reduction due to the high charge transfer resistance (RCT). The high RCT has a negative influence on the fill factor (FF) and shows poor energy conversion efficiency (η)[16].

Table 1. Photovoltaic parameters using conventional carbon of CE for dye-sensitized solar cell

CE	Voc/mV	Jsc/mA cm_2	FF (%)	η(%)
AC	0.581132	0.198157	0.897788	0.011516
BC	0.901621	0.220368	0.440098	0.008744
Graphite	0.701047	0.287001	0.524888	0.010561

[a]Jsc: short circuit current density, Voc: open circuit voltage, FF: fill factor, η: energy conversion efficiency

4. Conclusion

The study successfully demonstrated the counter electrode that was fabricated from activated carbon, black carbon, and graphite for DSSC applications. The activated carbon has the best performance compared to the black carbon and graphite electrodes due to the efficiency of dye-sensitized solar cells (DSSCs). The results showed that the activated carbon has more porous structure, smallest particle size, and highest fill factor compared to the other two. Due to this fact, we were able to deduce that the morphology of the surface on the counter electrode provides an effect on the performance of DSSCs.

References

[1] M Wu, X Lin, T Wang, J Qiu, and T Ma, *Low-cost dye-sensitized solar cell based on nine kinds of carbon counter electrodes Energy Env Sci* 2011.

[2] S Suhaimi, M M Shahimin, Z A Alamed, J Chysky, and A H Reshak, Material for Enhanced Dye-sensitized Solar Cell Performance: Electrochemical Application *electrochemsci* **vol. 10**, pp 2859–2871 2015.

[3] Z Huang *et al* Application of carbon materials as counter electrodes of dye-sensitized solar cells

elecom **vol. 9** pp 596–598 2007.

[4] E Ramasamy and J Lee Large-pore sized mesoporous carbon electrocatalyst for efficient dye-sensitized solar cells *ChemComm* **vol 46** pp 2136–2138 2010.

[5] J D Roy-Mayhew D J Bozym C Punckt and Ii A Aksay Fuctionalized Graphene as a Catalytic Counter Electrode in Dye-Sensitized Solar Cells *acsnano* **vol 4** pp 6203–6211 2010.

[6] Z Shi, K Deng, and L Li, Pt-free and efficient counter electrode with nanostructured $CoNi_2S_4$ for dye-sensitized solar cells **vol 5** p 9317 2015.

[7] Y-C. Wang *et al* FeS2 Nanocrystal Ink as a Catalytic Electrode for Dye-Sensitized Solar Cells *angewandte* **vol 52** pp 1–6 2013.

[8] G Veerappan, K Bojan, and S-W Rhee, Sub-micrometer-sized Graphite As a Conduvting and Catalytic Counter Electrode for Dye-sensitized Solar Cells **vol 3** pp 857–862 2011.

[9] F M Al-Marzouki, S Abdalla, and S Al-Ameer, Dye Sensitized Solar Cells with Low Cost Carbon Nanotubes Electrodes 2013.

[10] I Y Y Bu and J Zheng, A new type of counter electrode for dye sensitized solar cells based on solution processed SnO_2 and activated carbon *mssp* **vol. 39** pp 223–228 2015.

[11] Q MIAO, M Wu, W GUO, and T Ma, Studies of high-efficient and low-cost dye-sensitized solar cells **vol 4** pp 103–107 2011.

[12] J-K Lee *et al* Preparations of TiO_2 pastes and its application to light-scattering layer for dye-sensitized solar cells *J Ind Eng Chem* **vol 15** no 5 pp 724–729 2009.

[13] K Basu, D Benetti, H Zhao, L Jin, A Vetrone, and F Rosei, “Enhanced photovoltaic properties in dye sensitized solar cells by surface treatment of SnO_2 photoanodes.,” *Sci Rep* **vol 6** 2016.

[14] C Sampath Kumara Ranasinghe, E Nirmada Jayaweera, G Rajanya Ashoka Kumara, R Mudiyanselange Gamini Rajapakse, H Mudiyanselage Navarathna Bandara, and M Yoshimura, “Low-Cost Dye-Sensitized Solar Cells Based on Interconnected FTO-Activated Carbon Nanoparticulate Counter Electrode Showing High Efficiency” **vol 5** pp 361–368 2015.

[15] S-H Park, H-R Jung, and W-J Lee, “Hollow activated carbon nanofibers prepared by electrospinning as counter electrodes for dye-sensitized solar cells,” *Electrochimica Acta* **vol** 102 pp 423–428 2013.

[16] G Veerappan, K Bojan, and S-W Rhee, “Sub-micrometer-sized graphite as a conducting and catalytic counter electrode for dye-sensitized solar cells” *ACS Appl Mater Interfaces* **vol 3** no 3 pp 857–862 2011.

IOP Conf. Series: Journal of Physics: Conf. Series **853** (2017) 012045 doi:10.1088/1742-6596/853/1/012045

The effect of Cr_2O_3 doping on structures and dielectric constants of SiO_2-Bi_2O_3-B_2O_3-Na_2CO_3 glass based on silica gel of natural sand

M Diantoro[1,2,*], M B Zaini[1], Z Muniroh[1], Nasikhudin[1,3] and AHidayat[1,2]

[1]Department of Physics, Faculty of Mathematics and Natural Sciences, State University of Malang, Jl. Semarang 5 Malang, Indonesia
[2]Minerals and Advanced Materials Science Laboratory, State University of Malang, Jl. Semarang 5 Malang, Indonesia
[3]Department of Physics, Faculty of Mathematics, Gadjah Mada University, Jl. Sekip Utara Bulaksumur 55281, Yogyakarta, Indonesia

[*]Corresponding email address: markus.diantoro.fmipa@um.ac.id

Abstract: One of the abundant natural resources along the coastal lines of Indonesia is silica sand. One of the beaches which has a lot of silica content is Bancar-Tuban beach. Silica can be used as a raw material of glass that has multiple properties in optic, dielectric, and other physical properties by introducing specific dopants. Some oxides have been used as dopant e.g. Al_2O_3, Fe_3O_4, and NiO. However, there has not been any comprehensive study discussing the multiple properties of natural silica-sand-based glass with Cr_2O_3 dopant so far. A series of samples have been prepared, which mean two solid steps to state melting technique. Cr_2O_3 was selected as a dopant due to its potential to control its color and to increase the dielectric constant of the glass. The synthesis of silica (SiO_2) sand from BancarTuban beach was conducted through the sol-gel process. The composition varied as the addition of Cr_2O_3on $50SiO_2$-$25B_2O_3$-(6.5-x) Bi_2O_3-$18.5Na_2CO_3$-xCr_2O_3 (x = 0, 0.02, 0.04, 0.06 and 0.08mol), later called SBBN glass. The samples' characterizations of the structure and morphology were conducted through the use of XRD, and SEM-EDX. The measurements were done by using a DC capacitance meter in order to investigate the dielectric properties of the sample, under the influence of light. It is shown that addition of Cr_2O_3 did not alter the crystal structure but changed the structure of the functional bond formation. It is also revealed that the dielectric constant increased along with the increasing of Cr_2O_3. An interesting result was that the dielectric constant of the glass was quantized decreasingly as the increase of light.

1. Introduction

It is known that one of the abundant natural resources along the coastal lines of Indonesia is silica sand. One of the beaches which has a lot of silica content is Bancar-Tuban beach. The content of silica in BancarTuban, which is also called as natural sand beach, is 80.7% [1]This potential needs to be developed considering the relatively simple method of silica manufactur. In science, the use of silica is very diverse. One of silica utilizations is as the raw material of glass which exhibits multiple properties. The raw materials of typical glass consist of silica, soda ash, and limestone. Synthesis of pure-silica glass needs a relatively high temperature around 1760-1870°C. The use of high temperature for large scale production is not cost effectively, and needs more complicated equipment and technologies. Many efforts have been studied to lower the melting temperature, and to modify the glass properties itself. The addition of some oxides material such as Bi_2O_3, B_2O_3, and Na_2CO_3 has been used to reduce the melting temperature of silica. Another technique to lower the melting temperature is by reducing silica grains or by modifying in the form of amorphous. The sol-gelmethod can be implemented to prepare silica down to nanometer size as well as in the form of amorphous[2].

Glass is one of the essential material in human life. Rapid technological developments have also initiated the development of glass, e.g. the glass for household application to glass for sensors. Glass ceramics can be applied in materials science and biotechnology [3].Glass is widely used as window and drinking glasses for centuries. Nowadays, glass is used as touch screens, mobile phone screens, LCD screens, laser materials, memory devices, photonic devices, glass windshields of cars and others. A super hydrophobic glass has generated a widespread interest because of its applications in solar panel lighting [2].In solid glass, there is no contribution from dipole orientation. There is a distortion in the infrared region and an electronic vibration in UV region. In between of those regions, there is no strong coupling. This region is called a window transparency of the glass [4].

It was resulted from our previous work that oxidic dopant can change the glass properties[1]. An interesting outcome also showed that the light intensity rose to an interesting and prominent dielectric constant. The use of oxides material as dopant e.g. Al_2O_3, Fe_3O_4, and NiO has been studied with specific change of dielectric properties. A comprehensive study reducing melting temperature by silica sol-gel and the use of Cr_2O_3 dopant to modify the dielectric constant under visible light intensity has not been reported this yet so far. The use of Cr_2O_3 causes its stability in magneto-dielectric[5]. Chromium has a configuration of $3d^5\ 4s^1$ in paramagnetic ion transitio group [5]. It has several oxidation state as follows Cr^{3+}, Cr^{4+}, Cr^{5+}, Cr^{6+}. The most stabil valence is Cr^{3+} and Cr^{6+}[6]. $Cr_2O_3$is one of the oxide studied intensively due to its wide band gap of 3,3 eV[7]. The aims of this work is to study the effect of Cr_2O_3 and to intensity of visible light on the dielectric constant of the SBBN glass.

2. Research Methods

2.1 Tools and materials

The raw materials used in this works were Bancar beach natural sands as the source of silica, B_2O_3 Merck 100169, Bi_2O_3 Merck 101862, Na_2CO_3 and Cr_2O_3 Merck 102483. Several tools and equipment utilized in this study were digital balance HF-3000 (0.019), Brother Furnace XD-1700M, magnetic stirrer hot plate, beaker glass, ceramic mortarandpestel. For characterization, we operated Pan Analytical XRF and XRD, DTA, FEI SEM-EDX, FTIR, DC electric capacitance, standard Philips 100 W lamps and optical system.

2.2 Procedures

2.2.1 Preparation of SiO_2. Silica sand collected from BancarTuban beach was washed six times with distilled water, and then ingrained to 100 mesh. The fine grains sand that had been soaked with 32% HCl for 12 hours, then washed with distilled water to pH 7 and dried.

2.2.2 Sol-gel process. Sixty grams of silica sand and 1920 mL 5M NaOH was stirred by using a magnetic stirrer at a speed of 600 rpm and at 140°C until the water content was lost. 1920 mL of water was added while stirring for 3 hours and then, it was allowed to stand in order to precipitate. The filtrate was separated by sediment. The filtrate was stirred with a magnetic stirrer at a speed of 600 rpm while titrated with 5M HCl to pH 7. Then it formed a gel which was set aside to stand for up to 24 hours and dried at a temperature of 125°C. Furthermore, it was tested through the use of XRF to determine the percentage of purity of the silica.

2.2.3 Synthesis of SBBN glass.
The composition used in the synthesis is $50SiO_2$-$25B_2O_3$-(6.5-x) Bi_2O_3-$18.5Na_2CO_3$-xCr_2O_3, SBBNwith x = 0, 0.02, 0.04, 0.06, 0.08mol were mixed until it became homogeneous and heated via Brother Furnace XD-1700M at a temperature of 450 °C for 2 hours and was raised up to 950 °C for 1 hour in ceramics crucible, and then cooled to room temperature. Samples were separated from the crucible and their characterization was tested by using DTA, XRD, FTIR, SEM-EDX and capacitance meter.

3. Results and Discussion

3.1 Synthesis of SiO_2 via sol-gel
Silica powders were tested by using XRF. Table 1 shows the composition and content of silica sand before and after the synthesis via the sol-gel process. The silica quartz sand has 88.1% purity before the sol-gel process and increases up to 95.5% after the process. Based on the results, the sol-gel method was able to reduce some of the impurities contained in the silica sand, such as K, Ca, Ti and Fe. The small amount of impurity elements was still contained in the sol-gel due to insoluble of the compounds inNaOH and HCl or may need longer time and also smaller size of particles.

Table 1. The composition of silica before and after extraction by sol gel process.

Element	Si (%)	Ca (%)	Ti (%)	Fe (%)
before extraction	88.1	2.76	1.37	2.97
after extraction	95.5	1.2	0.44	1.92

3.2 The results of the DTA analysis
The DTA observation curves of SiO_2-B_2O_3-Bi_2O_3-Na_2CO_3 glass samples without doping (x=0) is demonstrated in Figure 1. From the DTA curve, it can be seen that the melted sample (Tm) is observed at 943.3 °C which is much lower than pure SiO_2 itself.

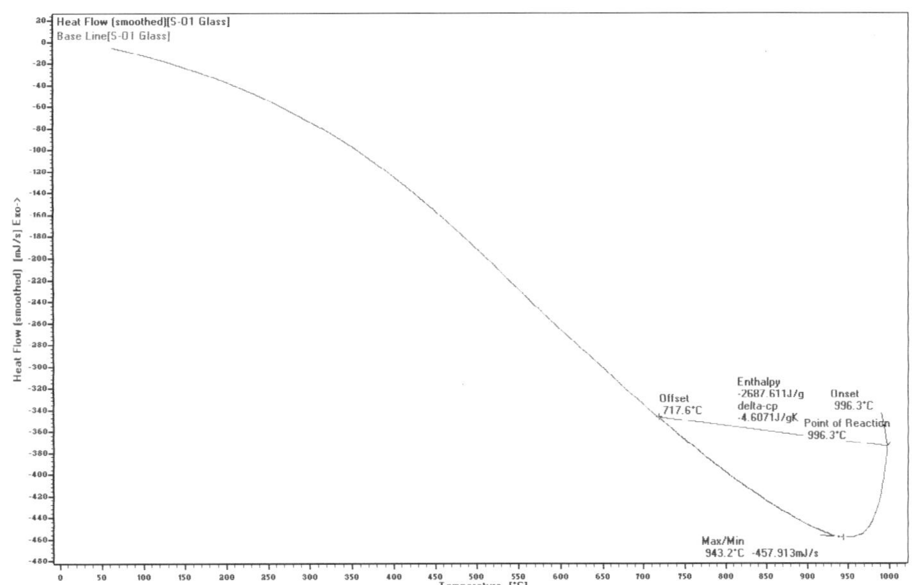

Figure1.DTA Test Results of SBBNGlass without Cr_2O_3 doping.

Regarding to Figure 2, in SBBN glass with Cr_2O_3 0.02 mol doping, there is a noticeable decrease in the DTA curve indicating the melting temperature of the sample glass. The optimum melting temperature increases at 949.2 °C. The introduction of.0,02 mol of Cr2O3 affects to rise the melting temperature of the glass as expected.

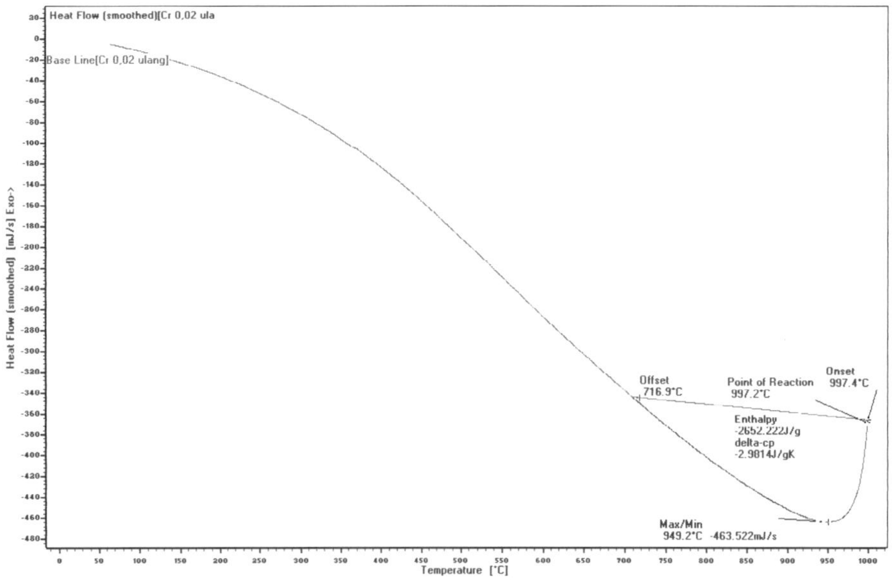

Figure 2.DTA Test Results of SiO_2-B_2O_3-Na_2CO_3-Bi_2O_3 Glass with 0.02 mol of Cr_2O_3 doping.

3.3 Results of XRD structure analysis

Based on the Figure 3, the diffraction pattern of the Cr_2O_3 sample doping of 0.02 mol, 0.04 mol, 0.06 mol and 0.08 mol indicates that the structure is in the glassy amorphous form (non-crystalline). The diffraction pattern shows only one weak peak at the 2θ position of about 27 degrees, illustrating the formation of a new phase of Bi_2O_3. One peak in the sample without doping at approximately 66 degrees (2θ position) is the culmination of boron oxide.

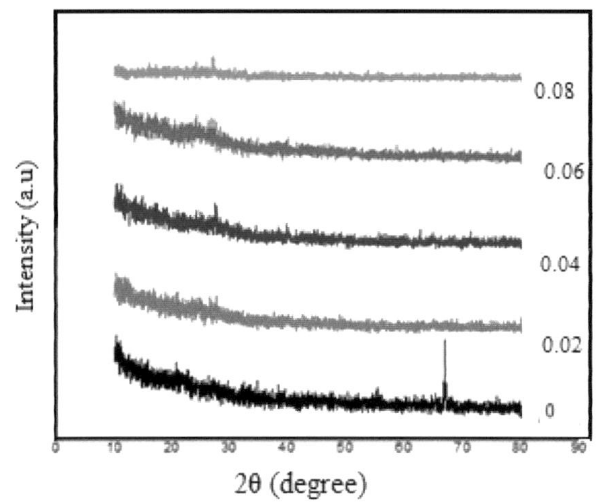

Figure 3. Comparison of XRD hystogramsofSBBN glass at various Cr_2O_3 doping.

3.4 FTIR Analysis Results
An FTIR transmittance spectra of un-doped and 0.02% Cr_2O_3 doped glass is displayed in Figure 4. The sample of glass without Cr_2O_3 doping (x=0) showed the Si-O-Si bond stretching vibration at a wavelength of 667 cm^{-1}, the vibration of Si-O-Si asymmetric at 1033 cm^{-1}, BO_3 unit at 1283 cm^{-1}, and the bending vibration is at 763 cm^{-1}.

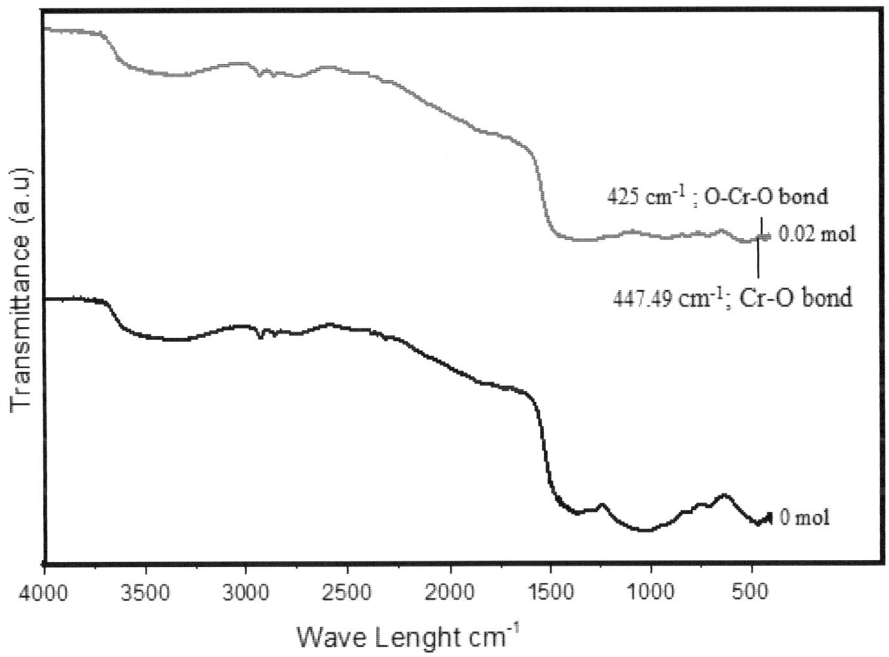

Figure 4.FTIR spectra of SBBN glass with x= 0and x=0.02mol doping.

In samples with x= 0.02 mol Cr_2O_3 doping at wave number 447.49 cm^{-1}, there is a Cr-O bond.The wave numbers are 472.56 to 698.23 Si-O-Si asymmetric vibration bond and the wavenumber at 425 cm^{-1} as O-Cr-O bond [8].The sample has also seen the modes of CrO_4^{2-} vibration at wave numbers of 862.18 and 879.54 cm^{-1}. Figure 4 describes that in the sample without doping in the range of wave number 800 cm^{-1} to 900 cm^{-1}, there is no peak detecting the presence of CrO_4^{2-} vibration. Also, in the range of wave number of 1300 cm^{-1} to 1400 cm^{-1}, there is no visible strong peak as well as samples with 0.02 mol Cr_2O_3 doping which should exist as the wave numbers are the BO_3 bond. It can be said that the doping sample with 0.02 molCr_2O_3 which generated a new peak in the range of wave number of 1300 cm^{-1} up to 1400 $cm^{-1}$is a composite material.

3.5 Analysis of SEM microstructure
Based on Figure 5, morphology of the glass with a 100-time magnification did not show any differences between samples of glass without Cr_2O_3 doping, and with 0.02 mol doping. There was no grains that was very similar in the whole area under inspection. However, overall, it was looked like a homogeneous sample. A several tiny grains at Figure 5a) and 5b) was actually not intrinsic in the samples. From the results of SEM-EDX above, even smooth small white granules can be seen, and white grain was detected from the Bi_2O_3.

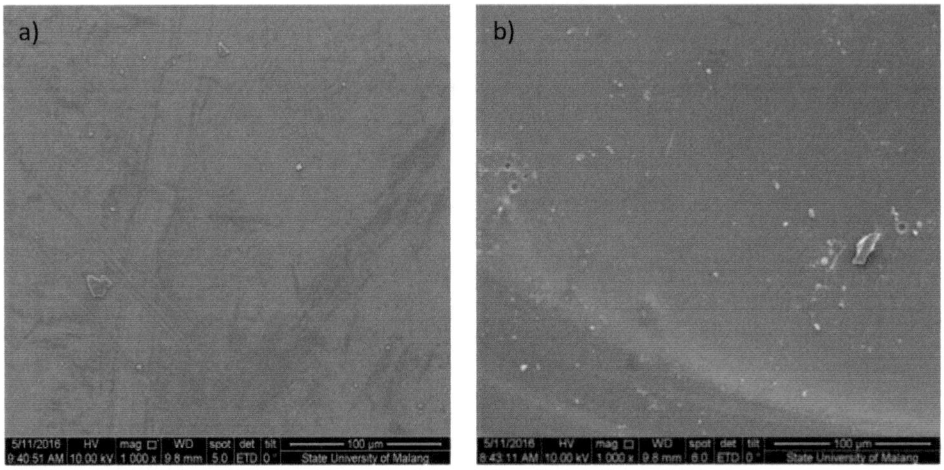

Figure 5. The result of SEM-EDX morphology SiO_2-B_2O_3-Bi_2O_3 Na_2CO_3 glass with Cr_2O_3 doping: a) x=0 mol and b) 0.02 mol.

3.6 Effect of the intensity of the glass dielectric constant

The dielectric constant influenced by the intensity of the glass was measured by using a series of LCR meters equipped with the lux meter. The lux meter is designated to measure the intensity of the visible light engaged. The value obtained from this measurement was electric capacitance (*C*) with respect to samples'surface area (*A*) and thickness (*d*). By using equation 1, the dielectric constant ($\kappa = \varepsilon/\varepsilon_0$) was presented in Figure 6. Figure 6 illustrates the effect of addition of Cr_2O_3 on the dielectric constant of $50SiO_2$-$25B_2O_3$-$(6.5\text{-}x)Bi_2O_3$-$18.5Na_2CO_3$-xCr_2O_3 glass on the intensity ranged from 0 to 2500 Lux.

$$\varepsilon_r = \frac{C.d}{\varepsilon_0.A} \qquad (1)$$

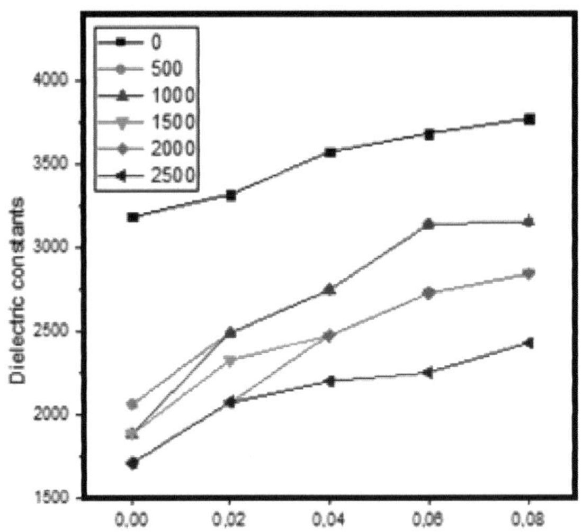

Figure 6. Dielectric constantof SiO_2-B_2O_3-Bi_2O_3 Na_2CO_3 glass against Cr_2O_3doping.

The addition of Cr_2O_3 doping significantly affected the dielectric constant. The greater the Cr_2O_3 were added, the greater the dielectric constant was. This was caused by a high dielectric constant in the Cr_2O_3 material, and since Cr_2O_3 has a centrosymmetric crystal structure[9], i.e. the high dielectric constant [10], so the addition of Cr_2O_3 doping into SiO_2-B_2O_3-Na_2CO_3-Bi_2O_3 glass resulted in the increase of dielectric constants. On the other hand, the intensity of light reduced the dielectric constant. The detail inferred dielectric constant measurement is shown in Figure 7.

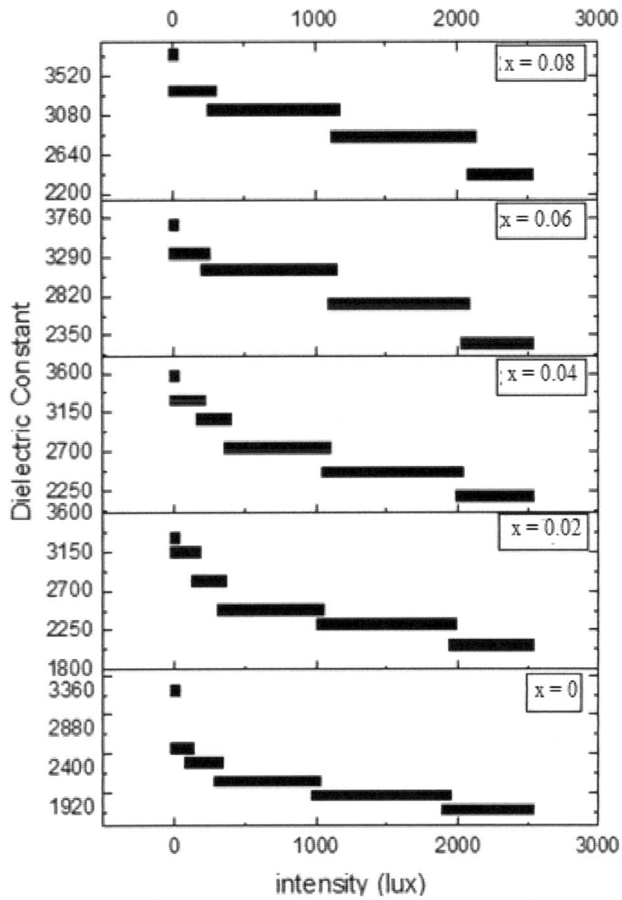

Figure 7. Dielectric Constant of SiO_2-B_2O_3-Bi_2O_3 Na_2CO_3 Glassas a Function of Intensity of Light.

The influence of the visible light intensity on the dielectric constant of the glass is depicted in Figure 7. It is significantly seen that the dielectric constant is inversely proportional to the intensity of use. The greater the intensity of use, the smaller the dielectric constant. Another distinct feature of Figure 7 is the quantization phenomena of the dielectric constant. This is in contrast to the influence of the increase of the Cr_2O_3 dopant. The decrease of the dielectric constant by light intensity as well as the quantization effect can be described as the increased volume of the crystal originated from the quantized of electron excitation by light. This mechanism may cause to the binding energy of the crystal and the further distance between atoms so that the dielectric constant decreases and quatized.

4. Conclusions

Based on the analysis and discussion in the previous sections, it can be concluded as follows; the addition of Cr_2O_3 does not alter the crystal structure of SiO_2-B_2O_3-Bi_2O_3-Na_2CO_3 yet functionally creates bond formation of the glass. It remains as a glass which is shown by XRD diffraction pattern.

On the other hand, its change on the structure of bond formation is indicated by the results of FTIR. The addition of Cr_2O_3 generally increases the dielectric constant of the glass. The most interesting fact is that the employment of light intensity causes the decrease of dielectric constant with some features of quantization phenomena at a specific intensity.

5. References

[1] M diantoro, P E Yuwita, D Olenka and Nasikhudin Sep. 2014 Fabrication of $CuAl_{1-x}MxO_2$ (M = Fe, Cr)/Ni film delafossite compounds using spin coating and their microstructure and dielectric constant *AIP Conf. Proc.* **1617**. 1, pp. 170–174.

[2] S Mor, C K Manchanda, S K Kansal and K Ravindra Feb. 2017 *Nanosilica extraction from processed agricultural residue using green technology J. Clean. Prod*, **143** pp. 1284–1290.

[3] S A M Abdel-Hameed and F H Margha Feb. 2012 Preparation, crystallization behavior and magnetic properties of nanoparticles magnetic glass–ceramics in the systems Fe2O3 CoO MnO2, Fe2O3 NiO MoO3 and Fe2O3 CoO V2O5 *J. Non-Cryst. Solids* **358**. 4 pp. 832–838.

[4] N K Mohan, M R Reddy, C K Jayasankar and N Veeraiah Jun. 2008 Spectroscopic and dielectric studies on MnO doped PbO–Nb2O5–P2O5 glass system *J. Alloys Compd.* **458**. 1–2 pp. 66–76.

[5] M M Abdullah, F M Rajab and S M Al-Abbas Feb. 2014 Structural and optical characterization of Cr2O3 nanostructures: Evaluation of its dielectric properties *AIP Adv.*, **4**. 2 p. 27121.

[6] G A Khater Sep. 2011 *Influence of Cr2O3, LiF, CaF2 and TiO2 nucleants on the crystallization behavior and microstructure of glass-ceramics based on blast-furnace slag Ceram Int.*, **37**. 7 pp. 2193–2199,

[7] Yuzheng Guo, Stewart J Clark and John Robertson 2012 Electronic and magnetic properties of Ti 2 O 3 , Cr 2 O 3 , and Fe 2 O 3 calculated by the screened exchange hybrid density functional *J. Phys. Condens. Matter*, **24**. 32 p. 325504.

[8] T Schneller, R Waser, M Kosec and D Payne 2013*Chemical Solution Deposition of Functional Oxide Thin Films springer*,

[9] R Vijay, P Ramesh Babu, V Ravi Kumar, M Piasecki, D Krishna Rao, and N Veeraiah, Jul. 2015 Dielectric dispersion and ac conduction phenomena of $Li_2O–Sb_2O_3–PbO–GeO_2$:Cr2O_3 glass system *Mater Sci. Semicond Process* **35** pp. 96–108,

[10] S A Makhlouf, Z H Bakr, H Al-Attar and M S Moustafa Apr. 2013 Structural, morphological and electrical properties of Cr2O3 nanoparticles *Mater. Sci. Eng. B*, **178**. 6 pp. 337–343.

Effect of temperature on the corrosion inhibition of iron in liquid lead using oxygen inhibitor: studied by MD simulation

Artoto Arkundato[1*], Fiber Monado[2] and Zaki Su'ud[3]

[1)]Physics department, Faculty of Mathematical and Natural Sciences, University of Jember, Jember
[2)]Physics department, Faculty of Mathematical and Natural Sciences, University of Sriwijaya, Palembang
[3)]Physics department, Faculty of Mathematical and Natural Sciences, ITB, Bandung

[*]Corresponding email address: a.arkundato@unej.ac.id

Abstract. For corrosion mitigation of steels used in a fast nuclear reactor power plant, oxygen gas is one of promising candidates of inhibitors. Many experiments have been conducted to reveal the mechanism of corrosion and mechanism of how to overcome the corrosion. In the previous work, we had shown computationally that the oxygen atom can be used to reduce the corrosion and we had predicted the oxygen contents. In the current work, not only to explore deeeper the ability of oxygen gas to reduce the corrosion, but also to include the variation of used temperature. We still used iron material to represent a real steels. Using MD (molecular dynamics) simulation based on the Lennard-Jones interaction potential, we sought to understand the concentration of oxygen gas as variation of temperature used in the reactor for the best corrosion mitigation. From this work, we conclude that the temperature does not give effect in related with how concentration of injected oxygen. The temparature merely affects to rise the diffusion coefficient of iron in liquid lead, yet it does not influence how much oxygen needed for corrosion mitigation. In this work, all simulations on different series of temperatures (1023^0K, 1073^0K, 1123^0K, 1173^0K) reveals that oxygen content of 0.1151wt% will cause the lowest corrosion level of iron in liquid lead.

1. Introduction

Molecular dynamics (MD) simulation is a powerful method in computational material sciences. Using MD simulation on many properties of materials under investgation may be predicted or calculated based on specific condition of temperature, pressure, concentration, and others. The purpose of this current study is to determine the corrosion phenomena and its material mitigation, especially the materials used in application of fast nuclear reactors (FNR) design. In FNR that uses liquid metals as a coolant material, for instance lead liquid, the coolant material may cause high corrosion level for steels besides of its important benefits. Many experiments showed the existing high corrosion on the surface of steel materials when they were immersed into high temperature liquid lead coolant [1]. Hence, the focus of the recent researches is to find the proper steels materials or a proper mitigation method to overcome the corrosion phenomena. One of the powerful methods to support the researches is MD simulation. In our previous works, we had shown that iron (representing steels) corrosion can be reduced or minimalized using injection of oxygen/nitrogen for specific and proper concentration [2-7]. We applied MD simulation to predict how much the proper oxygen concentration that would minimize

IOP Conf. Series: Journal of Physics: Conf. Series **853** (2017) 012046 doi:10.1088/1742-6596/853/1/012046

the corrosion at the lowest level. However, we attempted to examine it merely at 750 C in this research. The focus of the current research is to study the effect of temperature variation during oxygen injection for different concentrations (in %wt). It is designated to find the relation between temperature and required oxygen concentration resulting on the lowest corrosion level of iron. We are still studying the iron to simplify the problem.

1.1 Potential function

One of the important input of MD simulation is potential interaction supporting the interactions among atoms within the material. In this study, we found Fe-O-Pb system that is iron-oxygen-lead atoms system; where Fe is a solid BCC metal, O is oxygen gas and Pb is liquid lead. To simplify the study, we apply Lennard-Jones potential to support the whole atomic interaction inside material system. For liquid or gaseous state, the interaction of atoms is usually represented by Lennard-Jones potential which in general the form is as below,

$$u(r)=k\varepsilon\left[(\sigma/r)^{n}-(\sigma/r)^{m}\right] \tag{1}$$

where both σ and ε are the potential parameters and k is a proportional constant in the form of:

$$k=\frac{n}{n-m}\left(\frac{n}{m}\right)^{m/(n-m)} \tag{2}$$

For $n = 12$ and $m = 6$, then we have a popular (12-6) Lennard-Jones potential function,

$$u(r)=4\varepsilon\left[(\sigma/r)^{12}-(\sigma/r)^{6}\right] \tag{3}$$

Following the popular **Lorentz-Berthelot** mixing formula, the interaction among different atoms can be parameterized by:

$$\sigma_{AB}=(\sigma_{AA}+\sigma_{BB})/2 \tag{4}$$

$$\varepsilon_{AB}=\sqrt{(\varepsilon_{AA}\times\varepsilon_{BB})} \tag{5}$$

Table 1. The Value of Potential Interaction

Pair Interaction	ε (eV)	σ (Å)
Fe-Fe	0.4007	2.3193
Pb-Pb	0.1910	3.1888
Fe-Pb	0.2766	2.7541
Fe-Pb	0.0639	2.8737
Pb-O	0.0441	3.3084

1.2 Molecular dynamics program

Basically, the MD simulation will solve the Newton 2nd Law of motion, $\boldsymbol{F} = d\boldsymbol{p}/dt$ for specific potential function $u(r)$ numerically. The relation between F and $u(r)$ could be facilitated by $F = -du/dr$. Many MD simulation programs have been developed either simple MD or powerful MD that could be used to compute many properties of materials. In this work, we use Moldy molecular dynamics program created by Keith Refson[8].

2. Simulations and calculations

To understand the effect of oxygen for corrosion inhibition, we calculated the diffusion coefficient $D(T)$ of iron in liquid lead environment, using formulas:

$$MSD = \left\langle \left| \vec{R}(t) - \vec{R}(0) \right|^2 \right\rangle \quad \text{(Mean square displacement)} \tag{6}$$

$$D = \lim_{t\to\infty} MSD/6t \quad \text{(Einstein relation)} \tag{7}$$

Simulations ran for specific ensamble (NPT or NVT or NVE), temperature, pressure, number of atoms, simulation box, molecular structure of material, density of material, number of integration step and time step. In every simulation, we applied the different temperature on different concentrations of oxygen gas. Then, we calculated the diffusion coeficient of iron. We looked for the condition of the lowest diffusion coefficient of iron for certain oxygen concentration which was injected into liquid lead. In our work, we used NPT simulations.

Another important things of MD simulation is equilibration condition. All calculations should be done after simulation achieved equilibration condition. We determined this condition by plotting the energy of system vs integration step (see Fig.1).

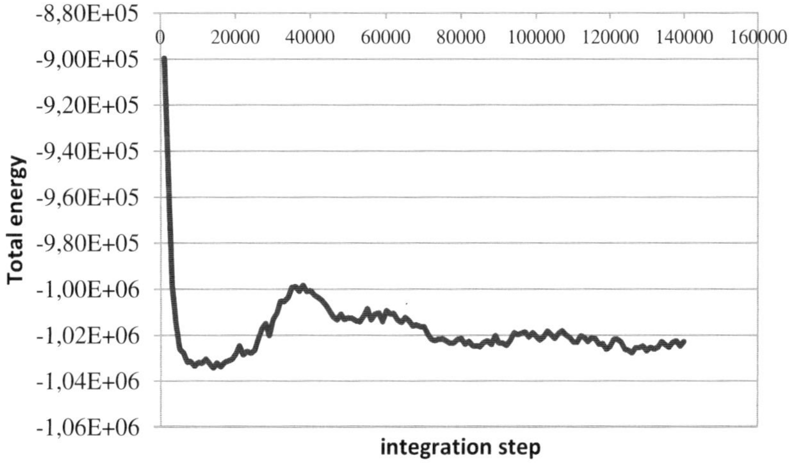

Figure 1. Equilibrium curve of total energy

With Moldy, we can find the information of equilibration condition by typing from terminal with the command =>$moldyext –f 4 name_of_Moldy's_output . From Fig. 1, we computed the properties after 40000 step of integration for better result.

2.1. The material system for moldy input

To realize the simulations, we prepared the material system (Fe-O-Pb) as in Fig. 2 where all the coordinates of material atoms was stored in specific file. The detail number of atoms for every simulation was illustrated in Table 2. Hence, we simulated the material system for every temperature (1023^0K, 1073^0K, 1123^0K, 1173^0K), for every wt% of oxygen (0.0, 0.038, 0.058, 0.0771, 0.1155, 0.1552, 0.1938 nad 0.231). Each simulation ran for pressure 0 atm, for 100000 integration step, for 0.0001 ps of time mesh and for begin-average 40000. We also used the Nosé–Hoover thermostat for constant-temperature molecular dynamics simulations, Andersen constant pressure molecular dynamics simulation, and Beeman algorithm.

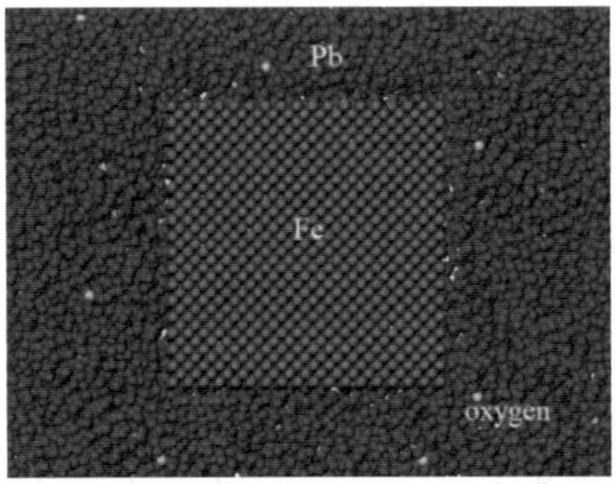

Figure 2. Preparation of material system before simulation

The number of atoms for material system of simulation are as follows:

Table 2. The Simulation Composition of Atoms in Material System before Simulations.

Iron (bcc)	10745	10745	10745	10745	10745	10745	10745	10745
Lead (liquid)	45006	45006	45006	45006	45006	45006	45006	45006
oxygen	0	226	340	450	674	906	1132	1348
Oxygen (in #%)	0.0	0.5	0.75	1.0	1.5	2.0	2.5	3.0
oxygen (in wt%)	0.0	0.038	0.058	0.0771	0.1155	0.1552	0.1938	0.231

3. Results and discussions

Simulations were done using Moldy MD program developed by Keith Refson [8]. This program could be downloaded from website shttp://ccl.net/cca/software/SOURCES/C/moldy/.

Figure 3 illustrates the graphs of iron diffusion D in liquid lead with oxygen injection for different temperatures T calculated by Equation 6 and 7. MSD values can be computed by using Moldy utilities as in manual [8].

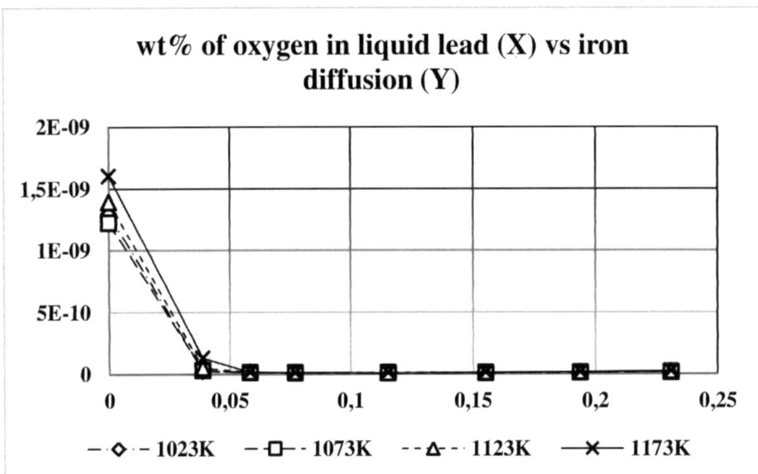

Figure 3(a). Graph of wt% of oxygen vs iron diffusion $D(T)$ in m^2/s.

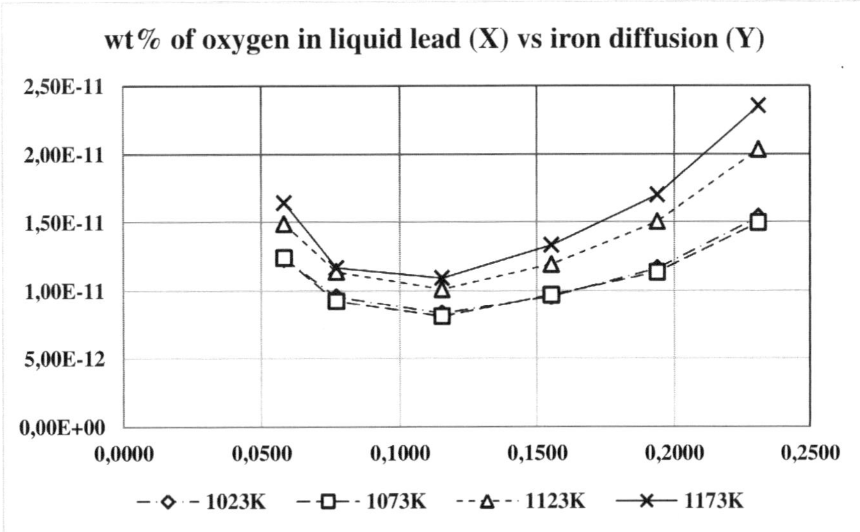

Figure 3(b). Zoom of graph 3(a) for wt% of oxygen more than 0.05wt%

By MD simulation, we can compute the diffusion coefficient of iron in liquid lead as in Fig. 3 (a) and (b) above. Experimental data [1] shows the iron diffusion in liquid lead is about ~ 2.8 × 10^{-9} m/s^2 for temperature 1023K. The use of 0.1155wt% of oxygen injected into liquid lead reduces the diffusion coefficient significantly from ~ 10^{-9} m/s^2 to ~ 10^{-11} m/s^2 and this occurs for all temperatures. This is about temperature does not effect for the best corrosion inhibition using 0.1155wt% of oxygen.

We could see from these figures that for all series of temperatures (1023^{0}K, 1073^{0}K, 1123^{0}K, 1173^{0}K) portray the similar trend, and the lowest iron diffusion (corrosion) occurs when the injected oxygen use concentration on 0.1155wt%. From these graphs, the condition in any temperature requires the oxygen content of 0.1155wt% to obtain the lowest iron diffusion (the lowest corrosion attack). The structures of iron after simulation without and with oxygen content demonstrating these combination (0wt%, 1023^{0}K), (0.038wt%, 1023K) and (0.1155wt%, 1023^{0}K) could be seen from Fig.4 as below:

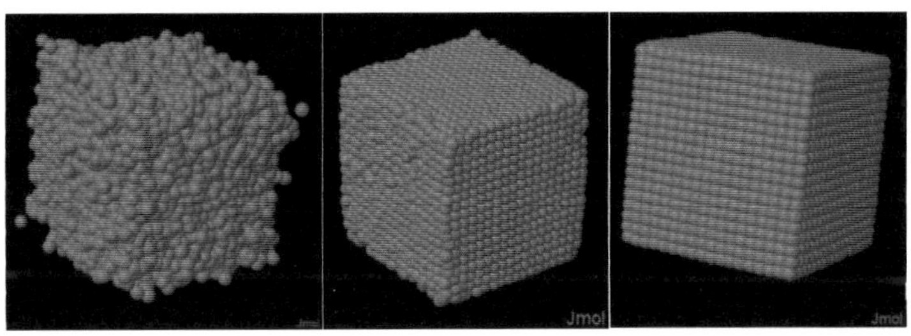

Figure 4. Effect of Oxygen on Iron BCC Structure Stability for Simulation at 1023K (by Jmol) [9]

From the Fig.4 (left), the iron BCC structure after simulation 100000 step without oxygen content shows heavy corrosion. The injection of 0.1151wt% of oxygen (right) shows the best reduction of iron corrosion, whereas the injection of 0.038wt% (middle) still shows relatively high corrosion. Quantitative representation about the effect of oxygen on temperature 1023K, reffering to Fig.4 can be summarized as Table 3.

Table 3. Quantitative Representation about Effect of Oxygen for Temperature 1023^0K

Oxygen content in liquid lead		# Fe before simulation		Fe in BCC after simulation	
# of atom	in %	# of atom	in %	# of atom	in %
0	0%	10745	100%	2739	25.5%
226	0.50%	10745	100%	6739	62.7%
674	1.50%	10745	100%	7471	69.5%

The number of atoms of iron after simulation by oxygen injection (Table 3) can be determined by Ovito code using facility of CNA common neighbour analysis [9].

4. Conclusions and remarks

By applying molecular dynamics simulation, this study proves that corrosion of iron in liquid lead can be reduced maximally using oxygen injection with content of 0.1151wt%. This oxygen content is independent from adjusted temperatures when the corrosion occurs. These simulations were conducted using Lennard-Jones potentials that may be not accurate for metal systems as iron in liquid lead. However for early prediction, we show that the oxygen contents for corrosion mitigation of iron in liquid lead are not affected by adjusted temperatures. In the other words, temperature has no effect towards the influence of the oxygen contents on corrosion mitigation. Eventually, this conclusion needs further and deeper studies for complete explanation of liquid corrosion inhibition and phenomena.

Acknowledgement

We thank to KEMENRISTEK DIKTI for a part of financial support of the research through the Hibah Fundamental grant.

References

[1] Zhang J and Li N 2008 Review of the studies on Fundamental Issues in LBE *J. Nucl.Mater.* 373 351-377

[2] Arkundato A, Su'ud Z, Mikrajudin A and Widayani 2012 Computational study: Reduction of iron corrosion in lead coolant of fast nuclear reactor *AIP Conference Proc.* **1454** pp.65 USA

[3] Arkundato A, Su'ud Z, Mikrajudin A, Widayani and Celino M 2012 Numerical Study: Iron Corrosion-Resistance in Lead-Bismuth Eutectic Coolant by Molecular Dynamics Method *AIP Conference Proc.* **1448** USA

[4] Arkundato A, Su'ud Z, Mikrajudin A and Widayani January 2013 Study of liquid lead corrosion of fast nuclear reactor and its mitigation by using molecular dynamics method, *International Journal of Applied Physics and Mathematics* **3**. 1 Singapore

[5] Arkundato A, Su'ud Z, Mikrajudin A and Widayani 2013 Molecular dynamic simulation on iron corrosion-reduction in high temperature molten lead-bismuth eutectic, *Turkish Journal of Physics*, DOI: 10.3906/fiz-1112-12.

[6] Arkundato A, Su'ud Z, Mikrajudin A, Widayani and Celino M 2013 Inhibition of iron corrosion in high temperature stagnant liquid lead: A molecular dynamics study **62** Pages 298–306, *Annals of Nuclear Energy* Elsevier

[7] Arkundato A, Su'ud Z, Sudarko, Hasan M and Celino 2015M Molecular dynamics simulation of corrosion mitigation of iron in lead-bismuth eutectic using nitrogen as corrosion inhibitor *Journal of Physics: Conference Series* **622** 012009 IOP Publishing.

[8] Refson K, 2000 *Moldy: a portable molecular dynamics simulation program for succesive and parallel computers* Comp Phys Commun, **126** (3), 309-328.

[9] Jmol: an open-source Java viewer for chemical structures in 3D. http://www.jmol.org/

[10] Stukowski A 2010 *Visualization and analysis of atomistic simulation data with OVITO - the Open Visualization Tool* Simul. Mater.Sci. Eng. **18.**

Microstructure and dielectric properties of cellulose acetate-ZnO/ITO composite films based on water hyacinth

M Diantoro[1,2,*], A A Mustikasari, NWijayanti, C Yogihati and A Taufiq[2]

[1]Universitas Negeri Malang, Faculty of Mathematics and Natural Science, Jl. Semarang 5 Malang, Indonesia
[2]Laboratory of Mineral and Advanced Materials, Universitas Negeri Malang, Jl. Semarang 5 Malang, Indonesia

*corresponding email address: markus.diantoro.fmipa@um.ac.id

Abstract. The electrical properties of Cellulose Acetate (CA), especially extracted from water hyacinth, is rarely informed. CA is generally more stable compared to its cellulose. It has a good potential for electronic application with specific modifications such as inducing metal oxide. A combination of intrinsic properties of Zinc Oxide (ZnO) and CA is expected as a great potential for electrical and optical applications. CA-ZnO/ITO composite film was investigated in relation with its structure, dielectric constant, and the effect of light intensity on their dielectric constant. CA-ZnO composite films were prepared with different mass of ZnO i.e. 0; 0,02; 0,04; 0,06 and 0,08 grams. CA-ZnO solution was synthesized via the mixing method with PEG:DMF solvents by using a magnetic hotplate stirrer with the rotation rate of 1500 rpm at 80°C. The CA-ZnO solution was then deposited onto ITO/glass substrate by using spin coating technique. The CA-ZnO/ITO films were annealed at 160 °C to remove the remaining solvents. The effects of ZnO composition on the structure (crystallinity and morphology) and dielectric constant properties were investigated by using X-Ray Diffractometer, Scanning Electron Microscopy, Fourier Transform Infrared Spectroscopy, and LCR meter. It was shown that cellulose can be isolated from water hyacinth with the yield of 67,72 % by Chesson method and can further be transformed into CA. The X-ray diffraction pattern showed that there are 2 phases formed i.e. CA and ZnO. Furthermore, greater ZnO amount increased the crystallinity of composite films. The CA-ZnO films exhibit porous films with ZnO distributed on the CA surface films. Therefore, ZnO increases the dielectric constant of CA-ZnO composite films.

1. Introduction

Water hyacinth is a common weed that grows rapidly in many areas of calm and shallow waters. The rapid growth of the weed can reduce its ecological functions such as silting rivers, decreasing biodiversity, inhibiting the flow of the river, and decreasing the solubility of oxygen in the water[1]. One way to overcome this problem is by using water hyacinth for traditional household products, such as traditional carpets, sacks, wallets and many other souvenirs. Water hyacinth may consist of 25-60% cellulose, 8–33% hemicellulose, and 10-17% lignin ([2];[3]). Cellulose can be used as emulsifier, adhesive, detergent, textile, cosmetic, paper materials, and drug releasing agent [4].

Cellulose has great mechanical and optical properties. On the other hand, cellulose is insoluble in many solvents, even in organic solvents. This characteristic limits its uses. However, it can be solved by a small modification which involves synthesizing the cellulose to its derivative called cellulose acetate (CA) [6]; [7].CA can be synthesized through acetylation process. CA exhibits dielectric properties which are originated from hydrogen bond[8]. The properties of CA are biodegradable and biocompatible as cellulose. Nevertheless, CA shows better optical and mechanical properties than cellulose [9]. CA can be applied as a fibre, film, laminate, adhesive, coating, and raw material for plastic production. The development of physical properties of this material can be done by adding polymers, metals, or oxides such as AgNPs, AuNPs, ZnO, and hydroxy gallium. This procedure may change its morphology, structure, hydrophobicity, and antibacterial properties[10];[11]; [12];[13]; [6];[14]; [15].

Many researchers have been widely using ZnO for many diverse studies and applications. ZnO has more unique properties than other semiconductor compounds due to its optoelectronic properties. Nowadays, ZnO is applied as a photovoltaic, gas sensor, clinic application, solar cell, piezoelectric nanogenerator, electric transducer, and LED[12]; [16]. ZnO is being widely used for electronic applications and CA is commonly used for medical applications. Currently, there are still very few studies concerned with synthesizing nanoCA-ZnO/ITO composite film for electronic applications which have been reported comprehensively.

In the present work, nanoCA-ZnO/ITO composite film was synthesized through several steps. The general aim of this work is to explore the information related to electronic properties especially its dielectric constant which is induced by light intensity. Firstly, CA was synthesized by extracting cellulose from water hyacinth. Furthermore, cellulose was acetylated to produce nano CA. Nano CA and ZnO were mixed in a DMF:PEG solvent by using a magnetic stirrer at 80°C. The Nano CA-ZnO solution was deposited onto ITO surface as a substrate by means of spin coating method. The Nano CA-ZnO/ITO composite film was characterized via FTIR, SEM-EDX, XRD, and LCR-meter which show its microstructure, morphology, and dielectric constant.

2. Materials And Methods

The raw materials used in the synthesizing process were water hyacinth, toluene 96%, ethanol 96%, sodium chloride 3 wt%, NaOH 4 wt%, acetic acid glacial p.a, sulphate acid p.a, DMF (dimethyl formamide) 99,8%, acetic anhydrate, zinc acetate dihydrate p.a, deionized water, methanol, and distilled water. The materials were provided by Merk and Aldrich, Germany. The instruments used were soxhlet apparatus, ultrasonic cleaner power sonic405 model LUC, SCILOGEX MS-H280-Pro hotplate stirrer, furnace 48000 thermolyne, and spincoater TC100 precision.

2.1. Synthesizing nano CA-ZnO/ITO composite film

CA-ZnO/ITO was synthesized with several steps. The process started with isolating cellulose from water hyacinth, transforming the cellulose into cellulose acetate, preparing nano-ZnO, and fabricating the CA-ZnO/ITO composite film.

2.2. Isolating cellulose from water hyacinth.

The water hyacinth was blended using a blender and was filtered by means of a 50-sized mesh filter. The dried water hyacinth is used as raw materials of isolating cellulose. Firstly, dewaxing process was done using toluene/ethanol (1:1) as solvent and used by soxhlet apparatus, followed by the bleaching 1 process which used $NaClO_2$ 3 wt% as the solvent for 4 hours at 50 °C. Then, the hemicellulose was removed by NaOH 4 wt% for 3 hours at the same temperature. The next process is de-lignification which uses NaClO 3 wt% for 4 hours and at the same temperature. Finally, the hydrolysis process was performed by using HCl. It was washed until the pH was neutral and dried at 80 °C.

2.3. Synthesizing cellulose acetate.

Cellulose acetate was synthesized through these following processes. Firstly, 10 gram of Cellulose and acetic acid glacial were mixed for 1 hour at 40°C. Then, sulfate acid and acetic acid glacial were added and mixed for 40 minutes. This mixing process was possible at 18 °C and acetic anhydrate acid, sulfate acid, and acetic acid glacial were added at 40 °C. Furthermore, it was hydrolyzed for 15 hours at room temperature. Then, DI water was added gradually to result in a precipitate. It was washed to neutral pH. Then, it was dried for 5 hours at 70 ˚C. Finally, this process produced cellulose acetate white powder.

Cellulose acetate was added to H_2SO_4 64 wt% and 200 mL distilled water at 45 ˚C for 20 minutes. This suspension was centrifuged at 4000 rpm. Finally, it was dried at 70 ˚C producing the nano CA.

2.4. Synthesizing ZnO.

ZnO was obtained through a 6.57 gram Zinc Acetate Dihydrate and DI water reaction. This solution was added by NaOH 3 M continually to reach the pH of 13 until it has a milky structure. Then, it was set aside for 30 minutes at 90 ˚C. Furthermore, it was washed by using methanol and then dried at room temperature. Finally, it produced ZnO white powder.

2.5. Fabricating CA-ZnO/ITO film.

CA-ZnO solution, the casting solution was synthesized by adding 0.3 gram CA powder and various amount of ZnO of 0; 0,02; 0,04; 0,06; and 0,08 gram to a solvent of PEG:DMF ratio (0,03 g: 3 mL). It was mixed for 5 hours at 80 ˚C with the speed of 1500 rpm. The casting solution was deposited on ITO/glass substrate via the spin coating method with the speed of 2500 rpm for 60 seconds. Finally, the CA-ZnO/ITO film was annealed at 160 ˚C.The CA-ZnO/ITO film properties were characterized by means of XRD, FTIR, SEM-EDX, and LCR-meter.

3. Result And Discussion

3.1. FTIR analysis

The result of the FTIR analysis was used to investigate the interaction between nano CA and ZnO. Figure 1 (a) shows the FTIR spectra of cellulose, nano CA film, and nano CA-ZnO/ITO. The main characteristic bands of cellulose are the hydroxyl group at 3271 cm-1 ([17]; [18]), C-H stretching at 2886 cm-1 ([19];[20]; [21]), and C-O-C stretching on β-glukosidic linkages at 1155, 1055, 896 cm-1 ([10]; [17]; [22]; [15]; [23]). The FTIR spectrum of the nano CA film on Figure1(a) shows the added peak at 1732 cm-1, 1239 cm-1 and 1367 cm-1 that respectively are –C=O stretching, acetyl –C-O stretching. In Figure1 (b), the Zn-O vibration confirms the formation of ZnO which is shown by the wave numbers of 400 - 1000 cm-1 on the nano CA-ZnO/ITO composite film.

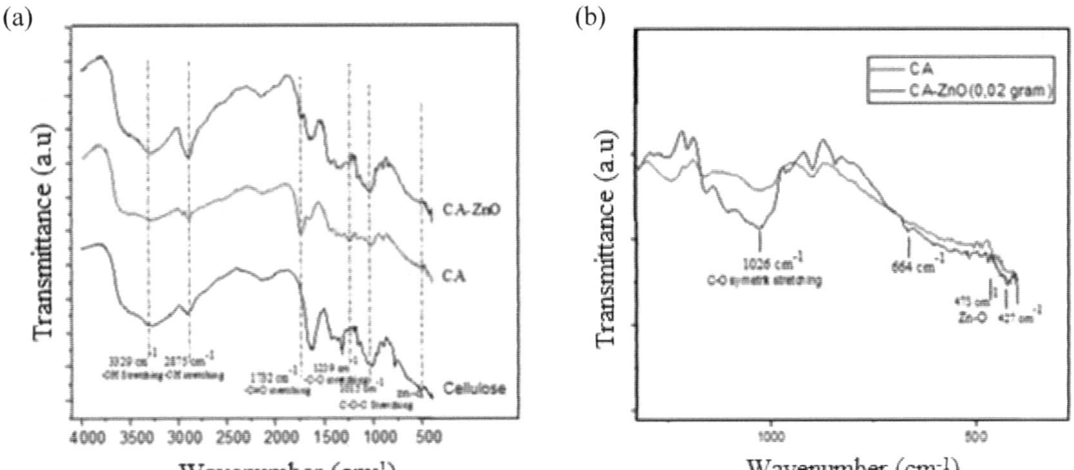

Figure 1. IR Spectrum on (a) Cellulose, CA, and CA-ZnO/ITOFilm,and (b) magnification scale 400 – 1000 cm^{-1} wave number in the CA and CA-ZnO film

IOP Conf. Series: Journal of Physics: Conf. Series **853** (2017) 012047 doi:10.1088/1742-6596/853/1/012047

3.2. XRD analysis

Based on Figure 2, the diffraction of the mass variety of ZnO which is added to nano CA indicates the constructed composite of nano CA-ZnO. It was shown on the new peak of nanoCA-ZnO/ITO film when ZnO was added. It is called composite since there are peaks that showed 2 phases. These phases are nano CA as matrix and ZnO as a dopant. The crystallinity of nano CA-ZnO/ITO film was obtained using the following equation (1).

$$Crystallinity = \frac{Crystalline\ Area}{Crystalline\ Area + Amorphous\ Area} x\ 100\ \% \quad (1)$$

It is shown in Table 1 that the Addition of ZnO has increased the crystallinity of ZnO and has decreased the crystallinity of CA which can also be seen in Figure 2.

Table 1. The Crystallinity of Nano CA-ZnO/ITOas a function of ZnO.

ZnO Mass (gram)	Crystallinity (%)	
	CA	ZnO
0	65.32	-
0.02	59.92	44.02
0.04	55.48	47.11
0.06	49.65	48.68
0.08	40.91	49.57

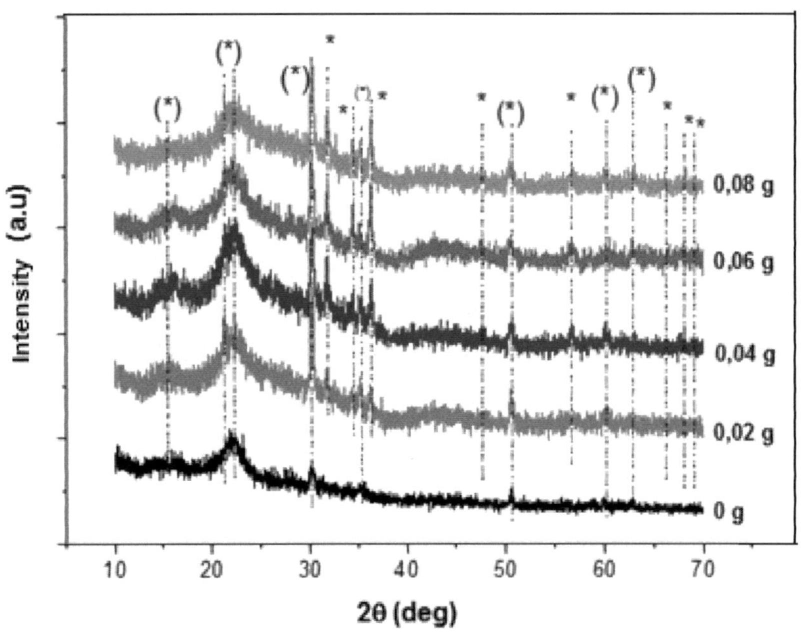

Figure 2. Diffraction Pattern of NanoCA-ZnO/ITO Composite Film for various ZnO which are 0; 0.02; 0.04; 0.06 and 0.08 grams. The symbol of (*) and * are designated respectively for CA and ZnO.

The crystal sizes of CA and ZnO on the Nano-CA-ZnO/ITO film were shown in Table 2. The crystal size was obtained by using equation 2. The crystal sizes of CA and ZnO are about 4,5 – 5.7 nm and 16.5 – 59 nm, respectively.

$$D = \frac{k\lambda}{\beta \cos\theta} \tag{2}$$

In formula 2, D is the crystal size, k is the crystal shape factor, λ is the wavelength of Cu (1.54056 Å), B is the FWHM (rad), and θ is the diffraction angle (^{0}).

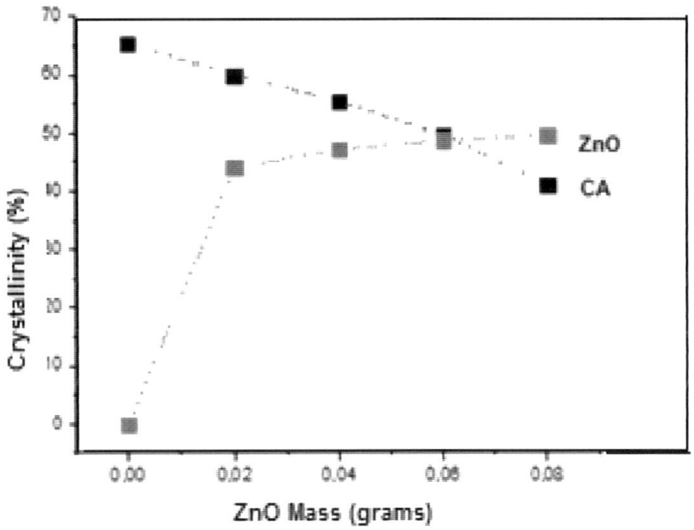

Figure 3. The Crystallinity of CA and ZnO at various Mass of ZnO of CA-ZnO/ITO Films.

Table 2. The crystalsize of CA and ZnO based on the addition of ZnO mass in the Nano CA-ZnO/ITO Films

ZnO (grams)	crystalsize (nm)	
	CA	ZnO
0	5.4	-
0.02	5.7	59.0
0.04	5.6	29.2
0.06	5.6	16.5
0.08	4.5	26.0

Table 3. Porous on the Morphology of Nano CA-ZnO Film

ZnO (grams)	porous (μm)
0	1.149
0.02	1.385
0.04	0.807
0.06	1.278
0.08	0.718

3.3. The Morphology of Nano CA-ZnO/ITO Film

The images of SEM show that nano CA-ZnO/ITO is the porous film. The porosity is related to the density of nano CA-ZnO film which can decrease or increase as shown in Figure4 (b,d) and Figure4 (c,e), respectively. The porosity of nano CA-ZnO/ITO can be found in Table 3.The porosity has been calculated using equation 3. In Figure4, ZnO spreads on the surface of nano CA-ZnO films. It shows in Figure 4f by the elemental mapping of nano CA-ZnO/ITO film

$$porous = \frac{\sum A_{porous}}{A_{total}} \tag{3}$$

where A is area of SEM result of film.

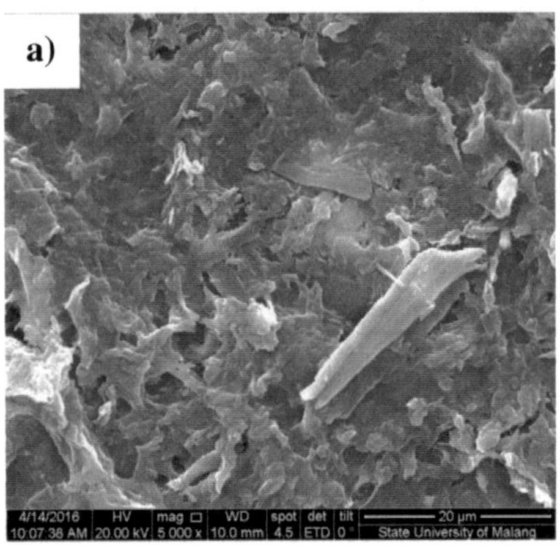

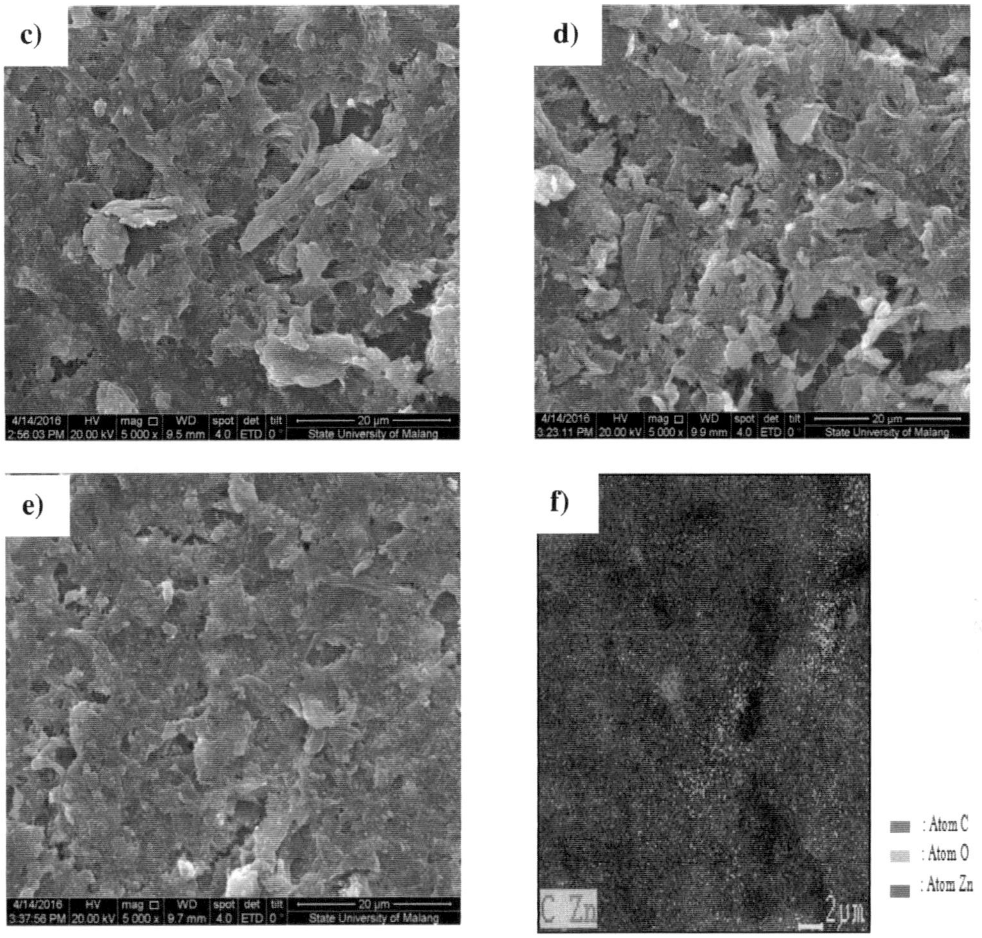

Figure 4. The Morphology of (a) CA; (b) CA-ZnO(0,02); (c) CA-ZnO(0,04); (d) CA-ZnO(0,06) and (e) CA-ZnO(0,08) and (f) the Elemental Mapping of nano CA-ZnO/ITO Film

3.4. Dielectric constant of nano CA-ZnO/ITO film

The addition of ZnO results in decreasing crystallinity of Nano CA-ZnO film as indicated in Figure 3. The decreased of crystallinity of CA significantly increase the dielectric constant of CA as shown in Figure 5. It is the result of its molecules being more polarized under the influence of the external field. The dielectric constant was calculated using Equation 4

$$\varepsilon_r = \frac{C.d}{\varepsilon_0.A} \tag{4}$$

where ε_r and ε_0 are relative dielectric constant and permittivity ≈8.854 x 10^{-12} $C^2/N\ m^2$. *A* is the area measured of the sample. *C* is capacitance and *d* is the distance between plate but in this case *d* is the thickness of the sample. The thickness is shown by the cross section of nano CA-ZnO/ITO film which was inspected by SEM.

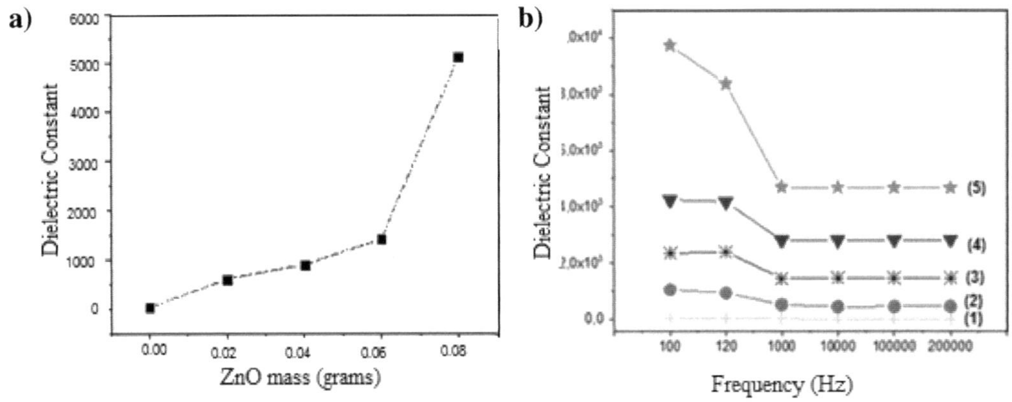

Figure 5. The dielectric constant of (a) The Nano CA-ZnO/ITO by The Mass Variety ZnO and By The Frequency of CA-ZnO (1) undoped nano CA film, (2) 0.02, (3) 0.04,(4)0.06, and (5) 0.08 of ZnO doped.

Figure5 (b) reveals that the frequency had influenced the dielectric of Nano CA-ZnO/ITO film. It investigated the polarization mechanism which was conducted. The dielectric decrease was followed by the increase of the frequency. Increasing frequency reduce the dielectric constant from ZnO, as well as the CA as a composite. The dielectric constant of ZnO and CA are mainly originated from ionic and weak dipole polarization mechanisms. The rise of electric field frequency may also produce vacancy before the full polarization occurs, which can reduce the dielectric constant [24].

4. Conclusion

The cellulose has been extracted from the water hyacinth which has 67,72% purity. ZnO-embedded nano CA film has been successfully prepared through the use of spin coating methods. From the analysis of FTIR spectra, it could be concluded that the nano CA-ZnO film is the composite film, which is also supported by XRD analyses. The grain size of CA and ZnO were 4.5 – 5.7 nm and 16.5 – 59.0 nm, respectively. The morphology of nano CA-ZnO/ITO film by analyzing SEM is categorized asporous film. Furthermore, inducing ZnO doped onto CA film inreasesits dielectric constant, while frequency tend to reduce its dielectric constant.

References

[1] R A Manish Batham and A Tiwari 2014 Time Efficient Co-composting of Water Hyacinth and Industrial Wastes by Microbial Degradation and Subsequent Vermicomposting, *J Bioremediation Biodegrad* **vol. 5**, no. 3, pp. 1–10.
[2] A F Abdel-Fattah and M A Abdel-Naby Feb 2012 Pretreatment and enzymic saccharification of water hyacinth cellulose *Carbohydr Polym* **vol. 87** no 3 pp. 2109–2113.
[3] T Istirokhatun, N Rokhati, R Rachmawaty, M Meriyani, S Priyanto and H Susanto 2015 Cellulose Isolation from Tropical Water Hyacinth for Membrane Preparation, *Procedia Environ. Sci.* **vol. 23** pp. 274–281.
[4] K J Murphy, Aquatic weed problems and their management: a review Oct. 1988 II Physical control measures, *Crop Prot.* **vol. 7**, no. 5, pp. 283–302.
[5] V Parra *et al* Mar 2007 New Hybrid Films Based on Cellulose and Hydroxygallium Phthalocyanine. Synergetic Effects in the Structure and Properties *Langmuir* **vol. 23** no. 7 pp. 3712–3722.
[6] K Ohkawa 2015 Nanofibers of cellulose and its derivatives fabricated using direct electrospinning *Mol. Basel Switz* **vol. 20**, no. 5, pp. 9139–9154.
[7] C Reid and E Breton, Water and Ion Flow Across Cellulosic Membrane - Google Scholar,

IOP Conf. Series: Journal of Physics: Conf. Series **853** (2017) 012047 doi:10.1088/1742-6596/853/1/012047

Journal of Applied Polymer Science, pp. 133–143, 1959.

[8] Gopiraman, M, Fujimori, K, Zeeshan, K, Kim, B S and Kim I S 2013 Structural and Mechanical Properties of Cellulose Acetate/Graphene Hybrid Nanofibers: Spectrocopic Investigations, *Express Polymer Letters* **vol. 7** pp. 354–563.

[9] E Fortunati, S Mattioli, I Armentano and J M. Kenny Nov 2014 Spin coated cellulose nanocrystal/silver nanoparticle films *Carbohydr. Polym.* **vol. 113** pp. 394–402.

[10] F Silva, R B Pinto, A Daniel-da-Silva, and T Trindade Oct. 2014 Cationic release behaviour of antimicrobial cellulose/silver nanocomposites, *Cellulose*, **vol. 21**, no. 5, pp. 3551–3560.

[11] A M Ferraria, S Boufi, N Battaglini, A M Botelho do Rego and M Rei Vilar Hybrid Systems of Silver Nanoparticles Generated on Cellulose Surfaces *Langmuir* **vol. 26**, no. 3, pp. 1996–2001, Feb. 2010.

[12] Y-Y Dong, S-M Li, M-G. Ma, K Yao, and R-C. Sun Jun. 2014 Compare study cellulose/Ag hybrids using fructose and glucose as reducing reagents by hydrothermal method, *Carbohydr. Polym.* **vol. 106,** pp. 14–21.

[13] N Bumbudsanpharoke, J Choi, I Park, and S K o2015 Facile Biosynthesis and Antioxidant Property of Nanogold-Cellulose Fiber Composite *J Nanomater*, **vol. 2015,** p. 9.

[14] S Anitha, B Brabu, D John Thiruvadigal, C Gopalakrishnan, and T S Natarajan Sep. 2013 Optical bactericidal and water repellent properties of electrospun nano-composite membranes of cellulose acetate and ZnO *Carbohydr. Polym.* **vol. 97** no. 2 pp. 856–863.

[15] K Shanmuganathan, J R Capadona, S J Rowan, and C Weder, Bio-inspired mechanically-adaptive nanocomposites derived from cotton cellulose whiskers," *J. Mater. Chem.*, **vol. 20**, no. 1, pp. 180–186, 2010.

[16] M Mariño, L Lopes da Silva, N Durán, and L Tasic 2015 Enhanced Materials from Nature: Nanocellulose from Citrus Waste, *Molecules*, **vol. 20,** no. 4, p. 5908.

[17] M Poletto and A J Zattera 2013Materials produced from plant biomass: part III: degradation kinetics and hydrogen bonding in lignin, *Mater. Res.*, **vol. 16**, pp. 1065–1070.

[18] Z-Y Yang, W-J Wang, Z-Q Shao, H-D Zhu, Y-H Li, and F-J Wang 2013 The transparency and mechanical properties of cellulose acetate nanocomposites using cellulose nanowhiskers as fillers *Cellulose* **vol. 20**, no. 1, pp. 159–168.

[19] J Zhang, T J Elder, Y Pu, and A J Ragauskas Jun. 2007 Facile synthesis of spherical cellulose nanoparticles *Carbohydr. Polym.* **vol. 69** no. 3 pp. 607–611.

[20] A Tercjak, J Gutierrez, H S Barud, R R Domeneguetti, and S J L Ribeiro Feb. 2015 Nano- and macroscale structural and mechanical properties of in situ synthesized bacterial cellulose/PEO-b-PPO-b-PEO biocomposites *ACS Appl. Mater. Interfaces* **vol. 7**, no. 7, pp. 4142–4150.

[21] J Rojas, A Lopez, S Guisao, and C Ortiz 2011 Evaluation of several microcrystalline celluloses obtained from agricultural by-products, *J. Adv. Pharm. Technol. Res.*, **vol. 2,** no. 3, pp. 144–150.

[22] V K Rastogi, D Stanssens, and P Samyn 2014 Mechanism for Tuning the Hydrophobicity of Microfibrillated Cellulose Films by Controlled Thermal Release of Encapsulated Wax, *Materials*, **vol. 7**, no. 11, p. 7196.

[23] N I Harun, R M Ali, A M M Ali, and M Z A Yahya 2012Dielectric behaviour of cellulose acetate-based polymer electrolytes, *Ionics*, **vol. 18**, no. 6, pp. 599–606.

[24] K Omar 2009Investigation on Dielectric Constant of Zinc Oxide, *Modern Applied Science*, **vol. 3**, no. 2.

IOP Conf. Series: Journal of Physics: Conf. Series **853** (2017) 012048 doi:10.1088/1742-6596/853/1/012048

Folate receptor mediated in vivo targeted delivery of human serum albumin coated manganese ferrite magnetic nanoparticles to cancer cells

A Zaidan[1,2,*], F Ilhami[1], M Z Fahmi[1,3], B Purwanto[1,4] and R Z Kharisma[1]

[1]Sekolah Pascasarjana Universitas Airlangga, Kampus B Universitas Airlangga, Jalan Airlangga 4-6, Surabaya, Indonesia.
[2]Department of Physics, Universitas Airlangga, Kampus C Universitas Airlangga, Mulyorejo, Surabaya, Indonesia.
[3]Department of Chemistry, Universitas Airlangga, Kampus C Universitas Airlangga, Mulyorejo, Surabaya, Indonesia.
[4]Faculty of Medicine, Universitas Airlangga, Kampus A Universitas Airlangga, Jalan Mayjen Prof. Dr. Moestopo 47, Surabaya, Indonesia.

*Corresponding email address: zaidan@fst.unair.ac.id

Abstract. Manganese ferrite nanoparticles ($MnFe_2O_4$) have received increasing attention due to their remarkable magnetic properties and have been used for various biomedical applications. They have potential applications in magnetic resonance imaging and hyperthermia for cancer. Both novel applications require a delivery system that will allow nanoparticle to move easily and localization of nanoparticle to the target tissue. In our work, we developed human serum albumin coated manganese ferrite magnetic nanoparticles (HSA-MF NPs). The nanoparticles were prepared using solvothermal method and modified with folic acid for targeted delivery. Structure and morphology of manganese ferrite nanoparticle were characterized by X-ray diffraction (XRD) pattern and transmission electron microscopy (TEM). The size of folic acid conjugated HSA-MF NPs (HSA-MF-FA NPs) were studied by dynamic light scattering (DLS). In the in vivo study, we used benzopyrene-induced cancer in mice. We successfully delivered HSA-MF-FA NPs through intravenous tail injection after induction of the tumour. We found that 54% of initial HSA-MF-FA NPs which previously injected localize in the target tissue. While obtained p-value from independent T-test is 0.013 which shows that there is a difference between the control group (HSA-MF NPs) and the treated group (HSA-MF-FA NPs)

1. Introduction

Cancer is one of the leading causes of mortality worldwide, it is also one of the most difficult puzzles to solve. Based on International Agency for Research on Cancer (IARC), there were more than 14 million new cases of cancer and 8 million cancer death worldwide in 2012 [1]. Traditional treatment method of cancer such as surgery, radiation therapy, and chemotherapy or a combination of these therapies. Over the years these methods have improved and managed to increase the number of patients healing. However, this method of treatment is still not perfect. The invasive nature of the surgery, providing a high degree of risk to patients both before and after the procedure. Radiation therapy have a short-term side effects such as nausea, itchiness, dryness and long-term side effects such as vascular damage, atrophy, and fibrosis due to radiation [2]. While chemotherapy causes systemic side effects. Drugs used in anti-cancer chemotherapy destroys cells that divide rapidly, either cancerous or healthy cells. Therefore, technology that can direct anti-cancer drugs right to the location of the tumour is strongly needed, thus the risk of chemotherapy side effects can be made as low as possible.

The current technological developments allow us to develop targeted therapy that can be an option for cancer treatment. Many scientists have done research to develop a drug delivery system (DDS) that can deliver the drug to the right place, localize the drug and reduce the risk of drugs side effects. One of the strong candidate for DDS is Magnetic nanoparticle with modified surface.

This study develops manganese ferrite nanoparticles ($MnFe_2O_4$) encapsulated in Human Serum Albumin (HSA). HSA was used because it can make nanoparticle system stable, it also acts as a phase transfer agent. The encapsulated $MnFe_2O_4$ magnetic nanoparticles system is expected to have stable, water-soluble properties and can be used for DDS.

Some targeting agent such as polyethylene glycol (PEG), lactic-co-glycolic acid (PLGA), human epidermal growth factor receptor (EGFR), folic acid, silica, and somatostatin / growth hormone (GH) has been developed. For cancer therapy, folic acid is very suitable for cancer targeting agent due to its low level of toxicity and long biodegradable, folic acid also has a high specificity towards cancer cells [3]. For these reasons, in this study manganese ferrite nanoparticles will use folic acid as targeting agent for cancer.

2. Materials and Methods

2.1. Nanoparticles and surface preparation

All materials used in this experiment were purchased from Sigma-Aldrich and used as-received without further purification. Manganese ferrite nanoparticles (MF NPs) were synthesized chemically using solvothermal method. For this method, an aqueous solution was prepared by dissolving 2 mmol Iron (II) acetylacetonate and 1 mmol Mangan (III) acetylacetonate in 15ml benzyl ether ($C_{14}H_{14}O$) and 15ml oleylamine ($C_{18}H_{35}NH_2$). Then the solution mixture was heated to 110 °C for 1 hour following by heated to 300 °C also for 1 hour. After solution temperature back to room temperature, the solution was washed with ethanol and separated by means of centrifugation at 6000 rpm for 10 minutes. MF NPs obtained from this procedure is in black powder form.

The HSA transfer was done by dissolve MF NPs with 1 ml of hexene. In another container, HSA were dissolved in 10 ml of aqua bidestilation. Both nanoparticles and HSA then mixed together and undergone ultrasonication process at 20 kHz for 2 minutes. The mixture is then left for 24 hours. From this procedure, we get human serum albumin coated manganese ferrite magnetic nanoparticles (HSA-MF NPs). The procedure is then repeated for the attachment of folic acid to obtain folic acid conjugated HSA-MF NPs (HSA-MF-FA NPs).

2.2. Samples characterization

The X-ray diffraction (XRD) patterns of prepare manganese ferrite nanoparticles sample was characterized by PAN Analytical Philips X'Pert diffractometer with CuKα (0.154 nm) radiation to generate XRD patterns from samples at room temperature in a 2θ range of 10°-70°. The morphology of the prepared sample is observed by transmission electron microscopy and size distribution of nanoparticles was determined using dynamic light scattering (DLS) method.

2.3. Cancer Cells Induction and in vivo Procedure

Benzopyrene-induced cancer in wistar mice was done by injecting benzopyrene solution to breast area of mice. Benzopyrene solution was prepared by dissolve benzopyrene powder in olive oil at a ratio of 1:1. Cancer induction was done by injecting mice with benzopyrene solution of 0.3ml/kg two times a week subcutaneously in the breast area for 4 weeks.

In vivo procedure consists of two group, one group of mice administered injection of HSA-MF NPs (positive control group) and other group of mice administered injection of HSA-MF-FA NPs (intervention group). Both of group administered injection of 1 cc solution intravenously from the tail. After 8 hours, cancer tissue of mice was taken for voltammetry characterization to determine the manganese metal content

3. Results and Discussion

3.1. X-ray diffraction, morphology, and size characterization

The XRD pattern of prepared MF NPs sample is shown in Fig. 1. The pattern of diffraction shows peaks which corresponds to Bragg reflections from (220), (311), (222), (400), (422), (333), (440) planes. Peaks of pattern in Fig. 1 is in good agreement with the standard XRD pattern of $MnFe_2O_4$ (JCPDS cards No. 74-2403, 10-0319) which indicate the samples had a face-cantered cubic crystal structure.

From XRD data, average crystallite size of MF NPs sample was calculated using Scherrer's formula [4]. The result of calculation using full width at half maximum intensity for (311) plane shows that average crystallite size of sample is 4.6 nm.

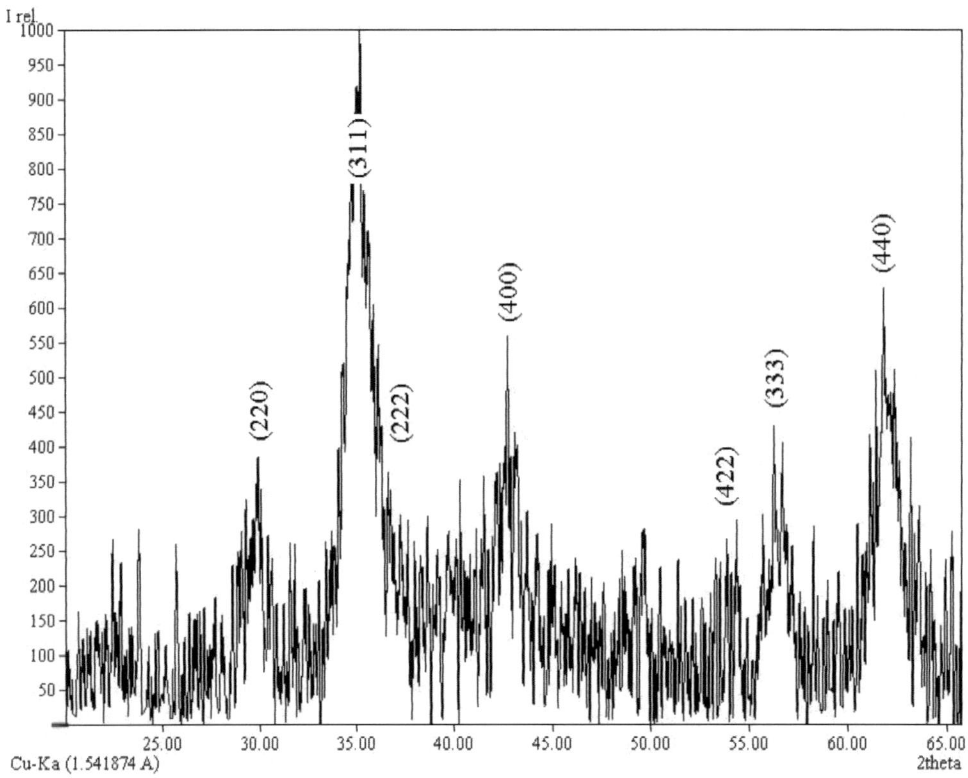

Figure 1. XRD pattern of prepared MF NPs.

The size of prepared MF NPs was studied by TEM which is shown in Fig. 2. The TEM result reveals that nanoparticles have almost spherical shape and quite homogeneous with a size between 10-20 nm. This result indicating that the synthesis method adopted is successful in obtaining homogeneous MF NPs.

Size distribution characterization result of HSA-MF-FA NPs using DLS show that coated nanoparticles has average size of 227.3 ± 114.5 nm. Addition size of MF NPs is due to coating and agglomeration. The size of HSA-MF-FA NPs is determined by concentration ratio of HSA with the nanoparticle. The greater the concentration of HSA used for coating procedure, the smaller size of HSA-MF-FA NPs [5].

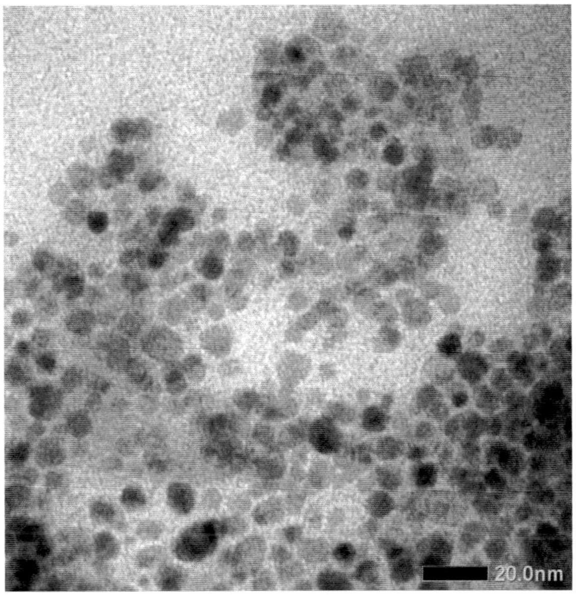

Figure 2. TEM micrograph of MF NPs.

3.2 In Vivo and voltammetry analysis

Voltammetry Analysis was conducted to determine the concentration of HSA-MF-FA NPs in cancer tissue. Standard model was generated to determine the concentration of manganese (Mn) using multiple solutions with different concentrations (1, 2, 4, and 8 ppm). The result shows that for intervention group, there is 54% of HSA-MF-FA NPs injected intravenously in wistar mice reached the cancer cells.

Based on the test results of Shapiro-Wilk, it is known that p-value in the positive control group was 0.081 and in the intervention group was 0.166. It shows that the data are normally distributed due to the p-value > α (0.05). Analyse using independent t-test show that p-value of 0.203, this is suggesting that the variance of the data is homogeneous due to the p-value> α (0.05). The complete statistical result is shown in Table 1

Table 1. Results of independent t-test analysis

Group	Mean	Median	SD±	P
Intervention group	0,0038	0,0038	0,0008	0,013
Positive control group	0,0024	0,0019	0,0011	

Based on the analysis results of independent t-test, p-value obtained was 0.013 <α (0.05) this shows that there are differences between (HSA-MF-FA NPs) and (HSA-MF NPs) as a delivery system media against cancer cells in wistar mice.

In vivo results in wistar mice show that HSA-MF-FA NPs have potential to be used as a DDS. Voltammetry analysis show that there was a 54% concentration of HSA-MF-FA NPs distributed in cancer cell of mice 8 hours after injected with 1cc HSA-MF-FA NPs intravenously. It happens due to the nano-size of HSA-MF-FA NPs making it possible to enter the cell through the leakage of blood vessels in the cancer cells. The nanoparticles also have a good ability to penetrate the interstitial part of blood vessel of the cancer cells with higher retention times than in healthy cells. Also, the uncontrollable cancer cells growth will always require faster vascularization and increases the permeability of that vascular so that it can easily be accessed by nanoparticles. This result is similar with study results by Hui S Huang in 2013 which revealed that the concentration of the nanoparticles identified in the tumour cells in the first 1 hour of 23%, 3 hours of 35%, 8 hours of 35%, 24 hours of 52%, and 96 hours of 20.58% [6].

In addition, HSA also play a role in the process of MF NPs entry into the cancer cells, it happens due to the differences in pH between healthy cells and cancer cells, where the high level of acidity in the environment of cancer cells will bring MF NPs move towards the cancer cells [7]. The change of pH in cancer cells will make the HSA structure opens and stimulates the release of folate from MF NPs so that MF NPs can enter cancer cells.

Delivery system media using HSA-MF-FA NPs are more effective accompanied by folic acid as a targeting agent that will move towards folate receptors on cancer cells. Some studies show promising results in accuracy and sensitivity detection of folate against cancer cells [3,8]. In our case folate receptor would normally distributed in the epithelium of healthy cells and increases in the epithelium of malignant cells. Folate receptor also activates the endocytosis progress so that HSA-MF-FA NPs can easily get into the cells' cytoplasm. This allows the use of targeted agents in the form of folic acid that have greater concentrations in cancer cells.

4. Conclusion

HSA-MF-FA NPs were successfully delivered through intravenous tail injection after induction of the cancer. It is found that 54% of initial HSA-MF-FA NPs which previously injected localize in the target tissue. Obtained p-value from independent T-test is 0.013 which shows that there is a difference between the control group (HSA-MF NPs) and the treated group (HSA-MF-FA NPs). It is indicating that HSA-MF NPs are less effective to be used as a delivery system media against cancer cells of wistar mice. From these results, it is concluded that the material of HSA-MF-FA NPs have potential to be used as a delivery system media against cancer cells in wistar mice.

Acknowledgments

This research was fully supported by Sekolah Pascasarjana Universitas Airlangga. We thank our colleagues from Department of Technobiomedics who provided insight and expertise that greatly assisted the research.

References

[1] Ferlay J, Soerjomataram I, Ervik M, Dikshit R, Eser S, Mathers C, Rebelo M, Parkin D M, Forman D and Bray F 2013 IARC CancerBase No. 11 (Available from: http://globocan.iarc.fr, accessed on 12/08/2016).
[2] Bentzen S M 2006 Nature Reviews Cancer 6 702-713.
[3] Zwicke G L, Mansoori G A and Jeffery C J 2012 Nano Reviews 3 18496.
[4] Langford J I and Wilson A J C, 1978 J. Appl. Cryst. 11 102-113.
[5] Fahmi M Z, Ou K L, Chen J K, Ho M H, Tzing S H and Chang J Y 2014 RSC Adv., The Royal Society of Chemistry 4 32762-32772.
[6] Huang H S and Hainfeld J F 2013 Int J Nanomedicine 8 2521-32.
[7] Elzoghby AO, Samy WM and Elgindy NA 2012. J Control Release 157(2) 168-82.
[8] Lu Y and Low P S 2002 Adv. Drug Deliv. Rev. 3;54(5) 675-93.